Expectations vs Realities of Information Privacy and Data Protection Measures

Editor
Indranath Gupta
Jindal Global Law School
O.P. Jindal Global University
Sonipat, Haryana, India

ISBN 978-981-99-6777-3 ISBN 978-981-99-6778-0 (eBook)
https://doi.org/10.1007/978-981-99-6778-0

This Springer imprint is published by the registered company Springer Nature Singapore Pte Ltd.
The registered company address is: 152 Beach Road, #21-01/04 Gateway East, Singapore 189721, Singapore

Indranath Gupta
Editor

Expectations vs Realities of Information Privacy and Data Protection Measures

A Machine-Generated Literature Overview

Springer

Acknowledgements

The editor is highly grateful to all those who have taken time out of their busy schedule and ensured the timely completion of this project. It has been made possible because of the extraordinary efforts of Paarth Naithani, Jindal Global Law School and the Springer team, who spent several hours on this project. The editor would also like to thank his family and colleagues at O.P. Jindal Global University (JGU) for their constant encouragement. The editor admires the unconditional support of Vice Chancellor Prof. C. Raj Kumar of O.P. Jindal Global University.

Introduction by the Editor

The issue of the privacy of individuals, the realm of informational privacy and the measures to protect informational privacy by way of implementing data protection measures by data controllers have been in the news in the last decade. Data controllers are those who process personal data of individuals or data subjects. Those processing personal data and those to whom personal data belong have been identified differently across jurisdictions. While the expectations associated with these stakeholders may differ, the overall meaning remains the same. The issue of data protection and informational privacy, which largely remained confined to the European Union, with the turn of the last century, gathered momentum in other jurisdictions. With rapid technological advancements on the horizon, countries began to harness the benefits of digitization. To operationalize digitization, processing of data, including personal data, became an unavoidable consequence. The wake of digitization somehow led many countries to implement a data protection framework to balance out the rights of data subjects vis-à-vis use and re-use of data in a meaningful way.

Although the data protection framework varies in different jurisdictions, it is primarily a cumulative outcome of data protection principles. These principles suggest that the processing of personal data of individuals, under all circumstances, should not obviate the basic norms of purpose and the virtue of minimal processing following the same purpose. Over time, consent in its various forms and types, informed, explicit and otherwise, ensured data subjects have a measured understanding of the purpose of such processing. The idea of consent with time has been challenging to implement with the rapid advancement of research in different areas. It remains the most critical fulcrum, yet there are instances when the implementation continues to challenge. Overall, not only consent but the inclusion of data protection requirements and their implementation in different services are rather cumbersome. Other societal rights in a democratic society must balance with the rights of those individuals to whom the personal data belong. There is the aspect of national security, the question of using and re-using the data in a data economy, and questions about fitting data protection framework within the rubric of other laws.

With these issues in the background, the author proposes two volumes of selective sources of work. These volumes have consciously tried to remain both jurisdictional and technology neutral while considering a range of data protection and privacy issues.

Owing to the nature of this sub-discipline, it remains a work in progress yet portrays a comprehensive range of issues. Towards that end, Volume 1 has chapters that capture overarching issues about data protection and privacy; conceptualizes data protection from different perspectives and its existing debates with other rights and developments in a democratic society; provides a snapshot of developments happening in various jurisdictions and how data protection framework engages with other laws. This volume also broaches the critical issue of consent and how consent as a requirement has evolved and integrated with health research and other allied areas. The subsequent volume, volume 2, would focus on different sectors and how these sectors have been tackling different expectations concerning data protection and privacy. It will also showcase how technology plays a catalyst in implementing data protection requirements.

Contents

About the Editor

Indranath Gupta is Professor at the Jindal Global Law School and Dean of Research at O.P. Jindal Global University, India. He is the Director of Jindal Initiative on Research in IP and Competition (JIRICO), and a Senior Fellow at the Jindal Institute of Behavioural Sciences (JIBS). He is currently the Jean-Monnet Chair- Academic Coordinator for the project 'Multi-dimensional Approaches to the Understanding of the E. U. Data Protection Framework'. He has been also invited as a Jean Monnet Visiting Research Fellow at the Centre of Excellence, Jean Monnet Centre for Intellectual Property & Management, Bournemouth University, UK. Prof. Gupta was earlier appointed as the research collaborator by the Università Bocconi, Milan, Italy, for a project funded by the European Commission under the 7th Framework Programme. He participated in a multi-jurisdictional project on Digital Era Competition: A BRICS View, whose report was published by the BRICS Competition Law and Policy Centre.

Professor Gupta holds two LL.M. degrees (taught and research) from the University of Aberdeen, UK, and the University of East Anglia, UK, and a Ph.D. from Brunel University, London. He also taught at the Brunel Law School during this Ph.D., which looked at the transnational influence of the US Supreme Court's judgement in the Feist case in the framing of the Database Directive in the European Union. Prof. Gupta specialises in Technology Law and aspects of Intellectual Property Law. Over the years, he has offered courses such as Data Protection Law, Information Technology Law, Music and Copyright, and EU Law and Approaches to New

Technology. He has contributed in reputed journals such as the *European Intellectual Property Review*, *Journal of World Intellectual Property*, *International Journal of Law and Information Technology*, *Journal of Intellectual Property Law & Practice*, *Journal of Data Protection and Privacy*, *Economic and Political Weekly*, and *European Journal for Law and Technology*. He has worked extensively with Springer on several book projects of all kinds, the latest being an in-process Major Reference Work titled 'Handbook on Originality in Copyright: Cases and Materials', currently available as a Living Edition on SpringerLink.

Chapter 1
Capturing Different Discourse to Privacy

Introduction by the Editor

There is often a bridging gap between prescribed norms and the operational steps that fulfil the objectives and goals of such norms. Data protection and privacy issues are no exception to this generality. Transposing overall goals is a challenge in a technology pervasive environment. There are other challenges in defining the expectations, either because of the absence of conceptual clarity or innate vagueness, leading to privacy roadblocks at a granular level. This chapter sets up the context of how privacy is understood and how data protection measures further percolate into technological advancements.

The discourse on privacy begins with the nature of the right associated with personal data and the sharing and distribution of such data. Should data have property rights assigned to it? There are arguments on both sides of this debate. The answer would decide the course of control over personal data. From a cultural perspective, the rights of an individual are, in fact, a reflection of the rights of a community or group of individuals. Privacy is not an absolute right and has to be balanced with other existing rights in a democratic society. The need for privacy is often contextualised in the backdrop of certain surveillance or other similar data gathering activities. These fundamental questions often get riddled with complexities that come with the use of advanced forms of technology. At times, with the emergence of a global phenomenon like the COVID-19 pandemic.

The ease of use of technology often pre-dates the goal of protecting the privacy of individuals. The use of technological interventions at appropriate times is not synchronised or sieved through with privacy goals. For instance, facial recognition technology or other forms of advancement in computing, automated or otherwise, raises legal, ethical and moral issues. The perception of having control over personal data and knowing the nature of processed personal data in advance seems difficult. The environment in which data is processed or advanced to other third parties is difficult to control. At times, users of different social networks do have the opportunity

I. Gupta (ed.), *Expectations vs Realities of Information Privacy and Data Protection Measures*, https://doi.org/10.1007/978-981-99-6778-0_1

to control the environment in which they share personal data. Yet, owing to awareness related challenges, the sharing of personal data happens in an unabated manner.

The urge to protect personal data also led to a discussion on information asymmetry, focusing upon an economic argument. Does the goal of privacy and protecting the rights of an individual extend to a situation where the flow of information is seriously aggravated? What sort of price would such action have on the data sharing digital economy? The anxiety is not anecdotal because the EU General Data Protection Regulation (GDPR) [1] recitals have mentioned the importance of free data flow.[2] There seems to be an underlying concern that the nature of protection may lead to over-protection. Following the same economic argument, sub-optimal transparency regarding the processes followed by service providers may result in sub-optimal collection and further use of personal data. At any given time, optimal information about processing may allow individuals to trust service providers in a digital environment driven by user-centric practices.

The COVID-19 pandemic signalled the final test for processing personal data in an environment enhanced by medical science and technology. It brought about that unexpected emotional change for humanity, not to mention its impact on the processing of personal data. The pandemic truly tested the objective enshrined under Recital 4 of the GDPR.

"The processing of personal data should be designed to serve humankind. The right to the protection of personal data is not absolute; it must be considered in relation to its function in society and be balanced against other fundamental rights, in accordance with the principle of proportionality".[3]

The intervention from the European Union has a long tradition- long before what we come across under GDPR. It showcases the extent of readiness for regulatory intervention in time of need. Following the principle of proportionality, the emphasis is on the point of balance. During COVID-19 times, public health and security issues became pivotal to governments across continents. The emergence of applications that track and trace individuals for public health reasons became a necessity. The issue of privacy and the data protection measures enabling the protection of sensitive health data became a paramount consideration. However, managing fundamental rights vis-à-vis a global health crisis became arduous.

The chapter discusses several of the issues mentioned above. It portrays the arguments that persist concerning the protection of privacy, adopting data protection measures and its effect on different circumstances.

The sources considered in this chapter are jurisdictional neutral. They cover a wide range of fundamental issues connected with privacy.

---

[1] Regulation (EU) 2016/679 of the European Parliament and of the Council of 27 April 2016 on the protection of natural persons with regard to the processing of personal data and on the free movement of such data, and repealing Directive 95/46/EC (General Data Protection Regulation) OJ L 119, 4.5. 2016, pp. 1–88. (GDPR).

[2] ibid, Recitals 4, 5 and 6.

[3] ibid Recital 4.

The sources have been arranged in a manner so that privacy for a wider audience does not remain as a unitary discipline centric concept. It goes much beyond security and allied risks and involves societal dimensions. Privacy is an amalgamation of concepts that transcends disciplinary boundaries, and hence, privacy requirements are a cumulative outcome meshed with disciplinary traits. As a term, it embodies different philosophies, and is not confined to a unidirectional understanding and includes social norms. There are efforts to place privacy and human interactions in connected environments in the context of various social psychological theories that showcase the prior needs before working on legislative interventions.[4]

Similarly, sources highlight the importance of understanding how the concept of private and public intertwines both in terms of institutional and informational aspects in the data economy, i.e. the public goals and private endeavours often match the desired outcomes.[5] The private and public spaces also have different dimensions when it comes to questioning the ownership of personal data and how private rights are balanced against public interests.[6] There are broad discussions concerning the economics of privacy, its impact on data flow and further dissemination, and the overarching value associated with customer information for companies.[7]

Quite a few selected sources have highlighted the issue of surveillance and other impacts that technology has on privacy, like the impact of social media networks[8] and the use of honeypots and their impact on privacy.[9] While surveillance is a common phenomenon across jurisdictions,[10] in the context of the UK, a submission has been made about following nine different principles. These principles help balance the rights of individuals vis-à-vis the need for surveillance and related policies.[11]

---

[4] For example, Mohamad Gharib & others, "An Ontology for Privacy Requirements via a Systematic Literature Review" [2020] Journal on Data Semantics 9 (4) 123–149; Jakko T Lehikoinen & others, "Understanding privacy regulation in ubicomp interactions" [2007] Personal and Ubiquitous Computing 12, 543–553; Diane P Michelfelder, "Philosophy, privacy, and pervasive computing" (2010) AI & SOCIETY 25, 61–70.

[5] Klaus Hoeyer, "Data promiscuity: how the public–private distinction shaped digital data infrastructures and notions of privacy" (2020) Humanities and Social Sciences Communications 7(37).

[6] Urbano Reviglio & Rogers Alunge "I Am Datafied Because We Are Datafied": an Ubuntu Perspective on (Relational) Privacy" (2020) Philosophy & Technology 33, 595–612.

[7] Benjamin E Hermalin & Michael L Katz, "Privacy, property rights and efficiency: The economics of privacy as secrecy" (2006) Quantitative Marketing and Economics 4, 209–239; Caleb S Fuller, "Privacy law as price control" (2017) European Journal of Law and Economics 45, 225–250; Carlos Lamela-Orcasitas & Jesús García-Madariaga, "How to really quantify the economic value of customer information in corporate databases" (2023) Humanities and Social Sciences Communications 10, 1–13.

[8] Francesca Cerruto & others, "Social network data analysis to highlight privacy threats in sharing data" (2022) Journal of Big Data 9(19).

[9] Pavol Sokol & others, "Honeypots and honeynets: issues of privacy" (2017) EURASIP Journal on Information Security 4.

[10] Qingxiu Bu, "The global governance on automated facial recognition (AFR): ethical and legal opportunities and privacy challenges" (2021) International Cybersecurity Law Review 2, 113–145.

[11] C N M Pounder, "Nine principles for assessing whether privacy is protected in a surveillance society" (2008) Identity in the Information Society 1, 1–22.

However, certain nuanced aspects are connected with the surveillance activity and its application. In fact, social stratification, which represents several socio-economic factors, has a role in deciding the collection of personal data, the extent of data collected, its use, and its recipients.[12] The idea of surveillance across jurisdictions has taken different imaginations with the exponential growth of technology. The fear of erosion of privacy rights is increasingly felt with the use of AI driven systems, and the introduction of new standards of communication (for instance, 5G),[13] and unstructured regulatory interventions. The threats to privacy from the use of an advanced form of technology raise various moral issues that are much deeper than merely protecting personal data or having such control over such data.[14]

Finally, the resources have been selected to showcase how privacy related issues and data protection measures introduced sit with public health situations. Handling COVID-19 matters has been a testament to the unprecedented challenges and has become a lesson for the future.[15]

Machine generated summaries

Disclaimer: The summaries in this chapter were generated from Springer Nature publications using extractive AI auto-summarization: An extraction-based summarizer aims to identify the most important sentences of a text using an algorithm and uses those original sentences to create the auto-summary (unlike generative AI). As the constituted sentences are machine selected, they may not fully reflect the body of the work, so we strongly advise that the original content is read and cited. The auto generated summaries were curated by the editor to meet Springer Nature publication standards. To cite this content, please refer to the original papers.

Machine generated keywords: app, pandemic, covid, ownership, resource, threat, network, surveillance, covid pandemic, computing, user datum, trace, social, see, deploy

---

[12] Marx, Gary T. Privacy and Social Stratification. Philosophy & Technology (2007) 20, 91–95.

[13] Samant Khajuria & Knud Skouby, "Privacy and Economics in a 5G Environment" (2017) Wireless Personal Communications 95, 145–154.

[14] Samant Khajuria & Knud Skouby, "Privacy and Economics in a 5G Environment" (2017) Wireless Personal Communications 95, 145–154.

[15] Emanuele Ventrella, "Privacy in emergency circumstances: data protection and the COVID-19 pandemic" (2020) ERA Forum 21, 379–393; Magdalena Kędzior, "The right to data protection and the COVID-19 pandemic: the European approach" ERA Forum (2020) 21, 533–543; Sjef van Erp, "COVID-19 apps, Corona vaccination apps and data "ownership"" (2022) China-EU Law Journal 8, 45-65 ; Majid Hatamian & others, "A privacy and security analysis of early-deployed COVID-19 contact tracing Android apps" (2021) Empirical Software Engineering 26(36).

1.1 Setting the Overview

Machine generated keywords: threat, ownership, surveillance, computing, social, see, privacy, network, democratic, informational, relational, social network, group, infrastructure, agreement

An Ontology for Privacy Requirements via a Systematic Literature Review [80]

This is a machine-generated summary of:

Gharib, Mohamad; Giorgini, Paolo; Mylopoulos, John: An Ontology for Privacy Requirements via a Systematic Literature Review [80].

Published in: Journal on Data Semantics (2020).

Link to original: https://doi.org/https://doi.org/10.1007/s13740-020-00116-5.

If you want to cite the papers, please refer to the original.

For technical reasons we could not place the page where the original quote is coming from.

Summary

"Much of existing work on privacy requirements deal with them as a special case of security requirements, thereby missing essential traits of such requirements."

"The main contribution of this paper is an ontology for privacy requirements that subsumes whatever proposals exist in the literature."

"The ontology was obtained through a systematic literature review concerning the literature on privacy, conducted in two phases, and followed by a consolidation phase where similar concepts were combined."

"The proposed ontology has been developed to be used by software engineers when dealing with privacy requirements, and since it is based on a systematic literature review, it is more comprehensive in coverage than all ontologies included in our systematic review."

Introduction

"A well-defined privacy ontology that captures privacy-related concepts along with their interrelations would constitute a great step forward in improving the quality of privacy-aware systems, since it can reduce the conceptual vagueness and terminological confusion by providing a shared understanding of privacy-related concepts between designers and stakeholders of system-to-be [1]."

"The privacy ontology should conceptualize privacy requirements in their social and organizational context."

"The ontology should consider not only the technical aspects of privacy but also its related social and organizational aspects."

"The ontology was obtained through a systematic literature review (SLR) whose main purpose is identifying key concepts and relationships for capturing privacy requirements."

"This allows for extending the ontology proposed in [2] with more refined concepts concerning both personal information and privacy requirements."

Review Process

"A systematic review consists of a systematic process for defining research questions, searching the literature for the best available resources to answer such questions, and collecting available data from resources for answering the research questions."

"Conducting the review through a search process after identifying the search terms and the literature resources, followed by study selection activity."

"This includes three main activities: 1.1 Identifying the need for the review which can be done by identifying and reviewing any existing systematic reviews related to privacy ontologies."

"Based on [3, 4], a review protocol should specify the following: the strategy that will be used to search for primary studies selection; study selection criteria; study quality assessment criteria; data extraction; and dissemination strategies."

"Identify the Search Terms Following [3, 4], we derived the main search terms from the research questions."

Review Results and Discussion

"We have identified 55 privacy-related concepts and relationships after analyzing the 34 selected studies in SLR1."

"We have identified 13 new privacy-related concepts and relationships after analyzing the 36 selected studies in SLR2."

"The main aim of selecting such variety of privacy-related concepts and relationships is having a comprehensive set of concepts that can be used for dealing with privacy requirements, which enable us to select the key ones for capturing privacy requirements (e.g., RQ2)."

"Each of the selected studies has been carefully investigated to identify concepts/relationships that capture privacy requirements."

"We have focused on identifying any concept/relationship that can be used for capturing privacy requirements in their social and organizational settings since many privacy-related vulnerabilities might manifest themselves into such settings."

"Privacy requirements Ten concepts and relationships have been identified for capturing privacy-related aspects (e.g., anonymity, confidentiality)."

A Novel Privacy Ontology

"Among the available concepts for capturing informational entities, e.g., data [5], a resource (physical or informational) [6–9], asset [10, 11], etc, we have adopted the following concepts and relationships: Information represents an informational entity."

"We define the own relationship that relates personal information to an identifiable legal entity, and we specify how information owner controls the use of its personal information relying on consent and permissions."

"Consents and Permissions a consent represents an agreement at the information owner side concerning the use of information it owns (e.g., personal information) for a specific reason [12–14]."

"Entity interactions: actors may not have the required capabilities to achieve their own objectives by themselves (e.g., achieve a goal, furnish information, etc)."

"Introduces concepts to capture data subjects' (information owners) privacy requirements/needs concerning their personal information."

Discussion

"While performing the systematic reviews, we were able to identify studies that propose concepts for capturing high-level privacy-related concepts (e.g., [15, 16]), and other studies propose security concepts that can capture some aspects of privacy in their social and organizational context (e.g., [9, 17])."

"We were able to identify mitigation and treatment concepts such as countermeasure and control (e.g., [18, 19]), which can be used to identify appropriate solutions for privacy vulnerabilities, threats, attacks, etc Nevertheless, we were not able to find any study that is able to deal properly with privacy requirements, i.e., capture all the main privacy concepts and relationships that have been identified in the review."

"The two reviews have identified 70 studies that cover a wide range of concepts capturing various organizational, risk, treatment, and privacy aspects of the system-to-be, which we consider essential for appropriately dealing with privacy requirements."

"These studies have been further analyzed to answer our research questions, and we used the identified concepts and relationships for proposing a privacy ontology."

Threats to Validity

"1—Construct threats is concerned with the extent to which a study measures what it claims to be measured [20]."

"We have identified the following internal threats: Potential publication bias is a common threat to the validity of systematic reviews, and it refers to a situation where positive research results are more likely to be reported than negatives ones [4]."

"3—External threats is concerned with to what extent the results of the study can be generalized [20]."

"It is almost impossible to capture all related studies, yet our review protocol and search strategy were very carefully designed to cover as much as possible of the related studies."

"The number of selected studies (70 studies) can raise concerns about the review results."

"4—Reliability threats is concerned with to what extent the study is dependent on the researcher(s), i.e., if another researcher(s) conducted the same study, the result should be the same."

Related Work

"There are few systematic reviews concerning privacy/securities ontologies."

"The authors considered several privacy-related concepts, but the main focus of this ontology was mainly on security and risk analysis."

"Although the ontology captures security requirements in their social and organizational setting, it does not consider privacy requirements."

"Thinakaran and others [21] propose a framework for modeling privacy requirements for IoT-based health applications."

"Several security approaches that either ignore privacy or deal with it as part of security requirements have been proposed."

"Liu and others [22] propose an agent-oriented modeling framework that allows for the identification of security and privacy requirements by employing a set of analysis mechanisms."

"Relying on Secure Tropos, Zannone [9] introduces the SI* methodology that propose concepts for modeling and analyzing security, privacy along with trust requirements."

Conclusions and Future Work

"To determine what are the right concepts for our ontology, we conduct a two-phase SLR concerning existing privacy/security literature with a main purpose of identifying the main concepts along with their interrelations for capturing privacy requirements."

"We have implemented, validated, and evaluated the ontology with the help of several software tools, lexical semantic experts, and privacy and security researchers [23]."

"We plan to better evaluate the completeness and validity of the ontology by deploying it to capture the privacy requirements for several case studies that belong to different domains (e.g., public administrations)."

"We are planning to develop a privacy assistant tool that exploits this ontology to support the identification and analysis of privacy requirements by analysts, and also the generation of a privacy requirements specification."

Data Promiscuity: How the Public–Private Distinction Shaped Digital Data Infrastructures and Notions of Privacy [81]

This is a machine-generated summary of:

Hoeyer, Klaus: Data promiscuity: how the public–private distinction shaped digital data infrastructures and notions of privacy [81]

Published in: Humanities and Social Sciences Communications (2020).

Link to original: https://doi.org/https://doi.org/10.1057/s41599-020-00535-6.

If you want to cite the papers, please refer to the original.

For technical reasons we could not place the page where the original quote is coming from.

Abstract-Summary

"This essay discusses the performative effects of the public–private distinction on digital data infrastructures in healthcare."

"This analysis focuses on how they are used both in an informational sense (what is kept secret or strictly controlled versus what is out in the open or shared) and an institutional sense (issues of ownership and purpose such as being state-owned and governed for the common good or privately owned and aimed at generating profit)."

"In the political construction of digital infrastructures, the two senses are deeply intertwined: changes in relation to ownership and purpose affect what is kept secret and what is shared."

"The informational interconnectedness facilitated by digital infrastructures produces a form of 'data promiscuity'."

"Data promiscuity is a condition where data are indiscriminate in the choice of partners: what is seen as data on a thing for one purpose can always become seen as data on another aspect of that thing and be used for another purpose and by another user."

"Data promiscuity is the contemporary condition of possibility for health research."

Introduction

"While the public–private distinction is used in many senses, two of them stand out in relation to digital data sharing (Ariès, 24; Johansen and Andrews, 25); one relating to informational aspects (what is kept secret or strictly controlled versus what is out in the open or shared) and the other to institutional issues of ownership and purpose (typically seen as a matter of being state-owned and governed for the common good, or privately owned and aimed at generating profit)."

"These two aspects are particularly interesting to look at in tandem because a datafied economy implies transformations in both the informational (who gets access to what) and institutional (who owns, operates and profits) aspects of emerging data infrastructures."

"To understand the conditions of possibility for health research in a digital age, it is therefore time to understand the wider transformations of data infrastructures through digitalisation."

The public–private distinction

"In a seminal article on the public–private distinction, feminist linguist Susan Gal once argued that though it is common to talk of a 'blurring of boundaries' between public and private, such talk relies on a basic misunderstanding (Gal, 26)."

"Public–private thinking operates as a 'fractal distinction', she argues, meaning that any segment that has been divided into either public or private can itself be subdivided again into its 'public' and 'private' aspects."

"As other fractal distinctions (e.g., nature/culture, female/male), the public–private divide is a sense-making device shaping the world that it is said to classify: public and private are performative concepts that align activities with meanings and thereby shape people's behaviour."

"The institutional transformations that I explore in this paper revolve around data infrastructures and therefore they have implications not only for ownership and moral attributions of rights to profit, but also for the informational aspects of the public–private distinction; for privacy."

'Privatising' infrastructures: Denmark as digital frontrunner

"Most Danes probably consider the CPR system 'public' and the whole-data-intensive form of government as indicative of the goals of the welfare state."

"The running of the register, however, has for years been outsourced to an international IT company, CSC, and private sector actors that use the register are not required to make their data available for public research."

"State ambitions of ensuring Denmark a position as digital frontrunner are not limited to the so-called 'public' sector."

"Danish citizens today consistently interact with what is described as 'public' services (authorities) through 'privately' (i.e., commercially) delivered digital interfaces."

"As digital infrastructures for 'public data' are outsourced to, run by or maintained by commercial companies, it makes little sense to see public and private as ontologically distinct domains."

Informational implications: the secret and the manifest

"The digitalisation of data infrastructures intervenes in intriguing ways in the dialectic between the informational aspects of private and public, or what Simmel (27) called the secret and the manifest."

"Digital data integration and the more complex networks of commercial suppliers open up access for many more potential users than the old closed filing cabinet did in the offices of authorised civil servants or health professionals."

"Even if it might be unusual, I believe it illustrates an interesting point about the kind of work that the public–private distinction performs as people seek to separate the secret and the manifest in the new digitally integrated data landscape: I have a feeling that the day I got diagnosed, many years ago, I became public property."

Institutional implications: 'public' and 'private' moral zones

"If the public–private distinction as a sense-making device implied incentivising public services (authorities) to operate in the same way as private markets (companies), the performativity of the moral zones of the distinction (how people use it to ascribe praise and blame) have proven to operate with reverse effect."

"The market value of the 'private' company NETS stems, as described above, from a series of 'publicly' sanctioned monopolies: NemID, E-Boks and digital payment options."

"The point is that when profits can be appropriated by these companies despite their failures, it is indicative of a culture where 'public' and 'private' have become associated with different moral zones."

"The digital infrastructure is not 'private' or 'public', but the moral zones of economic opportunity and political blame have come to operate in this way."

"The moral zones of public and private fundamentally shape the conditions of possibility for health research."

Discussion: searching for new solutions in an age of data promiscuity

"The GDPR gave individuals 'rights' to access, delete and sell (through the 'portability principle') data on themselves, but the very notion of 'portability' here reveals how this approach to privacy as matters of private property is as much about facilitating as retraining data exchange."

"Just as an ethics of monogamy fared better as a technology of blame than as a means of protection in relation to sexual promiscuity, responsibilisation of the individual does more to protect the institutions and companies thriving on data than the individuals who are asked to hand over data as a precondition for entering social relations and accessing services."

"Of proclaiming an individual right to privacy and suggesting it should be protected with property rights that cannot be upheld anyway, it is time for a much more thorough political debate about who may use which data for what purposes."

"I Am Datafied Because We Are Datafied": An Ubuntu Perspective on (Relational) Privacy [82]

This is a machine-generated summary of:

Reviglio, Urbano; Alunge, Rogers: "I Am Datafied Because We Are Datafied": an Ubuntu Perspective on (Relational) Privacy [82].

Published in: Philosophy & Technology (2020).

Link to original: https://doi.org/https://doi.org/10.1007/s13347-020-00407-6.

If you want to cite the papers, please refer to the original.

For technical reasons we could not place the page where the original quote is coming from.

Abstract-Summary

“A general overview of the debate on the ethics of privacy is provided by distinguishing between individual and relational privacy.”

“The challenges of ‘group privacy’ are discussed.”

“We introduce basic principles of Ubuntu focusing on how these may have affected privacy conceptions and related legal practices.”

“By outlining opportunities and risks of intercultural information ethics, we argue how Ubuntu—similarly to other communitarian moral philosophies—strengthens the development of the concept of relational privacy and, in particular, of group privacy.”

Introduction

“Cultures and philosophies that embrace a conception of privacy as fundamentally relational have already been analyzed in this context, notably Indian (Basu 28) and Japanese (Capurro 29) as well as Buddhist (Hongladarom 30, 31) and Confucian (Ma 32).”

“The conception of relational and group privacy can find ideological support in the Ubuntu philosophy which, in turn, can enrich the intercultural ethics debate and be used as a starting point for conceptualizing an African understanding of privacy.”

“In this article, we analyze how Ubuntu conceives relational autonomy and relational privacy, and how this would affect the protection of privacy.”

“The article is organized as follows: the first part reviews the literature on the ethics of privacy and the relationship between individual and relational privacy.”

“In the third part, after having briefly situated Ubuntu within the context of the intercultural information ethics debate, we discuss how it might have influenced privacy conceptions in Africa and how it can affect a new understanding of privacy, dealing in particular with the issue of group privacy.”

Emerging Challenges to Privacy

“Privacy in terms of “information control,” popularized by Alan Westin’s definition of privacy (33) as “the claim of individuals, groups, or institutions to determine for themselves when, how, and to what extent information about them is communicated to others.”

“The view that the individual self is considered constrained from the start of its own existence by social factors (relational privacy) and that people are not always in control of information regarding them (networked privacy) necessitates the consideration of norms which emphasize the dignity of humanity (as a community or group) over the dignity of individuals.”

“From a traditional Western perspective, such information, provided it does not identify any specific individual but refers to a given population, is not subject to privacy law However, there are reports of such information being (hacked and) used to study, locate, and launch attacks on targeted populations in warzones, or combined with other data points to (sometimes erroneously) infer, for instance, the Muslim faith of taxi drivers in New York city (Taylor and others 34)."

An Introduction to the Ubuntu Philosophy

"As cited by Mnyaka and Motlhabi (35, p. 217) for example, Sebidi, in his 1988 unpublished paper titled "Towards a Definition of Ubuntu as African Humanism," compares an attempt at defining Ubuntu to an attempt at defining time, based on the notion that Ubuntu is "something abstract, [a] non-perceptible…quality or attribute of human acts the presence or absence of which can only be intuited by the human mind," while another posits that a synthetic definition of the term will always be inadequate, as scholars will only address characteristics of the concept which appeal to them the most."

"Letseka (36) highlights the embodiment of these values in Section II of the South African Constitution which addresses human dignity, fairness, and equality before the law Kwangwangamulu's work on the definition of Ubuntu reveals a key feature: the concept advocates a society in which humans are expected to care for each other's needs."

An Ubuntu Perspective on (Relational) Privacy

"Studies of intercultural information ethics show how the concept of privacy is associated with the Western perspective of individual autonomy (Capurro 29), even though forms of privacy are found in every culture throughout the world."

"Following Ma (32), considering Ubuntu and other ethical perspectives such as Confucianism jointly could spark research contributing to the relational understanding of privacy in three ways: (i) to understand how cultural factors influence people's understanding of privacy, (ii) to assess whether there are some underlying shared social norms that might be more salient in one context/culture than another, and (iii) to find empirical evidence of how prevalent control-based belief of privacy is among individual, corporate, and/or government documents in other cultures other than Western ones."

"In the big data context, Ubuntu would therefore support a top-down regulatory paradigm of privacy, as a right which protects an individual through principles and rules regulating the collection and sharing of data relating to entire communities or groups of individuals."

Conclusions

"To explore potential alternative foundations for the conceptualization of privacy, this paper has presented and discussed the concept of relational privacy from the perspectives of the sub-Saharan ethno-philosophy of Ubuntu."

"Ubuntu problematizes Western philosophical foundations, especially the notion of individual autonomy and personhood, to eventually reconcile individualistic and communitarian perspectives."

"Despite the opportunity that Ubuntu presents to enrich and nuance the privacy debate, as a philosophy faces several limits and challenges in both theory and practice."

"African populations have to rediscover Ubuntu because many have not experienced it, and also because many live in two different cultures, practicing Ubuntu in the rural environments and Western values in the urban ones."

"Adopting a relational perspective to understand privacy, as embodied in Ubuntu, can stimulate new empirical research and provide a broader and more fruitful grounding for privacy laws and policies."

Nine principles for assessing whether privacy is protected in a surveillance society [83]

This is a machine-generated summary of:

Pounder, C. N. M.: Nine principles for assessing whether privacy is protected in a surveillance society [83].

Published in: Identity in the Information Society (2008).

Link to original: https://doi.org/https://doi.org/10.1007/s12394-008-0002-2.

If you want to cite the papers, please refer to the original.

For technical reasons we could not place the page where the original quote is coming from.

Abstract-Summary

"It argues that the current debate about the nature of a "surveillance society" needs a new structural framework that allows the benefits of surveillance and the risks to individual privacy to be properly balanced."

"The first part of this article sets out the reasons why reliance on the current framework of data protection or human rights legislation, or on the current regulatory regime does not necessarily protect privacy."

"The second part sets out nine principles that can be used to assess whether individual privacy is comprehensively considered when surveillance policy is developed."

"These principles are applied to surveillance in the UK to identify the structural improvements that could create an effective balance."

Part I: Why the current framework of privacy protection in the UK is deficient

"In such circumstances, it is going to be very difficult to claim that the First, Second, Third and Fifth Data Protection Principles have been breached because the enactment of surveillance legislation establishes that: the processing purposes are lawful and compatible (meeting the relevant First and Second Principle obligations in the Data Protection Act); specifies the duration of the processing of personal data (satisfying the Fifth Principle obligations over data retention) and specifies the items of personal data to be processed (satisfying the Third Principle obligations in connection with the relevance of the personal data to the purpose)."

"The National Identity Register (NIR) of the ID Card Act 2006 provides an extreme example of how a Government can minimise effective Parliamentary and public scrutiny of a proposal—the proposal in question being whether the NIR should be used as a general population register so that commonly held personal data (e.g. name, address) can be shared between most public authorities."

Part II: The nine principles

"In Part II, the nine Principles are presented in order to provide a framework that corrects the structural defects identified in Part 1."

"The Principles interact with each other and should be considered as a whole."

"It is argued that the application of these Principles to the development of surveillance policy, or to the bodies performing the surveillance, would help maintain public confidence in legitimate surveillance."

"Although the Information Commissioner is the most likely Regulator in relation to a surveillance activity, other specialist Regulators in the UK (usually dealing with law enforcement or national security) have to be included as being linked to these Principles."

Principle 1: The justification principle

"Information relating to any legislation or policy that involves surveillance (or extension to an existing surveillance policy) is provided so an assessment can be made to ensure that the surveillance can be justified in terms of pressing social needs and measurable outcomes; this information is provided prior to the approval of legislation or policy."

"Effective public scrutiny of legislation assumes that Parliament has access to all relevant documents produced in relation to a surveillance proposal (including legal background, a Human Rights Assessment as requested by the JCHR, policy options, cost benefit analysis, details about technical, operational and risk factors)."

"If these documents contain information that relates to actual surveillance operations (e.g. lessons learnt), then the Government can choose to summarise or redact this information (but such redaction to be indicated in the text)."

Principle 2: The approval principle

"Any relevant Regulator should have a role in assisting Parliament or informing public debate by commenting on the information provided by the application of the Justification Principle."

"Codes of Practice (or parts of Codes) or Statutory Instruments that concern surveillance matters should be subject to prior consultation with a Regulator."

"It is argued that if all Codes of Practices or Statutory Instruments dealing with surveillance could become subject to a requirement for affirmative action by Parliament, then civil servants would want to minimise possible difficulties during the approval process."

"Surveillance (in particular, data sharing or revealing of identity information) occurs with consent of the individuals concerned, the Approval Principle is satisfied if that consent is properly formulated; the Regulator has powers to modify improper consent procedures."

Principle 3: The separation principle

"As the Approval Principle (and Reporting Principle in the case of legislation that has been enacted) allows the Regulator to report to Parliament on any legislation or Code, this can include an independent view on the correct level of separation that is appropriate."

"The Regulator's recommendations on separation can thus be considered by Parliament when it considers the detail of a surveillance proposal contained in a Bill or Code in question."

"Separation is achieved by an authorisation officer or Single Point of Contact (SPOC), the Regulator could determine how separation procedures should apply, the nature of the records to be maintained by that officer or SPOC (Reporting and Adherence Principles)."

"The Regulator would be in a position to report to Parliament about any deficiency in the system of supervision (e.g. present an informed view as to whether judicial approval of warrants would provide better safeguards)."

Principle 4: The adherence principle

"Procedures which authorise a surveillance activity are professionally managed and audited; staff involved in a surveillance activity are fully trained to follow relevant procedures and that such training is assessed if appropriate; any malfeasance in relation to a surveillance activity can be identified and individuals concerned suitably punished."

"One would expect the Regulator to give advice in relation to the correct surveillance procedures to be followed whilst the Independent Supervision and Reporting Principles permits the Regulator to intervene on procedural matters if need be."

"The Adherence Principle provides an oversight mechanism that could include supervision of the privacy related obligations that are connected to the Government's Data Handling Reviews when these are completed (e.g. there is a senior board member responsible for the processing of personal data, that there is suitable training of staff, that procedures are reviewed and maintained, and that risk assessments are taken at regular periods)."

Principle 5: The reporting principle

"A Regulator shall determine what records, including statistical records, are retained and maintained concerning a surveillance activity, in order to ensure transparency and accountability to the Regulator, to the public and to Parliament."

"These records include details about authorisation, cost of surveillance, outcomes, training, management, procedures, any audit or information requirements as determined by the Regulator."

"It is essential to have an independent Regulator identifying all reporting requirements and the criteria which measure the success of a surveillance activity."

"Reports concerning a surveillance activity are produced by the Regulator and should be laid before Parliament and published."

"From an individual's perspective, accountability is achieved if he or she can refer relevant matters to a Regulator for investigation (e.g. after becoming aware of an unjustifiable surveillance activity that involves them)."

Principle 6: The independent supervision principle

"The more invasive the surveillance, the more important it is for the powers of the Regulator to be available."

"Referral to Parliament or public discussion of a surveillance issue (e.g. a report via the Reporting Principle) would only arise if there was a no agreement between Government and a Regulator as to surveillance procedures."

"If an organisation were to employ an independent Data Protection Officer, an idea that has resurfaced in the Information Commissioner's written evidence to Parliament, the Regulator could specify the procedures, records or reporting framework that Officer maintains in relation to a surveillance activity (the Reporting Principle)."

"To ensure independent supervision, a Regulator supervising a surveillance activity should be appointed by, removed by, and report to Parliament."

"The Regulator should be identified in law as being independent assessor of a surveillance activity with particular responsibility to protect the public."

Principle 7: The privacy principle

"Individuals should be granted a right to privacy of personal data which can be enforced by the Data Protection Commissioner and should possess a much simpler right to object to the processing of personal data in appropriate circumstances."

"A revised right to object should not interfere with that processing of personal data that has been undertaken by law enforcement etc, but the right would be easier to claim by the individual, in a context where the processing had been undertaken by a public authority on grounds such as administrative convenience."

"It is also noted that there could be a need to protect the public authority if the right to object to the processing of personal data was being exercised vexatiously (e.g. as part of a campaign to disrupt a public authority)."

Principle 8: The compensation principle

"An individual should obtain compensation if a surveillance activity has caused damage, distress or detriment that proves to be unjustified."

"A Regulator, following detailed investigation under the Independent Supervision Principle, should be able to award limited remedial compensation, to the level assessed in a small claims court, to each individual who has been damaged as a result of a surveillance activity."

"Balance could be provided by allowing the body undertaking the surveillance, to appeal against any award to the Courts."

"Larger sums of money are involved, the Courts would have to become involved immediately; in such cases, and where appropriate, the Regulator should be able to assist individuals with their claim."

Principle 9: The unacceptability principle

"It would be for Ministers and Parliament to decide what to do in the light of the report; however, the assumption is that the public interest generated by the Regulator's report would oblige Parliament or a Committee to consider the surveillance issue in detail."

"It is argued that there needs to be a far stronger "feed-back" loop which gives an informed Parliament a leading role in deciding public policy with respect to balancing the need to perform surveillance against the need to respect private and family life."

"Part of the feedback loop is an empowered Regulator who can investigate issues raised by data subjects and who can report policy problems to Parliament."

"If such conflict arises, the matter can be resolved by Parliament or the Courts; the former dealing with policy matters, the latter dealing with legalities."

Privacy and Social Stratification [84]

This is a machine-generated summary of:

Marx, Gary T.: Privacy and Social Stratification [84].

Published in: Philosophy & Technology (2007).

Link to original: https://doi.org/https://doi.org/10.1007/s12130-007-9009-5

If you want to cite the papers, please refer to the original.

For technical reasons we could not place the page where the original quote is coming from.

Abstract-Summary

"A number of factors that limit unleashing the full potential of privacy-invading technology, even in contexts of inequality, are considered: legal and moral normative constraints on power holders; the logistical and economic limits on total monitoring; the interpretive, contextual, and indeterminate nature of many human situations; system complexity and interconnectedness; human inventiveness; and the vulnerability of those engaged in surveillance to be compromised or responded to in kind."

Some Types of Surveillance

"Citizens can watch government through requirements for open hearings and meetings, freedom of information requests, and conflict of interest, and other disclosure statements."

"The corporation may require citizens to provide personal information in return for goods and services, but it does not offer equivalent information about those within its organization."

"On balance, asymmetry in formal surveillance and information rights within hierarchical organizations and other stratified settings remains."

"The development and use of information-avaricious technologies tends to reflect differential access to resources."

"The ability to control information is central to the borders of social groups."

"A legal oppositional political group (or indeed any group) needs to be able to control information about members, resources, and plans and to feel that freedom of expression within the group is respected."

Is Technology Neutral?

"George Orwell, the author of the novel 1984, was once asked, "Isn't technology neutral?""

"As with the animals in another of Orwell's novels, Animal Farm, surveillance is more neutral for some than for others."

"The actual use of a technology needs to be considered apart from its potential use."

"Part of the neutrality or equality-of-technology argument is equivalent to Anatole France's observation that the rich and the poor are both forbidden to sleep under bridges or steal a loaf of bread."

"This egalitarian potential of the new technology does not mean that all persons and settings have an equivalent chance of being surveilled."

"In the prison case noted above, in spite of the omnivorous potential of the lens, there is unlikely to be a camera or audio recording of what goes on in the warden's office or recreational room for guards, nor in the bathroom they use."

Complicating Factors

"A number of factors limit unleashing the full potential of privacy-invading technology, even in contexts of inequality: legal and moral normative constraints on power holders; the logistical and economic limits on total monitoring; the interpretive, contextual, and indeterminate nature of many human situations; system complexity and interconnectedness; human inventiveness; and the vulnerability of those engaged in surveillance to be compromised or responded to in kind."

"In spite of doomsday scenarios with respect to the death of privacy and liberty, in societies with liberal democratic economic and political systems, the advantages of technological and other strategic surveillance developments are sometimes short-lived and contain ironic vulnerabilities."

"As life comes to imitate art, contemporary electronic lifestyles reverse some aspects of the traditional relationship between stratification and surveillance, at least with respect to documentary records of behavior."

[Section 4]

"One insufficiently considered issue involves implications for stratification and equality."

"The control over information—the defining characteristic of privacy—is a resource related to, but with varying degrees of independence from, other scarce social resources such as class, status, and power."

"Rules regarding who can collect personal information, what is collected, the conditions under which it is gathered, and how it is used (and by whom) are very much connected to social stratification."

"Many contemporary concerns over privacy invasion involve large organizations and their employees and customers, police and suspects, guards and prisoners, and professionals and clients, as well as interpersonal relations as with parents and children."

"Situations involving power differences with respect to gender and ethnicity may also reflect information inequality."

"In the high visibility conditions of wired shopping malls, guards never know when they are being watched, overheard, or secretly tested."

Own Data? Ethical Reflections on Data Ownership [85]

This is a machine-generated summary of:

Hummel, Patrik; Braun, Matthias; Dabrock, Peter: Own Data? Ethical Reflections on Data Ownership [85].

Published in: Philosophy & Technology (2020).

Link to original: https://doi.org/https://doi.org/10.1007/s13347-020-00404-9

If you want to cite the papers, please refer to the original.

For technical reasons we could not place the page where the original quote is coming from.

Abstract-Summary

"We identify four conceptual dimensions of calls for data ownership and argue that these help to systematize and to compare different positions."

"Pluralism of data ownership claims, we introduce, spell out and defend a constructive interpretative proposal: claims for data ownership are charitably understood as attempts to call for the redistribution of material resources and the socio-cultural recognition of data subjects."

"We argue that as one consequence of this reading, it misses the point to reject claims for data ownership on the grounds that property in data does not exist."

"Data ownership brings to attention a claim to renegotiate such aspects of the status quo."

Introduction

"In this spirit, calls for data ownership demand that data can be property."

"We argue that all of these dimensions of data ownership matter for informational self-determination, understood as the ability of data subjects to shape how datafication and data-driven analytics affect their lives, to safeguard a personal sphere from others, and to weave informational ties with their environment."

"We need to take seriously the full range of dimensions of data ownership in order to understand how data subjects exercise informational self-determination, and how such exercises can be facilitated and promoted."

"We distinguish four dimensions of data ownership."

"Statements on data ownership touch upon and go back and forth between two different spheres: the redistribution of material resources and the socio-cultural recognition of data subjects."

"In view of our findings, the notion of data ownership can be understood as an expressive resource for articulating and negotiating claims concerning both spheres."

Legal Frameworks

"One primary reason for legal scholars to be critical of data ownership is that there are important differences between data and paradigm cases of property (Zech 37, 117–119)."

"This would raise at least two challenges: first, data ownership concerns data that relates to individuals, but unlike intellectual property, such data are typically not created or invented by them."

"This does not establish that individuals can or should have information as property, but it does challenge the suggestion that mere non-rivalry of information speaks against data ownership."

"Whether data or its usage are persistent, scarce, or rivalrous, and which significance this should have for justified ownership claims, are questions that have implications for law and relate to legal frameworks, but that importantly go beyond legal inquiry and discourse."

"This, in turn, suggests that doubts about the existence of data ownership relative to existing legal frameworks should not prevent us from exploring the notion in further detail."

Aspects and Indeterminacies of Data Ownership

"Evans argues that status quo regulations (in her case: US federal regulatory protections) and proposed data ownership models do not differ much with regards to balancing privacy with public access and non-consensual data use: "Creating property rights in data would produce a new scheme of entitlements that is substantively similar to what already exists, thus perpetuating the same frustrations all sides have felt with the existing federal regulations" (Evans 38, 75)."

"Other calls for data ownership share the conviction that certain fundamental resources need to be available to individuals for self-constitution, forming and implementing life plans, and participating in communal lifeforms."

"In these proposals, data ownership held by populations, society or mankind need not preclude further ownership or control rights at the individual level."

"We have suggested that calls for data ownership position themselves along four dimensions, specified by four pairs of poles: property versus quasi-property, marketability versus inalienability, protection versus participation, and individual versus collective claims and interests."

Data Ownership and the Prerequisites for Informational Self-determination in Digitization

"When property and quasi-property are discussed, the question concerns the extent to which the redistribution of data resources and power is called for, whether it shall take the form of assigning genuine ownership, or whether—and in which cases—this kind of redistribution stands in tension with commitments on recognition, for example the ascription of achievement to data processors."

"The dimensions of data ownership articulate stances on—and shift between matters of—distributive justice and cultural recognition."

"Proponents of data ownership argue that redistributions are necessary, whereas opponents maintain that from the cultural perspective of recognition and achievement, status quo distributions are appropriate."

"Data ownership thus expresses stances on the redistribution of material resources and socio-cultural recognition of data subjects in a datafied and data-driven lifeworld."

"Informational self-determination requires both redistribution and recognition, and the different dimensions of data ownership overlap with and across these spheres."

Conclusion

"Data ownership encodes a variety of different concerns."

"but also because commentaries on the topic oscillate between poles that concern primarily pre- or para-legal questions: whether data is rivalrous, whether property is the right category to articulate and recognize the claims of data subjects or whether surrogate notions are more suitable (3.1), whether individuals should be in a position to market their data (3.2), whether and how we shall consider both protective and participatory claims (3.3), and how the claims of individual data subjects relate to collective claims and interests (3.4)."

"Data subjects claiming data ownership are sometimes concerned primarily with issues of controlling data flows, inferences drawn about them, and the impact of data processing on their lives."

"Besides scoping out the conceptual space and highlighting how the dimensions of data ownership are entangled, we intend the present contribution as an invitation to relate endorsements and rejections of data ownership more explicitly and systematically to the proposed reference points of recognizing data subjects and redistributing resources throughout the datafied and data-driven lifeworld."

The global governance on Automated Facial Recognition (AFR): Ethical and legal opportunities and privacy challenges [86]

This is a machine-generated summary of:

Bu, Qingxiu: The global governance on automated facial recognition (AFR): ethical and legal opportunities and privacy challenges [86].

Published in: International Cybersecurity Law Review (2021).

Link to original: https://doi.org/https://doi.org/10.1365/s43439-021-00022-x

If you want to cite the papers, please refer to the original.

For technical reasons we could not place the page where the original quote is coming from.

Abstract-Summary

"Automated facial recognition (AFR) comes with many concerns as well as benefits."

"The way facial images are obtained and used, potentially without consent or opportunities to opt out, can have a negative impact on people's privacy."

"Laws on privacy vary across jurisdictions, which has an enormous effect on measures that could be taken to safeguard AFR-related ethical concerns."

"It is imperative to build an AFR infrastructure that incorporates society's legal and ethical commitments, and further address the challenges of governing the technology."

Introduction

"Automated facial recognition (AFR) is a biometric technology that uses cameras to match live footage of individuals with images from a database [39]."

"Despite its widespread use, AFR is operating in a legal grey area."

"The risk-intensive processes of AFR shed light on the limits of current legal and ethical frameworks, which entails a global challenge to alleviate the impending threats to privacy."

"This article proceeds as follows: Part I starts with a benefit–cost analysis and further ascertains whether the use of AFR is viable or untenable in terms of privacy protection."

"This part casts a challenge on a long-standing theory of moral and ethical supremacy and inferiority in the context of AFR deployment."

"The concluding remark affirms a momentous duty for both public and private actors to get the most out of AFR while still protecting privacy."

AFR's benefits and privacy challenges

"As a dual-use technology, AFR has a wide range of benefits and potential privacy implications."

"The threat of perpetual surveillance erodes fundamental rights, because there is a significant gap between the AFR and the laws regulating its use [40]."

"Even the attainment of privacy could be subject to limitations; values are modulated by circumstances with AFR being used circumstantially in both private and public spheres [41]."

"AFR should operate in an adaptive manner to ensure that the digital world has places where law-abiding people can enjoy privacy [42]."

"The question arises as to whether the use of AFR should be banned until the right legal framework along with privacy and security safeguards are in place [43, 44]."

"The central concern is that the deployment of AFR needs to be adequately regulated to preserve privacy."

Provisions under the general data protection regulation (GDPR)

"Under the EU law, biometric data is defined as "personal data resulting from specific technical processing relating to the physical, physiological or behavioural characteristics of a natural person, which allow or confirm the unique identification of that natural person, such as facial images"."

"The EU protects people from the threats of facial recognition by enforcing the General Data Protection Regulation (GDPR), which prevents the processing and sharing of biometric data without consent."

"The GDPR generally forbids the processing of biometric data for uniquely identifying purposes unless one can rely on one of the 10 exemptions."

"The law provides that collection and processing of biometric data including facial recognition is valid when "the data subject has given explicit consent to the processing of personal data"."

"The GDPR includes exemptions for the collection of biometric data like facial recognition by authorities, even though such information is considered "sensitive" and highly restricted in the hands of private companies."

AFR-related laws in the U.S.

"In Carpenter v. United States, the Supreme Court held that the government's warrantless access to an extensive compilation of cell phone user data violated the Fourth Amendment."

"The privacy-protective case helps frame the analysis, and the ruling in Facebook v. Patel would have shaped how courts view AFR."

"Although, in Riley v. United States, the Supreme Court's quantitative and qualitative analysis applies to the challenges of AFR surveillance, there has so far been no developed case law or constitutional precedent upholding the police use of facial recognition without a warrant."

"CCPA 2019 has substantial impact on privacy rights and consumer protection, which is sometimes considered as a model for a federal data privacy law On 14 March 2019, the Commercial Facial Recognition Privacy Act (CFRPA) was introduced by senators to offer legislative oversight on AFR's commercial application [45]."

Legal challenges against the use of AFR in the UK and China

"The UK is also one of the most surveilled countries in the world [46], where a plaintiff's challenge against policy is on the grounds that the use of AFR breaches the right to privacy and data protection laws."

"The legal framework which currently applies to the use of AFR by law agencies and private actors does not ensure those rights are sufficiently protected."

"The law imposes legal obligations on operators by stating the requirements for the collection, use and protection of personally identifiable information (PII), which includes biometric data [47]."

"There are currently neither laws governing the specific use of AFR, nor overarching principles of data protection being set at the national level."

"Despite the few regulations surrounding law enforcement's use of AFR, legislation should require that public agencies rigorously review biometric technologies for privacy concerns [48]."

Public enforcement authorities vis-à-vis private actors in deploying AFR: a theoretical analysis of social licence

"AFR is being used in public spaces, not only by law enforcement agencies but also increasingly by the private sector."

"Protecting privacy is a widely shared social preference [49], rigorous assessments of AFR use should be undertaken in the context of public and private partnership (PPP)."

"AFR use exposes consumers to the risk of their private information being shared with unintended recipients due to data breaches."

"Using the technology for security and surveillance reasons, public actors typically rely on private companies for procuring and deploying AFR."

"Data-sharing of surveillance material not only increases the potential of privacy harms, but also blurs the public and private boundaries given the differentiated rationales behind their access [50]."

"The asymmetric power over information between private AFR users and data subjects increases the potential for abuse."

"There is little justification and low public approval for the use of AFR systems operated by a private actor."

A paradoxical game in surveillance and data collection

"AFR raises a host of ethical and legal questions about privacy along with surveillance."

"As a world-leading AI-powered surveillance state, China has embraced AFR, using it to implement a national surveillance system."

"China has led the way in developing and deploying AFR, and set up the world's most sophisticated surveillance state."

"While domestic concerns grow, China's facial recognition companies are leading the global market for public surveillance systems [51]."

"Some commentators observed: While debate over the use of facial recognition in the EU and the U.S. is focused on the privacy threat of governments or companies identifying and tracking people, the debate in China is often framed around the threat of leaks to third parties, rather than abuses by the operators themselves [52]."

"Dominant in the global facial recognition market, Chinese companies have made every submission to the UN for international standards on surveillance technology in the past 3 years [53]."

Embed principles and rules into the AFR governance regime

"The use of AFR is subject to insufficient regulation that is specifically designed to protect human rights given the new challenges posed by emerging technology [54]."

"Pursuant to the GDPR, the processing of biometric data is only allowed where processing is: Necessary for reasons of substantial public interest, on the basis of Union or Member State law which shall be proportionate to the aim pursued, respect the essence of the right to data protection and provide for suitable and specific measures to safeguard the fundamental rights and the interests of the data subject."

"Initiating a much-needed and vigorous public debate about the proper balance between the AFR-related surveillance and privacy rights should be put high up on the agenda."

Conclusion

"The existing laws that protect individual biometric data are not adequate to respond to the challenges posed by AFR."

"The use of AFR calls for a new legal and regulatory framework to avoid a dystopian future."

"Enforcement authorities must be ensured of being able to make efficient use of AFR's powerful investigatory roles, while privacy should be taken into reasonable account."

"More factors need to be considered, including the legal basis, necessity, proportionality and justification, in order to address intrusive AFR processing."

"To ensure the respect of human rights in this new socio-technical context, both public and private actors need to be committed to striking the right balance when using AFR."

"Processing of personal data constitutes a limitation of privacy, it needs to be subjected to a strict necessity and proportionality test, including a clear legal basis to do so and a legitimate aim pursued."

Philosophy, privacy, and pervasive computing [87]

This is a machine-generated summary of:

Michelfelder, Diane P.: Philosophy, privacy, and pervasive computing [87].

Published in: AI & SOCIETY (2009).

Link to original: https://doi.org/https://doi.org/10.1007/s00146-009-0233-2

If you want to cite the papers, please refer to the original.

For technical reasons we could not place the page where the original quote is coming from.

Abstract-Summary

"Philosophers and others concerned with the moral good of personal privacy most often see threats to privacy raised by the development of pervasive computing as primarily being threats to the loss of control over personal information."

"One reason is that the parallels between pervasive computing and ordinary networked computing, where everyday transactions over the Internet raise concerns about personal information privacy, appear stronger than their differences."

"Another reason is that the individual devices which can become linked in a pervasive computing environment: PDAs, GPS sensors, RFID chips/readers, publicly-located video surveillance cameras, Internet-enabled mobile phones, and the like, each raise threats to individual privacy."

"Without discounting the value of this approach, this paper aims to propose an alternative; and, as a result of recasting the threat to individual privacy from pervasive computing, to identify other, and deeper, moral goods that pervasive computing puts at risk that otherwise might remain concealed."

Introduction

"I want to sketch out and defend an approach to thinking about the moral importance of protecting privacy in a world whose everyday lived-in environments reflect the presence of computing technology in a particular way."

"It is with this in mind that I want to take up one such question here: that of the moral value of protecting privacy in a world of pervasive computing."

"The moral value of privacy in an age of pervasive computing is best seen not by taking the protection of items of personal information as a starting point, but through looking more holistically at how pervasive computing affects one's very orientation to the world itself."

"In taking up the question of the moral value of privacy within an environment of pervasive computing, I will begin by considering those differences between these two worlds that are qualitative in nature."

Privacy as control of personal information in network-enhanced environments

"As the world of pervasive computing continues to develop along its present course, particularly through the convergence of forms of mobile communication with RFID technologies, we can expect that more and more information will be collected about individuals by the simple virtue of their presenting themselves within particular environments, and not because they are deliberately providing well-defined personal information via a computerized information network to another party."

"It is the presentational nature of the pervasive computing environment that makes it difficult to think of the moral value of protecting privacy within this context as first and foremost associated with control of personal information."

"If as I have been suggesting here it is the case that the presentational character of pervasive computing environments makes it very unlikely that I can know in advance what information about myself is being collected, then to find the moral dimension of privacy in a pervasive computing environment we need to extend our vision beyond the idea of having control over our personal information."

Privacy and existential autonomy

"This proposal, developed by the computer scientist Scott Lederer while a graduate student in the Group for User Interface Research at the University of California at Berkeley, represents an attempt to take the four traditional principles of fair information practice: notice, consent, security, and redress—principles upon which many Web sites rely in order to protect user privacy—and import them to the world of pervasive computing as a basis for designing mechanisms for individual privacy to be protected (Lederer 55, p. 14)."

"Lederer identified the principles of notice and consent as being crucial to the protection of privacy within a world of ubiquitous computing, as these principles offered individuals the "greatest everyday utility" (p. 48)."

"Combined with the provision of a data review log, Lederer initially hoped this would represent the pathway for an individual to find an effective way to manage who has access to his personal information in a world of pervasive computing."

Existential autonomy and human well-being

"There is a critical difference between having existential autonomy over being a provider of technological information, and having the autonomy to make decisions that contribute to the formation of the self and to the deep, personal narratives that emerge from this formation, as do the decisions involved in the constitutional-related cases."

"The spread of pervasive computing brings forcefully to our attention that while it might be possible to define surveillance quite generally so as to fit all situations and all purposes, distinguishing among different forms of surveillance might be helpful in recognizing another dimension to the risks to our well-being through the loss of privacy engendered by pervasive computing; and, in particular, through the loss engendered through compromises to our existential autonomy rather than through the collection of personal information."

Understanding Privacy Regulation in Ubicomp Interactions [88]

This is a machine-generated summary of:

Lehikoinen, Jaakko T.; Lehikoinen, Juha; Huuskonen, Pertti: Understanding privacy regulation in ubicomp interactions [88].

Published in: Personal and Ubiquitous Computing (2007).

Link to original: https://doi.org/https://doi.org/10.1007/s00779-007-0163-2

If you want to cite the papers, please refer to the original.

For technical reasons we could not place the page where the original quote is coming from.

Abstract-Summary

"We aim at analyzing this complex issue by applying and extending Altman's theoretical privacy framework, well known in social sciences, to privacy in ubicomp."

"Altman understands privacy as a two-way interactive process, which makes the approach promising in analyzing ubicomp where people, devices and the environment interact with each other."

"We point out similarities between the existing model and the features of ubicomp environment, and verify the results by applying and analyzing the resulting extended framework to typical ubicomp use cases."

Introduction

"For discussion on ubicomp challenges related to computer science in general, see also [56]."

"We focus on the class of human-related challenges, and emphasize one of the biggest threats usually associated with ubicomp: privacy."

"In this paper we combine the knowledge from both ubicomp research and privacy studies in social sciences."

"We aim at understanding the privacy in ubicomp via Altman's existing framework of privacy in social sciences [57]."

"Since Altman's theory is well recognized and significant in the field of social sciences and further, it emphasizes the roles of social interaction and an environment as essential elements in privacy regulation, it seemed to provide feasible grounds for this work."

"We review related work in ubicomp privacy, followed by a brief introduction to Altman's theory."

Related work

"They discuss the privacy threats regarding ubiquitous computing in the working environment, and continue by presenting a framework for designing privacy in ubicomp environments."

"They fail to refer to the works of Altman or Westin when discussing societal issues and privacy."

"Price and others [58] propose a model for user control of privacy."

"They have identified four layers related to privacy in ubicomp: the regulatory regime a user is currently on; the type of ubicomp service required; the type of data being disclosed, and personal privacy policy."

"Hong and others [59] propose privacy risk models as a tool for identifying concrete privacy issues in ubicomp, and prioritizing them."

"By applying the principle of minimum asymmetry, privacy-aware ubicomp spaces and system should decrease the asymmetry of information between data owners, data collectors, and data users."

A brief introduction to Altman's theory of privacy

"Altman [57] understands privacy as dialectic and dynamic boundary control process, which regulates interaction with others."

"Altman illustrates this process by the concepts of desired privacy, interpersonal control mechanisms, and achieved privacy."

"Altman's framework of privacy should be understood as two-way process, in which the self and the others interact with each other."

"Altman uses the concepts of crowding and its counterpart social isolation as resulting from breakdowns in achievement of desired levels of privacy."

"Social isolation means that achieved privacy is greater than desired level of privacy, which means that a person has been cut off from the social interaction, causing for example loneliness or boredom."

"All these interactions can potentially affect the level of achieved privacy for a person in the environment (or a person remotely present, for that matter)."

Analysis of Altman in ubiquitous computing environments

"Personal space, one of the control mechanisms discussed by Altman, can be understood in ubicomp environments as a combination of self, personal device(s) and digital self."

"The self may control the output of the personal device to other humans for instance by hiding it, adjusting volume of loudspeaker or brightness of display so that the information mediated via the device is more or less observable by other people."

"Context awareness can also be used as control mechanism in ubicomp environment; the self may manage the information that her device communicates to the environment or other people's devices based on pre-defined context attributes."

"The reason why Altman does not consider the amount of output as an potential threat for privacy is the different control mechanism in face-to-face and mediated communication: in face-to-face setting an individual has clear and well established means to control social interaction with others whereas in ubicomp environments they do not necessarily even notice that information about themselves has been communicated to others."

Examination of privacy regulation by typical ubicomp scenarios

"Due to the fact that Tom has no means to control the information exchange, this form of interaction is fairly close to the state of privacy what we call leaking."

"Lizzie utilizes also another way of practicing territorial control in what we call "homecomp" interactions; she limits the number of tunes to five that the guests can share from their mobile devices with the home audio system and thus with everyone present in the party."

"The following aspects have been identified in this scenario: private sphere allowing more control mechanisms for the owner, management of access rights, personal spacing, selection of material to be shared, semi-anonymity as control mechanism, and several levels of territorial responses."

"Personal spacing in ubicomp can be seen as a control mechanism that includes interaction between the self and a personal device in a social context."

Discussion

"The interactive approach fits well to the characteristics of a typical ubicomp environment, and allows analyzing them from the privacy point of view."

"With the identified extensions, like non-human actors, information as an object, and new control mechanisms, we are able to model interactions and thus privacy regulation in ubicomp to a greater degree."

"Of discussing interactions, the previous privacy research in field of ubicomp concentrates mostly on information—another important aspect of privacy."

"Many others have also studied privacy from the information point of view, including the actions performed on it."

"One could argue that the actions performed on the information and the interactions that take place between different actors in a ubicomp environment together form a rather holistic view on what affects ubicomp privacy and how."

"Considering telepresence from privacy regulation point of view is another interesting direction for further research."

Conclusions

"In order to better understand privacy in ubiquitous computing, we have examined whether Altman's well-known social psychological theory of privacy could be applied to ubicomp, extending it when needed."

"Altman considers privacy a two-way interactive process, making the approach promising from a ubicomp point of view where people, devices and services interact with each other in a rich manner."

"Our analysis is based on considering typical characteristics of a ubicomp environment that could potentially affect achieved privacy, and applying Altman's theory by utilizing illustrative use cases and pointing out significant features and their equivalencies to Altman's terms."

"Based on the analysis we argue that Altman's theory of privacy can be applied to ubicomp environments, provided that it can be extended when needed."

Social network data analysis to highlight privacy threats in sharing data [89]

This is a machine-generated summary of:

Cerruto, Francesca; Cirillo, Stefano; Desiato, Domenico; Gambardella, Simone Michele; Polese, Giuseppe: Social network data analysis to highlight privacy threats in sharing data [89].

Published in: Journal of Big Data (2022).

Link to original: https://doi.org/https://doi.org/10.1186/s40537-022-00566-7.

If you want to cite the papers, please refer to the original.

For technical reasons we could not place the page where the original quote is coming from.

Abstract-Summary

"The definitions of privacy requirements and default policies for safeguarding people's data are the most difficult challenges that social networks have to deal with."

"We have collected data concerning people who have different social network profiles, aiming to analyse privacy requirements offered by social networks."

"We have built a tool exploiting image-recognition techniques to recognise a user from his/her picture, aiming to collect his/her personal data accessible through social networks where s/he has a profile."

"We aim to analyse the amount of extrapolated data for evaluating privacy threats when users share information on different social networks to help them be aware of these aspects."

"This work shows how users data on social networks can be retrieved easily by representing a clear privacy violation."

"Our research aims to improve the user's awareness concerning the spreading and managing of social networks data."

Introduction

"With such a vast number of users, the need to analyse how they manage privacy over several social networks is becoming vital to detect which information is not privatised, by making users more aware of how to prevent privacy issues."

"Users tend to use social networks to share information massively; in most cases, they do not care about privatising data and are unaware of the privacy threats they can be exposed to."

"By exploiting face recognition and data analysis, we have built the social data analyzer (SODA), a tool for extrapolating users' information made available from social networks, aiming to perform an accurate analysis for revealing privacy threats linked to incorrect usage of data sharing in social networks."

"(SODA) is independent of the privacy settings offered by social networks; it simulates the search of a user retrieving data available in his/her social network profile."

Related work

"This section discusses relevant articles in which social networks privacy-preservation is addressed to evaluate risks connected to personal user data."

"In the context of privacy preservation for sharing data in social network platforms, several approaches define strategies to make users aware of the privacy issues linked to their posted data."

"In [60], the authors define a new approach for helping social media users to evaluate their privacy disclosure score (PDS)."

"The authors have analyzed data concerning 15 users registered over different social networks (Facebook, ResearchGate, LinkedIn, and Google+) to perform their analysis."

"Social network data represents a rich source of information, mainly when it characterizes users, and malicious users can jeopardize the user's privacy by performing targeted attacks to recover sensitive information."

"Authors explain that an attacker can violate the user's privacy by studying his/her social network."

Methodology

"If a match is found, the social network module yields all the user profile data available on his/her specific social network."

"The main difference is the exploitation of multiple social network modules for extrapolating several user profile data."

"Each module can extract user profile data from a specific social network."

"It possible to collect several user profile data from different social networks."

"All the user profile data associated with each social network feed the integration module for aggregating all collected user profile data."

Social data analyzer

"There are several issues related to extraction that yield specific choices for the components of the SODA tool: (1) the number of users involved in the analysis process can be large, (2) each social network relies on different implementation technologies, and (3) continuous updates of the social network platforms require continuous maintenance of system components."

"With respect to Social Mapper, the proposed tool SODA provides several novel functionalities that allow us to perform an in-depth analysis of the data shared by users, and extend the search on a large scale."

"It consists of all the initial user information, which enables Social Mapper to acquire data for user profiling; The Business layer containing the modules for extracting information from different social networks."

"To enable SODA to analyze the information that a user shares, it has been necessary to upgrade each component of Social Mapper in order to add new functionalities for crawling information from different social networks."

Experimental evaluation

"34% of them allows the reconstruction of this information from other social networks; Among 5000 analyzed users who have privatized their Date of birth on Facebook or are not registered to this social network, 3% shared it on VKontakte, and 3% on Linkedin."

"94% of analyzed users have privatized this information, since 6% of them shared it to other social networks; Among the 5000 analyzed users who have privatized the Email on Facebook or are not registered to this social network, only 1% of them shared it on Linkedin, while 1% on VKontakte."

"8% of analysed users shared Hometown or Residence on other platforms, so 92% have completely privatised it; Among the 5000 users analyzed, 4861 have privatized their Date of birth on Linkedin or are not registered to this social network."

Ethical discussion

"Processing and analyse data gathered by social networks could raise ethical discussions."

"According to the claims described above, we have collected social networks users data to perform a specific analysis with the only aim to improve user's awareness concerning privacy threats over different social networks."

"To comply with GDPR, only the statistics retrieved from the collected social network data will be made public without publishing the data itself."

"From the ethical point of view, we have not violated user's privacy because this was not our target; we have collected data with the only purpose to emphasise privacy issues related to social network data dissemination."

"This research could be the baseline to improve the user's awareness in terms of data privacy and also help to determine new strategies to privatise social network user's data"

Conclusion and future directions

"In our work, we have performed a single-social and a cross-social evaluation concerning users' data, to assess how easily they can be reconstructed from social networks."

"Through the cross-social analysis, we also reconstructed other significant users data by exploiting the combination of several social networks."

"We want to clarify that our analysis aims not to violate users' privacy over their social network accounts but only emphasise that users do not understand in deeply privacy threats linked to an incorrect sharing/privatisation of their data."

"We would like to collect more data concerning users, by integrating information over other social networks."

Honeypots and Honeynets: Issues of Privacy [90]

This is a machine-generated summary of:

Sokol, Pavol; Míšek, Jakub; Husák, Martin: Honeypots and honeynets: issues of privacy [90].

Published in: EURASIP Journal on Information Security (2017).

Link to original: https://doi.org/https://doi.org/10.1186/s13635-017-0057-4.

If you want to cite the papers, please refer to the original.

For technical reasons we could not place the page where the original quote is coming from.

Abstract-Summary

"We outline the privacy issues of honeypots and honeynets with respect to their technical aspects."

"The paper discusses the legal framework of privacy and legal grounds to data processing."

"The analysis of legal issues is based on EU law and is supported by discussions on privacy and related issues."

Introduction

"One of the most popular method of learning about attackers is using honeypots."

"A honeypot is deliberately allowed to be compromised, and the attack is then analyzed so that we can learn about the methods, procedures, and tools that the attacker used."

"We have to keep in mind that there are legal issues regarding honeypots that need to be addressed when deploying one, analysing the captured data, and sharing the results with others."

"This issue influences how a honeypot can be deployed, what data they are allowed to collect, and what we can do with the collected data."

"What data are legally allowed to be collected by honeypots?"

"This section focuses on previous literature related to the legal aspects of honeypots and honeynets, especially the issue of privacy."

State of the art

"The related work on honeypots and legal issues of honeypots will be presented."

"For the purpose of this paper, we classify honeypots according to their level of interaction and purpose."

"The papers dealing with the legal aspects of honeypots and honeynets focus on three fundamental legal issues of the deployment and usage of honeypots: privacy, liability, and entrapment."

"Most of papers focus on legal issues from the US law perspective."

"There are papers that at least outline the legal issues of honeypots from the perspective of the EU law For example, Dornseif and others [61] focus on legal issues of the usage of honeypots in the context of German laws."

"The paper also discusses the implications of data mining techniques and the issue of privacy protection with regard to the use of forensics methods."

Privacy and personal data protection

"EU Directive 2002/58/EC focuses on the processing of personal data and the protection of privacy in the sector of electronic communications."

"The Data Protection Directive recognizes in the Art 7 several legal grounds for data processing, from which the following are relevant for the case of honeypots and honeynets: The data subject has unambiguously given their consent (letter a)) The processing is necessary to comply with a legal obligation to which the controller is subject (letter c)) The processing is necessary for the purposes of legitimate interests pursued by the controller or by a third party or parties to whom the data are disclosed, except cases where such interests are overridden by the interests for fundamental rights and freedoms of the data subject (letter f)) The Data Protection Directive sets four conditions for the validity of consent which must be met."

Other privacy issues

"If we take a look at this in the light of data retention, the retention of data for considerably long periods of time leads to risk exposure, which will pose a long-term threat to privacy and security to guarantee privacy rights in practice."

"If data are stored, if all traffic is stored or monitored, then a leak of these data is a probable threat and a theft of these data is more probable the longer the period of time, because nothing is perfectly secure on the Internet and risks will become more salient as the time for their emergence and chance to occur increases."

"What if a honeypot worked as a support facility to a chat server or another electronic service and research and security would require specific data?"

"Network trace anonymization is an active area of research in the security community, as shown by the ongoing development of anonymization methods and the releases of network data that they enable [62]."

Conclusions

"The paper has outlined the concept of personal data protection in the EU law and focused on the issues of the purpose of data processing, the legal grounds for data processing, and the retention of data."

"It is also crucial to identify a relevant lawful purpose of personal data processing and choose a correct legal ground for such a processing."

"Administrators of honeypots and honeynets, who are seen as personal data controllers in the eyes of the law, because IP addresses are considered personal data within EU law, can rely on the legal ground of legitimate interest to collect and process personal."

"An adequate length of retention of personal data is also an important issue for the processing of honeypot data, since the data minimization principle applies."

"In situations where a researcher wants to publish data collected by honeypots and honeynets, anonymization is needed."

1.2 An Economic Argument

Machine generated keywords: business, company, advertising, value, revenue, economic, billion, customer, service provider, personally identifiable, personally, partner, asset, scholar, efficiency

Privacy, property rights and efficiency: The economics of privacy as secrecy [91]

This is a machine-generated summary of:

Hermalin, Benjamin E.; Katz, Michael L.: Privacy, property rights and efficiency: The economics of privacy as secrecy [91].

Published in: Quantitative Marketing and Economics (2006).

Link to original: https://doi.org/https://doi.org/10.1007/s11129-005-9004-7.

If you want to cite the papers, please refer to the original.

For technical reasons we could not place the page where the original quote is coming from.

Abstract-Summary

"A central element of privacy is the ability to control the dissemination of personally identifiable data to private parties."

"Others have argued that privacy issues are readily resolved by proper allocation of property rights to control information."

"We find: (a) privacy can be efficient even when there is no "taste" for privacy per se, and (b) to be effective, a privacy policy may need to ban information transmission or use rather than simply assign individuals control rights to their personally identifiable data."

Introduction

"There are also calls simply to establish appropriate property rights to information on the grounds that market forces will then lead to efficient privacy levels."

"Is there an assignment of property rights to personally identifiable information that leads to an optimal level of privacy or disclosure?"

"The Chicago School, most notably Posner (63) and Stigler (64), asserts that privacy is harmful to efficiency because it stops information flows that would otherwise lead to improved levels of economic exchange."

"Privacy can lead to informational asymmetries that destroy markets and prevent efficient exchange."

"We also establish conditions under which intermediate increases in information break what would have been efficient broad pooling of households types under privacy and lead to narrower pools that entail greater levels of inefficient exclusion."

A general model and results

"We are interested in comparing: equilibrium when firms have the right to compel revelation of the indicator variable with equilibrium when households have the right to conceal this information."

"If firms commit to contract offers before households reveal their indicator variables, then the set of equilibrium outcomes is independent of whether the property rights to personally identifiable data are granted to the firms or to the households."

"If there is a least-favored indicator value, then there exists an equilibrium when households hold the right to conceal their personally identifiable data that supports the identical outcome as the equilibrium that arises when firms possess the information property rights."

"Given the possible irrelevance of whether households have the right to conceal information or firms have the right compel its revelation, we are also interested in equilibrium when firms are legally prevented from making use of σ whether or not households would otherwise consent to its release and use."

Market power and price discrimination

"Suppose that there are two types of household, with linear demand functions having a common slope, and the indicator variable can take two values, one of which is good news (implying the other is bad news). (a) If low-type households would be excluded from the market under privacy, then total surplus is weakly lower under privacy than in the absence of privacy. (b) If, absent privacy, low-type households in both sub-populations would have positive consumption levels, then privacy leads to strictly higher total surplus than the absence of privacy. (c) In other cases, privacy may raise or lower total surplus relative to the absence of privacy depending on the parameter values."

"If the proportion of low-type households in the bad-news sub-population (e.g., $\sigma = 1$) is high enough, the seller will not exclude these households when conditioning on $\sigma = 1$, which results in welfare strictly greater than under privacy."

Competitive markets and adverse selection

"Observe, however, that once efficient pooling has been destroyed, further sorting raises total surplus back towards the efficient-pooling level—such sorting has no effect on the employment of low-ability workers but it moves high-ability workers from inefficient unemployment as members of sub-population 1 to efficient employment as members of sub-population 2 ($g_1 < g^* < g_0$ implies $g_2 > g^*$)."

"Once the workers are sufficiently sorted that $g_2 > g^*$, high-ability workers in sub-population 2 are efficiently employed."

"Additional sorting increases total surplus further because the number of (employed) high-ability workers in sub-population 2 rises and the number of (unemployed) high-ability workers in sub-population 1 falls."

"In the high-productivity case, low-ability workers are made increasingly worse off by more information when $g_1 < g^* < g_0$."

"Everyone is employed with or without privacy, but increased information means high-ability workers are less co-mingled with low-ability workers and, thus, appropriate a higher percentage of their revenue product."

Conclusion

"With so many people making extreme claims in discussions of privacy and related public policy, and with so little understanding of the underlying economics, it is important to identify the fundamental forces clearly."

"Can one identify conditions under which public policy should or should not promote privacy?"

"In those circumstances where privacy is desirable, how can public policy promote it?"

"As our analysis shows, discerning a privacy policy's distributional consequences is not always clear either; certainly in the case of employment, changes in privacy policy can make some households winners and others losers."

"In some situations, the only effective policy would be explicitly to block the dissemination or use of such information."

"Policy makers might allow dissemination and instead attempt to block the use of the protected information."

Privacy law as price control [92]

This is a machine-generated summary of:

Fuller, Caleb S.: Privacy law as price control [92].

Published in: European Journal of Law and Economics (2017).

Link to original: https://doi.org/https://doi.org/10.1007/s10657-017-9563-6.

If you want to cite the papers, please refer to the original.

For technical reasons we could not place the page where the original quote is coming from.

Abstract-Summary

"To address these fears, the European Union passed the Privacy Directive to regulate the common business practice of information collection."

"I extend this logic to argue that if a regulation enables consumers to stop supplying this information, while continuing to consume the site's content, it is equivalent to a price control."

"I discuss unintended consequences that this price control may generate: tie-in sales, investment flight, and altered exchange characteristics."

"My analysis suggests that one of the most well-researched policy interests of economics—the theory of price controls—can shed light on one of economists' newest interests: digital privacy."

Introduction

"Regulation of digital privacy stems from fears regarding two potential uses of consumer data: illegitimate, fraudulent activity as well as legitimate activity that consumers frequently decry, such as behavioral ad targeting and price discrimination (De Corniere and De Nijs 65)."

"Information asymmetry, some have concluded that digital privacy is inadequately protected and that the digital arena is a "classic example of a market failure," (Gertz 66)."

"How firms react to digital privacy legislation is considered to be among the most important questions in this burgeoning literature (Acquisti et al. 67, p. 479)."

"Despite fears regarding firms' privacy policies, information collection by firms serves a vital role in the so-called digital economy, a role so seminal that its absence would all but preclude some of the most popular sites on the Internet."

"Foremost among these differences in constraints is the EU Privacy Directive that curtails the information-collecting practices that are so common among US-based firms."

"Privacy price control" in theory

"By extension, if I visit an ad-supported website, the price I pay for consuming that site's content is the information collected about me upon my visit."

"To the extent that uncertainty or information asymmetry obscures the nature of the information being collected, those too, are a non-pecuniary component of the price that consumers pay."

"Because consumers can anonymize and thus forgo paying the full information price, it is true that the analogy to the price control may not be perfect, as one is not permitted to avoid paying the price of a good in traditional markets."

"If personal information is the price that consumers pay for a website's services, then regulation that sets a price other than the market price is an instance of price-fixing."

"Mandating opt-in as a solution to digital privacy differs from a policy that simply informs consumers about the information-price they must pay to access a site's services."

"Privacy price control" in practice

"The EU further updated its 2002 Directive, clarifying that the opt-in default also applies to the placement of "cookies," pieces of data stored in a web-user's browser and used to track browsers across sites (this is sometimes referred to as the "Cookie Directive") (Goldfarb and Tucker 68)."

"Many websites have responded to the Directive by attempting to exclude "non-payers," through use of technologies like the "tracking wall," which effectively prevents those who have not opted-in from accessing the site (Borgesius 69)."

"The logic of price control is applicable to many EU-based firms that are prohibited from exclusionary techniques: large firms, those primarily serving children, those judged as being monopolistic, and those where bans against exclusionary techniques like tracking walls are enforced."

Unintended consequences of "privacy price control"

"Investment may flow out of areas where the privacy price control has been imposed (ad-supported websites), and into non-price-controlled areas."

"Investment in digital ad-supported business may flow out of the privacy price-controlled area."

"The price control makes each banner ad less targeted, platforms might increase the total quantity of ads that they support, in an attempt to maintain revenue neutrality."

"Advertisers may shift to dynamic ads in the price-controlled region."

"The immediate effect of the privacy price control is to reduce the "revenue" that platforms are collecting because only some fraction of the visitors to their site pay an above-zero information-price."

"Theory suggests that the privacy price control will reduce the average quality of websites, particularly those that have fewer potential margins of adjustment."

The political economy of "privacy price control"

"For one, the price control may incentivize rent-seeking relative to other forms of government-sanctioned privacy provision."

"Firms that compete with ad-supported web services have an incentive to support the privacy price control."

"There is some, albeit scant, evidence that private interests may favor the privacy price control."

"It is apparent that the privacy price control hurts both platforms and advertisers, just as rent-control hurts landlords."

"The privacy price control may still be a useful tool for governments."

"One reason we may not see this policy is that, while likely less costly, it would also be less likely to confer the same reputational benefits on the administering officials as would the price control."

"Simple information disclosure may do less to enhance a government's reputation as compared with more stringent measures, such as the privacy price control."

Conclusion

"To admit that privacy law is a price control is not a conclusive case against privacy legislation."

"Regulatory issues are far from the only line of inquiry in the economics of digital privacy, the theory of price controls is useful in predicting the effects of the mandated opt-in."

"Seeing privacy protection as a form of price control also generates several testable predictions."

"If, in fact, a digital privacy regime of opt-in is a price control, we may see Internet platforms adjusting on myriad margins in an attempt to maintain profitability."

"Some consumers, namely the most privacy-sensitive, doubtlessly do benefit from a restriction that grants them enhanced protection of the personal information they generate in digital environments."

"Economists generally agree that price controls are welfare-reducing; they may benefit by applying their well-founded aversion regarding price controls to questions in digital privacy law."

How to really quantify the economic value of customer information in corporate databases [93]

This is a machine-generated summary of:

Lamela-Orcasitas, Carlos; García-Madariaga, Jesús: How to really quantify the economic value of customer information in corporate databases [93].

Published in: Humanities and Social Sciences Communications (2023).

Link to original: https://doi.org/https://doi.org/10.1057/s41599-023-01654-6

If you want to cite the papers, please refer to the original.

For technical reasons we could not place the page where the original quote is coming from.

Abstract-Summary

"The focus is on the concept of information and knowledge as an asset with intrinsic economic value and a variable to consider in order to achieve effective customer management."

"The proposed new customer relationship management (CRM) information value model (V_{ICRM}) offers significantly different results from other traditional models."

"Its main advantages include the possibility of carrying out new segmentations of customers and estimations of the company's value."

"The results of this research aim to serve any company whose business model is Business to Consumer (B2C) to calculate, in an effective way, the value of a large amount of information companies have about their customers."

Introduction

"In today's environment of a knowledge-based economy, it is not easy for businesses to assess the usefulness of acquiring or developing customer relationship management (CRM) tools to implement the corresponding strategies."

"Companies find it difficult to quantify this usefulness clearly and simply and to determine the economic value of the information generated and managed from the perspective of managing the customer base (Heidemann and others, 70)."

"The priority objective of this research is the conceptualization and empirical verification of an exploratory model that quantifies the economic value of the information and knowledge generated by the different computer tools that help implement CRM solutions."

"This study has three goals: (1) to propose and test a new model to measure the value of business information in CRM environments rooted in the information valuation theory."

Work proposals

"CKV $\in [-1;1]$: is an endogenous variable, a mediator of the CLV, which considers the value that customer knowledge brings to a company measured by the amount and recency of the available data based on the quality perceived by the user who manages the data."

"Based on the well-proven Metcalfe's law, we can estimate a customer's CSV as the potential value that their network of contacts can bring to the company due to the effect of their influence:where $Q_m \in [-1;1]$ = variable that measures the quality of contacts; as a coupling force between the size of the network and the value generated thanks to the client's authority in their network, measured by the level of influence exerted for the company and the volume of participation in the network, $Q_m = 0$: no authority and/or no social activity, $Q_m = 1$; maximum social activity and authority with a level of positive influence for the company and $Q_m = -1$: maximum social activity and authority with a level of negative influence."

Model construction

"The structure of the model can be explained mathematically as follows: From the above equation, it follows that the value of information in CRM environments (V_{ICRM}) is the result of the sum of the CLV functions of all the customers or segments that make up the company's portfolio."

"o Engagement value component measures the potential that a customer's engagement will have for the company, measured by the probability that they will continue generating all the CLV possible, or, on the contrary, will leave the company or not continue consuming in the next periods, with the consequent loss of profit."

"It is estimated by multiplying the absolute value of the CLV of a customer (i) or of a "typical" customer who is representative at the company/segment level (m)."

"It is estimated by multiplying the absolute value of the CLV by its corresponding CKV together with the retention rate of a "typical" customer who is representative at the company/segment level."

Main characteristics of the model

"Deterministic models may be suitable for many different systems and situations."

"These models accurately explain the results of the different parameters, relationship states, and initial conditions."

"This model is based on the multiplicative effect of each construct as a function of a potential transactional CLV that supports the economic value of a client's information."

"This formulation is proposed in recognition of the fact that there is a potential interaction effect between the different constructs in the model."

Methodology

"To apply this mathematical model, we have selected a company that is part of the financial services industry, specifically the insurance sector."

"This industry is an important part of the financial sector, not only because insurers are significant investors in the financial markets but also because they provide various important services to consumers and companies (Gaganis and others, 71)."

"To choose a paradigmatic company, we look for these characteristics: a high degree of organizational maturity, medium size, high availability of data, high willingness to commit the necessary resources for the investigation, and the lack of useful prior information about the value of its customers, which has led managers to receive new information with interest."

"We emphasize that although it is true that in terms of representativeness, no substantial differences were detected between those who answered and those who did not, there are significant differences in terms of other characteristics, as those who responded seem to have greater engagement with the company according to several metrics, including roughly double savings on average."

Results

"This implies that there are significant differences in the results depending on which segment the customers belong to and that the value of the different components of the model differs according to the construct analyzed."

"Other studies have found a relationship between models based on customer value and company value in the markets and their link with business results (Kumar and Reinartz, 72), it would be interesting to compare the data obtained with the available company results."

"There are differences in the results depending on whether customers belong to one segment or another and that the value of the models differs."

"In the new model, we found that there is not such a high relationship between income and final value due to the mediating effect of the f(CLV) compared to traditional models, which are much more dependent on operating margins."

Contributions

"This is the first research with a theoretical-practical approach that seeks to unify in a single deterministic model the different components that reflect the value of customer information in the financial industry."

"Thanks to this comparison, we could validate some theories about the overvaluation of some clients, the problems with retention rates, the distribution of value among customers, and how traditional models are too regressive."

"From a professional perspective, starting from the traditional knowledge of the financial sector on asset valuation, this model can be used in multiple B2C industries without many adaptations."

"It serves to assign a specific value to CRM solutions that, until now, were difficult to quantify, providing a tool that allows objectifying the decision-making processes at the managerial level (for example, improving regular segmentation processes, including the partial values of each component separately, offering a guide to know what and how to measure the information available from your customers and to enhance the accountability of IT investments in CRM)."

Conclusions

"Companies should improve flexibility in the use and distribution of information in order to make the appropriate decisions more safely and quickly, helping to provide a higher level of customer service by personalizing business–customer relationships, which results in greater satisfaction and loyalty over time (Chen and Tsou, 73)."

"It has been proven that the exclusive use of traditional assessment models may be insufficient to show a complete image of customers."

"The use of the V_{ICRM} model serves as a new, more holistic, and comprehensive approach that helps to capture the particularities of the information value of their customers that the companies possess in their databases and that many are still incapable of making profits."

"More field research should be done in other contexts that can help to fit the different metrics better or even investigate new mediating or moderating variables."

"Extending the customer value model to include new data-intensive and actionable metrics is a critical requirement for future research."

Privacy and economics in a 5G environment [94]

This is a machine-generated summary of:

Khajuria, Samant; Skouby, Knud Erik: Privacy and Economics in a 5G Environment [94]

Published in: Wireless Personal Communications (2017).

Link to original: https://doi.org/https://doi.org/10.1007/s11277-017-4421-y

Copyright of the summarized publication:

Springer Science+Business Media New York 2017.

If you want to cite the papers, please refer to the original.

For technical reasons we could not place the page where the original quote is coming from.

Abstract-Summary

"A generally accepted technology vision is that new services and applications especially related to social networking, and machine-to-machine communications will be accelerated, underlining the shift to a more user centric application-driven connectivity, transparently deployed over various technologies and infrastructures by users and devices."

"For the user centric and data centric 5G technologies to be accepted and function as a viable business model, user trust in their data privacy is paramount."

"The aim of the paper is to discuss user privacy concerns in today's and tomorrow's information society with a focus on economic aspects."

"The paper will further discuss the 'privacy economy' that will enable new business models focused on the economic value of private data"

Introduction

"The trend of exploiting the value of users' personal information by the service providers have already been here for a long time."

"When online users share their personal information with a service provider in order to use the service offered, the service providers gather the data to create customer profiles for targeted advertising and selling information to the 3rd parties "behind the scene"."

"The unbalanced information exchange and the lack of transparency related to the collection of users' personal data and its use, have made users reluctant to share information with their service providers."

"To build and deploy the user centric and data centric 5G technology, it is very crucial that the users trust their network and share accurate information with the services."

"The purpose of this paper is to discuss the role of users' privacy concerns and economics behind users' personal data as drivers or barriers for the emerging user centric and data centric 5G technology."

Notice and Choice: Privacy Concerns of Today and Tomorrow

"The notice from any service should be recognized as essential to ensure that users are properly informed before revealing any personal information, i.e., addressing identification of all the entities involved in the process; the nature of the data collected; the means by which it is collected; information about the requested data; whether it is voluntary or required that the data is given to the service; the consequences of a refusal and finally the steps taken to ensure the confidentiality, integrity and quality of the data [74]."

"As an answer to these issues of notice and choice, some privacy aware-tools are developed by privacy technologists for online users."

"These tools make users aware of their online data privacy."

"To this, the tool also tells the user how much the personal data is worth to Facebook, Google and LinkedIn."

The 5G Technology Environment

"The emerging network of the future will not only consist of a heterogeneous infrastructure provided by the service providers but also by the users, communities and a broad range of 3rd party access network providers."

"In 5G networks, all these different instances will be univocally understood as being a part of a "user domain" and will be orchestrated in order to provide the overall integrated services the user wants."

"The second aspect is the "Social aspect", which will impose transport of information, rather than the current moving of data in networking and computing, meaning that any data generated by the users will be used in the creation of new services."

"user context information from sources such as mobility and social metrics—are very important criteria for efficient design of 5G networks."

Stakeholders and Economics in a Privacy Oriented Ecosystem

"Similar to the 1995- data protection directive, the regulation also provides the legal framework for the new service environment and recognizes three distinct categories of parties for the processing of personal data."

"It does not specifically address the evolving market or the economic role of these categories in the market, e.g., not even their ownership to data and other assets emerging in the process."

"Information has been bought and sold since ancient times, but mainstream economic theory has had difficulties in handling information as a good or service, e.g., the example of personal information that can have peculiar characteristics related to non-rivalry and non-excludability; the data may easily be duplicated and accessed, and at the same time have privacy protection as a core tenet."

"Based on the GDPR which recognizes the above mentioned categories of parties for the processing of personal data, the stakeholders and their business objectives in the value chain are identified in the following [74]."

Conclusion

"For a successful deployment of user centric and data centric 5G technology, it is very crucial that the user trust their network and share accurate information with the services."

"The first part discusses the key challenges and two most essential principles "Notice and Choice" associated to the users' personal information when dealing with the online services."

"Part of the paper identifies the stakeholders in the value chain and their business objectives dealing with users personal data."

"It is generally agreed that transparent treatment of privacy related data is key to user acceptance of new 5G services and thereby to the development of the potentials of this system."

1.3 COVID-19 Pandemic

Machine generated keywords: app, pandemic, covid, covid pandemic, trace, contact, measure, complement, security, freedom, privacy security, employee, fundamental right, permission, fight

Privacy in emergency circumstances: data protection and the COVID-19 pandemic [95]

This is a machine-generated summary of:

Ventrella, Emanuele: Privacy in emergency circumstances: data protection and the COVID-19 pandemic [95].

Published in: ERA Forum (2020).

Link to original: https://doi.org/https://doi.org/10.1007/s12027-020-00629-3.

If you want to cite the papers, please refer to the original.

For technical reasons we could not place the page where the original quote is coming from.

Abstract-Summary

"A focus on the increase in cybercrime during the pandemic then provides insights on the relevant risks and remedies for the security of personal data."

Introduction

"The way we conceive our privacy and the importance which we attach to the protection of our personal data has also been heavily impacted by this ground-breaking event."

"As it has put into perspective other fundamental rights which until then we would never have accepted seeing restricted by state measures, the pandemic has required us to balance privacy with health and security."

"By first providing legal insights on the general discussion about the balance between the fundamental right to privacy and the general public interest, this article will describe the most critical and controversial processing operations employed by states to contain the pandemic and mitigate its effects."

Privacy in emergency circumstances

"Art 23(1) GDPR allows Member States to restrict data subject rights, as well as the data protection principles outlined in Art 5 GDPR, as long as this is done by way of a legislative measure and respects the essence of those same fundamental rights and freedoms."

"While the first legal basis allows processing personal data that is necessary to protect the vital interest of individuals (i.e., to save lives), the second can be relied upon to safeguard the public interest or in the exercise of official authority vested in the controller."

"According to the Regulation, the processing of special categories of personal data is permitted when it is necessary for reasons of public interest in the area of public health, "such as protecting against serious cross-border threats to health"."

Tracking individuals to contain the spread

"Insofar as concerns the GDPR, where the processing employed by contact tracing applications does not involve special categories of personal data, the recommended legal basis for such processing can be found in Art 6(1)(e) GDPR (i.e., public interest)."

"Where these applications involve the storage of health data (e.g., in order to monitor the health status of an infected individual), in addition to the above legal basis, Art 9(2) (h), or (i) can allow such processing on the basis of it being necessary for the purposes of preventive or occupational medicine and healthcare, or for reasons of public interest in the area of public health."

"Since the processing of personal data resulting from digital contact tracing applications is likely to produce high risk to the rights and freedoms of data subjects, a data protection impact assessment (DPIA) should always be carried out prior to their deployment."

The security of personal data during the pandemic

"Although taking various forms, phishing attacks share the common purpose of convincing individuals to give access to information (in most cases personal data), providing fraudulent opportunities both in the cyber and in the real world."

"Businesses and organisation, whether they be private or public entities, are required both to put in place procedures aimed at the protection of personal data and to implement cybersecurity measures at all levels."

"Preventative organisational measures showing consideration of the level of risk and the value of the processed data should be implemented in order to ensure a rapid response."

"Answering a question on how public organisations should implement measures to increase cyber resilience and mitigate the impact of attacks to the security of personal data, he stated: "Cyber security is a shared responsibility and—while technology can provide baseline protection—a strong focus should be put on human factors."

Conclusion

"They did however underestimate the power and effectiveness of the European data protection regime."

"Supervisory authorities have provided useful guidance regarding the development and deployment of invasive measures used to mitigate the effects of the pandemic."

"On the other, businesses and organisations may have discovered that compliance with the security-related requirements of the GDPR already provided the necessary technical and organisational measures to combat the rise in cybercrime during the pandemic."

"Although in many ways, the EU was unprepared for the management of the pandemic, it performed better than others at protecting the fundamental right to privacy of its citizens in a time of health crisis."

The right to data protection and the covid-19 pandemic: the european approach [96]

This is a machine-generated summary of:

Kędzior, Magdalena: The right to data protection and the COVID-19 pandemic: the European approach [96].

Published in: ERA Forum (2020).

Link to original: https://doi.org/https://doi.org/10.1007/s12027-020-00644-4

If you want to cite the papers, please refer to the original.

For technical reasons we could not place the page where the original quote is coming from.

Introduction and background

"Against the background of these developments since the outbreak of the coronavirus pandemic, Member States, backed by the European Commission, have been assessing the effectiveness, security, privacy, and data protection aspects of digital solutions to address the crisis."

"The reflections that follow touch upon major legal challenges that have been identified on the EU level with regard to safeguarding privacy when fighting the pandemic and the approach taken to these challenges by the EU institutions."

"This contribution aims therefore to map the EU's approach to COVID-19 with regard to privacy and data protection, to identify the areas that cooperation at European level has been focusing upon and to determine the added value thereof."

"Beyond this, the question of what might have been the possible obstacles to a more harmonised approach towards fighting the pandemic in the EU Member States will also be considered."

Legal challenges regarding data protection rights

"Discussions and controversies focused around several topics such as the processing of employees' personal data (including health data) by employers; workplace monitoring; the processing of health data by public authorities; the data protection requirements needed to be met by trustworthy and efficient apps; the tracing of location data; as well as data subjects' rights in connection with states of emergency in Member States."

"The European Data Protection Board in its Guideline 04/2020 on the use of location data and contact tracing tools in the context of the Covid-19 outbreak of 21 April emphasised that the GDPR and Directive 2002/58/EC both contain specific rules allowing for the use of anonymous or personal data to support public authorities and other actors at national and EU levels in monitoring and containing the spread of the virus."

Institutional responses to data protection in times of COVID-19

"The European Law Institute (ELI) also drafted principles for the COVID-19 crisis which address some of the most important legal issues arising in relation to the pandemic, offering guidance to the Data Protection Commissioner of the Council of Europe on how to steer through these unprecedented times."

"At EU level, the European Data Protection Board adopted a formal statement on the processing of personal data in the context of the COVID-19 outbreak on 19 March 2020."

"In subsequent steps, guidance on issues relating to data protection and the processing of health data in the context of the COVID-19 outbreak for scientific research purposes; a statement on restrictions on data subject rights in connection to the state of emergency in Member States; a statement on the data protection impact of the interoperability of contact tracing apps; and a statement on the processing of personal data in the context of the reopening of borders following the COVID-19 outbreak were all adopted."

EU data protection laws: an obstacle in fighting the pandemic?

"The European Data Protection Board recalled that the GDPR foresees derogations to the prohibition of processing of certain special categories of personal data, such as health data, where this is necessary for reasons of substantial public interest in the area of public health (Art 9.2.i), on the basis of Union or national law, or where there is a need to protect the vital interests of the data subject (Art 9.2.c), while recital 46 explicitly refers to the control of an epidemic."

"The European Data Protection Supervisor also confirmed that the GDPR allows for the processing of sensitive data when this is necessary for reasons of public interest in the area of public health, such as protecting against serious cross-border threats to health."

Conclusions

"Faced with challenges concerning the fundamental rights to privacy and to data protection, the EU institutions have taken note of some of the problems that have arisen in times of the COVID-19 pandemic and recalled the applicable standards."

"EU data protection experts have underlined how data protection rules, in particular the GDPR, should not hinder the measures that need to be implemented in the fight against the COVID-19 pandemic."

"By adopting numerous policy and guidelines documents, the EU institutions have proven themselves able to preserve privacy in emergency situations and even launch innovative technical solutions which fight the pandemic, while applying principles of personal data protection."

"The area of data protection in electronic communications is still governed by the E-Privacy Directive, subject to national implementation and the restrictions of the rights of data subjects may vary depending on the laws of the particular Member State."

Covid-19 Apps, Corona Vaccination Apps and Data "Ownership" [97]

This is a machine-generated summary of:

van Erp, Sjef: Covid-19 apps, Corona vaccination apps and data "ownership" [97].

Published in: China-EU Law Journal (2022).

Link to original: https://doi.org/https://doi.org/10.1007/s12689-022-00097-7.

If you want to cite the papers, please refer to the original.

For technical reasons we could not place the page where the original quote is coming from.

Abstract-Summary

“The rise of Covid-19 tracing (and/or tracking) apps and now to be followed by the advance of Corona vaccination apps has made us aware of the benefits which e-health may bring, particularly in a situation where distance means safety.”

“The apps contain very personal information and, consequently, have provoked questions as to whether the apps sufficiently protect a person’s right to privacy and data protection as safeguarded by the EU’s General Data Protection Regulation.”

“The nature of the data, however, is such that also questions as to the importance of access by public health authorities in the public interest can be asked.”

“The result is a conflict particularly between entitlement to privacy protection and the general interest, causing questions to be asked about which interest has priority.”

“Data are non-rivalrous and non-depletable, because they can be copied and copied, questions about which entitlement has priority cannot be answered in absolute terms.”

“Rights regarding data depend upon who at a particular time has control over the data, who else has control and what control between all those involved then means.”

“Calling data entitlement “ownership” is not a reference to ownership in the traditional sense of the word, but to management.”

“These exercises will change over time, as accordingly will the answer to the question who is “owner” of data in Covid-19 and Corona vaccination apps.”

Introduction: e-health

“Being monitored by an e-health app on your mobile phone allows treating patients in such a way that patients can stay at home while doctors continue working in their practice, without having to travel and getting into close physical contract, but still giving patients the care they need.”

“Several e-health tools are now available, but the focus of this contribution is on an e-health tool that was developed particularly in reaction to the Covid-19 pandemic: the Covid-19 tracing and/or tracking apps and an e-health tool that is now being developed, tested and already used in some countries that provides information on whether you have been vaccinated, tested negative or have Covid-19 antibodies: the Covid vaccine or immunity passport.”

“Both apps, as in fact all e-health apps, provide data about someone’s physical integrity and accordingly contain very personal information.”

Data: A mirror or creator of reality?

“This is an overarching question which provokes a debate about privacy (a data subject having certain public law rights against a data controller about that subject’s personal data), the nature of data sales contracts (what does selling data legally mean?) and how to attribute as well as distribute these entitlements and any possibly resulting economic benefits."

"In our quest for a data governance structure we seem to forget that by focussing on, what might perhaps be called, the design of rights, like we traditionally do when formulating a property right as part of the numerus clausus, we also define the object of that right."

"It is also economic, because by doing so at this initial stage of data law development we also distribute wealth, and it is that aspect that also makes this a political question, furthermore given that in the general interest also the government might be entitled to a data right."

Covid-19 apps: tracing and tracking

"Mobile phone providers already follow their users by tracking them, which can be done by using various technical means."

"If the Corona app would track mobile phone users this would be in violation of the EU's General Data Protection Regulation (GDPR), as the contacts data clearly would be seen as personal in the sense of article 4(1) GDPR and, consequently, the mobile phone user would qualify as a "data subject" in the sense of the GDPR and would be entitled to data protection."

"Initial storage of these (anonymous and pseudo-anonymous) Bluetooth data can be done centralised or decentralised, meaning that the Bluetooth codes either are stored locally on a user's phone or immediately on a central government server."

"In case the owner of the phone is tested positive, the owner will be asked if (s)he would be willing to upload those codes to a central (government) server and, if he agrees, will be given a code which triggers the app to upload."

Corona vaccination apps: identity and data certification

"It should not be forgotten that international air travel already now is only possible if a traveller shows a passport or, in some cases, an identity card, to which in this new app verified information is added about vaccination."

"From the perspective of who actually "owns" these data, the travel app developed by IATA, the e-Yellow Card by the WHO and the Digital Green Certificate present even more complicated problems than already exist regarding data in a Covid-19 app."

"With regard to vaccination passports, either as part of a travel app or as a self-standing app, a major extra problem is that the process of data collection does not begin with the phone exchanging anonymous codes with phones close by, in other words: the gathering of raw data, but starts with downloading an externally verified data object (information about, for example, vaccination) to an externally verified data subject (transforming the app into an e-passport)."

Data "ownership" and personal health data

"It matters whether the question is being asked from the viewpoint of personal data and privacy protection (public law) or from the standpoint of marketability (private law)."

"A public law approach will favour creating a distance from the market, not allowing parties to conclude contracts on personal data."

"The reason for creating such a distance is that, if a private law and, consequently, market transaction approach, would be followed, this would facilitate individual choices to contractually waive a right to privacy and data protection."

"A public law approach, therefore, would be more inclined to start by denying any "ownership" rights, whereas a private law approach would be more likely to begin by accepting that at least some type of ownership, and consequently trade in data, could exist."

"In other words: the above, briefly sketched debate, about whether to approach data ownership from the perspective of public law or private law, is pointless."

Data "ownership" as management

"Looking at their content, data rights provide control, both positive and negative, rather similar to traditional property rights."

"The data rights of a person will change depending upon other persons also gaining some form of control, which more often than not will be the case as several copies of data may exist to which a number of people at the same time may have varying degrees of access."

"Data rights resemble more a legal status than a primary property right such as "ownership" in the Civil Law tradition."

"Being in control, however, undeniably gives an illegitimate manager a bargaining position, but no legally enforceable rights will come into existence, unless perhaps when a third party in good faith gets access to such data."

Concluding remarks

"The apps contain very personal information and, consequently, have provoked questions as to whether the apps sufficiently protect a person's right to privacy and data protection as safeguarded by the EU's General Data Protection Regulation."

"The nature of the data, however, is such that also questions as to the importance of access by public health authorities in the public interest can be asked."

"The result is a conflict particularly between entitlement to privacy protection and the general interest, causing questions to be asked about which interest has priority."

"Data are non-rivalrous and non-depletable, because they can be copied and copied, questions about which entitlement has priority cannot be answered in absolute terms."

"These exercises will change over time, as accordingly will the answer to the question who is "owner" of data in Covid-19 and Corona vaccination apps."

A privacy and security analysis of early-deployed covid-19 contact tracing android apps [98]

This is a machine-generated summary of:

Hatamian, Majid; Wairimu, Samuel; Momen, Nurul; Fritsch, Lothar: A privacy and security analysis of early-deployed COVID-19 contact tracing Android apps [98].

Published in: Empirical Software Engineering (2021).

Link to original: https://doi.org/https://doi.org/10.1007/s10664-020-09934-4

If you want to cite the papers, please refer to the original.

For technical reasons we could not place the page where the original quote is coming from.

Abstract-Summary

"There has been a lot of speculations about the privacy and security aspects of these apps and their potential violation of data protection principles."

"The developers of these apps are constantly criticized because of undermining users' privacy, neglecting essential privacy and security requirements, and developing apps under time pressure without considering privacy- and security-by-design."

"We analyze the privacy and security performance of 28 contact tracing apps available on Android platform from various perspectives, including their code's privileges, promises made in their privacy policies, and static and dynamic performances."

"Our methodology is based on the collection of various types of data concerning these 28 apps, namely permission requests, privacy policy texts, run-time resource accesses, and existing security vulnerabilities."

"Based on the analysis of these data, we quantify and assess the impact of these apps on users' privacy."

"Our findings have revealed that the developers of these apps need to take more cautionary steps to ensure code quality and to address security and privacy vulnerabilities."

"They should more consciously follow legal requirements with respect to apps' permission declarations, privacy principles, and privacy policy contents."

Introduction

"We analyze the privacy and security aspects of COVID-19 contact tracing apps (a set of 28 apps available on Android platform) by applying a multilateral analysis method for Android apps introduced by Hatamian and others (75)."

"To address these questions, we develop a method that combines four metrics: (i) apps' data access potential from their permission requests, (ii) dynamic app-behavior analysis by monitoring their run-time resource accesses patterns, (iii) coverage of data protection principles by their corresponding privacy policy texts, and (iv) static app-code analysis to document existing security vulnerabilities."

"The collected data is used as a basis for a multi-perspective privacy and security analysis of COVID-19 contact tracing apps, enabling an understanding of privacy and security related quality indicators for rapidly deployed smartphone apps within the context of a health pandemic."

COVID-19 contact tracing apps

"We briefly discuss the existing technologies used in contact tracing apps along with the compatibility of such apps with privacy regulation."

"Followed by this, the European Data Protection Board published a public letter (EDPB-letter 76) where it was stated that "contact tracing apps do not require location tracking of individual users."

"Collecting an individual's movements in the context of contact tracing apps would violate the principle of data minimization."

"In a very similar effort, on 4th May 2020 the UK Information Commissioner's Office published (77) data protection expectations on contact tracing app development outlining nine data protection principles (transparency, data minimization, user control, data security, etc) which are linked to the core principles and provisions of data protection law and are designed to support design and development decisions of app developers."

Study design

"The privacy policy of an app is a statement, or a legal document that gives information about the ways an app provider collects, uses, discloses, and manages users' data."

"By law, service providers (including app providers) are required to be transparent about their data collection, sharing, and processing practices and specify how they comply with legal principles (Hatamian 78)."

"By Art 13 (2), 14 (2) of the GDPR, app providers are required to provide these rights to users to ensure fair and transparent data processing (principle of lawfulness, fairness and transparency Art 5 (1a))."

"To further ensure lawful, fair, and transparent processing of data, app providers should inform users in a transparent and understandable way about privacy policy changes."

"According to Art 13 (1a), 14 (1a) GDPR, users have the right to be informed about the actual identity of data collectors, i.e., app providers."

Multi-perspective privacy and security analysis

"In our analysis, we pay attention to privacy policy analysis of contact tracing apps and fulfillment of 12 fundamental legal principles proposed in Hatamian (78), the extent to which the privacy policy texts of COVID-19 contact tracing apps match what developers request (in manifest) and what they do in reality (actual permission usage), and ultimately the discrepancies/similarities in apps' privacy policies published by EU and non-EU bodies in terms of covering fundamental privacy principles."

"We also conducted a comparative analysis similar to the analysis that we carried out regarding the permission requests of apps across the EU and rest of the world to compare the privacy policy principles fulfillment of contact tracing apps developed and published by EU and non-EU bodies."

"Further, the app uses dangerous permissions for the purposes of contact tracing, which put a user's security at risk."

Impact assessment

"Similar to the Dangerous Permission Request Score, the Idle Dangerous Permission Usage Score has an upper limit of 8, since each app can access a maximum number of dangerous permissions as many as 8 permissions detected in our analysis (the higher the Idle Dangerous Permission Usage Score, the severe the impact for privacy)."

"Based on our analysis, the impact of an app on users' privacy can be deemed as more severe if it requests higher number of dangerous permissions, has less coverage of essential privacy principle in its privacy policy document, has higher number of dangerous permissions accesses at run-time, and has higher security vulnerabilities."

"We show dangerous permission requests score with green, absent specification in privacy policy text with red, idle dangerous permission usage score with orange, app security score with blue, and the Total Impact Score with a thick black line."

Discussion

"The privacy policy text analysis of contact tracing apps showed that these apps tend to be non-transparent regarding their data collection, processing, sharing and transfer practices."

"Such an extreme tendency in requesting location-related permissions can result in apps being over-privileged (asking for as much data as they can, even those that are irrelevant to their functionality) and invade individuals' privacy, such as the claim by a Minnesota law enforcement official confirming that the state was employing contact tracing to identify connections among protesters who were detained during the "Black Lives Matter demonstrations" (79)."

"We also observed that some of the contact tracing apps do not provide any privacy-by-default options to users to limit/isolate data sharing practices."

"In our analysis, we found a number of data transmissions to different countries by contact tracing apps that have not been justified in privacy policy documents."

Conclusion

"Our main findings are: Privacy policies provided for many of the apps were found incomplete or non-existent at the time of our study."

"Still, most apps fail to comply with one or more of GDPR's privacy principles Code vulnerability analysis showed several vulnerabilities in apps' codes."

"In terms of resource access pattern analysis, one could think of device instrumentation to jailbreak an iOS device to enable a run-time analysis, but this could then face the incompleteness of analyzed data set (6 out of 28 apps in our data set were not available for iOS)."

"This means although the results obtained from the legal compliance analysis can be easily generalized to all apps published within the EU, they may not be necessarily generalized to all non-EU countries as those countries may have less strict data protection requirements as offered by the GDPR, and therefore, they may not need to fully comply with all the studied requirements in this paper."

References

1. Uschold M, Gruninger M (1996) Ontologies: principles, methods and applications. Knowl Eng Rev 11(02):93–136
2. Gharib M, Giorgini P, Mylopoulos J (nov 2017) Towards an ontology for privacy requirements via a systematic literature review. In: International conference on conceptual modeling, vol. 10650 LNCS. Springer, pp 193–208
3. Kitchenham B (2004) Procedures for performing systematic reviews. Keele, UK, Keele University 33, TR/SE-0401 , 28
4. Kitchenham B, Charters S (2007) Guidelines for performing Systematic Literature reviews in Software Engineering Version 2.3. Technical Report, Keele University
5. Labda W, Mehandjiev N, Sampaio P (2014) Modeling of privacy-aware business processes in BPMN to protect personal data. In: Proceedings of the 29th annual ACM symposium on applied computing, ACM, pp 1399–1405
6. Giorgini P, Massacci F, Mylopoulos J, Zannone N (2005) Modeling security requirements through ownership, permission and delegation. In: 13th IEEE international conference on requirements engineering (RE'05)
7. Massacci F, Mylopoulos J, Paci F, Tun TT, Yu Y (2011) An extended ontology for security requirements. In: Advanced information systems engineering workshops, Springer, pp 622–636
8. Mouratidis H, Giorgini P (2007) Secure tropos: a security-oriented extension of the Tropos methodology. J Softw Eng Knowl Eng 17(2):285–309
9. Zannone N (2006) A requirements engineering methodology for trust, security, and privacy. PhD thesis, University of Trento
10. Elahi G, Yu E, Zannone N (2010) A vulnerability-centric requirements engineering framework: analyzing security attacks, countermeasures, and requirements based on vulnerabilities. Requir Eng 15(1):41–62
11. Kang W, Liang Y (2013) A security ontology with MDA for software development. In: Proceedings-2013 international conference on cyber-enabled distributed computing and knowledge discovery, CyberC 2013, IEEE, pp 67–74
12. Agostinelli S, Maggi FM, Marrella A, Sapio F (2019) Achieving GDPR compliance of BPMN process models. Lect Notes Bus Inf Process 350:10–22

13. Gerl A, Bennani N, Kosch H, Brunie L (2018) LPL, towards a GDPR-compliant privacy language: Formal definition and usage. In: Lecture notes in computer science, vol. 10940 LNCS. Springer, pp 41–80
14. European Parliament Regulation (EU) 2016/679 of the European Parliament and of the Council of 27 (April 2016) on the protection of natural persons with regard to the processing of personal data and on the free movement of such data, and repealing Directive 95/46/EC. Official Journal of the European Communities, vol 59, pp 1–88
15. Solove DJ (2006) A taxonomy of privacy. Univ Pennsyl Law Rev 154(3):477
16. Van Blarkom GW, Borking JJ, Olk JGE (2003) Handbook of privacy and privacy-enhancing technologies. Privacy Incorporated Software Agent (PISA) Consortium, The Hague
17. Paja E, Dalpiaz F, Giorgini P (2014) STS-tool: security requirements engineering for socio-technical systems. In: Engineering secure future internet services and systems. Springer, pp 65–96
18. Braber FD, Dimitrakos T, Stølen K, Gran B, Aagedal JØ, Lund (2003) The CORAS methodology: model-based risk assessment using UML and UP. UML and the Unified Process, pp. 332–357
19. Mayer N (2009) Model-based management of information system security risk. PhD thesis, University of Namur
20. Runeson P, Höst M (2009) Guidelines for conducting and reporting case study research in software engineering. Empir Softw Eng 14(2):131–164
21. Thinakaran K, Dhillon JS, Gunasekaran SS, Chen LF (2017) A conceptual privacy framework for privacy- aware IoT health applications. In: 6th international conference on computing and informatics, no. October, pp 175–183
22. Liu L, Yu E, Mylopoulos J (2003) Security and privacy requirements analysis within a social setting. In: 11th International requirements engineering conference, IEEE, pp 151–161
23. Gharib M, Mylopoulos J, Giorgini P (2020) COPri: a core ontology for privacy requirements engineering. In: Lecture notes in business information processing, vol. 385 LNBIP, Springer, pp 472–489
24. Ariès P (1989) Introduction: passion of the renaissance. In: Chartier R (ed) A history of private life, vol 3. The Belknap Press of Harvard University, Cambridge, pp. 1–11
25. Johansen VF, Andrews TM (2016) On challenges to the private-public dichotomy. Soc Theor Health 15(1):66–83
26. Gal S (2002) A semiotics of the public/private distinction. J Femin Cult Stud 13(1):77–95
27. Simmel G (1950a) Secrecy. In: Wolff KH (ed) The sociology of Georg Simmel. The Free Press, Glencoe, pp. 330–345
28. Basu, S. (2012). Privacy protection: a tale of two cultures. Masaryk UJL & Tech., 6, 1.
29. Capurro, R. (2005). Privacy. An intercultural perspective. Ethics and Information Technology, 7(1), 37–47.
30. Hongladarom, S. (2016a). A Buddhist theory of privacy. In A buddhist theory of privacy (pp. 57–84). Singapore: Springer.
31. Hongladarom, S. (2016b). Intercultural information ethics: a pragmatic consideration. In Information cultures in the digital age (pp. 191–206). Wiesbaden: Springer VS.
32. Ma, Y. (2019). Relational privacy: where the east and the west could meet. Proceedings of the Association for Information Science and Technology, 56(1), 196–205.
33. Westin, A. F. (1967). Privacy and freedom. New York: Atheneum.
34. Taylor, L., Floridi, L., & Van der Sloot, B. (Eds.). (2016). Group privacy: new challenges of data technologies (Vol. 126). Springer.
35. Mnyaka, M., & Motlhabi M. (2015). The African concept of and its socio-moral significance. Black Theology, 3(2):215–237.
36. Letseka, M. (2012). In defence of Ubuntu. Studies in Philosophy and Education, 31(1), 47–60.
37. Zech, H. (2012). Information als Schutzgegenstand. Tübingen: Mohr Siebeck.
38. Evans, B. J. (2011). Much ado about data ownership. Harvard Journal of Law & Technology, 25(1), 69–130.

39. Hirose M (2017) Privacy in public spaces: the reasonable expectation of privacy against the dragnet use of facial recognition technology. Conn Law Rev 49(5):1591
40. Hamann K, Smith R (2019) Facial recognition technology: where will it take us? Crim Justice 34(1):9–13
41. (2019) Beyond face value: public attitudes to facial recognition technology' (Ada Lovelace Institute). https://www.adalovelaceinstitute.org/wp-content/uploads/2019/09/Public-attitudes-to-facial-recognition-technology_v.FINAL_.pdf (Created 09.2019). Accessed 21 Jan 2021
42. 'Perfected in China, A Threat in the West' Economist (2 June 2018)
43. Crawford K (2019) Halt the use of facial-recognition technology until it is regulated. Nature 572:565
44. House of Commons Science and Technology Committee (2019) The work of the biometrics commissioner and the forensic science regulator contents. House of Commons Science and Technology Committee, London (https://publications.parliament.uk/pa/cm201719/cmselect/cmsctech/1970/197003.htm)
45. Schatz, blunt introduce bipartisan commercial facial recognition privacy act. https://www.schatz.senate.gov/press-releases/schatz-blunt-introduce-bipartisan-commercial-facial-recognition-privacy-act (Created 14 Mar 2019). Accessed 18 March 2021
46. Stephanie Hare, 'Facial Recognition is Now Rampant, The Implications for Our Freedom are Chilling' The Guardian (18 August 2019) "With more than 6 million CCTV cameras in the UK, and 420,000 in London."
47. Lee S Coming into focus: China's facial recognition regulations' (Washington DC, centre for strategic and international studies (CSIS)). https://www.csis.org/blogs/trustee-china-hand/coming-focus-chinas-facial-recognition-regulations (Created 4 May 2020). Accessed 8 Feb 2021
48. Ringrose K (2019) Law enforcement's pairing of facial recognition technology with body-worn cameras escalates privacy concerns. Virginia Law Rev 105:57–66
49. Welinder Y (2012) A face tells more than a thousand posts: developing face recognition privacy in social networks. Harv J Law Technol 26(1):165–239
50. Wright E (2019) The future of facial recognition is not fully known: developing privacy and security regulatory mechanisms for facial recognition in the retail sector. Fordham Intellect Prop Media Entertain Law J 29(2):611–685
51. Yuan Yang and Nian Liu, 'China Survey Shows High Concern over Facial Recognition' Financial Times (5 December 2019)
52. Yuan Yang and Madhumita Murgia 'Facial Recognition: How China Cornered the Surveillance Market' Financial Times (6 December 2019)
53. Anna Gross and Madhumita Murgia, "China Shows its Dominance in Surveillance Technology" Financial Times (27 December 2019)
54. Olatokun M Automated facial recognition and the rule of law (London, BIICL). https://binghamcentre.biicl.org/comments/69/automated-facial-recognition-and-the-rule-of-law (Created 8 Oct 2019). Accessed 8 March 2021
55. Lederer S (2003) Designing disclosure: interactive personal privacy at the dawn of ubiquitous computing. Retrieved 10 Dec 2008 at http://www.cs.cmu.edu/~assist/publications/old-pubs/privacy-lederer-msreport-1.01.pdf
56. Weiser M (1993) Some computer science issues in ubiquitous computing. Commun ACM 36(7):75–84
57. Altman I (1975) The environment and social behavior. Brooks/Cole Publishing, Monterey
58. Price B, Adam K, Nuseibeh B (2005) Keeping ubiquitous computing to yourself: a practical model for user control of privacy. Int J Hum Comput Stud 228–253
59. Hong JI, Ng JD, Lederer S (2004) Privacy risk models for designing privacy-sensitive ubiquitous computing systems. In: Proceedings of Designing Interactive Systems (DIS) 2004, pp 91–100
60. Aghasian E, Garg S, Gao L, Yu S, Montgomery J. Scoring users' privacy disclosure across multiple online social networks. IEEE Access. 2017;5:13118–30.
61. M Dornseif, FC Gärtner, T Holz, Vulnerability assessment using honeypots. Praxis der Informationsverarbeitung und Kommunikation. 27(4), 195–201 (2004).

62. P Defibaugh-Chavez, R Veeraghattam, M Kannappa, S Mukkamala, A Sung, in 2006 IEEE Information Assurance Workshop. Network based detection of virtual environments and low interaction honeypots (IEEE, 2006), pp. 283–289.
63. Posner, R. A. (1981). The economics of privacy. American Economic Review, 71(2), 405–409.
64. Stigler, G. J. (1980). An introduction to privacy in economics and politics. The Journal of Legal Studies, 9(4), 623–644.
65. De Corniere, A., & De Nijs, R. (2016). Online advertising and privacy. The RAND Journal of Economics, 47(1), 48–72.
66. Gertz, J. D. (2002). Purloined personality: Consumer profiling in financial services. The San Diego Law Review, 39, 943.
67. Acquisti, A., Taylor, C., & Wagman, L. (2016). The economics of privacy. Journal of Economic Literature, 54(2), 442–492.
68. Goldfarb, A., & Tucker, C. E. (2011). Privacy regulation and online advertising. Management Science, 57(1), 57–71.
69. Borgesius, F. J. Z. (2015). Improving privacy protection in the area of behavioural targeting. Philadelphia: Wolters Kluwer Law and Business.
70. Heidemann J, Klier M, Landherr A, Zimmermann S (2013) The optimal level of CRM it investments: an economic model and its application at a financial services provider. Electron Mark 23(1):73–84
71. Gaganis C, Hasan I, Pasiouras F (2019) Cross-country evidence on the relationship between regulations and the development of the Life Insurance Sector. Econ Model 89:256–272
72. Kumar V, Reinartz W (2016) Creating enduring customer value. J Mark 80(6):36–68
73. Chen J, Tsou H (2012) Performance effects of IT capability, service process innovation, and the mediating role of customer service. J Eng Technol Manag 29(1):71–94
74. Reidenberg, J. R., et al. (2015). Disagreeable privacy policies: Mismatches between meaning and users' understanding. Berkeley Technology Law Journal, 30(1), 39.
75. Hatamian M, Momen N, Fritsch L, Rannenberg K (2019) A multilateral privacy impact analysis method for android apps. In: Annual Privacy Forum. Springer, pp 87–106
76. Edpb letter concerning the european commission's draft guidance on apps supporting the fight against the covid-19 pandemic (2020). European Data Protection Board. https://edpb.europa.eu/sites/edpb/files/files/file1/edpbletterecadvisecodiv-appguidance_final.pdf
77. Covid-19 contact tracing: data protection expectations on app development (2020). UK Information Commissioner's Office https://ico.org.uk/media/for-organisations/documents/2617676/ico-contact-tracing-recommendations.pdf
78. Hatamian M (2020) Engineering privacy in smartphone apps: A technical guideline catalog for app developers. IEEE Access 8:35429–35445
79. (2020). Calling police investigations 'contact tracing' could block efforts to stop covid-19 (2020). One Zero https://onezero.medium.com/calling-police-investigations-contact-tracing-could-block-efforts-to-stop-covid-19-349cdc27766e
80. Gharib, Mohamad; Giorgini, Paolo; Mylopoulos, John An Ontology for Privacy Requirements via a Systematic Literature Review. Journal on Data Semantics (2020). https://doi.org/10.1007/s13740-020-00116-5
81. Hoeyer, Klaus Data promiscuity: how the public–private distinction shaped digital data infrastructures and notions of privacy. Humanities and Social Sciences Communications (2020). https://doi.org/10.1057/s41599-020-00535-6
82. Reviglio, Urbano; Alunge, Rogers "I Am Datafied Because We Are Datafied": an Ubuntu Perspective on (Relational) Privacy. Philosophy & Technology (2020). https://doi.org/10.1007/s13347-020-00407-6
83. Pounder, C. N. M. Nine principles for assessing whether privacy is protected in a surveillance society. Identity in the Information Society (2008). https://doi.org/10.1007/s12394-008-0002-2
84. Marx, Gary T. Privacy and Social Stratification. Philosophy & Technology (2007). https://doi.org/10.1007/s12130-007-9009-5
85. Hummel, Patrik; Braun, Matthias; Dabrock, Peter Own Data? Ethical Reflections on Data Ownership. Philosophy & Technology (2020). https://doi.org/10.1007/s13347-020-00404-9

86. Bu, Qingxiu The global governance on automated facial recognition (AFR): ethical and legal opportunities and privacy challenges. International Cybersecurity Law Review (2021). https://doi.org/10.1365/s43439-021-00022-x
87. Michelfelder, Diane P. Philosophy, privacy, and pervasive computing. AI & SOCIETY (2009). https://doi.org/10.1007/s00146-009-0233-2
88. Lehikoinen, Jaakko T.; Lehikoinen, Juha; Huuskonen, Pertti Understanding privacy regulation in ubicomp interactions. Personal and Ubiquitous Computing (2007). https://doi.org/10.1007/s00779-007-0163-2
89. Cerruto, Francesca; Cirillo, Stefano; Desiato, Domenico; Gambardella, Simone Michele; Polese, Giuseppe Social network data analysis to highlight privacy threats in sharing data. Journal of Big Data (2022). https://doi.org/10.1186/s40537-022-00566-7
90. Sokol, Pavol; Míšek, Jakub; Husák, Martin Honeypots and honeynets: issues of privacy. EURASIP Journal on Information Security (2017). https://doi.org/10.1186/s13635-017-0057-4
91. Hermalin, Benjamin E.; Katz, Michael L. Privacy, property rights and efficiency: The economics of privacy as secrecy. Quantitative Marketing and Economics (2006). https://doi.org/10.1007/s11129-005-9004-7
92. Fuller, Caleb S. Privacy law as price control. European Journal of Law and Economics (2017). https://doi.org/10.1007/s10657-017-9563-6
93. Lamela-Orcasitas, Carlos; García-Madariaga, Jesús How to really quantify the economic value of customer information in corporate databases. Humanities and Social Sciences Communications (2023). https://doi.org/10.1057/s41599-023-01654-6
94. Khajuria, Samant; Skouby, Knud Erik Privacy and Economics in a 5G Environment. Wireless Personal Communications (2017). https://doi.org/10.1007/s11277-017-4421-y
95. Ventrella, Emanuele Privacy in emergency circumstances: data protection and the COVID-19 pandemic. ERA Forum (2020). https://doi.org/10.1007/s12027-020-00629-3
96. Kędzior, Magdalena The right to data protection and the COVID-19 pandemic: the European approach. ERA Forum (2020). https://doi.org/10.1007/s12027-020-00644-4
97. van Erp, Sjef Covid-19 apps, Corona vaccination apps and data "ownership". China-EU Law Journal (2022). https://doi.org/10.1007/s12689-022-00097-7
98. Hatamian, Majid; Wairimu, Samuel; Momen, Nurul; Fritsch, Lothar A privacy and security analysis of early-deployed COVID-19 contact tracing Android apps. Empirical Software Engineering (2021). https://doi.org/10.1007/s10664-020-09934-4

Chapter 2
Data Protection and Competing Interests

Introduction by the Editor

The objective of regulatory interventions is to provide a framework of data protection measures that protect privacy, which has emerged in the background of the burgeoning need for data exchange in the midst of technological advancements.[1] Often, there are myths surrounding technology advancements as an antithesis to privacy. It is crucial to locate the ethos and identify the philosophical foundations behind structuring data protection and privacy measures.[2] Equally, it is important to recognise the different phases in the life cycle of technological growth. Arguably, the final objective of regulatory interventions is not to put a check on technical development but to place balanced perspectives in its progression.

Similarly, technology is not evolving to bypass security or intrude upon the privacy of individuals but is serving for the betterment of humankind, which is not different from the broader objective of data privacy legislation. There may not be an inherent problem with a particular technology, but its use and how it is supervised.[3] There could be possible contradictions at the cusp where areas of law, policy and technology meet but are not limited to these areas only and, at times, extend to the space created by different legislative interventions at different times. For instance, collective interest does not always go against or oppose individual interest. The free flow of personal data is not an antithesis of the act of protecting personal data. It brings us to the overarching cost attached to data privacy legislation. Economists and scholars

---

[1] Oskar J Gstrein & Anne Beaulieu, "How to protect privacy in a datafied society? A presentation of multiple legal and conceptual approaches" (2022) Philosophy & Technology 35(3).

[2] Tore Hoel & Weiqin Chen, "Privacy and data protection in learning analytics should be motivated by an educational maxim—towards a proposal" (2018) Research and Practice in Technology Enhanced Learning 13(18).

[3] Amitai Etzioni, "Are New Technologies the Enemy of Privacy?" (2007) Philosophy & Technology 20, 115–119.

I. Gupta (ed.), *Expectations vs Realities of Information Privacy and Data Protection Measures*, https://doi.org/10.1007/978-981-99-6778-0_2

studying the anti-competitive effect have pointed to the perils of over-protection. There are concerns about privacy legislation affecting the startup economy.[4]

Several contradictions exist in jurisdictions that rely on their local laws as well as Regulations like GDPR. With so many changes in the making, EU member states have developed a niche for legislation and practices that have evolved over the years. For instance, the European Union has a rich history of data protection legislation. Their endeavours have been initiated at different levels with varied expectations. Therefore, the member states had to adjust local laws to match the broader goals.[5]

If there are debates with data protection legislation, such as the GDPR, one needs to examine how the legislation has been structured and its basis. Of course, it is early days to observe the negative impact of this legislation with progressive technological advancements. The questions are related to the envisaged processes protecting personal data and whether they would survive the extraordinary ramifications of technology.

Further, there are contradictions associated with using data in a data-driven economy and delivering administrative services. The quality of data, its reliability, integrity, and authenticity, as well as the nature of data and its use, have been questioned. Further, with the trend in data monetisation through digitally empowering technologies, regulators should account for the possible harm these practices may cause individuals. It is a rather arduous task to have legal and policy interventions that are essentially neutral in nature.[6]

This chapter portrays the over-arching differences that exist in the digital ecosystem. It considers the immense value that data has in economic terms[7] for other administrative purposes and how the use of technology facilitates all these transactions. It showcases the possible mutual coexistence of rights vis-à-vis proper and effective use and reuse of data.

The theoretical premise that lays the foundation for legal intervention has been questioned. The privacy framework and the data protection measures must have an underlying theoretical basis, without which the legal intervention may sound arbitrary.[8]

---

[4] Nicholas Martin & others, "How Data Protection Regulation Affects Startup Innovation" (2019) Information Systems Frontiers 21, 1307–1324.

[5] Zoltán Alexin, "Hungary's unorthodox approach to personal privacy" (2017) Health and Technology 7, 423–440; András L Pap, "Human Rights and Ethnic Data Collection in Hungary" (2007) Human Rights Review 9, 109–122; Jonathan Mance, "Human Rights, Privacy and the Public Interest—Who Draws the Line and Where?." (2009) Liverpool Law Review 30, 263–283.

[6] Sara Quach & others, "Digital technologies: tensions in privacy and data" (2022) Journal of the Academy of Marketing Science 50, 1299–1323.

[7] Janja Komljenovic, "The future of value in digitalised higher education: why data privacy should not be our biggest concern" Higher Education (2020) 83, 119–135.

[8] Dag Elgesem, "The structure of rights in Directive 95/46/EC on the protection of individuals with regard to the processing of personal data and the free movement of such data" (1999) Ethics and Information Technology 1, 283–293.

Scholars have found that the two worlds of data privacy and data security have come a lot closer due to the advancement of technology. These concepts must not be read in isolation.[9]

Regulatory interventions point to data security practices as a vital norm in protecting data privacy. The emphasis on a design-level technical approach brings security measures to the forefront, reimagining the means to protect data privacy.[10]

Sources have identified the role of legislative interventions in relation to the general use of personal data and the issue of personal autonomy. There are certain trade-offs between interests that intersect different stakeholders. We have personal interests that relate to the data subject, yet there are other interests that connect with the different facets of a democratic society. Therefore, the impact of data processing and use of data can never be unidimensional but rather multidimensional. It relies upon the scope of control and the autonomy that the data subject exercises.[11] As a reference point, Article 20(4) [GDPR] is under consideration, which primarily covers the data portability right, drawing upon the need to balance the rights of a data subject and that of others.[12]

As a continuation of legislative intervention and the question of balance, scholars have been critical of legislation protecting digital privacy. There is a cost attached to such intervention, and they believe that it creates multiple barriers in the market where stakeholders rely on the use of consumers' data.[13] For startups, empirical research suggests that GDPR simultaneously has the potential to stimulate and impede innovation. It may take a little while to understand the actual impact, yet those relying heavily on the use of data may change their business model, ultimately impacting innovation. Startups may respond to data protection legislation depending on several external and internal issues they may face.[14]

Sources suggest that techniques like anonymisation and pseudonymisation used to protect individual privacy may go against them and risk surveillance, especially in the context of the advancement of technology. Apparently, it may lead to higher surveillance compared to cases where no such techniques have been used. Trading of anonymised data should not preclude individuals from exhorting their right- not

---

[9] Emanuele Ventrella, "The symbiotic relationship between privacy and security in the context of the general data protection regulation" (2019) ERA Forum 20, 445–469.

[10] Ibid.

[11] Inge Graef & others, "Conceptualizing Autonomy in an Era of Collective Data Processing: From Theory to Practice" (2023) Digital Society 2(19).

[12] Regulation (EU) 2016/679 of the European Parliament and of the Council of 27 April 2016 on the protection of natural persons with regard to the processing of personal data and on the free movement of such data, and repealing Directive 95/46/EC (General Data Protection Regulation) OJ L 119, 4.5.2016, pp. 1–88. (GDPR).

[13] Caleb S Fuller, "The perils of privacy regulation" (2017) The Review of Austrian Economics 30, 193–214.

[14] Nicholas Martin & others, "How Data Protection Regulation Affects Startup Innovation" (2019) Information Systems Frontiers 21, 1307–1324.

to share.[15] As an offshoot of technological advancements, there are concerns about using big data for multiple reasons, including policing. "Data-driven policing" has been a point of concern in the US. There are challenges, including certain biases in such data and how it would further contaminate decision-making. Other questions surrounding effective transparency, accountability and the utility of using such data point to ethical and legal issues.[16]

At the end, there are existing debates about the extent of data use and the scope of rights enshrined under legislative interventions.[17] Solutions have been proposed in the healthcare domain based on the use of data and not necessarily on the type of data.[18] Similarly, in the healthcare care domain, suggestions have been made to bring about the idea of a fiduciary relationship, which would offer better protection to sensitive data as well as improve the trust quotient in all privacy and data protection matters.[19]

Machine Generated Summaries

Disclaimer: The summaries in this chapter were generated from Springer Nature publications using extractive AI auto-summarization: An extraction-based summarizer aims to identify the most important sentences of a text using an algorithm and uses those original sentences to create the auto-summary (unlike generative AI). As the constituted sentences are machine selected, they may not fully reflect the body of the work, so we strongly advise that the original content is read and cited. The auto generated summaries were curated by the editor to meet Springer Nature publication standards. To cite this content, please refer to the original papers.

Machine generated keywords: digital, value, privacy, collective, interest, firm, notion, big datum, argue, data driven, health datum, human right, conflict, analytic, product service.

The structure of rights in Directive 95/46/EC on the protection of individuals with regard to the processing of personal data and the free movement of such data [50]

This is a machine-generated summary of:

Elgesem, Dag: The structure of rights in Directive 95/46/EC on the protection of individuals with regard to the processing of personal data and the free movement of such data [50].

---

[15] Jane Andrew & Max Baker, "The General Data Protection Regulation in the Age of Surveillance Capitalism" (2019) Journal of Business Ethics 168, 565–578.

[16] Jeremy Davis & others, "Five ethical challenges facing data-driven policing" (2022) AI and Ethics 2, 185–198.

[17] Xiaolan Yu, "Allocating Personal Data Rights: Toward Resolving Conflicts of Interest over Personal Data" (2021) Fudan Journal of the Humanities and Social Sciences 14, 549–563; Lucas Introna & Athanasia Pouloudi, "Privacy in the Information Age: Stakeholders, Interests and Values" (1999) Journal of Business Ethics 22, 27–38.

[18] Amelia Fiske & others, "Value-creation in the health data domain: a typology of what health data help us do" (2022) BioSocieties 18, 473–497.

[19] Chirag Arora, "Digital health fiduciaries: protecting user privacy when sharing health data" (2019) Ethics and Information Technology 21, 181–196.

Published in: Ethics and Information Technology (1999).

Link to original: https://doi.org/10.1023/a:1010076422893.

If you want to cite the papers, please refer to the original.

For technical reasons we could not place the page where the original quote is coming from.

Abstract-Summary

"A survey and analysis is given of the structure of individual rights in the recent EU Directive on data protection."

"Of the paper it is argued that theories of privacy popular among philosophers are not able to shed much light on the issues treated in the Directive, which are, arguably, among the central problems pertaining to the protection of individual rights in the information society."

"In the third part of the paper, some suggestions are made for a richer philosophical theory of data protection and privacy."

The symbiotic relationship between privacy and security in the context of the general data protection regulation [51]

This is a machine-generated summary of:

Ventrella, Emanuele: The symbiotic relationship between privacy and security in the context of the general data protection regulation [51].

Published in: ERA Forum (2019).

Link to original: https://doi.org/10.1007/s12027-019-00578-6.

If you want to cite the papers, please refer to the original.

For technical reasons we could not place the page where the original quote is coming from.

Abstract-Summary

"The purpose of this article is to demonstrate how technological development, instead of worsening the cleavage between privacy and security, allows considering the two principles to be inter-related and to affect each other reciprocally."

Introduction

"By impacting every aspect of modern human society, technology has detached both privacy and security from real world situations, allowing them to find their field of application in the arena of cyberspace."

"The purpose of this article is to demonstrate how technological development, instead of worsening the cleavage between privacy and security, allows considering the two principles as inter-related and reciprocally affecting each other."

"By theorising this relationship, the article will first focus on how the evolution of the concept of privacy led by the advancement of technology has moved the two values closer together."

The Evolution of Privacy

"In accordance with the Fourth Amendment of the Bill of Rights, Warren and Brandeis' view derived a right to privacy from the entitlement to individual freedom from unwarranted intrusion and focused on the harm caused by physical access to a person and his or her possessions."

"Privacy being the expression of the core value of security, Moor justifies its interpretation in terms of protection of personal information in this way: "Without protection species and cultures don't survive and flourish."

"In a large, highly computerised culture in which lots of personal information is greased it is almost inevitable that privacy will emerge as an expression of the core value, security."

"In the context of the cyber-revolution, the increased connectivity derived from the explosive use of computer technologies has conditioned both quantitatively and qualitatively the dissemination of personal information, making the relationship between privacy and security even more concrete."

The Revolution of the General Data Protection Regulation (GDPR)

"In five different sections of the Regulation it appears that: (1) pseudonymisation may facilitate the processing of personal data beyond their original collection purpose; (2) pseudonymisation may reduce the possibility of identifying individuals when personal data breaches occur, decreasing the risk of harm to data subjects and positively impacting the process of risk evaluation which is functional to notify personal data breaches to data protection authorities; (3) pseudonymisation represents a relevant safeguard for processing personal data for archiving purposes in the public interest, as well as for scientific or historical research purposes or statistical purposes; (4) pseudonymisation may avoid controllers providing a data subject with access to data, with rectification and erasure of data, with restrictions on processing or with data portability; and5.pseudonymisation may be considered a mitigating factor for supervisory authorities calculating potential fines."

Conclusion

"Being concrete expressions of the principle of accountability and the risk-based approach, innovative data protection measures like DPIA and data breach notifications show how the GDPR puts emphasis both on the security of data subjects and on the security of processing operations."

"The employment of ICT solutions that—"by design"—integrate security measures implementing data protection safeguards has been prescribed as a way to ensure compliance with the obligations of the Regulation."

"Through detaching data subjects from personal data and resulting in the minimisation of risks to the rights and freedoms of individuals, the adoption of security measures like pseudonymisation may allow a "relaxation" of some of the data protection requirements of the GDPR."

"By acknowledging that the proper implementation of data protection always requires data security measures, professionals should question whether data protection should be considered a separate area of responsibility and expertise to data security."

Conceptualizing Autonomy in an Era of Collective Data Processing: From Theory to Practice [52]

This is a machine-generated summary of:

Graef, Inge; Petročnik, Tjaša; Tombal, Thomas: Conceptualizing Autonomy in an Era of Collective Data Processing: From Theory to Practice [52].

Published in: Digital Society (2023).

Link to original: https://doi.org/10.1007/s44206-023-00045-3.

If you want to cite the papers, please refer to the original.

For technical reasons we could not place the page where the original quote is coming from.

Abstract-Summary

"While literature has already recognized the relational and collective impact of data processing, there is still limited understanding of how this affects the design of legislative instruments."

"We submit that legislators must recognize trade-offs between one's own interests, the interests of other individuals, and collective or societal interests more explicitly in regulating data."

"We examine how legislators can make trade-offs between different interests explicit and thereby bring the regulation of data more in line with the current societal reality that is increasingly dominated by relational and collective effects of data processing."

Introduction

"Moving beyond the GDPR, this paper will illustrate that two other related European legislative initiatives, namely the EU Digital Markets Act [1] and the EU Data Act proposal (Proposal for a Regulation of the European Parliament and of the Council on harmonised rules on fair access to and use of data), may in fact equally disregard the trade-off between individual and collective interests in some of their provisions."

"As a core message, this paper submits that legislators must recognize trade-offs between one's own interests, the interests of other individuals, and collective or societal interests (such as democracy, public health, fostering competition and innovation, or the protection of the environment) more explicitly in developing regulatory mechanisms relating to the use of data in order to better account for the relational and collective impact of data processing."

Autonomy and Data Processing: Perspectives from the Literature

"Said otherwise, the individuals' right to autonomy and informational self-determination should not only be understood as their ability to decide which data they share with whom, but also, and more fundamentally, as their right to understand and exercise control on who has their data, what is being done with it and how this impacts their life and their possibility to exercise their autonomy by making their own choices (and acting upon them); as opposed to being subject to decisions made about them on the basis of personal data used as proxies and over which they might not have control [2: 56, 3: 82]."

"There can be a tension between retaining one's own autonomy over whether to share personal data, and other individuals' and collective interests that may be harmed by an individual's decision relating to data processing."

Autonomy and Data Processing: Perspectives from Regulation

"To identify to what extent the regulatory initiatives reflect the multi-dimensional nature of autonomy, we will analyse three aspects, namely: (1) the extent of control and responsibility an individual has over the protection of her personal data (the inward-looking dimension of autonomy); (2) the limits that other individuals' and collective interests pose to the exercise of control and responsibility by an individual (the outward-looking dimension of autonomy); and (3) if relevant, how such balancing of interests takes place and by whom."

"While the choice to let users decide on whether a gatekeeper can combine personal data is welcome from the perspective of the inward-looking dimension of autonomy, it does not leave any room for considering how such a decision of an individual can affect the other individuals' and collective interests in contestable markets."

"The GDPR's right to data portability arguably considers both the inward- and outward-looking dimensions of autonomy, as it makes room for considering other individuals' interests."

Conclusion

"This thus requires balancing the interests of the individual who takes a decision about her personal data, with other individuals' and collective interests."

"Other areas, such as the GDPR's right to data portability, already partly acknowledge the need to balance the different interests and dimensions of autonomy, as Article 20(4) GDPR provides that the individual's interests must be balanced with other individuals' interests."

"The design of these data spaces, among other things, also requires balancing individual and collective interests by opening up more data for use in the economy and society, while letting individuals and business retain some level of control."

"This demonstrates the relevance of balancing individual control against other, societal considerations within the European Health Data Space as another legislative area beyond the ones studied in this paper."

The perils of privacy regulation [53]

This is a machine-generated summary of:

Fuller, Caleb S.: The perils of privacy regulation [53].

Published in: The Review of Austrian Economics (2016).

Link to original: https://doi.org/10.1007/s11138-016-0345-0.

If you want to cite the papers, please refer to the original.

For technical reasons we could not place the page where the original quote is coming from.

Abstract-Summary

"Advocates of digital privacy law believe it is necessary to correct failures in the market for digital privacy."

"Legislators allegedly craft digital privacy regulation to protect consumers, some advocates have understated the dangers that digital privacy law may engender."

"My research suggests that policy-makers should consider a more holistic accounting of the costs before imposing additional digital privacy regulation."

Introduction

"Hoofnagle, a law professor and outspoken commentator on digital privacy topics, argues [4] that the "[Federal Trade Commission] has to move into the twenty-first century and meaningfully address Internet privacy," that "the FTC should abandon its faith in self-regulation," and that "the FTC is certainly capable of protecting privacy online.""

"Other examinations of the costs of digital privacy regulation [5] have focused extensively on the ways that these laws constrict the free flow of information."

"As this paper shows, digital privacy law generates perils by failing to provide an analogue to the market's disciplinary corrective of profit-loss accounting, by stifling the market's discovery process, and by creating superfluous avenues of discovery."

"Government-mandated digital privacy rules are not neutral, however, with respect to the workings of the market; as Kirzner [6] argues, all intervention stifles and redirects the market process."

Literature Review

"The second is that the market for digital privacy fails due to the presence of information asymmetry; consequently, regulation must correct the deficiencies of the unhampered market."

"They identify several costs of digital privacy legislation: reduced ability to match between consumers and producers as well as a likely reduction in innovation by Internet firms."

"This analysis notes that to the extent digital privacy law raises barriers to entry (to take just one example of many), it impedes the ceaseless churning of the market as entrepreneurs continually discover the prices, quantities, and qualities that satisfy consumer demands."

"That regulatory agencies purport to ensure consumers' best interests by imposing food safety mandates, groups like the FTC maintain they are correcting market deficiencies by passing digital privacy law Such market-failure reasoning appears dubious, however."

"Regardless of the soundness of market failure claims, market process theory illuminates several regulatory dangers that advocates of digital privacy law overlook."

Some Overlooked Perils of Digital Privacy Regulation

"There is evidence to suggest that digital privacy law, for instance, has stifled the discovery process in industries ancillary to those that collect consumer data directly."

"To the extent that digital privacy law has this effect, it stifles the discovery process undertaken by small or entrant firms."

"It states that controllers must be able, at all times, to demonstrate compliance with EU digital privacy law to data subjects, the general public, and to regulators."

"Directly imposing fixed costs is relatively easier to identify than is the fact that these strictures may impede the market's discovery process in ancillary industries, such as firms that specialize in third-party certification of other companies' privacy practices."

"Because legislation creates new, "superfluous" avenues for entrepreneurial discovery, firms can leverage digital privacy law to further their interests."

Conclusion

"The current literature under-values the market process perspective on digital privacy problems."

"Economists working in this tradition might find it worthwhile to examine the ways entrepreneurs have responded to alleged failures in the market for digital privacy."

"This paper briefly explored "web seals" as a way for firms to signal the quality of their privacy protection."

"This paper should inform that debate by highlighting some of the perils that such government-enforced privacy protection may entail."

"This research indicates that even seemingly "innocuous" legislation, such as digital privacy law, creates newly profitable avenues in the form of rent-seeking opportunities."

"Bibas [7], in a prescient article on digital privacy law, anticipates the knowledge problems that regulatory solutions impose."

"This paper, employing insights from the market process perspective, suggests that the problems of digital privacy regulation may be even more pervasive than Bibas anticipated."

How Data Protection Regulation Affects Startup Innovation [54]

This is a machine-generated summary of:

Martin, Nicholas; Matt, Christian; Niebel, Crispin; Blind, Knut: How Data Protection Regulation Affects Startup Innovation [54].

Published in: Information Systems Frontiers (2019).

Link to original: https://doi.org/10.1007/s10796-019-09974-2.

If you want to cite the papers, please refer to the original.

For technical reasons we could not place the page where the original quote is coming from.

Abstract-Summary

"While many data-driven businesses have seen rapid growth in recent years, their business development might be highly contingent upon data protection regulation."

"We therefore study how data protection regulation affects startup innovation, exploring this question during the ongoing introduction of the EU General Data Protection Regulation (GDPR)."

"Our results show that the effects of data protection regulation on startup innovation are complex: it simultaneously stimulates and constrains innovation."

"We identify six distinct firm responses to the effects of the GDPR; three that stimulate innovation, and three that constrain it."

Introduction

"Previous research that has studied country or industry-wide effects of data protection regulation on firm performance has often taken data protection regulation effects as a monolithic black box, providing high-level insights but without identifying concrete impacts and responses at the firm level, or identifying concrete causes of these effects."

"We ask the following research questions: How does data protection regulation affect innovation among startups?".

"We seek to answer these questions with a two-stage approach consisting of semi-structured interviews: First we interviewed lawyers and other intermediaries of the startup ecosystem to obtain insights into the regulatory framework conditions (enforcement levels, compliance strategies, etc.) as well as a broad overview of data protection regulation's apparent effects on innovation."

"Building on this, Section 3 develops a basic conceptual framework to explain how firms may respond to data protection regulation and how this would impact innovation."

Effects of Regulation on Firm Innovation

"Empirical research on the impact of social regulation on innovation and economic performance has focused primarily on the effects of environmental regulation, but has been unable to derive definitive predictions [8, 9]."

"The time frame may also be important: Negative effects may predominate in the short-term as new regulation disrupts existing structures of production and ways of doing things, while positive, innovation-stimulating effects might only show up in the medium term, since the time to market of innovations starting from research and development take time and depend heavily on the characteristics of the technology [8, 10]."

"There has been little research on data protection regulation's effects on innovation."

"These studies often also miss a firm level perspective when assessing the effects of privacy regulation on innovation, or they do not link the effects of data protection regulation to firm performance."

Developing a Conceptual Framework for Firm Responses to Privacy Regulation

"Strategic Non-compliance: They can deliberately contravene the regulation, at the risk of running afoul of the authorities and facing punitive consequences (fines, closure, etc.) Which of these three responses firms choose is likely to depend on a mix of internal and external factors, including their technological and financial capabilities, the relative technological and financial easiness of the innovation needed to achieve compliance, the expected level of market demand for regulation-compliant products (which may be low if achieving compliance impairs functionality or raises prices), and the expected level of regulatory enforcement."

"We expect little strategic non-compliance (due to enforcement), little compliance innovation and little regulation-exploiting innovation (due to low demand for compliant products), but also high levels of abandonment (again due to enforcement)."

Methodology

"As they advise many startups in diverse sectors, we expected lawyers and consultants to be able to provide a broader overview than individual company executives, particularly concerning what kinds of innovation-obstructing or -stimulating effects data protection regulation commonly produced in startups, and how frequent and severe these effects tended to be."

"We sought startups for which processing of personal data was at the heart of their products and business models, in particular companies where the processing was likely to pose a "high risk" with regards to the GDPR, as these companies could be expected to be particularly affected by data protection regulation."

"As the interviews with the intermediaries indicated that data protection regulation's effects may vary depending on whether companies are situated in B2B or B2C markets, we sought to obtain a mix of both across different industries for the second stage."

Results and Discussion

"VC Partner 1, University Accelerator Manager 2, Lawyer 5 and Startup Executive 1 all noted that complete data protection compliance was a basic prerequisite to work in these markets, putting real pressure on startups to innovate fully compliant products and services."

"Lax enforcement and low market pressure (due to low end-user pressure) for fully data protection-compliant products and services limited data protection-related innovation in these sectors, at least before the GDPR."

"Lawyers 5, 6 and 7 and Startup 3, 6 and 9 described a lively and innovative market developing for products to enable companies achieve compliance and solve obstacles from data protection regulation."

"Our interviews found evidence for abandonment and also identified two further innovation-constraining responses: entrepreneurial discouragement, whereby concerns that data protection law will make realization of their ideas impossible might discourage would-be founders from starting firms, and what we label "data minimization", whereby the cumulative impact of privacy regulation reduces innovators' access to data to such an extent that certain products and technologies, especially in the field of big data and artificial intelligence, become hard to develop for lack of input data."

Implications

"Our interviews reveal data protection law to have both innovation-stimulating, and innovation obstructing effects and suggest that both have only become pronounced once the GDPR changed the enforcement environment and market demand for compliant products and processes."

"This finding, that hitherto data protection law simply had not had much effect because it had gone unenforced, has important implications for debates about the causes of the relative weakness of the European digital economy: at least for startup innovation in the pre-GDPR era, data protection law would not seem to have been a significant cause."

"These innovation efforts seem to mostly still be young and it still needs to be seen how effective the innovated solutions to the compliance challenges posed by data protection law really will be."

Conclusion and Limitations

"Despite frequent company complaints about regulatory burdens, little was known about data protection regulation's actual effects on particular corporate functions."

"Our work underscores the need for researchers to dig deep into the individual stipulations of regulation and how these interacts with specific business models and technologies, and to pay close attention to whether regulation is in fact enforced, rather than taking that as given."

"Our theoretical framework and research focus primarily concerned how regulation affects product innovation and strategy."

"We ensured that all interviewees had substantial relevant experience, and chose startups that were particularly likely to be affected by data protection regulation—i.e. those, where data protection regulation's effects on innovation were likely to manifest most clearly."

"Due to the rather small size of the interviewed companies, the potential gap in perspective between founders and CTOs concerned with innovation/product strategy and lower-level engineers' focus on practical design questions was, in our case, probably fairly small."

Hungary's unorthodox approach to personal privacy [55]

This is a machine-generated summary of:

Alexin, Zoltán: Hungary's unorthodox approach to personal privacy [55].

Published in: Health and Technology (2017).

Link to original: https://doi.org/10.1007/s12553-017-0181-7.

If you want to cite the papers, please refer to the original.

For technical reasons we could not place the page where the original quote is coming from.

Abstract-Summary

"The fundamental rights defined in the European Convention of Human Rights were included in it and, in addition, the right to the protection of personal data."

"The main characteristic of the legal system is that politicians understand the constitution so that it would always require passing a new law whenever they want a public authority to collect personal data for some purpose."

"Hungary has several hundred laws and decrees in effect on the obligatory collection and transfer of personal data."

"Turning our attention to healthcare, we see that medical authorities and researchers are authorized by the law to collect personal medical data without providing information and the right to object to data subjects."

"He gives an account of the key characteristics of the Hungarian legislation and provides a comparison with the relevant European Union legislation in relation to medical data."

"When the EU General Data Protection Regulation came into force, the conflict between the two became obvious."

"This paper gives the reader an expert's view of the Hungarian data protection policy, while attempting to pave the way to stimulate social debate concerning the necessity for better, and more privacy aware legislation."

Introduction

"Article 54 of the amended Constitution declared that Hungary respects human rights; human rights must be executed in accordance with the interests of the socialist society; the execution of rights is inseparable from the performance of duties; and regulations on rights and duties are laid down by Hungarian laws."

"They have no opportunity before the Constitutional Court because the Fundamental Law gives the Hungarian Parliament a free hand on how it should implement these fundamental rights."

"In the case of personal data, on the question of fundamental rights, this means that every time the Hungarian Parliament wants to create a new database containing information about people, the only action the Hungarian Parliament has to take is to pass a law (or an amendment to an existing law) that restricts the right to the protection of personal data."

The Rise and Decline of the Right to Protection of Personal Data

"The bipolar nature of the regulation (consent or law) remained the same, and Section 5 of the new Data Protection Act says: Personal data may be processed under the following circumstances: (a) when the data subject has given his consent, or (b) when processing is necessary as decreed by law or by a local authority based on authorization conferred by law concerning specific data defined therein for the performance of a task carried out in the public interest (hereinafter referred to as "mandatory processing") [11]. The new Data Protection Act established a renewed authority called the Hungarian National Authority for Data Protection and Freedom of Information."

"The new Data Protection Act no longer contains the prohibition of a unique personal identifier and Hungarian Parliament invalidated all decisions that the Constitutional Court had made, referring to the old Constitution."

Medical Privacy Regulations

"The author challenged the decree concerning the obligatory examination made without medical indication before the Constitutional Court, referring to the new Act CLIV of 1997 on Health and privacy rights of the patients."

"Although the new Data Protection Act Section 21, paragraph 1b) grants the right to object [11] when the purpose of the processing is scientific research, this right is revoked by another law Medical research databases raise concerns even if they do not contain direct identifiers to patients such as name, mother's name, or resident address."

"Patients can restrict access to the documents except where an authority (court, police, security service, public health, health insurance fund etc.) requires the data in accordance with a law, or a medical doctor in the case of an obligatory medical fitness exam, or in the case of an emergency."

Court cases concerning the Regulation of Privacy

"3110/2013 (4th June, 2013) that in the case of obligatory medical data processing, everybody is a potential subject of such a regulation and it need not be certified since they can get ill at any time, and the court recognised that there is no possible remedy before the normal courts, hence applicants may turn directly to the Constitutional Court."

"The author also has a pending case from 2011 before the court challenging the Health Data Processing and Protection Act, referring to the decisions C-468/10 and C-469/10 of the Court of Justice of the EU, which applies to Hungary as well."

"The Strasbourg Court in the Copland vs. United Kingdom, 62617/00, in paragraph 43 of the decision stated that the storing of personal data relating to the private life of an individual also falls within the application of Article 8 § 1."

The Adoption of the New EU Data Protection Regulation May Open Ways to Legal Dispute

"This was one reason why the European Parliament decided that the new data protection regulation would be the type of community legal act that must be applied directly in all member states."

"This is described in Article 10 of the 680/2016 Directive: Processing of personal data revealing racial or ethnic origin, political opinions, religious or philosophical beliefs, or trade union membership, and the processing of genetic data, biometric data for the purpose of uniquely identifying a natural person, data concerning health or data concerning a natural person's sex life or sexual orientation shall be allowed only where strictly necessary, subject to appropriate safeguards for the rights and freedoms of the data subject, and only: (a) where authorised by Union or Member State law; (b) to protect the vital interests of the data subject or of another natural person; or (c) where such processing relates to data which are manifestly made public by the data subject."

Conclusions

"Three decades have elapsed from the time of the insertion of the right to protection of personal data in the Constitution since 1989, but Hungary still does not have data protection case law The courts are uncertain of how to apply even the definition of personal data, and how data subjects can execute their right to access to data relating to them."

"Other medical privacy issues, as in the matter of obligatory examinations, and the substantially restricted living will, may remain unsolved [12]. Since Hungary excluded the case law of the ECtHR, it means that the implementation of privacy rights has stalled at the 1950s level."

"Hungary should completely revise the regulation concerning the right to protection of personal data."

"In such cases, the regulation must be amended in such a way that data subjects may object to the processing of personal data relating to them."

Human Rights and Ethnic Data Collection in Hungary [56]

This is a machine-generated summary of:

Pap, András L.: Human Rights and Ethnic Data Collection in Hungary [56].
Published in: Human Rights Review (2007).
Link to original: https://doi.org/10.1007/s12142-007-0032-4.
Copyright of the summarized publication:
Springer Science + Business Media B.V. 2007.
All rights reserved.

If you want to cite the papers, please refer to the original.

For technical reasons we could not place the page where the original quote is coming from.

Abstract-Summary

"The article analyzes ethnic data collection pertaining to criminal justice in Hungary."

"With such a sensitive and delicate issue at hand, Hungary has decided on an evasive approach, resisting ethnic data collection by law enforcement authorities."

"The first part of the article describes general problems relating to ethnic data collection and analyses of the Hungarian minority protection framework, in particular, the minority self-government structure (a unique constitutional institution)."

Introduction

"Even the European Commission against Racism and Intolerance (ECRI) has expressed concern "at evidence that severe problems in the administration of justice exist as regards discrimination against members of the Roma/Gypsy community [...]" This article will analyze ethnic data collection pertaining to criminal justice in Hungary."

"It will be demonstrated that Hungary is one of the (many) countries in which extensive legal restrictions on the collection of nonanonymous data concerning ethnic, national, or religious identity have prompted law enforcement authorities to simply deny that ethnicity is of significance in their actions."

"In 1999 the European Commission against Racism and Intolerance reported that "while acknowledging the fact that the collection and utilisation of data on ethnic origin is restricted in Hungary for valid reasons, [the] ECRI is concerned that the lack of reliable information about the situation of various minority groups living in the country makes evaluation of the extent of possible discrimination against them or the effect of the actions intended to fight such discrimination difficult.""

When is a Minority a Minority?

"The 1993 Act defines national and ethnic minorities as groups which have been present in the territory of Hungary for more than 100 years and "constitute a numerical minority within the population of the country, whose members hold Hungarian citizenship and differ from the rest of the population in terms of their own tongue, cultures, and traditions, and who prove to be aware of the cohesion, national or ethnic, which is to aim at preserving all these and at articulating and safeguarding the interests of their respective historically developed communities" (§ 1)."

"Some commentators claim that the creation of the above described homogenous legislation for national and ethnic minorities may help promoting rights of ethnic Hungarians in the neighboring countries; it cannot, however, provide an effective institutional framework to deal with the specific and robust Roma problem."

Ethnicity and Data Collection

"It should also be stressed that the anonymous collection of data relating to one's perceived ethnic origin for research purposes is not explicitly prohibited by the Data Protection Act, as it defines sensitive data as one that relates to racial origin and national and ethnic minority affiliation—not perceived racial origin."

"The case law pertaining to penalizing 'violence against a member of an ethnic group,' suggests that Hungarian criminal law recognizes the difference between self-identification and perceived ethnic origin and attaches the same criminal liability to violence committed on either ground."

"As Lilla Farkas points out, with Hungarian law allowing for the handling of data on racial and ethnic origin only with the consent of the person concerned, the effect is a severe impediment on the prospect of litigation against indirect discrimination or institutional racism."

Criminal Justice and Ethnic Data

"For example, in the case of "violence against members of national, ethnic or racial minorities and religious groups," the following number of instances had been registered: 1999, 3, 2000, 8, 2001, 12, 2002, 5, 2003, 11."

"Referring to data protection regulations, official statistics have no reliable data on the ethnicity of race crime victims, either, which is entirely absurd, given that the existence of racially motivated crimes logically presupposes membership in the given (racial or ethno-national) community."

"Other evidence for demonstrating that such a narrow interpretation of a superficially minority-friendly principle is actually contrary to the victimized groups' interest is that the abovementioned criminal statistics operate with a homogenous victim category, and statistics will not contain figures broken up into the various protected categories (racial, national, ethnic, religious groups)."

Police Profiling and Ethnic Data

"Should this (maximum 12-h) arrest not be sufficient, another type of detention ("public order detention") may be ordered, which (including the time spent in arrest) may take as long as 24 h. For these stop and search procedures, no suspicion whatsoever is needed, no probable cause standards are set forth, and, as demonstrated above, unsuccessful identification itself may lead to up to 24 h of detention."

"The researchers found that Roma offenders and suspects were significantly more likely to have been identified via police stops and searches, whereas in the case of nonminority suspects, other investigatory methods—particularly being caught in the act—were the dominant causes that lead to the suspects' capture."

"Another empirical comparative research, conducted on behalf of the Open Society Justice Initiative between 2005 and 2006, indicated that in Hungary, the Roma are discriminated against in the context of stops and searches by the police, especially in the case of pedestrians."

Conclusion

"As for the first, we may confidently state that the Hungarian legislator did not err in providing a narrow framework for ethnic data processing; as whatever the reasons for ethnic profiling and the ill treatment of ethno-national minorities may be, it is certainly not related to ethno-national data collection."

"Maybe it is also due to the superficially satisfying legislative framework for ethnic data collection that indirect and direct discrimination in policing and other fields of life can exist."

"After all, the preferential treatment of minorities should never be compulsory but must always remain the prerogative of the rightful recipients, and no victim of ethnic discrimination should be deprived of legal remedy because of data protection arguments that actually fail to serve their initial goals."

The General Data Protection Regulation in the Age of Surveillance Capitalism [57]

This is a machine-generated summary of:

Andrew, Jane; Baker, Max: The General Data Protection Regulation in the Age of Surveillance Capitalism [57].

Published in: Journal of Business Ethics (2019).

Link to original: https://doi.org/10.1007/s10551-019-04239-z.

If you want to cite the papers, please refer to the original.

For technical reasons we could not place the page where the original quote is coming from.

Abstract-Summary

"Prior research has investigated the impact of Big Data on individual privacy rights, however, there is also growing recognition of its capacity to be mobilized for surveillance purposes."

"We postulate that efforts at controlling Big Data may create a trade-off of risks rather than an overall improvement in data protection."

"We posit that safeguards advocated by the law, anonymization and pseudonymization, while representing effective counter measures to privacy concerns, also incentivize the use, collection, and trade of behavioral and other forms of de-identified data."

"We consider the legal status of these ownerless forms of data, arguing that data protection techniques such as anonymization and pseudonymization raise significant concerns over the ownership of behavioral data and its potential use in the large-scale modification of activities and choices made both on and offline."

Introduction

"With reference to Zuboff's [13] latest book, The Age of Surveillance Capitalism, we argue that while efforts to de-identify data subjects are no doubt beneficial in terms of privacy, these same processes, which disconnect subjects from their personal data, free up the use and trade of other types of behavioral data."

"It is important to note here that these techniques represent a new form of surveillance that is less to do with a centralized state-led panopticon targeting individuals as private citizens and more concerned with promoting various data marketplaces engaged in the trade of both behavioral data and its associated predictive algorithms and products."

"The next section defines Big Data and explores it as a potential site of privacy and surveillance risks."

"We then conceptually delineate privacy and surveillance as two separate but related data protection issues, both of which are always present in various Big Data practices."

"We first outline the improvements to privacy and surveillance risks, created by the GDPR, in terms of identified data."

What is Big Data?

"The term 'Big Data' refers to the binding of advanced predictive tools with large data sets on human activity, with the aim of tracking, monitoring, analyzing, and disseminating this information [14–16]."

"Ironically, the defining feature of Big Data is not its size alone but a capacity to search and integrate large data sets in order to more accurately predict future events, activities, and behaviors [14]."

"The properties that define Big Data, such as volume, variety, and veracity [17], contain insights into the advances of these new technologies and practices."

"To volume, the variety of text, images, and video produced around human activities and their digital footprints is an important feature of Big Data."

"This includes the production of new types of metadata (data that describe and give information about other data), which has begun to change the shape and usefulness of large data sets."

The Risks of Big Data

"We review two ways that ethics-based research has engaged with Big Data, that is, through exploring privacy and surveillance issues."

"Privacy work primarily deals with the impact of Big Data practices on individual rights, including the extraction of data, the use of disclosures, and consent, access, and privileges of users."

"We refer to these concepts as privacy risk, which focuses on the adverse effects of Big Data practices on the rights of individuals as private citizens, and surveillance risk, which relates to how Big Data is used by the state and companies to monitor and control individuals as participants in politics and markets."

"It is important to understand that surveillance capitalism, as facilitated by Big Data, is not a centralized state-controlled program that observes individual subjects but rather a market place, wherein behavioral data and "prediction products that anticipate what you will do now, soon, and later" are exchanged [13, p. 14]."

A Data Protection Trade-Off: Privacy Risk Versus Surveillance Risks

"While both concerns relate to possible harms that may be inflicted on individuals as a result of new forms of visibility, privacy and surveillance risks should not be conflated."

"Privacy risks are thought to increase with the extraction of more detailed and more personal data because the depth of these data can be used to intimately target individuals as subjects."

"The conceptualization of surveillance risk extends our understanding of these risks to include the broader control and governance implications of gathering and trading in this personal data."

"There are a range of scenarios that include varying levels of both privacy and surveillance risks, but while data surveillance might benefit from access to personal information, it does not need this information to be effective."

"As the accumulation of a wide variety of data has grown, new regulatory efforts have emerged to address both surveillance and privacy risks."

The General Data Protection Regulation

"We will first explore the main principles of the GDPR in cases where persons or data subjects are identifiable under the law We will briefly explore how the law may offer an improvement to both privacy and surveillance risks."

"Article 4.5 of the GDPR defines pseudonymization as "the processing of personal data in such a way that the data can no longer be attributed to a specific data subject without the use of additional information.""

"The relaxation of the purpose specification principle means that pseudonymization is an attractive option for controllers, allowing for regulatory compliance with the GDPR, while permitting an expansion of data collection and the subsequent uses of this data."

"While pseudonymized data provide certain additional protections for individual privacy, it introduces higher surveillance risk than situations in which no identification techniques are used and where data controllers are required to observe the principles of purpose specification and data minimization."

De-identification Techniques in the Age of Surveillance Capitalism

"Zuboff's [13] The Age of Surveillance Capitalism details and traces the activities of a new breed of data capitalist; one driven to acquire ever more predictive sources of behavioral data on our choices, activities, behaviors, personalities, and emotions."

"For Zuboff [13], the system of 'surveillance capitalism' does not operate as a centralized panopticon that identifies and targets individuals, but as a de-centralized marketplace that exchanges both behavioral data (as a raw commodity), and its associated predictive models to analyze this data."

"The law, in approving the use of complex modification techniques like anonymization and pseudonymization may be inadvertently crystallizing the power of tech elites, like Google and Facebook, which have already established vast economies of scale in the collection and analysis of behavioral data."

"The value of the advertising services of tech platforms like Facebook is the access to data, predictive algorithms, and behavioral modification techniques, the valuable combination of which has been described as an emerging "behavioral futures market" by Zuboff [15, p. 15]."

Recommendations

"Our recommendation is not necessarily wedded to the establishment of new property rights over res nullius data but may encompass alternative approaches, such as a stakeholder perspective that would ask the question: 'if Big Data can predict and modify the activities of people with great accuracy then should it not stand to reason these same individuals have a stake in its function?'".

"For the GDPR to be robust, data that are collected with the intention to be anonymized should still be subject to the consent and agreement of the individual."

"While pseudonymization requires consent processes, under the current law, anonymized data do not."

"Perspective, it is important to ensure all individuals know if their data can be traded, even if the data are completely anonymized, and to grant them the right not to make their anonymous data available for this purpose."

Conclusion

"This emphasis placed on individual privacy does not provide sufficient protections to citizens against the surveillance risks associated with the collection, analysis, and trade of their behavioral data."

"The way the law has been constructed has created space for a behavioral futures market to emerge with very few protections for individuals."

"There is little doubt that the GDPR represents a bold attempt by the EU to introduce new data protection regulations, and it will go some way towards addressing emergent ethical concerns about data collection and processing."

"The law creates the space for a behavioral data market in which commercial self-interest is likely to flourish."

"Whether deliberately or otherwise, the GDPR will make it possible for the behavioral data market to continue to function unencumbered."

Five ethical challenges facing data-driven policing [58]

This is a machine-generated summary of:

Davis, Jeremy; Purves, Duncan; Gilbert, Juan; Sturm, Schuyler: Five ethical challenges facing data-driven policing [58].

Published in: AI and Ethics (2022).

Link to original: https://doi.org/10.1007/s43681-021-00105-9.

If you want to cite the papers, please refer to the original.

For technical reasons we could not place the page where the original quote is coming from.

Abstract-Summary

"This paper synthesizes scholarship from several academic disciplines to identify and analyze five major ethical challenges facing data-driven policing."

"The challenges have been articulated in the context of related discussions, but their distinctively ethical dimensions have not been explored in much detail."

"Our goal here is to articulate and clarify these ethical challenges, while also highlighting areas where these issues intersect and overlap."

"Responsible data-driven policing requires collaboration between communities, academics, technology developers, police departments, and policy makers to confront and address these challenges."

Introduction

"After years of local resistance and national criticism about racial bias and other civil liberties concerns, the Los Angeles Police Department, in the spring of 2020, ended its decade-long experiment with PredPol, an algorithmic crime forecasting system designed to predict the location of property crimes [18, 19]."

"That some municipalities are erecting barriers to the use of data-driven methods in policing, more departments use data-driven methods than ever before."

"Given the constantly evolving landscape of data-driven policing methods and the inconsistent publicity about the use of such technologies, it is difficult to know precisely how many departments employ data-driven methods, but it is safe to say that most major cities across the U.S., as well as several smaller cities and towns, use one or more forms of data-driven policing technology."

Methods

"This paper synthesizes scholarship from several academic disciplines to identify and analyze five major ethical challenges facing data-driven policing."

"Because the term "data-driven policing" encompasses a broad swath of technologies, we first outline several data-driven policing initiatives currently in use in the United States."

"Responsible data-driven policing requires collaboration between communities, academics, technology developers, police departments, and policy-makers to confront and address these challenges."

"Before beginning, however, two caveats are in order: first, our focus in this paper is the deployment of data-driven policing tools in the US context."

"What we provide in this article is neither intended to be comprehensive nor to be the final word on data-driven policing."

"We hope what follows gives the reader a sense of some of the obstacles and open questions facing any attempt at a holistic assessment of data-driven policing."

What Is Data-Driven Policing?

"What characterizes big data systems, of which data-driven policing is one instance, is that they make use of vast quantities of data from a variety of data sources, and, using rapid computer processing and algorithms, identify correlations or patterns in the data—often patterns that humans would otherwise overlook—to classify new instances in the domain of interest [20–22]."

"Premised on a criminological phenomenon known as the "near repeat effect," wherein one crime event in an area tends to give rise to a spike in similar crimes in that area, PredPol was used by the LAPD to forecast future property crime—vehicular theft and theft from a vehicle—at highly specific locations and times during a police officer's patrol shift [23]."

"The London Gangs Matrix has been used by the London Metropolitan Police since 2011 to identify individuals at high risk of involvement in the city's gang activity and includes both individuals who are associated with committing gang-related crime as well as victims."

Challenge #1 Racial Bias and Civil Liberties

"If, as differential selection theory suggests, police target people differently on the basis of race, then arrest data, particularly for drug and nuisance crimes, is significantly influenced by racial bias in police officers' choices about whom to investigate."

"If arrests for drug and nuisance crimes are a reflection of racial bias on the part of police, and arrests are used to generate the forecasts of predictive policing systems, then these systems will tend to forecast greater crime than actually exists in communities where racial minorities are concentrated."

"Marda and Narayan have identified several potential sources of bias within the New Delhi place-based predictive policing system CMAPS, which makes predictions solely on the basis of data from past emergency calls for service and police reports [24]."

"Something else that has been lost in the discussions of bias and discriminatory impacts is how to better include members of minority communities in decision-making about data-driven systems and additional police attention being directed toward their community."

Challenge #2 What Data for Which Purposes?

"Our second challenge concerns two assumptions made by proponents of data-driven policing that its critics have called into question."

"Advocates of data-driven policing often tacitly assume that once a department adopts certain data-driven methods, these methods can be used in a value-neutral way."

"Choices about which data to use in data-driven policing implicitly or explicitly involve choices about the proper role of police in crime prevention."

"Much of the discussion surrounding the implementation of data-driven policing systems has thus far ignored this question."

"Insofar as data-driven policing systems encourage more aggressive enforcement-oriented policing tactics than other systems, community members have a significant stake in participating in the deliberative process about which systems are adopted."

"Data-driven technologies are a tool of the police."

"If existing data-driven policing technologies are tools of oppression and racial control, then we should not be using data at all in the service of these technologies [25]."

Challenge #3 The Standard of Success

"Is it good enough for data-driven policing systems to make more accurate classifications, offer greater transparency with respect to the factors that play a role in institutional decision-making, and produce criminal justice outcomes that better align with societal commitments to fairness than the status quo in policing?".

"We note in Challenge #4 below various obstacles to inspecting the inner workings of data-driven systems for discriminatory potential, but these obstacles are not obviously greater in the case of data-driven systems than they are for human decision-making."

"Is it worth investing in an algorithmic system that predicts 2.5% more crime than human analysts?".

"Brantingham and Valasik's study examining discriminatory arrest patterns when using PredPol found no significant differences in the proportion of arrests by racial-ethnic groups when comparing police patrols allocated by PredPol to patrols allocated by human crime analysts [26]."

Challenge #4 Explanation and Transparency

"Like many other algorithmic systems, data-driven policing systems have been criticized for their relative lack of transparency—that is, their opacity [20, 27–29]."

"Data-driven policing systems are opaque in several ways."

"Data-driven systems developed using "deep learning" tools only amplify technical obscurity."

"Most data-driven systems are shrouded in secrecy."

"Whether it originates in the technical features or proprietary features of data-driven technology, opacity in the use of data-driven policing technologies is problematic on several fronts."

"Many police departments have adopted data-driven systems in response to concerns about racial bias in policing that occurs when police departments rely entirely on officers' individual judgment and discretion."

"As we discuss in Challenge #1, race-blind data-driven systems can reflect and reinforce discriminatory practices, if the data used by those systems reflects discriminatory practices."

Challenge #5 Accountability and Community Oversight

"Privatization thus poses a significant obstacle to the efforts of concerned citizens, advocacy groups, and those directly impacted by these systems to hold law enforcement accountable."

"Without full access to these systems, any attempts at holding law enforcement accountable will be incomplete."

"It is widely thought that accountability can be improved by subjecting data-driven systems to 'algorithmic audits' and 'algorithmic impact assessments', in which an independent organization evaluates the system for bias, unfair treatment, improper data, corrupt source codes, harmful impacts and other problems [30, 31]."

"Such a requirement should extend to those private algorithms being leased by U.S. law enforcement agencies and courts."

"Another encouraging example of algorithmic auditing in practice comes from a collaboration between ShotSpotter Technologies and NYU's Policing Project."

"These stories are promising, and algorithm audits are, to be sure, an essential step in ensuring law enforcement accountability; however, several key issues remain."

Conclusion

"Each of the ethical challenges we discussed raises many more complex and nuanced issues than we have addressed here."

"There are many other urgent technical, legal, and ethical challenges facing data-driven policing."

"Other challenges, such as those relating to bias, data selection, standards of success, and deeper questions about the social value of police require more collective deliberation."

"Those who wish to continue on the path to better data-driven policing must work to address the challenges we have raised above, and many other issues that are raised by this technological shift."

"Those who oppose data-driven policing practices ought to be aware of and sensitive to the full range of challenges, their complexity, and the various possible pathways forward."

"Socially responsible data-driven policing requires collaboration between academics, technology developers, police departments, and policy-makers to confront and address these challenges."

How to protect privacy in a datafied society? A presentation of multiple legal and conceptual approaches [59]

This is a machine-generated summary of:

Gstrein, Oskar J.; Beaulieu, Anne: How to protect privacy in a datafied society? A presentation of multiple legal and conceptual approaches [59].

Published in: Philosophy & Technology (2022).

Link to original: https://doi.org/10.1007/s13347-022-00497-4.

If you want to cite the papers, please refer to the original.

For technical reasons we could not place the page where the original quote is coming from.

Abstract-Summary

"This article explores privacy by mapping out different legal and conceptual approaches to privacy protection in the context of datafication."

"The resulting overview provides insight into the main strengths and limitations of the different approaches arising from specific traditions."

"This analytic overview therefore serves as a key resource to analyse the usefulness of the approaches in the context of the increasing datafication of both private and public spheres."

"We contrast the approach focusing on data subjects whose data are being 'protected' with others, including Fair Information Practice Principles, the German right to 'informational self-determination', and the South American 'habeas data' doctrine."

"We also present and contrast emerging approaches to privacy (differential privacy, contextual integrity, group privacy) and discuss their intersection with datafication."

Living in the Digital Age

"This article explores venues to the protection of privacy by mapping out and contrasting different legal and conceptual approaches in the context of datafication."

"This observation leads to a first key point in this discussion of datafication and privacy: many insights about individuals rely on assumptions about populations and groups, whereas most privacy and data protection regimes focus on atomistic agents and their individual rights, often labelling them as autonomous individuals [32]."

"To better understand the tensions between approaches to the protection of privacy, the objective of this article is to concisely explain and analyse multiple approaches in the context of datafication."

"It is important to make this variety of approaches towards the protection of privacy in a datafied society clearly visible for at least two reasons."

"We add to this regulatory overview by presenting recently emerging approaches to the protection of privacy from different areas."

Datafication

"As soon as one considers the various environments and practices involved in making and interpreting data, it becomes clear that datafication is not only about data and related computational techniques."

"Datafication is a practice that links four essential elements: First, the community of actors (and related institutions) who engage with the data."

"(4), we consider several ways in which datafication changes the very grounds from which privacy-related concepts arise."

"Possession—or lack—of digitally verifiable credentials therefore translates personal data into economic value, shapes who we can be on the internet, and connects the private and the public in new ways. (On the internet, platforms know exactly what kind of dog you are and which segment of dogfood they can sell you.) Potentially, these many changes brought about through datafication can affect individual and collective self-determination, as well as human dignity [33]."

Approaches to the Protection of Privacy

"The demand for transparency increased, and with it, the duty to inform an individual upon request whether one is subject to data processing, what use is made of personal data, who has access to it, and for what reason [34]."

"The development of this 'data protection law' must be considered in parallel with the development of frameworks based on the protection of privacy as an individual right."

"Informational self-determination is a specific and independent individual right that is not based on traditional privacy provisions and has no direct connection to data protection, or the general right to personal development."

"There are three important parameters that need to be considered: the accuracy of the data as initially collected, the addition of noise to manipulate the set and conceal characteristics of individuals, as well as the number of queries available to protect individuals and keep the dataset useful (often called the 'privacy budget')."

Discussion: Strengths, Limitations and Remaining Issues Around the Approaches to Privacy

"Using the argument of personalisation, many platforms have eroded the boundaries between what would be private, or what would belong in different individual or collective spheres, by increasingly combining and circulating personal data with the justification of providing more personalised services to customers."

"Privacy becomes an issue of circulation, combination and evaluation of data, rather than of labelling data for a specifically definable and static purpose, which is typical for data protection law The correspondence of personalisation and of privacy as associated with a subject is not accidental: these dynamics share deep roots in systems of ownership and rights."

"Privacy is then not tied to a special place, but to the possibility for the data subject to modulate, be informed, and keep control on what one allows to circulate, to whom and for what purpose."

"Data protection frameworks have the advantage of clearly defined roles, which can then be associated with concrete individual rights and duties."

Conclusion

"Is it the case that several different approaches thrive or are we moving towards a monoculture of privacy focused on the least common denominator?".

"Data protection might be a useful approach from a legal perspective, whereas approaches such as habeas data and informational self-determination might be especially strong when it comes to establishing governance principles and political guidelines."

"Group privacy might be an approach that has its merits in the context of social sciences and the development of ethical frameworks for the use of PII."

"Rather than trying to find one single solution that works across all these settings and worldwide, it might be more useful to identify synergies and stumbling blocks to make sure privacy is respected, protected and promoted where it is necessary and desirable."

"It also seems that some approaches to privacy may be less amenable to certain technological configurations."

The future of value in digitalised higher education: why data privacy should not be our biggest concern [60]

This is a machine-generated summary of:

Komljenovic, Janja: The future of value in digitalised higher education: why data privacy should not be our biggest concern [60].

Published in: Higher Education (2020).

Link to original: https://doi.org/10.1007/s10734-020-00639-7.

If you want to cite the papers, please refer to the original.

For technical reasons we could not place the page where the original quote is coming from.

Abstract-Summary

"Universities around the world are increasingly digitalising all of their operations, with the current COVID-19 pandemic speeding up otherwise steady developments."

"This article focuses on the political economy of higher education (HE) digitalisation and suggests a new research programme."

"I argue that current changes in digitalising HE cannot be studied in isolation from broader changes in the global economy."

"HE digitalisation is embedded in the expansion of the digital economy, which is marked by new forms of value extraction and rentiership."

"The emerging research on the intersection of marketisation and digitalisation in HE seems to follow the theories of marketisation qua production and commodification."

"I argue that we need theories with better explanatory power in analysing the current digitalisation dynamics."

"The current pandemic has led to 'emergency pedagogy', which has intensified overall digitalisation in the sector and is bypassing concerns of data value redistribution."

Introduction

"The commonality among these forms is a shift from a business model of selling a commodity, which would involve a transfer of ownership rights for an exchange of a product or a service, to a business model based on a digital platform, which enables extracting value through ownership and control rights of digital data, and monetisation via rent."

"The political economy of digitalisation in and of HE thus far developing implies that it is most likely proprietary platforms who profit from the business models emergent in the digital economy by owning and controlling the rights over platforms and extracted data."

"Quite possibly, when universities deliver modules and courses in partnerships with the platform companies, they might be selling their services as commodities; while platform companies might benefit from assetising digital data and charging rent."

Conclusion

"This article focuses on the political economy of EdTech and extracted digital data in the HE sector."

"This is crucial as the digital economy brings new forms of value construction based on data extraction, analysis, and intelligence, and governed by IP rights."

"I then argued that there are different forms and types of HE digitalisation, most of which include proprietary platforms."

"Digital platforms and extracted data are constructed as assets and besides monetary rent, also collect data rent."

"I suggested a new research agenda involving a careful, in-depth, case-to-case analyses of different assetisation processes, the kinds of assets that are created, and the kinds of monetary and data rents that are collected."

"I explained the importance of analysis of extracted data ownership, and contracts between platform owners and universities, as well as between platform owners and individual students and staff."

Digital technologies: tensions in privacy and data [61]

This is a machine-generated summary of:

Quach, Sara; Thaichon, Park; Martin, Kelly D.; Weaven, Scott; Palmatier, Robert W.: Digital technologies: tensions in privacy and data [61].

Published in: Journal of the Academy of Marketing Science (2022).

Link to original: https://doi.org/10.1007/s11747-022-00845-y.

If you want to cite the papers, please refer to the original.

For technical reasons we could not place the page where the original quote is coming from.

Abstract-Summary

"With a comprehensive analysis of digital technologies and data strategy informed by structuration theory and privacy literature, the authors consider privacy tensions as the product of firm–consumer interactions, facilitated by digital technologies."

"This perspective in turn implies distinct consumer, regulatory, and firm responses related to data protection."

"By consolidating various perspectives, the authors propose three tenets and seven propositions, supported by interview insights from senior managers and consumer informants, that create a foundation for understanding the digital technology implications for firm performance in contexts marked by growing privacy worries and legal ramifications."

"On the basis of this conceptual framework, they also propose a data strategy typology across two main strategic functions of digital technologies: data monetization and data sharing."

Firms' Digital Technology Use and Consumer Privacy Tensions

"Digital technologies allow firms to access vast amounts of data, which they might leverage to increase their profitability (i.e., data monetization) or improve the performance of their broader business networks (i.e., data sharing)."

"With vast digital technologies, they generally engage in little sharing but monetize data through extended data wrapping (e.g., game development services) or sales of information or analytics (e.g., data brokering services)."

"Digital technologies offer data monetization and sharing benefits to firms but have concomitant costs for consumers, especially with respect to privacy."

"Data aggregation, processing, and storing technologies empower data monetization practices by establishing novel insights about customers from IoT-enabled devices and big data, facilitated by cloud technologies, which can inform consumer profiling, behavior prediction, and targeting efforts."

A Structuration Approach to Digital Technology–Privacy Tensions

"Firms and consumers inherently belong to social systems that establish structures, such as regulatory frameworks or strongly held social norms about privacy."

"Considering the novelty of this structuration theory application to privacy topics, as well as the complexity of our proposed model, which involves interplays of institutions (regulators), groups (firms), and individuals (consumers), we focus here on the duality of structure and social practices in an effort to clarify privacy tensions among firms, consumers, and regulators, rather than test the original analytical dimensions of structuration theory."

"We can explain the regulatory framework associated with privacy (i.e., structure) and predict both reactive and proactive responses by consumers and firms (i.e., actors)."

"According to structuration theory, structures such as regulatory frameworks (i.e., rules) can both constrain and enable consumer and firm actions in the digital landscape, exacerbating or offsetting privacy tensions."

Consumer Privacy Protection Behavior

"If consumers realize greater privacy risks due to the nature of the data being collected or increased breach likelihood in a firm relationship, they become more likely to engage in privacy protective behaviors, manifested as future responses to the structures and resources available within that relationship."

"Some privacy-protecting strategies increase consumers' control over personal information (e.g., decrease disclosures, minimize their digital footprint) or establish requirements for explicit permission for uses of their personal data (information access and use) [35]."

"With a proactive strategy, consumers preemptively address privacy threats; with a reactive strategy, they act as explicitly advised by a firm or in response to an immediate threat."

"In response to regulatory and consumer actions, firms might comply with privacy rules (reactive response) or go beyond them to engage in privacy innovation (proactive response), which potentially shapes new structures [36]."

Integrated Framework of the Structuration of Privacy

"Drawing from the notion in structuration theory that structure can both shape and be shaped by social practices, we note that the inherent privacy tensions of data monetizing practices provoke consumer and regulatory privacy responses, which have direct implications for firm performance."

"We propose the following tenet and propositions: Enabled by digital technologies, data monetization creates a trade-off between firm profitability and privacy tensions (information, communication, and individual privacy)."

"Data monetization negatively influences firm performance through increased privacy tensions (information, communication, and individual privacy), which trigger consumer data protection behaviors and privacy regulations."

"Data sharing negatively influences firm performance through increased privacy tensions (information, communication, and individual privacy), which trigger consumer data protection behaviors and data privacy regulations."

"Compared with reactive privacy responses, firm proactive privacy responses reduce consumer privacy protection behavior by mitigating the negative effects of data monetization and data sharing strategies on privacy tensions."

Implications for Firms with Different Data Strategies

"Our novel framework explains that digital technologies empower firms' data monetization or data sharing strategies, while also creating privacy tensions."

"The effect of data privacy regulation on data sharing and monetization thus should vary across firms which rely on data to varying levels."

"The more firms rely on data monetization and data sharing, the more pronounced the effects of regulatory changes and consumer data protection behaviors become."

"Data harvesters engage in limited, internal data monetization and data sharing practices, leaving them less exposed to consumer privacy behaviors and regulations."

"The head of digital marketing of an apparel firm shared: We don't have as much access to that sort of [third-party] data but yet we do have first-party data, which will be those [data] of those people within our leads database in our customer base."

"Data harvesters likely comply closely with government regulations and tend to adopt reactive privacy strategies to protect consumer privacy."

Recommendations for Policymakers

"On the basis of our integrative framework and typology of data strategy, which establish a comprehensive view of privacy tensions linked to emerging digital technologies, we offer policymakers several recommendations for drafting, implementing, and monitoring effectively such regulations."

"Although future-proofing privacy regulations is untenable, regulatory parameters that govern fundamental data exchanges, rather than specific technological techniques for gathering or processing data, can protect consumers more broadly, even as technologies change."

"It is thus imperative that regulatory frameworks adequately protect consumer data, regardless of technological advances, with legal and protective parameters drafted with a technology-neutral approach that avoids regulatory obsolescence and prevents innovative subversion by firms."

"Even with comprehensive data protection regulations in force, multinational corporations appear to adopt reactive privacy strategies for the most part, while continuing to engage in behaviors and practices that put them and their customers at risk for data breaches [37]."

Conclusion and Research Directions

"We begin by identifying the pressing tensions between firms' data monetization and sharing practices enabled by digital technologies, as well as their implications for consumer privacy."

"We overlay these predictions with a typology of firm strategies for data monetization and data sharing to highlight how firms' privacy responses are influenced by resources (i.e., digital technology), rules (i.e., regulations), and other actors' actions (i.e., consumers' privacy protection behaviors)."

"To stimulate further research, we propose an agenda across three broad areas: firm data and privacy strategies, regulatory impacts on firms and ecosystems, and consumer responses to the digital technologies."

"Research thus might investigate consumers' perceptions of firms' data strategy (i.e., data harvester, informant, patron, and expert), actual data privacy actions, and regulation stringency, which could help construct long-term, technology-proof outcomes for all parties involved."

[Section 8]

"Such reliance prompts increasing privacy concerns about firms' data behaviors and actions among both consumers and regulators."

"To develop new marketing insights into digital technologies and privacy, we explicitly consider this firm–consumer–regulatory intersection and work to disentangle the data strategies and embedded technologies that firms use to create mutual value for themselves and their customers."

"We investigate the consumer implications of firms' digital technology use, with a particular focus on their privacy responses."

"To relying on effective regulations for data protection, consumers exhibit other privacy protection behaviors and demands, which then intensify the pressure on firms to respond to privacy tensions."

"We begin with a description of the consumer privacy tensions that emerge from firms' digital technology uses in four areas: data capture; data aggregation, processing, and storage; data modeling and programming; and data visualization and interaction design."

Value-creation in the health data domain: a typology of what health data help us do [62]

This is a machine-generated summary of:

Fiske, Amelia; Degelsegger-Márquez, Alexander; Marsteurer, Brigitte; Prainsack, Barbara: Value-creation in the health data domain: a typology of what health data help us do [62].

Published in: BioSocieties (2022).

Link to original: https://doi.org/10.1057/s41292-022-00276-6.

If you want to cite the papers, please refer to the original.

For technical reasons we could not place the page where the original quote is coming from.

Abstract-Summary

"Important epistemic and political questions remain unexamined, such as how the value of data is created, what data journeys are envisioned by policies and regulation, and for whom data types are (intended to be) valuable."

"Drawing on two empirical cases, (a) interviews with physicians on the topic of digital selfcare, and (b) expectations of stakeholders on the use of Real-World Data in clinical trials, as well as existing literature, we propose a typology of what health data help us to do."

"This typology is intended to foster reflection about the different roles and values that data use unfolds."

Introduction

"In the health domain, the value of data is historically contingent, purpose-specific, and relational."

"Some actors in the health data space have sought to establish themselves as platforms, with business models built around the idea of creating different types of value with the same kind of data."

"In reflecting on the Covid-19 pandemic, and increasing calls for the need for 'more' data, or the inherent value of data in devising appropriate public health responses, we argue it is productive to shift the focus away from generalised notions of data that are considered sensitive or not, or data that can 'facilitate research' or 'improve healthcare' to instead consider data in terms of roles and uses, within a suite of relationships and practices that can, in turn, affect value."

Methods

"The first one revolved around reflections on digital self-care (DSC) by physicians in Northern Germany, and the other on expectations of the use of Real-World Data (RWD) in clinical trials by those working for institutions that run or participate in these trials."

"Of the interview, participants were asked to share their views about the role of data in clinical research, and in particular on data collection."

"Following this, each author discussed coding results, and refined sub-themes to ensure a high degree of inter-coder reliability, with the aim of identifying different roles that data play in health research and medical practice, and extrapolating different types of value that they yield."

"The case studies, combined with our reflection and review of literature relating to the uses of health data during the Covid-19 pandemic, helped us to produce the first iteration of the typology."

Results: Typology of Roles of How Health Data Become Valuable

"We also observe this trend of datafication facilitating observation and interpretation in the context of Covid-19, e.g. with regard to self-testing: a new type of health data becomes individually and socially meaningful when the SARS-CoV-2 infection status is relevant knowledge for participation in public life."

"At the population level, for instance, as the institutional context of some of the authors as well as the background research in the context of the RWD study showed, routinely collected real-world health data (e.g. from medical health records or administrative data) can be used to predict healthcare demand, for instance as related to the prevalence of a specific disease at a certain point in the future."

"Patient-collected data, e.g. on lifestyle or health registry data, can make the process of identifying possible participants in a clinical study more efficient, reducing the cost of clinical research."

Discussion and Conclusions

"We argue that to assess the value of specific instances of data use, the typology provides a missing link between types of data (survey data/registry data; personal data/non-personal data; etc.) and purpose of use (pandemics management, insurance fraud detection, drug research, medical treatment, etc.)."

"Focusing on the use and not the type of data (such as format or source) can help researchers, regulators, and the public to advance a more nuanced discussion of the risks, harms and benefits that emerge from the context of use."

"As conversations continue around the increasing value of health data in our societies, we hope that this typology, which brings together empirical examples as well as collective reflection on the Covid-19 pandemic, can help to support a more structured reflection of thedifferent roles and values through which data use unfolds in other health contexts."

Human Rights, Privacy and the Public Interest—Who Draws the Line and Where? [63]

This is a machine-generated summary of:

Mance, Jonathan: Human Rights, Privacy and the Public Interest—Who Draws the Line and Where? [63]

Published in: Liverpool Law Review (2009).

Link to original: https://doi.org/10.1007/s10991-008-9043-8.

If you want to cite the papers, please refer to the original.

For technical reasons we could not place the page where the original quote is coming from.

Abstract-Summary

"It considers the significance and extent of the margin of appreciation which that Court allows to individual states in respect of the protection of privacy and the balancing of privacy against other interests such as press freedom."

In Brief Conclusion

"There remain differences in attitudes and laws regarding privacy in different European states."

"It is unclear that the margin of appreciation allowed by the European Court of Human Rights in this area of privacy is in practice very large, despite different attitudes in different European countries."

"The internet is also a phenomenon with the potential to disseminate information and affect privacy on a world-wide basis (and, as Editions Plon v. France illustrates, so to undermine national attempts to enforce privacy)."

"There are obvious benefits in this connection in terms of speed and costs in the potential redress offered by more informal dispute resolving mechanisms, but they do not of course offer damages or indeed enforceable injunctive relief."

"There have in the past been calls for domestic legislation regulating the area of privacy."

[Section 2]

"The European Court of Human Rights in Strasbourg has held it can be incumbent on a subscribing state to take positive steps to protect the Convention interests of those within its jurisdiction by having and enforcing appropriate laws providing horizontal protection."

"In the light of later Strasbourg and English case-law, it is interesting to note that various members of the House stressed their view that the mere taking of photographs of a celebrity or anyone else in a public place would not constitute any infringement of privacy."

"58148/00), by a different section of the court, echoes interestingly the observation in paragraph 64 of the von Hannover case that the public right to be informed can "in certain special circumstances" even extend to aspects of the private life of public figures."

Allocating Personal Data Rights: Toward Resolving Conflicts of Interest over Personal Data [64]

This is a machine-generated summary of:

Yu, Xiaolan: Allocating Personal Data Rights: Toward Resolving Conflicts of Interest over Personal Data [64].

Published in: Fudan Journal of the Humanities and Social Sciences (2021).

Link to original: https://doi.org/10.1007/s40647-021-00330-w.

If you want to cite the papers, please refer to the original.

For technical reasons we could not place the page where the original quote is coming from.

Abstract-Summary

"This article attempts to resolve such conflicts of interest by examining the possibility of allocating data rights which balance competing interests."

"It suggests that the allocation of personal data rights should follow certain rules to ensure that the interests of relevant data subjects are protected."

"By examining the reasons for conflicts of interest over personal data, as well as the subjects and substance of such conflicts, the article puts forward an appropriate approach for allocating personal data rights by balancing the interests of different data subjects."

Introduction: Data and Wealth

"The fact that Internet users' personal data have become an important commodity today vividly illustrates how data have emerged as a new form and maker of wealth in today's information society."

"The term "data" have been used to describe this new form of wealth more than "information" in, for example, the EU's General Data Protection Regulation (GDPR) and almost all state-level legislation in the U. S. [38]."

"Value is embedded in personal data–i.e., data are valuable because of their possible actual use."

"This article suggests that the realization of personal data's value happens in four stages."

"The rapid development of the digital economy provides a driving force for the use of personal data and realization of their value."

Conflicts of Interest in the Use of Personal Data

"Conflicts of interest arise between derivative data subjects and data users over property interests."

"This means that there will be conflicts of interest between the different subjects in the data flow chain listed above, who pursue their personal interests, and data regulators, who act in the social and public interest."

"Personal interests sometimes come into conflict with public interests—for example, there is often a conflict between protecting source data subjects' personality interests and ensuring the free flow of data."

"Conflicts often arise when governments try to balance protection of private rights (e.g., personality and property interests) and the exercise of public power; for example, when governments utilize user data in the name of crafting effective public policies or when governments supervise citizens' and enterprises' private data [39, pp. 487–490]."

Potential Resolutions for Conflicts of Data Interest: Allocating Interests and Rights

"The basic principles of private law should be followed when allocating personal data rights."

"The rest of this section examines the allocation of personal data rights from the perspective of civil legislation and theory in order to illustrate the possibility (perhaps even necessity) of applying the theory of interest measurement in this domain of private law Both civil legislation and theory base the allocation of rights on some kind of measurement of interests."

"If we view rights as the interests of justice that conform to the social value orientation which the law protects and aspires to, then the process of allocating rights in legislation is the same as the process of interest measurement."

"There are many different data-relevant interest claims from various subjects, and it is essential that we balance these subjects' interests when allocating rights."

Conclusion

"The data-relevant rights framework has a solid basis in the theory of legal persons' ownership of property–that is, the right to possess, use, profit from, and dispose of data."

"This article has considered how to resolve the various conflicts among data subjects by adopting the perspective of data rights allocation in the field of private law Leaning on interest measurement theory, this article has examined the relevant interests of relevant data subjects, explored appropriate strategies for resolving conflicts of interest among different data subjects, and put forward a possible technical route for the allocation of personal data rights."

"It has suggested that conflicts of interest between data users, enterprises, and regulators even data users can be resolved through proper allocation of their rights to personal information, possession/use/profiting from/disposing of data, and right to supervision and utilization, respectively."

Privacy and data protection in learning analytics should be motivated by an educational maxim—towards a proposal [65]

This is a machine-generated summary of:

Hoel, Tore; Chen, Weiqin: Privacy and data protection in learning analytics should be motivated by an educational maxim—towards a proposal [65].

Published in: Research and Practice in Technology Enhanced Learning (2018).

Link to original: https://doi.org/10.1186/s41039-018-0086-8.

If you want to cite the papers, please refer to the original.

For technical reasons we could not place the page where the original quote is coming from.

Abstract-Summary

"Privacy and data protection are a major stumbling blocks for a data-driven educational future."

"This paper is based on an actual conundrum in developing a technical specification on privacy and data protection for learning analytics for an international standardisation organisation."

"Legal arguments vary a lot around the world, and seeking ontological arguments for privacy does not necessarily lead to a universal acclaim of safeguarding the learner meeting the new data-driven practices in education."

"This paper explores the legal and cultural contexts that make it a challenge to define universal principles for privacy and data protection."

"The different justifications for privacy, the legal obligation to separate analysis from intervention, and the way learning and teaching works makes it necessary to argue data privacy from a pedagogical perspective."

"The paper concludes with three principles that are proposed to inform an educational maxim for privacy and data protection in learning analytics."

Introduction

"In preparing for the digital futures of learning analytics (LA), adaptive education, multimodal learning support and other data-driven approaches educationalists need to develop what we have termed an educational maxim for privacy and data protection in this field."

"The question raised in this paper, is whether there also are pedagogical ideas that should be brought to bear when designing privacy policies and solutions for educational big data."

"To construct the foundation for an educational maxim for privacy related to educational big data this conceptual analysis builds on issues and concerns identified in design of LA applications."

"The questions leading up to the proposal of grounding privacy for LA in an educational maxim are the following: Is reference to privacy as an individual universal right the answer to data management and control?".

"These questions describe the structure of this paper, which concludes with three principles that could be used to further develop an educational maxim for data privacy in learning analytics."

Research Background Through a Practical Case

"The editors of the draft specification suggested that privacy problems should be looked at 'in a LET [learning, education and training] context to be able to specify privacy and data protection principles for LET that address specific problems and support a good learning environment for the individuals involved'."

"The editors suggested the following principle for development: 'The educational context of LA requires that the right to be informed is not interpreted restrictively; it is a pedagogical value of its own to be as open as possible about data collection and processing.'."

"The educational context is, however, an opportunity to clarify [for the students] privacy and data protection issues related to use of LA' (T. Hoel, personal communication, August 2018)."

"Working draft, it is clear that the authors of the standard try to carve out an educational argumentative space that would allow for certain policy principles regarding privacy."

A Universal Right to Privacy Using Educational Technologies?

"If this is so, then managers have an obligation to protect personal information privacy in every system and in every country, regardless of distinctions in national levels of concern or of regulatory approaches' [40, p. 73]."

"Trade-offs between ideals and reality may not be the best way to understand how privacy and related interests with regard to the processing of personal data are protected internationally."

"The formal normative basis for the data protection laws may well be derived 'mainly from the catalogues of fundamental human rights' [41, p. 180]; however, when it comes to applying these principles in international instruments one should note that an important motivation for developing international privacy frameworks are promoting the exchange of goods and services across borders."

Educational Data Privacy by Asking for Consent

"The new European Data Protection Regulation also proposes restricting the opportunities for the processing of personal data on legal grounds other than consent' [42, p. 49]."

"What does it mean when the Norwegian Data Protection Authority states as their primary recommendation to meet the challenges of big data: 'consent [is] still the point of departure' [42]?".

"That means the individual as an actor with rights to decide on data management, not as an object in need of protection by others."

"The Norwegian data protection authority points to claims, 'that the constant demand for consent on the Internet paradoxically may result in poorer protection for the individuals' (ibid, p. 50)."

"The problem with consent, we would argue, does not so much lie in hollowing out the consent mechanism as with the fact that consent is not the sole legal ground for access to personal data."

Balancing Interests for Educational Big Data Analysis

"In terms of legitimate interests, Cormack [43] sums up how European law specifies requirements for this justification to be used: Where personal data are processed for legitimate interests, there must be a clearly stated purpose, the processing must be necessary for that purpose, the impact and risk for the individuals whose data are processed must be minimised, and any remaining impact or risk must be justified by a balancing test against the claimed interest."

"Interests, even though legitimate, cannot justify processing that involves an inappropriate risk to the individuals whose data are processed [43]."

"The approach proposed by Cormack [43, 44] dividing the monolithic 'big data' process into two stages (analysis, to find patterns; and intervention, to identify and affect relevant individuals) opens up a need for examining the educational specific consequences and opportunities when applied on LA."

Pedagogical Opportunities Arising from LA Data Privacy

"Does data for analysis come only from institutional systems, like Student Information Systems or Learning Management Systems?".

"In order to be able to do analysis one needs to decide upon which learning activities to monitor; to collect the data that serve as proxies for the activities under study; and as a last important step also imbued with a host of privacy and data protection issues, decide upon how the data should be stored and processed before analysis."

"It is the risks to the learner, caused by the institution acting upon the knowledge from analysis that make it necessary to ask the learner to consent to processing of personal data, giving him or her the opportunity to opt out when the nature of the proposed intervention is clarified."

"Both the Tongzhou and the Oslo teachers had strong pedagogical motivation for their interest in student activity data."

Conclusions and Future Work

"It would help to build consensus about privacy and data protection policies if these also could be argued from an educational perspective, not only from universal or individual rights perspectives."

"This result of our explorations, the positioning of privacy and data protection for LA in an argumentative space of continuous negotiations, gives hope for achieving an international consensus on educational privacy policies."

"Privacy and data protection in LA are achieved by negotiating data sharing with each student."

"We would suggest that these principles are further developed and expressed in a LA privacy maxim for education."

"We will therefore follow up this research with more empirical studies of how different countries develop privacy policies in education, and how we can develop solutions for privacy in LA that could be accepted across cultures."

"This paper has shown that there is a need to understand how data privacy policies for LA connect to pedagogical practices."

Privacy in the Information Age: Stakeholders, Interests and Values [66]

This is a machine-generated summary of:

Introna, Lucas; Pouloudi, Athanasia: Privacy in the Information Age: Stakeholders, Interests and Values [66]

Published in: Journal of Business Ethics (1999).

Link to original: https://doi.org/10.1023/a:1006151900807.

Copyright of the summarized publication:

Kluwer Academic Publishers 1999.

If you want to cite the papers, please refer to the original.

For technical reasons we could not place the page where the original quote is coming from.

Abstract-Summary

"We conclude that privacy as the freedom or immunity from the judgement of others is an extremely useful concept to develop ways in which to understand privacy claims and associated risks."

"We develop a framework of principles that explores the interrelations of interests and values for various stakeholders where privacy concerns have risen or are expected to rise."

"We argue that conflicts between the interests and values of different stakeholders may result in legitimate claims of privacy/transparency being ignored or underrepresented."

"We argue that stakeholders are persons or groups with legitimate interests, of intrinsic value, in the procedural and/or substantive aspects of the privacy/transparency claim and subsequent judgements on that basis."

Are New Technologies the Enemy of Privacy? [67]

This is a machine-generated summary of:

Etzioni, Amitai: Are New Technologies the Enemy of Privacy? [67].

Published in: Philosophy & Technology (2007).

Link to original: https://doi.org/10.1007/s12130-007-9012-x.

If you want to cite the papers, please refer to the original.

For technical reasons we could not place the page where the original quote is coming from.

Abstract-Summary

"The merits or defects of particular technologies are not inherent to the technologies, but rather, depend on how they are used and above all, on how closely their use is monitored and accounted for by the parties involved."

"In order to reassure the public and to ensure accountability and oversight, a civilian review board should be created to monitor the government's use of surveillance and related technologies."

"Proper accountability requires multiple layers of oversight, and should not be left solely in the hands of the government."

An Arms Race Versus a Luddite Imagery

"Although these alarmist critics recognize that there is no way to turn back the clock to a pre-digital age, they bemoan the rise of privacy-invading technologies—criticisms similar in tone and terms to the complaints that the Luddites lodged against the development of industrial equipment, from the loom to the steam engine, in the nineteenth century."

"As far as the relationship between technology and privacy is concerned, it is best viewed as an arms race between advancements that diminish privacy and those that better protect it—as opposed to a one-sided development in which those who invade privacy overrun those who seek to protect it."

"True, the balance between privacy-invading and privacy-securing technology is changing all of the time."

"If one compares routine communications today to those of say 1975, one finds a significant net increase in privacy, largely due to the development of high-powered encryption."

Accountability

"One should note first of all that, like many technologies cited by civil liberty groups as sources of privacy violations, such as programs enabling credit card orders over the Internet, E-ZPass participation is voluntary."

"The main issue with E-ZPass is not the technology itself but the ways in which it is used—and that use supervised."

"As with other technologies, E-ZPass can be employed both in ways that most would find unproblematic and quite beneficial and in a manner that many would find very troubling indeed."

"To return to the example at hand, assume we are told that E-ZPass is used merely in the minimal way described above—information is immediately erased after verification that the toll has been paid, and the information about who traveled when and where is not otherwise available."

How Can Accountability Be Provided?

"The record shows that, on their own, these committees cannot provide the needed countervailing force to government agencies hell bent on following their own course in the name of national security."

"The press, the next line of defense, has been doing a sound job of regularly reporting about a variety of abuses and about programs with absurd designs, leading the government to send many ideas back to the drawing board and to greatly reform aspects of the no-fly lists, the tracking of foreign students, and airport scanning methods, among others."

"To further strengthen oversight for law enforcement authorities and to reassure the public that they are not running amok, we need a civilian review board."

"The board would issue regular reports about its generalized findings without revealing specifics about sources and methods."

"Such reports should lead to internal reforms in government agencies, as they will have to expect future rounds of similar audits."

Reliable E-ID

"Referred to as 'digital certificates,' such E-IDs can be provided through a 'certificate authority' or CA, such as GeoTrust or VeriSign."

"The electronic postmark (EPM) extension launched by the US Post Office in partnership with Microsoft and Authentic Date in October of 2003 is just one example of a technological development that seeks to provide reliable E-IDs."

"Such E-IDs make it much more difficult for unauthorized persons to gain access to a variety of information, and they will help to minimize identity theft."

"I simply wish to outline the basic concepts of E-IDs as a means by which to provide another example of how some advancements in technology are helping to safeguard privacy as others infringe upon it."

Conclusion

"Privacy is under attack by new technologies, but it is also benefiting from new technologies."

"Those concerned about privacy should work to improve the regulations controlling the use of these technologies rather than adopt a semi-Luddite position, hoping these technologies will go away or be suppressed."

"How carefully the use of various technologies is monitored is, as a rule, more important than the capabilities of the technologies themselves."

Digital health fiduciaries: protecting user privacy when sharing health data [68]

This is a machine-generated summary of:

Arora, Chirag: Digital health fiduciaries: protecting user privacy when sharing health data [68].

Published in: Ethics and Information Technology (2019).

Link to original: https://doi.org/10.1007/s10676-019-09499-x.

If you want to cite the papers, please refer to the original.

For technical reasons we could not place the page where the original quote is coming from.

Abstract-Summary

"They also create a conflict between individual interests of avoiding privacy harms, and collective interests of assembling and using large health data sets for public benefits."

"This paper argues for fiduciary relationships, which put deliberative demands on digital health data controllers to keep the interests of their data subjects at the forefront as well as cater to the contextual nature of privacy."

"These deliberative requirements ensure that users can engage in collective participation and share their health data at a lower risk of privacy harms."

Introduction

"I, therefore, further argue that digital health data controllers should be recognized as fiduciaries, such that they are required to keep the interests of the users at the forefront in making decisions about processing of health data."

"This discussion establishes the background for the next section, where I argue that the relationship between digital health data controllers and users should be recognized as fiduciary for three reasons: (a) the relationships shares key features with traditional fiduciary relationships; (b) it involves circumstances similar to those that have led to establishing fiduciary relationships in the past; and (c) fiduciary law is better suited than contractual or statutory law to protect user privacy and enable trust required for sharing health data with health data controllers."

"In the same section, I then propose an account of the scope of fiduciary duties that digital health data controllers should owe to their beneficiaries (users sharing health data)."

Transparency

"The success of transparency, in alleviating individual privacy concerns, as well as in promoting sharing of data for beneficial purposes (such as research), however, seems limited."

"The HIPAA Privacy Rule, for example, has been criticized both for allowing too much access to data (as individual concerns regarding sharing of their data for research were not addressed) as well as for restricting sharing of data for useful research (as consent requirements impeded sharing of data) [45]."

"While data controllers may fulfil their legal obligations related to transparency by providing 'notice and consent' forms and privacy policies, individuals may still be uncertain about what they are consenting to Barocas and Nissenbaum [46]."

"Rather than being empowered by transparency, the individual (consenting to share their data) has to take a leap of faith and rely on the assumption that the actors involved in the use of individual's data will be committed to a set of ethical principles, professional commitments, and guiding norms and regulations which protect the individual from privacy harms."

Fiduciary Relationships

"Fiduciary law (through assigning fiduciary duties and obligations) counters the asymmetrical power relationship between fiduciaries and beneficiaries and protects the beneficiary against opportunism [47]."

"A number of non-fiduciary economic actors, for example, seek self-interest with some guile, without requiring the courts to intervene through fiduciary law Secondly, as argued before, if an agent is entrusted with power or resources to warrant opportunism, the said agent could also be restricted through contract law Why then do we require the category of fiduciary relationships?".

"Fiduciary law, thus, provides the legal system with a way to counter opportunism which cannot be completely dealt with through contract law The open-ended approach required to guard against opportunism, that cannot be detected ex-ante, also partially explains the lack of consensus among courts and scholars regarding a common definition for fiduciary relationships and why fiduciary law seems like a "concept in search of a principle" [48]."

Digital Health Data Controllers as Fiduciaries

"I present three main arguments for recognizing the relationship between digital health data controllers and users sharing their health data as fiduciary: (a) the relationship shares features with traditional fiduciary relationships; (b) the relationship involves circumstances similar to those that have led to establishing fiduciary relationships in the past; and (c) fiduciary law is better suited than contractual law in protecting user privacy and enabling trust required for sharing health data with data controllers."

"An alternative possibility to the scope of fiduciary loyalty proposed here, would be to require that fiduciaries go beyond protection of privacy, and also ensure that a broader or general set of interests of the users are kept at the forefront when sharing health data with third parties, even in anonymized and de-identified form."

Gaps in and Limits of Fiduciary Law

"While the application of information fiduciary status to health data controllers will address user concerns about privacy when sharing health data with digital health data controllers, there are other problems that remain unsolved with this proposal and would need to be addressed by other methods."

"One challenge for the proposal to give information fiduciary status to health data controllers is the diverse nature of legal systems and regulations across the globe."

"The unique challenge for the proposal to have fiduciary relationships between health data controllers and data-subjects, however, is that fiduciary law is explicitly defined in only a few legal systems, in particular in systems of common law tradition (for example, in legal systems of countries such as USA, Australia, England, Canada) [49]."

"Governments, for example, may use such latitudes and exceptions within fiduciary law, by forcing health data controllers to share information with them on grounds of public safety."

Conclusion

"This potential revolution, enabled by collection, use, and analysis of large amounts of health data, however, requires collective participation and poses threats to the individuals, exposing intimate information to privacy related harms."

"To ensure the protection of privacy of users sharing health data, I have argued that the relationship between users sharing health data and digital health data controllers should be recognized as a fiduciary relationship, such that health data controllers would keep the interests of the users at the forefront."

"A fiduciary relationship between health data controllers and users is also better suited than alternative approaches for protecting user privacy and thus, enabling users to trust data controllers with their health data."

References

1. Regulation (EU). (2022a). 2022/1925 of the European Parliament and of the Council of 14 September 2022 on contestable and fair markets in the digital sector and amending Directives (EU) 2019/1937 and (EU) 2020/1828 (Digital Markets Act), 12 October 2022, OJ L265/1.
2. Rouvroy, A., & Poullet, Y. (2009). The right to informational self-determination and the value of self-development: Reassessing the importance of privacy for democracy. In S. Gutwirth, Y. Poullet, P. de Hert, C. de Terwangne, & S. Nouwt (Eds.), Reinventing data protection: Proceedings of the International Conference (Brussels, 12–13 October 2007) (pp. 45–76). Springer.
3. Delacroix, S., & Veale, M. (2020). Smart technologies and our sense of self: Going beyond epistemic counter-profiling. In M. Hildebrandt & K. O'Hara (Eds.). Life and the Law in the Era of Data-Driven Agency (pp. 80–99). Edward Elgar Publishing. https://doi.org/10.4337/978 1788972000
4. Hoofnagle, C. J. (2005) Privacy self regulation: A decade of disappointment. In J. K. Winn (Ed.) Consumer Protection in the Age of the 'Information Economy' (Ashgate 2006).
5. Lenard, T. M., & Rubin, P. H. (2009). In defense of data: information and the costs of privacy. Technology Policy Institute Working Paper 9–44.
6. Kirzner, I. M. (1985). Discovery and the capitalist process. University of Chicago Press.
7. Bibas, S. A. (1994). Contractual approach to data privacy. Harvard Journal of Law & Public Policy, 17, 591.
8. Ambec, S., Cohen, M. A., Elgie, S., & Lanoie, P. (2013). The porter hypothesis at 20: Can environmental regulation enhance innovation and competitiveness? Review of Environmental Economics and Policy, 7(1), 2–22.
9. Koźluk, T. & Zipperer, V. (2014). Environmental policies and productivity growth: A critical review of empirical findings. OECD Journal: Economic Studies. 2014(1), 155-185
10. Bourke, J., & Roper, S. (2017). Innovation, quality management and learning: Short-term and longer-term effects. Research Policy, 46(8), 1505–1518.
11. Act CXII of 2011 on the Right of Informational Self-Determination and Freedom of Information, in English http://naih.hu/files/Act-CXII-of-2011_EN_23June2016.pdf (retrieved 31st August 2016).
12. Julesz M. Passive euthanasia and living will, in orvosi Hetilap (Medical Weekly), 1057–1062, 155, 27. In: Hungarian 6th July; 2014. https://doi.org/10.1556/OH.2014.29950.
13. Zuboff, S. (2019). The age of surveillance capitalism: the fight for the future at the new frontier of power. New York: Profile Books.

14. Boyd, D., & Crawford, K. (2012). Critical questions for big data: Provocations for a cultural, technological, and scholarly phenomenon. Information, Communication & Society, 15(5), 662–679.
15. Lohr, S. (2012). The age of big data. New York Times. Retrieved February 11, 2012, from https://www.nytimes.com/2012/02/12/sunday-review/big-datas-impact-in-the-world.html .
16. Warner, R., & Sloan, R. H. (2014). Self, privacy, and power: Is it all over. Tulane Journal of Technology and Intellectual Property, 17, 61–108.
17. Geng, H. (2017). Internet of things and data analytics handbook. Chichester: Wiley.
18. Haskins, C.: Academics Confirm Major Predictive Policing Algorithm is Fundamentally Flawed. Vice.com. (2019). https://www.vice.com/en/article/xwbag4/academics-confirm-major-predictive-policing-algorithm-is-fundamentally-flawed . Accessed 25 May 2021.
19. Moravec, E.R.: Do Algorithms Have a Place in Policing? The Atlantic. https://www.theatlantic.com/politics/archive/2019/09/do-algorithms-have-place-policing/596851/ (2019). (Accessed 15 Feb 2021)
20. Ferguson, A.G.: The Rise of Big Data Policing: Surveillance, Race, and the Future of Law Enforcement. NYU Press, New York (2017)
21. Brayne, S.: Predict and Surveil: Data, Discretion, and the Future of Policing. Oxford University Press, New York (2020)
22. Lazer, D., Radford, J.: Data ex machina: introduction to big data. Annu Rev Sociol 43, 19–39 (2017)
23. Short, M.B., D'Orsogna, M.R., Brantingham, P.J., et al.: Measuring and modeling repeat and near-repeat burglary effects. J. Quant. Criminol. 25, 325–339 (2009)
24. Marda, V., Naravan, S.: Data in New Delhi's predictive policing system. In: Proceedings of the 2020 Conference on Fairness, Accountability, and Transparency, pp 317–324
25. Noble, S.U.: Algorithms of Oppression: How Search Engines Reinforce Racism. NYU Press, New York (2018)
26. Brantingham, J., Valasik, M.: Does predictive policing lead to biased arrests? Results from a randomized controlled trial. Stat. Public Policy 5, 1–6 (2018)
27. Goodman, B., Flaxman, S.: European Union regulations on algorithmic decision-making and a "right to explanation." AI Mag. 38, 50–57 (2017)
28. Ferguson, A.G.: Policing predictive policing. Wash Univ. Law Rev. 94, 1109–1189 (2017)
29. Bakke, E.: Predictive policing: the argument for public transparency. Ann. Surv. Am. Law 74, 131–172 (2018)
30. Barocas, S., Selbst, A.D.: Big data's disparate impact. Calif. Law Rev. 104, 671–732 (2016)
31. Reisman, D., Schultz, J., Crawford, K., et al.: Algorithmic Impact Assessments: A Practical Framework for Public Agency Accountability. AI Now Institute https://ainowinstitute.org/aiareport2018.pdf (accessed 24 May 2021)
32. Floridi, L. (2014). Open Data, Data Protection, and Group Privacy. Philosophy & Technology, 27(1), 1–3. https://doi.org/10.1007/s13347-014-0157-8
33. Zwitter, A., Gstrein, O. J., & Yap, E. (2020). Digital Identity and the Blockchain: Universal Identity Management and the concept of the 'Self-Sovereign' Individual. Frontiers in Blockchain, 3, 26. https://doi.org/10.3389/fbloc.2020.00026
34. van der Sloot, B. (2014). Do data protection rules protect the individual and should they? An assessment of the proposed General Data Protection Regulation. International Data Privacy Law, 4(4), 307–325. https://doi.org/10.1093/idpl/ipu014
35. Walker, K. L. (2016). Surrendering information through the looking glass: Transparency, trust, and protection. Journal of Public Policy & Marketing, 35(1), 144–158.
36. Luo, Y. (2006). Political behavior, social responsibility, and perceived corruption: A structuration perspective. Journal of International Business Studies, 37(6), 747–766.
37. Norwegian Consumer Council (2020), Out of Control: How Consumers are Exploited by the Online Advertising Industry, 177–83, viewed 20/09/2021, from https://www.forbrukerradet.no/out-of-control/
38. Mao, Mark, Ronald Raether Jr., Yanni Lin, Sheila Pham, Jonathan Yee, Sadia Mirza, Julia Hoffmann, Molly DiRago, Melanie Witte, and Julie Hoffmeister. 2018. Data privacy: the current legal landscape. Annual Compendium, Ver. 1: 1.

39. Stevens, Tim. 2018. Cyberweapons: power and the governance of the invisible. International Politics 55 (3–4): 487–490.
40. Milberg, S. J., Burke, S. J., Smith, H. J., & Kallman, E. A. (1995). Values, personal information privacy, and regulatory approaches. Communications of the ACM, 38(12), 65–74. https://doi.org/10.1145/219663.219683
41. Bygrave, L. A. (2010). Privacy and data protection in an international perspective. Scandinavian Studies in Law. Online: http://www.scandinavianlaw.se/pdf/56-8.pdf , http://www.scandinavianlaw.se/pdf/56-8.pdf. Accessed 2 Aug 2017.
42. Datatilsynet. (2013). Big Data – privacy principles under pressure. September 2013. Online: https://www.datatilsynet.no/globalassets/global/english/big-data-engelsk-web.pdf. Accessed 24 July 2017.
43. Cormack, A. N. (2016a). Downstream consent: A better legal framework for big data. Journal of Information Rights, Policy and Practice, 1(1). https://doi.org/10.21039/irpandp.v1i1.9.
44. Cormack, A. (2016b). A data protection framework for learning analytics. Journal of Learning Analytics, 91–106 https://doi.org/10.18608/jla.2016.31.6.
45. Evans, B. J. (2011). Much ado about data ownership. Harvard Journal of Law & Technology, 25, 69.
46. Barocas, S., & Nissenbaum, H. (2009). On notice: The trouble with notice and consent (SSRN Scholarly Paper No. ID 2567409). Rochester, NY: Social Science Research Network. Retrieved from https://papers.ssrn.com/abstract=2567409.
47. Licht, A. N. (2016). Motivation, information, negotiation: Why fiduciary accountability cannot be negotiable (SSRN Scholarly Paper No. ID 2811237). Rochester, NY: Social Science Research Network. Retrieved from https://papers.ssrn.com/abstract=2811237.
48. Miller, P. (2011). A theory of fiduciary liability. McGill Law Journal/Revue de Droit de McGill, 56(2), 235–288. https://doi.org/10.7202/1002367ar .
49. Gelter, M., & Helleringer, G. (2018). Fiduciary principles in European Civil Law Systems (SSRN Scholarly Paper No. ID 3142202). Rochester, NY: Social Science Research Network. Retrieved from https://papers.ssrn.com/abstract=3142202.
50. Elgesem, Dag The structure of rights in Directive 95/46/EC on the protection of individuals with regard to the processing of personal data and the free movement of such data. Ethics and Information Technology (1999). https://doi.org/10.1023/a:1010076422893
51. Ventrella, Emanuele The symbiotic relationship between privacy and security in the context of the general data protection regulation. ERA Forum (2019). https://doi.org/10.1007/s12027-019-00578-6
52. Graef, Inge; Petročnik, Tjaša; Tombal, Thomas Conceptualizing Autonomy in an Era of Collective Data Processing: From Theory to Practice. Digital Society (2023). https://doi.org/10.1007/s44206-023-00045-3
53. Fuller, Caleb S. The perils of privacy regulation. The Review of Austrian Economics (2016). https://doi.org/10.1007/s11138-016-0345-0
54. Martin, Nicholas; Matt, Christian; Niebel, Crispin; Blind, Knut How Data Protection Regulation Affects Startup Innovation. Information Systems Frontiers (2019). https://doi.org/10.1007/s10796-019-09974-2
55. Alexin, Zoltán Hungary's unorthodox approach to personal privacy. Health and Technology (2017). https://doi.org/10.1007/s12553-017-0181-7
56. Pap, András L. Human Rights and Ethnic Data Collection in Hungary. Human Rights Review (2007). https://doi.org/10.1007/s12142-007-0032-4
57. Andrew, Jane; Baker, Max The General Data Protection Regulation in the Age of Surveillance Capitalism. Journal of Business Ethics (2019). https://doi.org/10.1007/s10551-019-04239-z
58. Davis, Jeremy; Purves, Duncan; Gilbert, Juan; Sturm, Schuyler Five ethical challenges facing data-driven policing. AI and Ethics (2022). https://doi.org/10.1007/s43681-021-00105-9
59. Gstrein, Oskar J.; Beaulieu, Anne How to protect privacy in a datafied society? A presentation of multiple legal and conceptual approaches. Philosophy & Technology (2022). https://doi.org/10.1007/s13347-022-00497-4

60. Komljenovic, Janja The future of value in digitalised higher education: why data privacy should not be our biggest concern. Higher Education (2020). https://doi.org/10.1007/s10734-020-00639-7
61. Quach, Sara; Thaichon, Park; Martin, Kelly D.; Weaven, Scott; Palmatier, Robert W. Digital technologies: tensions in privacy and data. Journal of the Academy of Marketing Science (2022). https://doi.org/10.1007/s11747-022-00845-y
62. Fiske, Amelia; Degelsegger-Márquez, Alexander; Marsteurer, Brigitte; Prainsack, Barbara Value-creation in the health data domain: a typology of what health data help us do. BioSocieties (2022). https://doi.org/10.1057/s41292-022-00276-6
63. Mance, Jonathan Human Rights, Privacy and the Public Interest—Who Draws the Line and Where?. Liverpool Law Review (2009). https://doi.org/10.1007/s10991-008-9043-8
64. Yu, Xiaolan Allocating Personal Data Rights: Toward Resolving Conflicts of Interest over Personal Data. Fudan Journal of the Humanities and Social Sciences (2021). https://doi.org/10.1007/s40647-021-00330-w
65. Hoel, Tore; Chen, Weiqin Privacy and data protection in learning analytics should be motivated by an educational maxim—towards a proposal. Research and Practice in Technology Enhanced Learning (2018). https://doi.org/10.1186/s41039-018-0086-8
66. Introna, Lucas; Pouloudi, Athanasia Privacy in the Information Age: Stakeholders, Interests and Values. Journal of Business Ethics (1999). https://doi.org/10.1023/a:1006151900807
67. Etzioni, Amitai Are New Technologies the Enemy of Privacy?. Philosophy & Technology (2007). https://doi.org/10.1007/s12130-007-9012-x
68. Arora, Chirag Digital health fiduciaries: protecting user privacy when sharing health data. Ethics and Information Technology (2019). https://doi.org/10.1007/s10676-019-09499-x

Chapter 3
Conceptualizing Data Protection Provisions

Introduction by the Editor

The significant provisions about personal data processing are mostly defined. Definitions on their own or in association with other related provisions may not always underline the path to operationalising expectations of the law. Mere definitions will not help those processing personal data.[1] The concepts surrounding the legislative framework may not always be succinct, leaving room for ambiguity at the operational stage.[2] For instance, the disclosure requirements from the data controllers under GDPR for automated data processing apparently need improvement because it raises doubt about the responsibilities bestowed upon data controllers. Equally, it may fall short of the expectations that data subjects have of data controllers regarding automated processing.[3]

Further challenges surface with technological advancements and when the processing of personal data is not linear. Therefore, it is crucial to ascertain meanings and unearth legislative intents to set the expectations threshold for the data controllers. For instance, the extent of liability when data controllers may be handling third-party activities concerning the processing of personal data. It is not just limited to the data controllers, but it is essential to ascertain how data subjects perceive their rights under data protection legislation. They must be measured against various provisions that guide data controllers.

---

[1] Jan von Hein & Anna Bizer, "Social Media and the Protection of Privacy: Current Gaps and Future Directions in European Private International Law" (2018) International Journal of Data Science and Analytics 6, 233–239; Lachlan Urquhart & others, "Realising the right to data portability for the domestic Internet of things" (2018) Personal and Ubiquitous Computing 22, 317–332.

[2] Elena Gil González & Paul de Hert, "Understanding the legal provisions that allow processing and profiling of personal data—an analysis of GDPR provisions and principles" (2019) ERA Forum 19, 597–621.

[3] Alexander J Wulf & Ognyan Seizov, " "Please understand we cannot provide further information": evaluating content and transparency of GDPR-mandated AI disclosures" (2022) AI & SOCIETY 39, 235–256.

I. Gupta (ed.), *Expectations vs Realities of Information Privacy and Data Protection Measures*, https://doi.org/10.1007/978-981-99-6778-0_3

There are certain question marks associated with the scope of different definitions, such as profiling and whether legal interventions can do justice in capturing the activity of profiling in its entirety in the context of technological advancements. The suggestion is to use "technological enhancing tools (TET)"[4] instead to match the rapid technological changes. It will help to increase transparency and the correct use of the activity and cover for any shortcomings that legal interventions may have at present.

Across jurisdictions, the roles and responsibilities under data protection legislation are set against the framework of stakeholders. For instance, jurisdictions may not have followed the exact terminology of a controller, processor, and joint controller, but they have classified roles based on the decision-making power of a particular stakeholder. It has been aligned with the stakeholder that decides the purpose of processing personal data. To assume the role of a joint controller, an entity needs to influence the purpose and means of processing. It becomes important to locate the influence of the entity in the overall decision-making for the purposes of processing.[5] The idea of joint controllership is also floated to cover social network sites and users who upload the personal information of other individuals.[6] Recital 18 of the GDPR, extends to "social networking and online activity"[7] within the scope of personal or household activities, which falls outside the material scope of GDPR.[8] The issue of personal or household activities and their scope may require certain further clarifications in the future.

Irrespective of jurisdictions, data protection principles form the crux of data protection law. These principles ensure that personal data are processed within the limited purpose of data collection and on a lawful basis. The purpose is to decide the time limit for storing personal data and further encourage the secure processing of such data. Cumulatively, all these add up to transparent and fair processing of personal data and make the data controller accountable. The principles of transparency and fair processing can introduce the necessary balance when the data controllers rely upon the different legal grounds of data processing. They augur well with the proportionality and necessity concepts that are embedded in GDPR.[9] It also extends to

---

[4] Mireille Hildebrandt, "Profiling and the rule of law" (2008) Identity in the Information Society 1, 55–70.

[5] Veronique Cimina, "The data protection concepts of 'controller', 'processor' and 'joint controllership' under Regulation (EU) 2018/1725" (2021) ERA Forum 21, 639–654.

[6] Brendam Van Alsenoy, "Social networks and web 2.0: are users also bound by data protection regulations?" (2009) Identity in the Information Society 2, 65–79.

[7] Regulation (EU) 2016/679 of the European Parliament and of the Council of 27 April 2016 on the protection of natural persons with regard to the processing of personal data and on the free movement of such data, and repealing Directive 95/46/EC (General Data Protection Regulation) OJ L 119, 4.5.2016, pp. 1–88. (GDPR).

[8] Ibid.

[9] Elena Gil González & Paul de Hert, "Understanding the legal provisions that allow processing and profiling of personal data—an analysis of GDPR provisions and principles" (2019) ERA Forum 19, 597–621.

processing by public authorities in the member states by virtue of the legal interventions in consonance with GDPR and personal data processing within the scope of employment.[10]

We should consider three specific situations that would safeguard personal data. They are about adopting different masking techniques, implementing rights like the 'right to be forgotten' and safeguards adopted while transferring personal data. Anonymisation and pseudonymisation techniques protect personal data from unauthorised use and allow the sharing of data in a digital economy. It is difficult to achieve a state of proper anonymisation, which essentially reflects a situation where data directly or indirectly cannot identify a natural person.[11] At times, the existing technology would not facilitate the identification process, but that situation would not lead to a state of anonymisation. There cannot be different grades for anonymisation. It assumes a binary form—yes or no. Anonymisation techniques also improve with technology, thereby reducing the possibility of re-identification that may give rise to data breach cases.

Reiterating these ideas, the concepts of anonymisation and pseudonymisation have been traced back with a further investigation into how these terms have evolved over time. 'Reasonably likely' is the key term that decides the outcome of processes that would rely on techniques like anonymisation and pseudonymisation. Of course, other than the general idea of the likelihood of deidentification, such a threshold may change based on the context and the nature of the data used. It is suggested that there is a difference in how these terms have been approached in legal instruments in the pre and post GDPR era. Much reliance has been placed on re-identification risk, which is a relative term in the context of different parties.[12]

The second situation concerns the right to be forgotten or the right to prevent personal data accessibility for the general public. A natural person may be willing to exercise this right for multiple reasons, starting from the sensitivity of personal content to causes emanating from social reputation. Scholars have identified categories like "information disclosure, content sensitivity, social reputation, control over further processing, system/process, and sociality".[13] There are exciting conversations happening about data subjects' rights and exercising these rights vis-à-vis their limitations, scope, and applicational challenges.[14] The right to be forgotten seems like an all-encompassing right, yet conflicting situations are bound to surface with its

[10] Dirk Bieresborn, "The impact of the General Data Protection Regulation on Social Security" (2019) ERA Forum 20, 285–306.

[11] John S Davis & Osonde Osoba, "Improving privacy preservation policy in the modern information age" (2019) Health and Technology 9, 65–75.

[12] Zhicheng He, "From Privacy-Enhancing to Health Data Utilisation: The Traces of Anonymisation and Pseudonymisation in EU Data Protection Law" (2023) Digital Society 2(17).

[13] Chanhee Kwak & others, "Could You Ever Forget Me? Why People Want to be Forgotten Online" (2022) Journal of Business Ethics 179, 25–42.

[14] George Christou & Imir Rashid, "Interest group lobbying in the European Union: privacy, data protection and the right to be forgotten" (2021) Comparative European Politics 19, 380–400; André Habisch & others, "Unpleasant Memories on the Web in Employment Relations: A Ricoeurian Approach" (2022) Humanistic Management Journal 7, 347–368.

sweeping application. Scholars have pointed to different situations where the applications of the right to be forgotten may pose certain questions, like deleting medical and genetic records in cases of gender transitions[15] and situations in the context of different traditions.[16]

The third point would be safeguarding personal data by ensuring personal data transfer following a certain protocol. Transferring personal data is a sensitive issue owing to the different nature of data protection laws in different countries.[17] Any data transfer should precede safety norms in the recipient country.[18] For instance, following the GDPR norms, there are foreseeable challenges concerning the transfer of personal data for health research purposes.[19] The general issues, like providing greater protection to transferred data, that prevailed during the old regime continued after the passage of the GDPR.[20]

This chapter showcases the extensive need for a comprehensive interpretation of the data protection framework. There are also additional global implications depending on how these provisions are interpreted. Although countries have certain uniformity regarding data protection principles, there are differences in their approach to protecting the personal data of individuals. It would be interesting to see how data protection decisions in a particular jurisdiction impact global business.

Machine generated summaries

Disclaimer: The summaries in this chapter were generated from Springer Nature publications using extractive AI auto-summarization: An extraction-based summarizer aims to identify the most important sentences of a text using an algorithm and uses those original sentences to create the auto-summary (unlike generative AI). As the constituted sentences are machine selected, they may not fully reflect the body of the work, so we strongly advise that the original content is read and cited. The auto generated summaries were curated by the editor to meet Springer Nature publication standards. To cite this content, please refer to the original papers.

Machine generated keywords: transfer, controller, gdpr, right, international, delete, article gdpr, processing, health datum, third, health, country, social medium, transfer personal, identity.

[15] Mónica Correia & others, "Gender Transition: Is There a Right to Be Forgotten?" (2021) Health Care Analysis 29, 283–300.

[16] Amr Osman, "The Right to be Forgotten: an Islamic Perspective" (2022) Human Rights Review 24, 53–73.

[17] Kornbeck, Jacob Athlete consent as a legal base for data transfers to third countries for anti-doping purposes, under EU and German law. The International Sports Law Journal (2017). https://doi.org/10.1007/s40318-017-0112-9.

[18] Paul Van den Bulck, "Transfers of personal data to third countries" (2017) ERA Forum 18, 229–247.

[19] Dara Hallinan & others, "International transfers of personal data for health research following Schrems II: a problem in need of a solution" (2021) European Journal of Human Genetics 29, 1502–1509.

[20] Paul Van den Bulck, "Transfers of personal data to third countries" (2017) ERA Forum 18, 229–247.

3.1 Basic Concepts in Data Protection

Machine generated keywords: processing, article gdpr, controller, gdpr, article, operation, health datum, domestic, healthcare, check, art, attribute, section, real, privacy risk.

From Privacy-Enhancing to Health Data Utilisation: The Traces of Anonymisation and Pseudonymisation in EU Data Protection Law [94]

This is a machine-generated summary of:

He, Zhicheng: From Privacy-Enhancing to Health Data Utilisation: The Traces of Anonymisation and Pseudonymisation in EU Data Protection Law [94]

Published in: Digital Society (2023).

Link to original: https://doi.org/10.1007/s44206-023-00043-5

If you want to cite the papers, please refer to the original.

For technical reasons we could not place the page where the original quote is coming from.

Abstract-Summary

"Taking a forward-looking perspective, this paper aims to contribute to the scholarly debate around anonymisation and pseudonymisation by extending the discussion to the contexts of forthcoming EU data laws, with a focus on the draft European Health Data Space (EHDS) Regulation."

"Starting with a positivist enquiry, the paper investigates the traces and evolution of anonymisation and pseudonymisation in EU data protection instruments from both before and after the entry into force of the General Data Protection Regulation."

"It then shifts focus to future EU data laws and examines the roles of anonymisation and pseudonymisation in these instruments, including the draft EHDS Regulation, the newly adopted EU Data Governance Act and the draft EU Data Act."

"The paper makes preliminary remarks and recommendations on the draft EHDS Regulation and questions to what extent its current incorporation of anonymisation and pseudonymisation can be reconciled with the health data sharing arrangements proposed in this Regulation."

Introduction

"Taking a forward-looking perspective, this paper aims to contribute to the scholarly debate around anonymisation and pseudonymisation by extending the discussion to the contexts of forthcoming EU data laws, with a focus on the draft European Health Data Space (EHDS) Regulation [1]."

"Starting with a positivist enquiry, the paper first investigates the traces and evolution of anonymisation and pseudonymisation in both pre- and post-GDPR EU data protection instruments."

"It then shifts focus to future EU data laws and examines the roles of anonymisation and pseudonymisation in such instruments, including the draft EHDS Regulation, the newly adopted EU Data Governance Act and the draft EU Data Act."

"The paper makes preliminary remarks on the draft EHDS Regulation and questions to what extent current legal definitions of anonymisation and pseudonymisation can be reconciled with the health data sharing arrangements proposed in this Regulation."

Traces from the Pre-GDPR Era

"In this Opinion, 'anonymous data' are defined as: "any information relating to a natural person where the person cannot be identified, whether by the data controller or by any other person, taking account of all the means likely reasonably to be used either by the controller or by any other person to identify that individual" (Article 29 Data Protection Working Party, 2) This definition basically affirmed the notion of 'data rendered anonymous' in the Directive."

"On this note, the Opinion further provided that 'anonymised data' constitute anonymous data that have previously been personal data, but where identification has been rendered 'no longer possible' (Article 29 Data Protection Working Party, 2)."

"Regarding the definition of anonymisation, Opinion 05/2014 stated that 'anonymisation results from processing personal data in order to irreversibly prevent identification' (Article 29 Data Protection Working Party, 3)."

Traces from the Post-GDPR Era

"It is indicated that if the means to re-identify the natural person are not reasonably likely because of time, money or technology, the data could be considered anonymous in the sense of the GDPR, despite a theoretical risk of re-identification."

"While no definition of 'anonymisation' is given, the GDPR specifically defines the term 'pseudonymisation' as meaning 'the processing of personal data in such a manner that the personal data can no longer be attributed to a specific data subject without the use of additional information' [4]."

"According to Convention 108 + : "Data is to be considered as anonymous only as long as it is impossible to re-identify the data subject or if such re-identification would require unreasonable time, effort or resources, taking into consideration the available technology at the time of the processing and technological developments." [5] As seen above, Convention 108 + 's definition of 'anonymous data' broadly aligns with the GDPR's description of anonymous information."

Normative Tensions and Scholarly Debates

"This is supplemented by Article 29 Data Protection Working Party's Opinion 05/2014, which interprets 'likely reasonably' and in doing so essentially sets a relatively high standard for achieving anonymisation, i.e. anonymisation can only be achieved if no person can re-identify the data subject."

"Post-GDPR instruments point to a more relative and contextual understanding of anonymisation, indicating that individual-level anonymity can be achieved if the re-identification risk by a specific party is relatively low considering the context, meaning that the data can be treated as anonymous relative to that party."

"Groos and van Veen argue that the Breyer decision in fact deviated from the Article 29 Data Protection Working Party's rather absolute approach by leaning toward a more relative approach, i.e. a contextual approach [6]."

"The Dutch Data Protection Authority leans toward the approach described in Opinion 05/2014 and seems to believe that anonymisation must be absolute [6]."

Traces in Future EU Data Laws

"There is a clear trend that anonymisation and pseudonymisation play important roles in the data sharing schemes proposed in future EU data laws, including the newly adopted Data Governance Act, the forthcoming Data Act and the EHDS Regulation."

"Apart from the pseudonymisation requirement, the draft Data Act does not provide further guidance on the use of anonymisation and pseudonymisation techniques."

"The purpose of the data user's processing cannot be achieved with anonymised data, taking into account the information provided by the data user, the health data access bodies shall provide access to electronic health data in pseudonymised format." [1] Where health data are provided on a pseudonymised basis, there are extra requirements."

"It empowers public sector bodies to impose anonymisation or pseudonymisation requirements on the re-use of personal data [7]."

Critical Aspects of the Draft EHDS Regulation

"While the draft EHDS Regulation proposes a health data sharing system on the basis of anonymisation and pseudonymisation techniques, there are aspects that require further clarification or careful monitoring in the further drafting and implementation of this Regulation."

"The draft EHDS Regulation does not clarify what anonymisation and pseudonymisation entail in the health data sharing context, beyond making a reference to the GDPR definitions."

"The draft EHDS Regulation imposes similar requirements on health data shared on an anonymised and/or pseudonymised basis."

"Anonymised and pseudonymised data are treated similarly under the draft EHDS Regulation, in the sense that they are all subject to the requirements of the EHDS and the data permits under which they are granted."

"On the plus side, the EDHS's data sharing approach better protects data subjects by assuming that neither anonymised nor pseudonymised data are risk-free and subjecting them to the requirements of the data permit and the EHDS Regulation."

Concluding Remarks

"The paper explores the interaction of anonymisation and pseudonymisation with EU data protection law by examining the past, present and future of these techniques in various EU data protection legal instruments."

"While all three instruments attach great importance to the use of anonymisation and pseudonymisation techniques, none gives further clarification on these terms in the data sharing context."

"If a legal obligation to anonymise or pseudonymise health data was to be introduced under the EHDS, such an obligation should also be considered for the processing of health data outside the scope of the EHDS."

"This may be achieved by introducing a general legal obligation under the GDPR requiring the anonymisation or pseudonymisation of health data before processing, so that the privacy-enhancing powers and data sharing facilitation potentials of these techniques can be unleashed to a fuller extent."

Improving privacy preservation policy in the modern information age [95]

This is a machine-generated summary of:

Davis, John S.; Osoba, Osonde: Improving privacy preservation policy in the modern information age [95]

Published in: Health and Technology (2018).

Link to original: https://doi.org/10.1007/s12553-018-0250-6.

If you want to cite the papers, please refer to the original.

For technical reasons we could not place the page where the original quote is coming from.

Abstract-Summary

"Anonymization or de-identification techniques are methods for protecting the privacy of human subjects in sensitive data sets while preserving the utility of those data sets."

"The efficacy of anonymization methods has come under repeated attacks and several researchers have shown that anonymized data can be re-identified to reveal the identity of the data subjects via approaches such as "linking.""

"Even given these deficiencies, many government privacy policies depend on anonymization techniques as the primary approach to preserving privacy."

"We review the European Union's General Data Protection Regulation (GDPR) and show that it takes a more goal-oriented approach to data privacy."

"It defines data privacy in terms of desired outcome (i.e., as a defense against risk of personal data disclosure), and is agnostic to the actual method of privacy preservation."

"GDPR goes further to frame its privacy preservation regulations relative to the state of the art, the cost of implementation, the incurred risks, and the context of data processing."

Introduction

"The value to platforms often comes in the form of some combination of higher user engagement, more attention to ads, more product sales etc But this data-for-service exchange process raises privacy concerns and forces society to balance the utility of the delivered services against the risk that the data may be used against the will of the individuals from which it is collected."

"Given these privacy risks, many government agencies have mandated policies for constraining the exchange of personal data."

"HIPAA is one of the most extensive privacy regulations in the U.S. and governs health data that is directly tied to an individual's identity by virtue of containing personally identifiable information (PII)."

"The purpose of this paper is to facilitate a nuanced discussion of anonymization for the privacy policy community to help better consider the options for balancing data privacy and data utility."

The Privacy of Tabular Data

"The second problem with this approach is that the quasi-identifiers often provide enough information for an observer to infer the identity of a given individual via a "linking attack," in which entries from multiple, separate data sets are linked together based on quasi-identifiers that have been made public."

"Most data anonymization schemes attempt to preserve subject anonymity (non-identifiability) by obfuscating, aggregating, or suppressing sensitive components of the data set."

"The anonymization work has led to two related research areas: (i) privacy-preserving data publishing (also referred to as non-interactive anonymization systems) and (ii) privacy-preserving data mining (also referred to as interactive anonymization systems) [8, 9]."

"To the dichotomy of privacy preserving data publishing and data mining, there are two general algorithmic approaches to anonymization: syntactic anonymization and differential privacy."

"We discuss both syntactic anonymization and differential privacy and show their relationship to privacy preserving data publishing and data mining."

Syntactic Anonymization

"Unlike syntactic anonymization, differential privacy modifies the actual values of the sensitive attributes (as opposed to the quasi-identifiers) by adding random noise with a judiciously chosen distribution."

"Differential privacy works best with query functions that have a small global sensitivity since this implies that privacy will be maintained without distorting the data too much."

"In order for the database to be differentially private, we need to select a randomized function, a mechanism M, that adds noise to the datasets that will produce a randomized result R. Since D_1 and D_2 are adjacent, the probability that $M(D_1) = R$ should be close to the probability that $M(D_2) = R$. More formally we can write For small ε, note that $e^{\varepsilon} \sim 1 + \varepsilon$ and if our probabilities are identical, we get The amount and kind of noise that M adds is constrained by the global sensitivity of the query function, f, that will be applied to the data."

Syntactic Versus Differential Privacy

"In comparing the syntactic and differential privacy approaches, it is important to keep in mind the trade-off between privacy preservation and data utility."

"On the other extreme, not releasing any data will maximize privacy preservation but the shared data (an empty dataset) will be useless."

"An organization wishing to share sensitive data must consider the logistical challenges associated with the sharing process in addition to carefully balance the privacy and utility of released data."

"In order to estimate the amount of anonymization noise (via a value for ε) differential privacy requires an understanding of the space of possible attribute values (even those not contained within the dataset in question) and the space of possible query functions that will be used to process the data (and that imply a value for global sensitivity) [10]."

Anonymity in practice: existing standards and legislation

"The importance of privacy and anonymization for data sets is recognized in US and EU law In the United States, much early work on the anonymization of data about individual respondents and statistical disclosure control was motivated by constitutionally-mandated privacy requirements for Census data-collection activity."

"There have been efforts to further enshrine privacy guarantees for sensitive data sets such as those needed for public health research or census efforts."

"It places non-trivial constraints on the use of data sets containing PII, but it allows for the use of de-identified health information (DHI) without any regulatory constraints."

"An alternative approach to privacy regulation might compel data publishers to demonstrate that their privacy-preservation measures reasonably account for state-of-the-art re-identification techniques."

"Title 13 USC §9 directs the Census Bureau to safeguard the privacy of the data they collect."

"The third class of non-consenting disclosure allows a school to release education records from which personally identifiable information (PII) has been removed."

The European Union's general data protection regulation (GDPR)

"The regulations recognized a fundamental right for the protection of data relating to natural persons (Article 2)."

"More pertinent to privacy, it makes a distinction between pseudonymized and anonymized data (Article 4(5), Recitals 26-30)."

"GDPR is novel in the way it identifies the individual natural person as the primary holder of rights to most data related to the person."

"Right to rectify errors in user data (Article #16) 3."

"Right to data erasure or "to be forgotten" (Article #17) 4."

"It is a goal-oriented regulation; GDPR identifies privacy and other data rights as goals or ends-in-themselves and tries to define a regulatory infrastructure to safeguard these rights subject to considerations of the technological state-of-the-art, implementation costs, and the severity of potential disclosure risk."

"The Article 29 Working Party Opinion on anonymization techniques, a non-binding commentary associated with GDPR, did provide advisory recommendations for de-identifying data."

Alternatives to anonymity

"They suggested confidentiality policies that compel researchers with full data access to uphold the privacy of the human subjects [11]."

"There are several models that can inform alternatives to data anonymization for preserving privacy."

"One approach is to grant systematic researcher access on the premises of the data owning entity."

"The U.S. Census Bureau has established 23 Research Data Centers for accessing census data as well as data from over 50 other partner organizations."

"Would treat sensitive research data similar to the way the U.S. federal government handles classified information [12]."

"This approach would require designated researchers to undergo a clearance process in order to gain access to sensitive data and could benefit from a well-established set of procedures for validating would-be researchers."

Conclusion

"The utility of data often stands in sharp contrast to the privacy of its subjects."

"A key assumption in our development as a data-based society is the assumption that we can preserve both privacy and utility."

"Many government policies assume that although it is easier to privilege utility over privacy, there are de-identification techniques that will effectively preserve privacy when necessary."

"We provided an overview of recent anonymization approaches designed to protect our growing data stores, but the commensurate growth of powerful algorithms and hardware makes it increasingly easy for a motivated interloper to violate any reasonable assumption of privacy, even when attempts to anonymize data are employed."

"The GDPR is a recent policy that takes a more sophisticated approach at addressing privacy preservation, by explicitly acknowledging the weaknesses associated with de-identification techniques and by avoiding regulatory language that presupposes particular technologies or de-identification techniques."

The data protection concepts of 'controller', 'processor' and 'joint controllership' under Regulation (EU) 2018/1725 [96]

This is a machine-generated summary of:

Cimina, Veronique: The data protection concepts of 'controller', 'processor' and 'joint controllership' under Regulation (EU) 2018/1725 [96]

Published in: ERA Forum (2020).

Link to original: https://doi.org/10.1007/s12027-020-00,632-8.

If you want to cite the papers, please refer to the original.

For technical reasons we could not place the page where the original quote is coming from.

Abstract-Summary

"Following the entry into force of the General Data Protection Regulation (GDPR) and Regulation (EU) 2018/1725, multiple questions were raised in relation to the novelties introduced to the concepts of 'controller', 'processor' and 'joint controllership' and their roles in the processing of personal data."

Introduction

"Following the entry into force of the General Data Protection Regulation ('GDPR') and Regulation (EU) 2018/1725 ('the Regulation'), numerous questions were raised on the novelties brought to the concepts of controller and processor and their respective roles."

"When processing personal data, EU institutions and bodies ('EUI' or 'EUIs') must comply with specific data protection rules."

"This article aims to summarise such novelties with regard to the different roles EUIs may assume both in terms of specific processing operations and in terms of obligations under the Regulation."

The Concept of 'Controller'

"Article 3(8) of the Regulation, similarly to Article 4(7) of the GDPR, defines a controller as: "(…) the Union institution or body or the directorate-general or any other organisational entity which, alone or jointly with others, determines the purposes and means of the processing of personal data; where the purposes and means of such processing are determined by a specific Union act, the controller or the specific criteria for its nomination can be provided for by Union law"."

"As also confirmed by the CJEU, each of the actors involved have specific obligations under data protection law Thirdly, the Article refers to the determination (of the purpose and means) that a controller may have on a specific processing operation, by virtue of its decision-making power."

The Concept of 'Processor'

"When a processor goes beyond the agreed mandate set by a contract or another legal act, thus deciding on the purpose and essential elements of the means of the processing operation, it may qualify as controller (or joint controller)."

"According to Article 29(1) of the Regulation, a controller shall only use processors who provide sufficient guarantees to implement appropriate technical and organisational measures in a way that the processing operation will meet the requirements set out by the Regulation and protect data subjects."

"Clear data protection obligations must be set out in a contract, other legal act or binding arrangement between the controller and the potential processor."

"Among a list of specific recommendations, the EDPS firstly suggests EUIs "(...) to only use processors providing sufficient guarantees to implement appropriate technical and organisational measures that the processing will meet the requirements of the Regulation and ensure the protection of the rights of the data subjects"."

The concept of 'joint controllership'

"The fact that one of the joint controllers may in practice have no access to personal data may be of particular importance when establishing its level of responsibility within the processing operation, but not for the purpose of establishing its joint controllership role."

"Article 28(1) of the Regulation states that joint controllers: "(...) shall in a transparent manner determine their respective responsibilities for compliance with their data protection obligations, in particular as regards the exercising of the rights of the data subject (...), by means of an arrangement between them unless, and in so far as, the respective responsibilities of the joint controllers are determined by Union or Member State law to which the controllers are subject (...)"."

"Article 28 of the Regulation clearly specifies that joint controllers need to enter into a specific arrangement laying down their roles and responsibilities, particularly towards data subjects."

EU case law on joint controllership

"By defining specific parameters relating to its target audience and objectives or for the promotion of its own activities, the CJEU considered that such administrator would be participating in the determination of the purposes and (essential elements of the) means of the processing of personal data of the fan page's visitors."

"The CJEU also clarified that the absence of written guidance and/or instructions from the Community to the members is not enough to exclude the exertion of influence over the processing of personal data, thus assuming the role of controller by participating in the determination of the purposes and (essential elements of the) means of the processing operation in place."

"The CJEU clarified that the operator, such as Fashion ID, in its role as joint controller in the processing operations in which it is involved, must provide, at the time of the data collection, specific information to the visitors of the page, such as its identity and the purpose of the processing operation."

Concluding Remarks

"While the concept of joint controllership is not a novelty, the Regulation and the GDPR have included a specific Article on the subject, providing for specific details in relation to the roles and responsibilities of the joint controllers."

"If an entity has specific influence over the processing operation for its own purposes, thus participating in the decision on purposes and (essential elements of the) means, this may be sufficient to trigger a situation of joint controllership."

"One entity may be considered a joint controller only in relation to the processing operations to which it participates by influencing the purpose and (essential elements of the) means of those specific processing operations."

"Specifying roles and responsibilities of the entities involved in a processing operation, whether in a situation of joint controllership or not, not only fosters the protection of data subject's rights, entailing their complete and full protection, but also underlines the importance of the principle of accountability, central to data protection legislation."

Understanding the legal provisions that allow processing and profiling of personal data—an analysis of GDPR provisions and principles [97]

This is a machine-generated summary of:

Gil González, Elena; de Hert, Paul: Understanding the legal provisions that allow processing and profiling of personal data—an analysis of GDPR provisions and principles [97]

Published in: ERA Forum (2019).

Link to original: https://doi.org/10.1007/s12027-018-0546-z.

If you want to cite the papers, please refer to the original.

For technical reasons we could not place the page where the original quote is coming from.

Abstract-Summary

"This contribution looks at the legal grounds for data processing ('when is one allowed to collect and use data on others?') according to the General Data Protection Regulation (GDPR)."

"Data protection principles such as transparency and overall fairness as enshrined in Article 5 GDPR may in this case serve as a resort to identify appropriate checks and balances."

"Additional understanding can be found outside data protection legislation—for instance, in competition law."

Introduction and Structure

"Regulation (EU) 2016/679 (General Data Protection Regulation, GDPR) aims to harmonise data protection legislation across EU member states, enhancing individuals' rights."

"The GDPR applies to organisations established in the EU as well as organisations operating outside the EU which offer goods or services to, or monitor the behaviour of, individuals in the EU."

"It applies to organisations processing personal data which have an establishment within the EU and also those organisations which operate outside the EU but offer goods or services to, or monitor the behaviour of, individuals in the EU."

"We begin by reviewing the provisions in the GDPR that identify a limited set of (six) legal grounds for lawful processing of personal data."

"In a last section, we conclude that the vagueness and subjectivity of certain concepts in the GDPR can undermine subjects' rights."

Legal Grounds for the Processing of Personal data According to the GDPR

"Article 6 expands on the first principle, that of lawfulness, and provides six grounds for lawful processing: the consent of the data subject (6(1)(a)), the necessity of the performance of a contract (6(1)(b)), the necessity to comply with a legal obligation on the controller (6(1)(c)), the necessity to protect the vital interests of the data subject (6(1)(d)), the necessity to perform a task carried out in the public interest or the exercise of official authority (6(1)(e)), and necessity in the legitimate interest of the controller or another third party (6(1)(f))."

"Article 6(1)(f) GDPR provides for the lawfulness of processing when it 'is necessary for the purposes of the legitimate interests pursued by the controller or by a third party, except where such interests are overridden by the interests or fundamental rights and freedoms of the data subject which require protection of personal data, in particular where the data subject is a child'."

Profiling, Definition and Distinctions

"Article 4(4) GDPR defines profiling as 'any form of automated processing of personal data consisting of the use of personal data to evaluate certain personal aspects relating to a natural person, in particular to analyse or predict aspects concerning that natural person's performance at work, economic situation, health, personal preferences, interests, reliability, behaviour, location or movements' (emphasis added)."

"Profiling is defined as 'applying to a particular individual the profile of a group with which he or she can be identified through the data collected on him or her."

"If a person lives in certain location, buys beer more than twice a week, and his rent is under X euros per month, he can be categorised under the below-35-single-male profile, and the data shows that people in that group have a probability of 27% for falling behind with a payment."

Solely automated decision-making (profiling) with legal or similarly significant effects

"These articles establish the duty of informing beforehand about 'the existence of automated decision-making, including profiling, referred to in Article 22(1) and (4) and, at least in those cases, meaningful information about the logic involved, as well as the significance and the envisaged consequences of such processing for the data subject' (emphasis added)."

"When one of the exceptions applies, the data controller is required to implement 'suitable measures to safeguard the data subject's rights and freedoms and legitimate interests, at least the right to obtain human intervention on the part of the controller, to express his or her point of view and to contest the decision' (Article 22(3) GDPR)."

"Automated decisions can never be based on the processing of special categories of personal data, unless there is explicit consent or substantial public interest, and there are suitable measures to safeguard the data subject's rights and freedoms and legitimate interests (Article 22.4)."

Not solely automated decision-making (profiling)

"The controller is the one carrying out its own self-assessment to decide if Article 6(1)(f) may apply, as well as self-assessing whether the profiling or decision is automated or (even more relevantly, due to the degree of subjectivity), the effects are significant."

"' In addition, if the profiling activities are challenged and subsequently found not to be grounded in a lawful basis, controllers face the highest tier of sanctions under the GDPR."

"This can be the case when the necessity test is not passed (for every Article 6 ground apart from consent), when the balancing test is not fairly done (for the legitimate interests) or when users have not provided free, informed and unambiguous consent."

"For all that, it states 'it would be difficult for controllers to justify using legitimate interests as a lawful basis for intrusive processing and tracking practices for marketing or advertising purposes, for example those that involve tracking individuals across multiple websites, locations, devices, services or data-brokering'."

Bringing in principles (transparency and fairness) and competition law to get the balance right

"Data protection and competition law can be linked, specifically in relation to the concept of fairness and the legal grounds for processing personal data."

"Kalimo and Majcher argue that the digital economy companies usually provide a service in exchange for user consent to process personal data, which creates a commercial relationship between the provider and the individual, subject to the Treaty on the Functioning of the European Union (TFEU)."

"The concept of fairness would allow, for instance, the argument that where a company provides information about its terms of service in a way that is difficult to understand, consent has not been obtained in accordance with the GDPR."

"Fairness also influences the concepts of proportionality and necessity found in data protection law, which, as already mentioned, are vital in assessing the best lawful basis for processing under the GDPR."

"Therefore, typical competition law concepts such as market share were considered by the authority to construe its arguments under data protection laws."

Conclusion: reading principles together with provisions to apply the legal grounds and the profiling mechanisms correctly

"We have first looked at the legal grounds for data processing according to Article 6 of the 2016 General Data Protection Regulation (GDPR), namely, the data subject's consent, the necessity for the performance of a contract/for a compliance with legal obligation/to protect vital interests, for the performance of a task carried out in the public interest or exercise of official authority/necessity for a legitimate interest pursued by a controller or by a third party."

"This prohibition is followed by three exceptions as set in Article 22(2): the data subject's explicit consent, the necessity for the performance of or entering into a contract, and the authorisation by Union or Member State law to which the controller is subject and which also lays down suitable measures to safeguard the data subject's rights and freedoms and legitimate interests."

"Regardless of whether the processing falls in Article 22 or not, we have argued that the vagueness and subjectivity of the concepts used in the GDPR can undermine subjects' rights and data protection principles in practice."

The impact of the General Data Protection Regulation on Social Security [98]

This is a machine-generated summary of:

Bieresborn, Dirk: The impact of the General Data Protection Regulation on Social Security [98]

Published in: ERA Forum (2019).

Link to original: https://doi.org/10.1007/s12027-019-00,565-x.

If you want to cite the papers, please refer to the original.

For technical reasons we could not place the page where the original quote is coming from.

Abstract-Summary

"Public authorities must have a legitimate reason for processing data; the legal grounds are given by the GDPR, some have to be specified by domestic law The processing of special categories of data is strictly prohibited, if there is not an exemption."

"Consent is not always necessary in order to process personal data, but when it must be freely and unambiguously given and can be just as easily withdrawn."

"In respect to the rights of the data subject it has always to be checked, if the rights are applicable or if there is a restriction by the GDPR itself or by domestic law Infringement of the GDPR can cause damage compensation, fines and inadmissibility of evidence."

Purpose of Data Protection?

"Data Protection is about avoiding harm to individuals by misusing or mismanaging their personal data."

"The aim of the GDPR is to strengthen and unify personal data protection for all individuals living in the European Union."

"There are two extreme opinions existing on data protection: Some allege the protection of personal data by law as the highest level a state of law obliged to human rights can achieve."

What are Personal Data?

"Personal data is defined as any information about a living individual which is capable of identifying that individual (Art 4 Nr 1 GDPR)."

"Special rules apply to special categories of (sensitive) personal data, which are defined as 'any information relating to an individual's racial or ethnic origin, political opinions, religious beliefs, trade union membership, physical or mental health or condition, sexual life, alleged or actual criminal activity and criminal record' (Art 9 par 1 GDPR)."

"Briefly summarized Personal data are not only Sur- and Family-Name, but any information that is capable of identify a person, as e.g. the IP-Address (Internet Protocol address), which used to identify computers communicating via the internet, an eMail even if there is no name included and blood-type."

"It doesn't matter if that data is already in the public domain—you still have to comply with the GDPR in the way in which you collect, use and store it."

Social Security and Processing of Data?

"Employers and public authorities have to collect, save, use and sometimes to transfer data to other authorities or private bodies or medical experts."

"He can ask the applicant himself, he can ask third persons as former employers, what can be an infringement of the principle of data minimization (Art 5 par 1 lit b) GDPR) and of proportionality, because so a third person besides the controller and the data subject gets to know personal data of the latter."

"Is it allowed to collect data from tax authorities?"

"They are, if there is a (national) law which allows the transfer of those personal data."

"Is it allowed for employers or authorities to collect medical reports by asking the physician, who treated the person?"

"May e.g. the health-insurance-authority transfer medical data to a pension insurance?"

History

"The Charter of Fundamental Rights of the European Union applies to everyone, but—since it didn't became a European constitution—only after ratification of the Lisboa treaty in respect to European matters involved (Art 51 ChFR)."

"The first draft of The General Data Protection Regulation was published 25th of January 2012."

"It was intended to create a new legal frame of a general data protection within the EU and follows the principle, to regulate almost all life issues which request processing of data besides some exceptions."

"It applies directly since 25th of May 2018 (Art 99 GDPR) in all member states and repealed Directive 95/46/EC."

Important Facts of the GDPR

"Art 6 par 3 GDPR says, that Member States may maintain or introduce more specific provisions to adapt the application of the rules of this Regulation ... by determining more precisely specific requirements for the processing and other measures to ensure lawful and fair processing including for other specific processing."

"Some regulations of the GDPR contain opening clauses, which allow own domestic regulations by member states within the given frame."

"This rule has to be applied reluctantly, since the Court of Justice of the European Union restricts the repetition of European Law given in regulations in domestic law Reason is, that the user has to know, if he is dealing with European or National law Otherwise the role of the Court of Justice of the European Union could be diminished."

Material Scope of the GDPR

"The GDPR follows the idea of a broad scope of European Data protection law This Regulation applies to the processing of personal data wholly or partly by automated means and to the processing other than by automated means of personal data which form part of a filing system or are intended to do so. (Art 2 par 1 GDPR)."

"The GDPR applies to a single written question brought by a client to the authority only if it is intended to be put in a file."

"It applies to the health insurance authority, which has to find a legal ground, usually provided by domestic law, which regulates more precisely the rules in Art 6 par 1/Art."

Basic principles of the GDPR

"The basic principles of the GDPR as laid down in Art 5 GDPR are Lawfulness (lit a), means that personal data may only be processed if there is a legal basis pursuant to Art 6, 9 or 10 and in addition the controller has lawful processing purposes."

"Purpose limitation (lit b), which means that personal data may only be collected for specific purposes and then not used for other purposes (provided there are no exceptions); the legitimacy of purposes requires not only the compliance with a legal basis in Art 6, 9 and 10, but also with other legal obligations like employment law The purposes have to be explicit, if they are documented and expressed in an intelligible manner for controller, data-subject and supervisory authorities."

"Accountability (Art 5 par 2 GDPR), means that the data controller must be able to prove that he complied with the above."

Most important definitions

"Processing of data is anything done with/to personal data, including storing it (Art 4 Nr 2 GDPR)."

"The data controller is the person or organisation who determines the how and what of data processing (Art 4 Nr 7 GDPR)."

"The data subject is the person about whom personal data is being processed (Art 4 Nr.1 GDPR)."

"It is not clear, if 'to collect data' (Art 4 Nr."

Lawfulness of Data Processing

"By Art 6 par 1 GDPR the processing of Data requires a legal ground, which means that Processing shall be lawful only if and to the extent that at least one of the following applies: The data subject has given consent to the processing of his or her personal data for one or more specific purposes (lit a) Processing is necessary for the performance of a contract to which the data subject is party or in order to take steps at the request of the data subject prior to entering into a contract (lit b) Processing is necessary for compliance with a legal obligation to which the controller is subject (lit c) Processing is necessary in order to protect the vital interests of the data subject or of another natural person (lit d) Processing is necessary for the performance of a task carried out in the public interest or in the exercise of official authority vested in the controller (lit e) Processing is necessary for the purposes of the legitimate interests pursued by the controller or by a third party, except where such interests are overridden by the interests or fundamental rights and freedoms of the data subject which require protection of personal data, in particular where the data subject is a child (lit f) The subparagraph lit f) shall not apply to processing carried out by public authorities in the performance of their tasks."

"By Art 6 par 2 and 9 GDPR member states may maintain or introduce more specific provisions to adapt the application of the rules of this Regulation with regard to processing for compliance with points (c) and (e) of par 1 by determining more precisely specific requirements for the processing and other measures to ensure lawful and fair processing including for other specific processing situations as provided for in Chapter IX."

"If the data subject has given explicit consent to the processing of those personal data for one or more specified purposes, except where Union or Member State law provide that the prohibition referred to in paragraph 1 may not be lifted by the data subject (lit a); explicit consent means that the consent has to mention the certain data which the consent refers to, If the processing is necessary for the purposes of carrying out the obligations and exercising specific rights of the controller or of the data subject in the field of employment and social security and social protection law in so far as it is authorized by Union or Member State law or a collective agreement pursuant to Member State law providing for appropriate safeguards for the fundamental rights and the interests of the data subject (lit b); this subparagraph mainly applies this in the field of employment contracts."

"If processing is necessary to protect the vital interests of the data subject or of another natural person where the data subject is physically or legally incapable of giving consent (lit c) If processing is carried out in the course of its legitimate activities with appropriate safeguards by a foundation, association or any other not-for-profit body with a political, philosophical, religious or trade union aim and on condition that the processing relates solely to the members or to former members of the body or to persons who have regular contact with it in connection with its purposes and that the personal data are not disclosed outside that body without the consent of the data subjects (lit d) If processing relates to personal data which are manifestly made public by the data subject (lit e) If processing is necessary for the establishment, exercise or defence of legal claims or whenever courts are acting in their judicial capacity (lit f) If processing is necessary for reasons of substantial public interest, on the basis of Union or Member State law which shall be proportionate to the aim pursued, respect the essence of the right to data protection and provide for suitable and specific measures to safeguard the fundamental rights and the interests of the data subject (lit g) If processing is necessary for the purposes of preventive or occupational medicine, for the assessment of the working capacity of the employee, medical diagnosis, the provision of health or social care or treatment or the management of health or social care systems and services on the basis of Union or Member State law or pursuant to contract with a health professional and subject to the conditions and safeguards referred to in paragraph 3 (lit h) If processing is necessary for reasons of public interest in the area of public health, such as protecting against serious cross-border threats to health or ensuring high standards of quality and safety of health care and of medicinal products or medical devices, on the basis of Union or Member State law which provides for suitable and specific measures to safeguard the rights and freedoms of the data subject, in particular professional secrecy (lit i) If processing is necessary for archiving purposes in the public interest, scientific or historical research purposes or statistical purposes in accordance with Article 89(1) based on Union or Member State law which shall be proportionate to the aim pursued, respect the essence of the right to data protection and provide for suitable and specific measures to safeguard the fundamental rights and the interests of the data subject (j) Again it is up to the Member states to lay down, maintain or introduce more specific provisions to adapt the application of the rules."

Doctor-client privilege

"This strict confidentiality between the doctor and its client is based on the oath of Hippocrates that protects communications between a patient and his or her doctor from being used against the patient in court."

"By protecting the confidentiality between physicians and clients the latter are more willing to speak freely with a doctor who is treating him or her to ensure that the patient receives the best and fullest medical care available under the circumstances."

"The doctor-client-privilege is not as strict as often alleged, since a doctor may betray things at least to public authorities, if it is declared by law, or if he has to defend himself in court or if he wants to rescue higher values."

"The GDPR doesn't contain any provision in respect to the doctor-client privilege, but it accepts the existence of an obligation of professional secrecy regulated by Union or Member State law (Art 15 par 5 GDPR; rec."

"It doesn't automatically allow to breach the doctors-client privilege!"

The Consent

"You can process personal data without consent where it is necessary: For the performance of a contract (Art 6 par 1 lit b); this happens seldom in public law!"

"Recital 43 says, that in order to ensure that consent is freely given, 'consent should not provide a valid legal ground for the processing of personal data in a specific case where there is a clear imbalance between the data subject and the controller, in particular where the controller is a public authority and it is therefore unlikely that consent was freely given in all the circumstances of that specific situation'."

"The authority investigates further on, asks the physician for a medical report and denies a pension because of other reasons (e.g. because the applicant doesn't provide enough time of insurance).The person now wants damage-compensation (Art 82 GDPR), by complaining that his health-data weren't processed on an legal basis."

Forward Processing

"The latter requirement is specified by Art 6 par 4 GDPR: Forward processing is legitimate, provided it is based either on a consent or an European or a domestic law or if 'the controller ascertains, that the processing for another purpose is compatible with the purpose for which the personal data are initially collected by 5 criteria: (a) any link between the purposes, (b) the context in which the personal data have been collected, (c) the nature of the personal data; (d) the possible consequences of the intended further processing, (e) the existence of appropriate safeguards'."

"This—nor really convincing—understanding of Art 6 par 4 GDPR as a broad authorization in favor of the controller to decide about forward processing can only be restricted by domestic law So e.g. German law allows to transfer personal data to private persons/bodies nowadays only, if the recipient declares, that he will use them only for the purpose, why they are transferred to him."

Physicians as Processors?

"The GDPR defines in Art 4 Nr 8 that a 'processor' means a natural or legal person, public authority, agency or other body which processes personal data on behalf of the controller."

"7 GDPR it defines a 'controller' as the natural or legal person, public authority, agency or other body which, alone or jointly with others, determines the purposes and means of the processing of personal data; where the purposes and means of such processing are determined by Union or Member State law, the controller or the specific criteria for its nomination may be provided for by Union or Member State law Art 4 Nr.10 GDPR says, that 'third party' means a natural or legal person, public authority, agency or body other than the data subject, controller, processor and persons who, under the direct authority of the controller or processor, are authorized to process personal data."

Transfer of social-security-data abroad?

"The transfer of data within the EU is always possible, because all member states are supposed to be safe in respect to data protection."

"By Art 49 par 1 GDPR a transfer of personal data to a third country or an international organisation is allowed where the Commission has decided that the third country, a territory or one or more specified sectors within that third country, or the international organisation in question ensures an adequate level of protection."

"Since Privacy Shield as agreement following to Safe Harbour, which the CJEU held not complying with EU-standards, is now enlisted by the commission (White List), it is possible to transfer social-security data to the U.S.A. e.g. by using cloud-computing-offers."

Rights of data-subjects

"By Art 14 par 1 to 4 GDPR the duty to inform the data subject shall not apply where and insofar as: (a) The data subject already has the information; (b) The provision of such information proves impossible or would involve a disproportionate effort, in particular for processing for archiving purposes in the public interest… (c) Obtaining or disclosure is expressly laid down by Union or Member State law to which the controller is subject and which provides appropriate measures to protect the data subject's legitimate interests; or (d) Where the personal data must remain confidential subject to an obligation of professional secrecy regulated by Union or Member State law, including a statutory obligation of secrecy."

"– provided that the medical report contains personal data—Art 21 GDPR only applies, if the processing is based on Art 6 par 1 lit e) or f)!"

"Art 77 GDPR provides, that every data subject shall have the right to lodge a complaint with a supervisory authority, which each Member State shall provide by law (Art 54 GDPR), if it considers that the processing of personal data relating to him or her infringes the GDPR."

Consequences of violating the GDPR

"By Art 82 par 1 GDPR any person who has suffered material or non-material damage as a result of an infringement of the GDPR shall have the right to receive compensation from the controller or processor for the damage suffered."

"Art 82 par 2 GDPR stipulates, that any controller involved in processing shall be liable for the damage caused by processing which infringes the GDPR."

"A processor shall be liable for the damage caused by processing only where it has not complied with obligations of the GDPR specifically directed to processors or where it has acted outside or contrary to lawful instructions of the controller."

"Art 83 par 7 GDPR allows Member States to lay down the rules on other penalties applicable to infringements of this Regulation in particular for infringements which are not subject to administrative fines pursuant to Art 83, and shall take all measures necessary to ensure that they are implemented."

A glance into the future

"Big data can be analyzed for insights that lead to better decisions and strategic business moves."

"Using of Big Data is interesting for social security companies (public and private insurances) not only for economic reasons."

"The processing of data can as well be used to improve the service of health."

"What may be now called using of 'Small Data' could be expanded to the storing of diseases, prescriptions."

"The using of big data could also help to discover and fight epidemics and finally it could cause better research-possibilities (e.g. in respect to cancer)."

"If it is once possible to decipher the human genome in an economical way and stored on the card, a hacking would mean even a discrimination of future generations."

"The new possibilities of technology and Big Data will lead to demands a broad discussion in society."

Summary

"Data-controllers like public authorities must have a legitimate reason for processing data, some are specified by domestic law By judging questions of data-protection it is necessary to regard the GDPR-rules on one hand and national regulations—if existent—on the other hand."

"Art 9 par 4 GDPR provides an opening clause, which enables the member states to restrict the processing of health-data, biometric data and genetic data by domestic law 4."

"Art 6 par 4 GDPR might be misinterpreted in a way, that—even if there is a domestic law regulating forward processing—the controller can always decide himself to process data for other purposes, than they were collected originally."

"In respect to the rights of the data subject (Art 13–21 GDPR) it has always to be checked, if the rights are applicable or if there is an restriction by the GDPR itself or by domestic law (Art 23 GDPR)."

[Section 19]

"The General Data Protection Regulation ('GDPR') applies since 25."

"It is the core and overall aim of the GDPR to ensure a consistent and high level of protection of natural persons and to remove the obstacles to flows of personal data within the EU."

"The GDPR strives to establish a harmonized level of protection of the rights and freedoms of natural persons regarding the processing of personal data on an equivalent level in all EU Member States."

Profiling and the rule of law [99]

This is a machine-generated summary of:
Hildebrandt, Mireille: Profiling and the rule of law [99]
Published in: Identity in the Information Society (2008).
Link to original: https://doi.org/10.1007/s12394-008-0003-1.

If you want to cite the papers, please refer to the original.

For technical reasons we could not place the page where the original quote is coming from.

Abstract-Summary

"Potential customers are profiled to detect their habits and preferences in order to provide for targeted services."

"Such intelligent networked environments will entirely depend on real time monitoring and real time profiling, resulting in real time adaptation of the environment."

"The author will assess the threats and opportunities of such autonomic profiling in terms of its impact on individual autonomy and refined discrimination and indicate the extent to which traditional data protection is effective as regards profiling."

Introduction: The Age of Identification

"E-government and e-health that aim to provide targeted services reiterate this quest for identification, although in this case the identification needed is more sophisticated and resembles what business undertakings seek when they develop targeted servicing and reinvent customer relationship management (CRM)."

"It seems that they are less interested in unique identification of any particular customer than in a refined type of categorisation that allows them to provide targeted servicing at the right time and in the right place."

"It is these threats that touch some of the fundamental tenets of democracy and rule of law, being the particular mélange of positive and negative freedom that allows citizens to develop their relative autonomy (section 5)."

"To counter such threats the focus of legal scholars and practitioners should be extended from the protection of personal data to the protection against the undesired application of profiles and the creation of transparency rights regarding group profiles (section 7)."

What else is new? autonomic behaviour and autonomous action

"One could say that the difference between living and lifeless material is the fact that living organisms are capable of self-constitution over and against an environment which is constituted as such by the act of self-constitution [13]."

"Profiling is thus a crucial sign of life, because it consists of a repeated identification of risks and opportunities by an organism in its environment [14]."

"Enough, such organic profiling is not dependent on conscious reflection."

"One could call it a cognitive capacity of all living organisms, without thereby claiming consciousness for an amoeba."

"What matters is our capacity for conscious reflection on the profiles that we have unconsciously generated, because this gives us the freedom to deliberate on them, to reject or to reinforce them and deliberately to apply them."

[Section 3]

"Automated profiling can be described as the process of knowledge discovery in databases (KDD), of which data mining (DM, using mathematical techniques to detect relevant patterns) is a part (Fayyad and others 15)."

"Applying the profiles Only the third step is what is called data mining in the sense of using mathematical algorithms to locate correlations, clusters, association rules and other patterns."

"Both Zarsky [16] and Custers [17] emphasize that the knowledge generated by profiling machines is new."

"He especially focuses on the difference between classification based on predefined classes and data mining techniques, which provoke unexpected clusters."

"He points out that in the case of KDD the hypothesis emerges in the process of data mining and is tested on the population rather than a sample."

"We have two differences with autonomic organic profiling: (1) the profiling is performed by machines and (2) the type of knowledge it generates differs from 'classical' empirical statistics."

A vision of ambient intelligence and 'the internet of things'

"Both the European Commission [18] and, for instance, Philips [19], have invested heavily in what is called the vision of Ambient Intelligence (AmI), vaguely defined by its 'key elements' [19], being: (1) embeddedness, meaning that networked devices are integrated into the environment; (2) context-awareness, since these devices can recognize you and your situational context; (3) personalisation, as they can be tailored towards your needs; (4) adaptiveness, meaning that they may change the environment in response to your behaviours; and (5) anticipatory, since they should anticipate your preferences without your deliberate input, the environment will always be one step ahead of you."

"We may conclude that profiling technologies are the crucial link between an overdose of trivial data about our movements, temperature, and interaction with other people or things and applicable knowledge about our habits, preferences and the state of the environment."

"To adapt the environment seamlessly we cannot afford to wait for a human interpreter but need profiling machines that draw their own conclusions about what we prefer when and where, hoping we can thus solve the problem of endless choice and deliberation."

Democracy and the rule of law

"This negative freedom is matched with a need to act, to anticipate and participate in the public space, for which some measure of transparency is needed."

"I would claim that negative freedom is an illusion as long as transparency is absent, as we may think that we are facing up to reality in the privacy of our own thoughts, while in fact we have no access to the knowledge needed to assess this reality."

"Profiling may endanger the intricate combination or negative and positive freedom whenever we (1) think we are alone, but are in fact watched by machines that observe our online behaviour, and in an AmI world any move we make; and (2) think we are making private decisions based on a fair idea of what is going on, while in fact we have no clue as to why service providers, insurance companies or government agencies are dealing with us the way they do."

Threats: Knowledge is Power

"As in relation to the so called trade-off between privacy and security two points can be made: (1) a loss of privacy may imply a loss of security, because it exposes the vulnerability of human identity, which renders even more complicated the idea of a trade-off between the two; and (2) trading of personal data implies co-modification and with Schwartz [20] we may expect a market failure due to the unequal access to information about the consequences of trading one's personal data (especially in the case of profiling)."

"The threats can be summarised as concerning (1) privacy, which, however, must not be reduced to hiding one's personal data; (2) security, which, however cannot be traded with privacy since a loss of the one may cause the loss of the other); (3) unfair discrimination, meaning that power relations must be balanced to provide equal bargaining positions; and (4) autonomy, meaning that our negative and positive freedom to act must be established and maintained, since manipulation on the basis of knowledge that we are not aware of violates our autonomy."

Legal pitfalls and challenges

"One can argue that autonomic application of profiles falls within the scope of article 15 of the Data Protection Directive (D 95/46 EC)."

"Paragraph 1 of this article reads: Member States shall grant the right to every person not to be subject to a decision which produces legal effects concerning him or significantly affects him and which is based solely on automated processing of data intended to evaluate certain personal aspects relating to him."

"This is the case even if after application the profile may in fact be a personal data and fall within the scope of articles 11 and 12, which obligate the relevant decision maker to notify the data subject and provide access to the data."

"In as far as the directive does grant legal rights of access to relevant group profiles we need a technological and organisational infrastructure to enable the exercise of these rights."

Transparency enhancing tools: from PETs to TETs

"The Commission is of the opinion that the response to the challenges posed by RFID technologies should include the 'adoption of design criteria that avoid risks to privacy and security, not only at the technological but also at the organisational and business process levels'."

"Multi-stability means that upstream involvement of potential end-users and others who may be affected by the technologies, will broaden the scope of technical design and increase the opportunities to construct a socio-technical infrastructure that does not obstruct the flourishing and autonomy of individual citizens."

"Which technological and organisational infrastructure will provide the transparency of profiles that we argued above?"

"This should provide citizens, whether as consumers, patients or targets of government investigations, with the legal and technological tools to understand which profiles may impact their life in which practical ways."

Closing Remarks

"Advanced profiling technologies answer questions we did not raise."

"These asymmetries challenge the relative autonomy of individual citizens and allow an unprecedented dynamic segmentation of society, especially if the vision of Ambient Intelligence is realised: based on refined real time monitoring, followed by proactive adaptation of our smart environment."

"As long as we lack the legal and technological infrastructure to counter the emerging asymmetry we may find ourselves in a gilded cage: an environment that anticipates our preferences before we become aware of them."

"This contribution argues that we urgently need to develop legal and technological transparency enhancing tools (TETs) to match the proactive dimension of our smart environments."

Social networks and web 2.0: are users also bound by data protection regulations? [100]

This is a machine-generated summary of:

Van Alsenoy, Brendan; Ballet, Joris; Kuczerawy, Aleksandra; Dumortier, Jos: Social networks and web 2.0: are users also bound by data protection regulations? [100]

Published in: Identity in the Information Society (2009).

Link to original: https://doi.org/10.1007/s12394-009-0017-3.

If you want to cite the papers, please refer to the original.

For technical reasons we could not place the page where the original quote is coming from.

Abstract-Summary

"21 and implementing legislation define the respective obligations and liabilities of the different actors that may be involved in a personal data processing operation."

Introduction

"The rapid growth in recent years, of the amount of both social network sites (SNS) (Facebook, LinkedIn, Netlog, MySpace, ...) and their users, shows how important these new fora for social interaction have become."

"Due to their continuously growing technical capabilities and infrastructural setting (as internet applications), SNS offer individuals the opportunity to share information and develop vast social networks with greater ease and speed than previously possible."

"The privacy risks that linger on the dark side of SNS, feed themselves by the (personal) data that are being processed through the services SNS provide."

"What happens when SNS users include information relating to others in their social networking activities?"

"Given the fact that most SNS are designed to house virtual circles of friends and personal profiles of its users (and are therefore primarily intended for recreational use), we must also examine the extent to which processing operations or entities may be exempted from complying with data protection regulations."

Misuse of SNS

"Privacy implications and risks typically associated with the potential misuse of personal data exchanged on SNS range from exposure to direct marketing, re-identification, profiling, identity theft, online and physical stalking, blackmailing and embarrassment. (See Gross and Acquisti 22: 78–79) Contemporary developments in case law have confirmed that privacy may also be jeopardized by 'ordinary' SNS users."

"Such conduct will be actionable on the basis of other grounds than the provisions of a Member States' data protection act (e.g. defamation, right of personal portrayal)."

"When there is no pre-existing relationship of confidence between the plaintiff and the defendant, a claim for misuse of private information under UK law requires that the defendant acquired the information by 'unlawful or surreptitious means'. (See Brimsted 23: 466) It is in light of such possible restrictions that we seek to analyze to what extent the processing activities performed by SNS participants are subject to data protection regulations."

Role Definition

"In order to be qualified as a controller, an entity must exercise at least some level of decision-making power with regards to both the purposes and means of a particular processing operation."

"The controllership of the user does not extend to the SNS as a whole, but only to those processing operations for which he can actually determine the purposes and means."

"The user of an SNS therefore only acts as a controller with regards to the content he chooses to provide and the processing operations he initiates."

"The Belgian Data Protection Act for instance provides that the controller of the processing shall be exempted of liability 'if he can prove that the injurious fact can not be attributed to him'."

"Only insofar as the SNS processes the data in accordance with the terms of use, can it be sustained that user and SNS act as co-controllers."

Personal Use

"As to the question of whether the activities of Mrs. Lindqvist were covered by the exception for personal use, the ECJ replied that the exception must 'be interpreted as relating only to activities which are carried out in the course of private or family life of individuals, which is clearly not the case with the processing of personal data consisting in publication on the internet so that those data are made accessible to an indefinite number of people'. [24] The Court appears to have put forward two elements to determine whether the exception of personal use is applicable."

"The exception in fact concerns processing activities carried out 'in the course of' a personal or private activity."

"3.2 of the Directive exempts data processing in the context of 'purely personal or household' activities, the ECJ makes reference to activities which are carried out in the course of 'private or family life'."

Implications

"They remain obligated to obtain the individual's prior informed consent, and must remove the data immediately if the data subject chooses to exercise his right to object. [25] The Dutch Data Protection Authority has announced that their Ministry of Justice is considering an exemption to the notification obligation for 'personal publications'."

"Two additional conditions are expected to apply: the processing shall only be exempted if (1) all personal data is removed as soon as the data subject objects to the processing of his personal data and (2) the web pages containing personal data are protected against indexing by search engines. [26] In other Member States, e.g., in Belgium, there has not yet been any indication of the intent to provide exemptions to the notification obligation for personal websites, blogs or social network activities."

Conclusion and Outlook

"Seeing as this creates a setting in which the interests of third parties may easily be damaged, it is necessary that a clear legal framework is in place which governs the activities of SNS users."

"Many of the core provisions of the data protection directive could substantially benefit third parties who feel they have been wronged by information posted on an SNS profile."

"We hope that the Article 29 Working Party includes them in their analysis when they adopt their opinion on on-line social networks. [27] In this regard notice should also be taken of the report of the International Working Group on Data Protection in Telecommunications regarding privacy in social networks, which stated in its 'guidance towards regulators' that regulators should rethink the 'current regulatory framework with respect to controllership of (specifically third party-) personal data published on social networking sites, with a view to possibly attributing more responsibility for personal data content on social networking sites to social network service providers'. [28] Whether or not more responsibility needs to be attributed to SNS providers is also still open for debate."

"It is our opinion that further clarification, guidance and perhaps some revision of the existing framework is needed to accommodate the new realities of web 2.0 and social networking applications."

"Please understand we cannot provide further information": evaluating content and transparency of GDPR-mandated AI disclosures [101]

This is a machine-generated summary of:

Wulf, Alexander J.; Seizov, Ognyan: "Please understand we cannot provide further information": evaluating content and transparency of GDPR-mandated AI disclosures [101]

Published in: AI & SOCIETY (2022).

Link to original: https://doi.org/10.1007/s00146-022-01,424-z.

If you want to cite the papers, please refer to the original.

For technical reasons we could not place the page where the original quote is coming from.

Abstract-Summary

"The General Data Protection Regulation (GDPR) of the EU confirms the protection of personal data as a fundamental human right and affords data subjects more control over the way their personal information is processed, shared, and analyzed."

"Where data are processed by artificial intelligence (AI) algorithms, asserting control and providing adequate explanations is a challenge."

"Due to massive increases in computing power and big data processing, modern AI algorithms are too complex and opaque to be understood by most data subjects."

"Articles 15 and 22 of the GDPR provide a modest regulatory framework for automated data processing by, among other things, mandating that data controllers inform data subjects about when it is being used, and its logic and ramifications."

"By means of an online survey (N = 835), we investigated how data subjects expect to be informed about the automated processing of their data."

"The combined findings reveal that current GDPR-mandated disclosures do not meet the expectations and needs of data subjects."

Introduction

"In a qualitative study on transparent online disclosure in the EU, Wulf and Seizov found that consumers habitually click away cookie notices because they find them too tedious— so what hope is left for a much more complex explanation of AI applications for data processing and automated decision-making?"

"Wulf and Seizov reviewed a number of cases related to Articles 15 and 22 GDPR or to their precursors in German national law and identified a number of instances in which German, Dutch and other European courts limited data subjects' rights of access to information, be it to reduce administrative burden, to keep companies' proprietary algorithms and internal processes hidden (i.e., trade secrets), or to narrow the very definition of what constitutes 'personal data'."

"The court, therefore, ruled that no automated decision-making within the meaning of Article 22 GDPR takes place and thus rejected the data subjects' request for information about this algorithm."

A survey of consumers' knowledge and experience of AI in e-commerce

"In a sub-section of a large-scale online survey on the transparency of online contracts we explored the respondents' awareness of AI usage and their experience and knowledge of the GDPR as related to their rights to data access and to explanation of decisions made by AI algorithms."

"Our findings reveal another knowledge gap, namely the divide between those who believed in the need for human review over AI decisions (96 per cent) vs. those who did not know they could or did not wish to exercise their right to request an explanation about how their personal data is being processed, by humans and by algorithms alike (90 percent)."

"Almost 80 percent of the respondents required a substantive AI disclosure and more than half of them expected information not only on the logic but also on the personal data that go into the decision-making process."

Exercising the right to information and explanation: a content analysis of 100 AI disclosures

"While the sample thus reflects the range of companies and organizations offering goods and services online to consumers in Germany and we took care to sample within industries with highly standardized products and similar data processing practices (e.g., banking, insurance, air travel, e-commerce, etc), we cannot claim that the sample is fully representative, so any findings derived from it cannot be generalised beyond the specific set of companies that we investigate and the results should thus be regarded as exploratory."

"While not as opaque as the above example, an additional 45 responses also failed to provide a full explanation of their AI data processing (for example, they mentioned automated services or processes but did not reveal the mathematical or other logic behind them)."

"Such responses stated that the company uses automated processing for one or more purposes but included no further details about the process or its practical and legal implications for the consumer."

Conclusions

"Almost 80 percent of the respondents wanted a substantive disclosure on AI, and the majority of them expected information not only about the logic but also on the personal data that go into the decision-making process."

"As previously pointed out by legal commentaries, the findings from our content analysis of the disclosures on AI use confirm that the requirements of the GDPR are too vague and leave substantial room for interpretation."

"These findings raise concerns about the effectiveness of the GDPR-mandated AI disclosures, both in terms of the data subjects' ease of access to information and of the quality and completeness of the information provided."

"We evaluated the transparency of the AI disclosures according to two standards: (a) the requirements of the GDPR and (b) the recommendations made by Seizov and Wulf."

Discussion

"Although the GDPR is the most developed piece of legislation on data protection currently in force, the guidance that it provides on how companies should inform data subjects about AI and automated decision-making appears to be too superficial and open to interpretation."

"The modest and rather vague regulatory framework of the GDPR, the fact that consumer interest in automated data processing has only recently begun to emerge, and the lack of enforcement of the relevant articles in the GDPR seem to create an environment in which businesses do not have sufficient incentive to provide comprehensive information on their use of AI."

"While it is beyond the scope of this paper to discuss solutions for all factors contributing to this problem, we now review how standardising GDPR-mandated AI disclosures could improve the way companies inform data subjects about AI and automated decision-making."

Social Media and the Protection of Privacy: Current Gaps and Future Directions in European Private International Law [102]

This is a machine-generated summary of:

von Hein, Jan; Bizer, Anna: Social Media and the Protection of Privacy: Current Gaps and Future Directions in European Private International Law [102]
Published in: International Journal of Data Science and Analytics (2018).
Link to original: https://doi.org/10.1007/s41060-018-0108-z.

If you want to cite the papers, please refer to the original.

For technical reasons we could not place the page where the original quote is coming from.

Abstract-Summary

"This article considers the current situation in European private international law regarding the protection of privacy and personality rights in social media."

"The article addresses these problems taking into account the recent case law of the Court of Justice of the European Union."

Introduction

"Social media platforms such as Facebook or Twitter operate in an international environment; frequently, the habitual residence of a user of social media in Europe and the seat of a service provider diverge, as the latter is often based in the USA."

"Social media platforms are in most cases open to users from different jurisdictions."

"The answer may be found in international contract law (regarding claims against the service provider), international tort law (regarding the relations among social media users), or even international successions law (concerning the right to terminate the Facebook account of a deceased person, for example)."

"This article will analyse these challenges from the perspective of European private international law After identifying the objects of protection according to European law (2) and the existing legal sources (3), this text focusses firstly on the legal relationship between social media providers and their (potential) customers (4)."

The object(s) of protection

"The right to privacy is recognized as a fundamental right and therefore explicitly protected (article 7 EU Charter of Fundamental Rights; article 8 European Convention on Human Rights [ECHR]) as is the right to protection of personal data (article 8 EU Charter of Fundamental Rights)."

"According to the European Court of Human Rights, respect to privacy does not only consist of a right to be left alone, but it also obliges the Member States to ensure the effective protection of privacy against infringements by private parties [29]."

"Privacy as a legal term comprises various aspects worthy of protection that essentially determine the personality of a human being, such as a person's name [30]"

"This proves a high awareness of the significance of personal data protection in European politics (also: [31]"

Legal Sources

"It applies when the defendant is domiciled in a member state (article 4 Brussels Ibis), in cases ruled by a jurisdiction agreement selecting a member state court (article 25 Brussels Ibis) and in cases regarding EU consumer contracts [article 18 (1) Brussels Ibis], while the remaining cases are governed by the national rules of civil procedure."

"In social media cases, however, bringing a claim on contractual grounds in the state where the plaintiff is domiciled will usually be possible because the relation between a user and a social media provider can be classified as a consumer contract (article 17, 18 Brussels Ibis)."

"In cases between users, a contractual relation between the parties is usually lacking; thus, violations of privacy and personality are, in those cases, subject to general jurisdiction and tort jurisdiction [article 7 (2) Brussels Ibis]."

The relationship between social media providers and (potential) users

"The standard terms as well as any claim between Facebook and its users are governed by the laws of the State of California."

"It is doubtful whether these clauses are valid under European law The jurisdiction agreement in favour of a third-state court violates article 25 in conjunction with article 19 Brussels Ibis when the user is a consumer (cf. [32])"

"In the case of Facebook, the invalidity of the choice-of-law agreement may be the result of the opaque phrasing of the clause, making it unfair towards the consumer."

"The CJEU recently decided that the applicable law to the validity of standard terms is determined by article 6 (1) Rome I regardless of the situation in which the clause is being attacked [33]"

"The court also decided that a clause failing to inform the consumer about the fact that mandatory provisions favourable to the user remain applicable irrespective of a choice-of-law agreement is unfair [33, para."

The relations among users of social media

"As a contractual relation between users does usually not exist, claims for violation of privacy, personality rights and defamation of a social media user against another are mostly governed by tort law The Brussels Ibis Regulation applies to most of the cases with regard to jurisdiction, whereas the applicable law in privacy cases is still governed by national laws."

"One could argue that the connection that justifies a special jurisdiction should be reason enough to allow the court to simply apply its own law Thus, the general choice-of-law rule could be: if a case falls within the courts special jurisdiction, the lex fori is the tort law applicable."

"This desirable parallelism of jurisdiction and choice of law presupposes that special jurisdiction is applied in a restrictive way to avoid an excessive application of the lex fori to cases that are merely connected to the state of the forum."

Conclusion and Outlook

"Existing European law only partially covers the specific questions concerning PIL issues arising in the context of social media."

"It is to be welcomed that the CJEU interprets the exclusion of privacy and personality rights from the scope of Rome II in a restrictive manner so that it does not hinder the application of the regulation to claims relating to unfair competition."

"As social media are characterized by their worldwide accessibility, it is reasonable to face these challenges by creating uniform rules on jurisdiction and choice of law In particular, the Rome II Regulation should be amended by a choice-of-law rule determining the law applicable to violations of privacy and personality rights, including data protection."

"The EU should develop comprehensive common rules on jurisdiction concerning third-state defendants."

Realising the right to data portability for the domestic Internet of things [103]

This is a machine-generated summary of:

Urquhart, Lachlan; Sailaja, Neelima; McAuley, Derek: Realising the right to data portability for the domestic Internet of things [103]

Published in: Personal and Ubiquitous Computing (2017).

Link to original: https://doi.org/10.1007/s00779-017-1069-2.

If you want to cite the papers, please refer to the original.

For technical reasons we could not place the page where the original quote is coming from.

Introduction

"Bringing the new right to data portability (RTDP) from an abstract legal provision in Article 20 of the EU General Data Protection Regulation (GDPR) 2016 into practice requires a greater role for the IT design community."

"Many IoT services maintain an ongoing relationship with users where their personal data is mined and analysed with the goal of providing value-added, contextually appropriate, services—for example automating routine tasks like room heating management."

"Legally, the lack of scope users have to control data flows after it is collected is a concern for ethical and sustainable growth of the emerging IoT market."

"We challenge the current zeitgeist that monetisation of data, with its incumbent legal obligations, is the best business model for many personal domestic IoT systems."

"PIMS support realisation of legal rights by giving users greater control over their personal data."

Legal perspective

"The RTDP only relates to personal data processed by automated means (i.e. by a computer, not paper records), where the lawful basis for doing so is either: The data subject has given consent, or explicit consent (for special categories of personal data), or Where processing is necessary to fulfil performance of a contract the data subject is party to, or at their request, prior to entering a contract."

"To only relating to data based on contracts or consent, RTDP should be applied 'without prejudice' to other rights, like the right to erasure but not to 'processing necessary for the performance of a task carried out in the public interest or in the exercise of official authority vested in the controller' or when it will 'adversely affect the rights and freedoms of others'."

RTDP for domestic IoT

"Consider how data from different domestic devices can be pieced together to form occupancy models: Increase in CO_2 levels (could be the cat though [34]) Lighting and heating control Passive Infrared (PIR) sensors and door contacts (i.e. traditional sensors for burglar alarms) In home, or more likely access point, cameras Energy monitoring indicating appliance use (e.g. a kettle is easy to detect) Internet traffic flows (e.g. streaming audio services) Increasing access to such sensed data for users, through the RTDP, could enable greater literacy around how devices use their data, and lead to more creative, useful responses or applications."

Realising the RTDP with PIMS

"We are interested in the role new personal data processing architectures can play in supporting increased user control, to enable the right to data portability, particularly for the IoT. Accordingly, we evaluate the current technical state of the art in PIMS."

"Mydex allows the users to authorise services access to specific sets of data and control the third parties with whom this data is shared [35]."

"They provides an API that allows for attribute exchange services that include sending data to an individual's Mydex store, allowing secure access to the store (with their consent), and developing applications that run across multiple devices, accessing personal data from users."

"Higgins is a cloud-based approach that provides storage and a locus of control to the user to manage their personal data, particularly addresses, profiles, interests, contacts, friends, affiliations, etc It allows bidirectional flow of data between the store and service providers and with the user's friends and other social connections."

Discussion and conclusions

"Positively, services could find mechanisms to provide functionality without collecting personal data, mitigating the need for users to rely on DP rights."

"Negatively, in that scenario, users may still be subject to profiling and the impacts that it creates, but as personal data is not being processed, they cannot rely on their DP rights to exercise control through the RTDP."

"There are a number of commercial drivers that require continued UK engagement with EU DP law and privacy standards post Brexit: As a precondition for market access, UK needs to consider DP law Under GDPR Article 3(2), UK data controllers, like companies, targeting goods or services towards EU citizens need to comply with GDPR standards to obtain access to the EU market."

"Data portability has scope to address many problems of users needing greater control, but more fundamentally, business models need to change so we never need to use this right in the first place."

3.2 Right to Be Forgotten

Machine generated keywords: delete, right, search, request, usa, union, tradition, online, individual privacy, employee, success, social medium, google, web, institutional

Gender Transition: Is There a Right to Be Forgotten? [104]

This is a machine-generated summary of:

Correia, Mónica; Rêgo, Guilhermina; Nunes, Rui: Gender Transition: Is There a Right to Be Forgotten? [104]

Published in: Health Care Analysis (2021).

Link to original: https://doi.org/10.1007/s10728-021-00,433-1.

If you want to cite the papers, please refer to the original.

For technical reasons we could not place the page where the original quote is coming from.

Abstract-Summary

"In gender transition, this right represents a powerful form of control over personal data, especially health data that may reveal a gender with which they do not identify and reject."

"It is pertinent to discern whether the right to have personal data deleted—in particular, health data—is ethically acceptable in gender transition."

"Towards addressing the ethical dimensions of the right to be forgotten in this case, this study presents relevant concepts, briefly outlines history, ethics and law of records considering the evolution from paper to electronic format, the main aspects of identity construction and gender identity, and explores the relationship between privacy, data protection/information control and identity projection."

"Conclusions on the ethical admissibility of the 'right to be forgotten' to control gender-affirming information are presented."

Introduction

"In gender transition, the right to be forgotten represents a powerful form of control for the owner over personal data, especially health data that may reveal a gender with which they do not identify and reject."

"It is pertinent to discern whether the right to delete personal data—in particular, health data—is ethically acceptable in gender transition."

"Notwithstanding other motives to modify registries, their legal impact, or other significant branches of the topic, this article's scope is to discuss the ethical admissibility of the right to be forgotten to control gender-affirming health information."

"Regardless of transition-related surgery, as long as gender transition has legal recognition in public registries and critical documents, we aim to provide an ethical assessment of possible claims to have gender-affirming health information deleted."

"Conclusions on the ethical admissibility of the right to delete to control gender-affirming health information."

Definitions

"An individual's sex at birth (birth-assigned sex) is usually determined based on genital appearance, with those present usually assuming that other components of sex are consistent with the newborn's genital sex." [36] "Cisgender person"—"A person who has a gender identity that corresponds with the gender assigned at birth." [37] "Gender"—"The attitudes, feelings, and behaviours linked to one's biological sex's experience and expression." [36] "Gender identity"—"refers to each person's deeply felt internal and individual experience of gender, which may or may not correspond with the sex assigned at birth, including the personal sense of the body (which may involve, if freely chosen, modification of bodily appearance or function by medical, surgical or other means) and other expressions of gender, including dress, speech and mannerisms." [37, 38] "Gender expression"—"The expression of one's gender identity, often through appearance and mode of dress, and also sometimes through behaviour and interests."

"Gender stereotypes often influence gender expression." [36] "Gender transition" or "Gender affirmation"—"refers to the social process of being recognised or affirmed in one's gender identity, expression, and/or role."

"Some trans individuals pursue social but not medical gender affirmation; some pursue medical but not legal gender affirmation, and so on."

"Gender affirmation does not require linearly following a discrete series of "transition" events—it is conceptualised as an ongoing process throughout the life course." [39] "Transgender and gender-nonconforming people (trans or gender minority)"—"People who have a gender identity that is different from the gender assigned at birth."

A Brief Outlook of History, Ethics and Law of Identity Records

"According to Durbach [40], the history of registers in the United Kingdom, for example, suggests that, although people's documentation primarily addressed the needs of the state, this technique was also favourable to some social classes that participated in the process to the extent that their economic interests benefited."

"In the UK, since the sixteenth century, people have understood that the registration of vital information could guarantee their property rights and their rights to state support in times of economic deprivation [40]."

"Registration has a double role: it is an organisational measure of the state; and is regarded as a means of personal affirmation or self-determination, since individualisation through registration facilitates respect for human dignity."

"Given the technological revolution of the last two decades, the state has advanced with registering and archiving personal identification data and health and genetic information."

Gender and Identity Construction

"Although the right to delete may cover registered realities other than gender, namely, challenges to paternity or maternity established at birth or adoption, gender identity—a personality-related fundamental value—represents an independent scope for ethical analysis given its complexity."

"It should be acknowledged that the pathology-based approach has been subject to substantial critique because the establishment of gender does not depend on physical characteristics but identity."

"Gender expression is another concept related to gender identity, and a person's gender expression may, or may not, follow their gender identity [36]."

"Gender identity is based on self-determination [37, 41], so people have the right to maintain primary and secondary sexual characteristics even if they chose to transition their gender."

"The state could change the last digit of the identification document number (even for female; odd for male), which reflected an outdated conceptualisation of gender based on a dualistic understanding of biological sex [42]."

Privacy, Data Protection and Identity Projection

"According to this author, two conceptions of privacy exist, sustained by two main sets of values: in Europe, privacy is an aspect of personal dignity, understood as the right to image, name, reputation, threatened mainly by the media; in the USA privacy is an aspect of freedom, endangered mainly by the government."

"Perspective, he claims that it is defenceless to consider human dignity as a universal value, equally felt by any society, regardless of its culture and history."

"Angle, human dignity is imposed on the individual himself and considered a universal value."

"Privacy is a condition of human dignity."

"Technology is why we have gone from protecting intimacy and privacy to protect personal data, a fundamental human right."

"Modern technology has created a "remembering-by-default" environment where personal data is recorded and stored indefinitely and shared quickly, resulting in the right to privacy violated, identity itself endangered [43]."

The Right to Be Forgotten and Trans Identity Information

"It is necessary to discuss the limits of the right to be forgotten regarding gender transition information, considering the principle of liberty and the proportionality principle as a foundation in human rights law Based on these arguments, we argue that the right to be forgotten may raise ethical and legal difficulties in the following circumstances: a. The right to erase health and genetic information related to gender transition makes tracing genetic diseases impossible, especially relevant when dealing with their biological offspring."

"e. Also, in case of repentance: changing ideas is also an exercise of autonomy, which is why deleting gender-affirming health data can harm the owner, preventing them from exercising the most varied rights if data recovery is not possible; f. The right to be forgotten so far as gender transition is concerned might create the inability to trace the previous identity, and so criminal responsibilities could be avoided, as it would allow identification forgery; moreover, civil responsibilities could be jeopardised, as it might cause contractual and inheritance problems, for example."

Conclusion

"Recognising gender identity as a fundamental human right—in the sense of full development of an individual's personality through self-determination and self-realisation—is accepted as an integral part of respect for human dignity."

"Forgetting gender-affirming health data should be regulated at the international level to protect the human person's inalienable rights and future generations."

"After examining some inter-related and conflicting issues regarding the right to delete personal data, specifically medical and genetic records, in the context of gender transition, we conclude that the admissibility of this right should be limited and regulated."

"Regardless of the future of information about the past, it urges a discussion about specific features associated with the right to delete health and genetic data in gender transition from an ethical perspective."

The Right to be Forgotten: an Islamic Perspective [105]

This is a machine-generated summary of:

Osman, Amr: The Right to be Forgotten: an Islamic Perspective [105]

Published in: Human Rights Review (2022).

Link to original: https://doi.org/10.1007/s12142-022-00,672-2.

If you want to cite the papers, please refer to the original.

For technical reasons we could not place the page where the original quote is coming from.

Abstract-Summary

"Following an overview of some aspects of the right to be forgotten and why it is endorsed in the European Union with its emphasis on privacy but not in the USA with its exaltation of free speech, the paper discusses two related issues: (1) elements in the Islamic tradition potentially conducive to endorsing a particular understanding of the right to be forgotten and (2) some possible obstacles from that tradition that could challenge the recognition of this right from an Islamic perspective."

"The paper concludes that despite some challenges, including Qur'anic verses and certain views, activities, and institutions, the right to be forgotten is defendable from within the Islamic tradition by reference to the principle of satr, which breaches that one should not seek to publicize personal information that causes harm to other people or even to oneself."

"The paper seeks to contribute to discussions on this new right by showing what a religious tradition such as Islam can offer to them."

Introduction

"This paper asks a straightforward question: is the right to be forgotten consistent with the Islamic tradition?"

"In part 3, the paper discusses two questions: Are there ideas in the Islamic tradition that can lend support to any understanding of the right to be forgotten, and under what conditions?"

"What are some possible obstacles from that tradition that could impede the recognition of this right from an Islamic perspective?"

"The paper concludes that the right to be forgotten can be defended from within the Islamic tradition by relying on the notion of satr, namely, refraining from searching for and publicizing whatever could harm other people's reputation, with harm aversion being another principle of the Islamic worldview."

The Right to be Forgotten

"Simply put, the right to be forgotten is the right of individuals to request the deletion of search engine links to information about them when this information is no longer "relevant" or is "excessive.""

"Put differently, whereas under the right to be forgotten links containing certain personal information remain operative but are removed from the search results using a specific name, under the "right to erasure," the original publisher of that information would be required to remove links to this data altogether [44]."

"The right to be forgotten does not actually delete information from the Internet altogether, but only makes it harder to find through search engines [45]."

"Not only did the court not request that information be removed from the Internet altogether, but it even failed, to the dismay of the supporters of the new right, to order search engines to delist links to the information worldwide."

The Right to Be Forgotten and the Islamic Tradition

"To demonstrate the status of satr as an eminent and meritorious Islamic value, Muslim scholars have used Qur'anic verses and Prophetic Hadith (that is, statements attributed to the Prophet Muhammad or reports about his actions)."

"A Muslim who does expose himself either by committing sins in public or revealing his transgressions (a practice called mujāhara) loses the right of moral immunity from backbiting, even if by revealing his transgressions he does not intend to show contempt to the rules of Islamic ethics or defy God's laws."

"This notwithstanding, there are other elements in the Muslim tradition that may pose a challenge to the endorsement of the right to be forgotten from an Islamic perspective."

"A special group of scholars known as the Hadith critics—who were typically transmitters themselves—took it upon themselves to ascertain that the individuals who transmitted traditions were reliable, which, once more, required that they be morally upright as well as accurate in their transmission."

Conclusion

"It is not hard to see why some countries and scholars insist that the right to be forgotten is inconsistent with the right to free speech, which requires, among other things, free access to information, including information that some people may consider private, irrelevant, or even embarrassing."

"The rule should perhaps be that individuals have the right to ask to be forgotten and that others who want to challenge this rule in a specific case have to prove that that would lead to a harm that is greater than the harm that the concerned individual may endure (such as in cases where a society needs to be protected from a particular individual with a criminal record)."

"It may be tempting indeed to argue that since Islam does emphasize the duty of respecting others' privacy [46, 47], and since the right to free speech, as currently understood, was virtually absent in the Islamic tradition, then the right to be forgotten is actually Islamic."

Interest group lobbying in the European Union: privacy, data protection and the right to be forgotten [106]

This is a machine-generated summary of:

Christou, George; Rashid, Imir: Interest group lobbying in the European Union: privacy, data protection and the right to be forgotten [106]

Published in: Comparative European Politics (2021).

Link to original: https://doi.org/10.1057/s41295-021-00,238-5.

If you want to cite the papers, please refer to the original.

For technical reasons we could not place the page where the original quote is coming from.

Abstract-Summary

"The issue of trust and control of data online has become critical for many European Union (EU) citizens in an era where we are increasingly reliant on digital platforms across a plethora of everyday activities."

"The future of the EU's Digital Single Market Policy is reliant on developing trust through robust legislation that ensures explicit control of data by EU citizens."

"This article explores the extent to which interest groups have been able to successfully achieve their goals through actions in the European Union institutional spaces that construct privacy and data protection legislation."

Introduction

"We seek to extend and add value to the work on privacy and data protection lobbying through a focus on how interest groups lobbied in relation to the specific provisions of the RTBF within the GDPR."

"We argue that the story of preference attainment and success in relation to the RTBF is more nuanced than the predominant industry versus CSO storyline found in the privacy lobbying literature, when it comes to relative winners and losers across the multiple dimensions of the RTBF issue."

"Our findings in relation to the RTBF confirm those in other lobbying studies on issue salience and strange bedfellows [48–52) in demonstrating that diverse or strange bedfellows have significantly higher preference attainment and that EU institutional policy-makers were sensitive to finding compromise that had broad support."

"We conclude by offering thoughts on the implications of interest group success relating to the RTBF and more broadly, debates on privacy, data protection and freedom of expression."

Interest group lobbying success and the European Union

"Among the scholars that have focused on lobbying success, numerous factors have been identified as significant in achieving interest group preferences across a number of studies: individual, institutional, organizational and policy or issue related [53–61)."

"We argue that a single case study of this sort is insightful in terms of: (a) consideration of representation and influence in relation to the RTBF and GDPR and providing further evidence on the factors that allow interest groups to achieve their preferences in the digital rights field, (b) providing a deeper exploration of EU policy outcomes and legitimacy and the implications for digital rights in Europe and (c) offering a more nuanced understanding of the public interest in relation to digital rights, where such an interest can be represented by enhancing privacy and data protection; maintaining freedom of expression; or, indeed, providing a platform through which industry (data controllers in particular in this case) can balance the necessary level of consent with growth and innovation."

Discussion: The RTBF and lobbying success in the EU

"Pro-privacy and anti-surveillance oriented CSOs such as the German Action Alliance Freedom Not Fear and consumer protection groups such as the European Consumers' Organisation (BEUC) supported and were in alignment with the Commission's (more specifically, Commissioner Reding's) preferences to create and promote a RTBF clause in order to strengthen and maximize the privacy of data subjects."

"Industry actors as represented by major online players and EuroISPA, in particular, favoured a minimalist liability and obligation model in relation to data controllers, which conflicted with the maximalist obligation preferences of Commissioner Reding in the first draft of the GDPR and that of consumer protection groups and certain pro-privacy groups."

"In terms of the RTBF, however, MEP Jan Philipp Albrecht (Greens/EFA), appointed as the leading rapporteur for the GDPR, was in favour of stronger individual rights in terms of privacy and data protection, but also backed the freedom of expression argument by certain CSOs and industry players [62]."

Conclusions

"Our study confirmed that policy compromise in the case of RTBF came as a result of factors such as issue salience and conflict, as well as the congruence of interest groups with EU institutional actors, and certain CSOs with industry, throughout the institutional process."

"Whilst those CSOs representing consumers and advocating for more robust privacy rights and maximal obligations for data controllers did not achieve their over arching preference, industry actors and certain CSOs lobbying for balanced rights, a right to erasure and reduced obligations for data controllers were relatively successful in attaining their preferences."

"Our study adds to the general civil society versus industry storyline in the EU privacy lobbying literature, by highlighting the alignment of interests on certain RTBF conflict dimensions, between pro-freedom of expression groups and industry actors calling for technical change."

Could You Ever Forget Me? Why People Want to be Forgotten Online [107]

This is a machine-generated summary of:

Kwak, Chanhee; Lee, Junyeong; Lee, Heeseok: Could You Ever Forget Me? Why People Want to be Forgotten Online [107]

Published in: Journal of Business Ethics (2021).

Link to original: https://doi.org/10.1007/s10551-021-04,747-x.

If you want to cite the papers, please refer to the original.

For technical reasons we could not place the page where the original quote is coming from.

Abstract-Summary

"Capturing the difficulties in terms of a cultural lag between technological advancements and regulations on individuals' data privacy needs, we identify six motives for individuals wishing to be forgotten online and investigate its expected effects on online content generation through a qualitative content analysis of 222 responses from open-ended surveys in Korea."

"Our findings provide implications for the literature on individual privacy and the right to be forgotten employing the cultural lag, as well as, elaborate further on the relationship between being forgotten online and the legitimacy of such requests of individuals."

"Implications for data providers, data controllers/processors, and governments to address this lag and build a balanced system of personal information are provided."

Introduction

"Individuals may not be able to delete disclosed personal data because of technological and legal obstacles."

"The right to be forgotten, a recent solution to reduce the cultural lag regarding individual privacy, should also be based on social consensus."

"This right's primary purpose is to provide a legal foundation for individuals to manage their personal data when such data manipulations do not harm public interests (European Parliament 63)."

"Social consensus can resolve the cultural gap that has emerged from the difference between the speeds of technology and regulation in the context of individual privacy."

"The findings have implications for the literature on individual privacy and the right to be forgotten with the cultural lag, and can elaborate further on the relationship between being forgotten online and the legitimacy of such requests of individuals."

Background

"In Article 17, the General Data Protection Regulation (GDPR) defines the right to be forgotten as the right "to obtain from the controller the erasure of personal data concerning him or her without undue delay and the controller shall have the obligation to erase personal data without undue delay [63, p. 4].""

"Following the guidelines of the right, data processors/controllers need to prepare their criteria and procedures to conform to the EU's regulation, otherwise, they cannot continue their business on the continent."

"We argue that the missing link of implementing the right to be forgotten can be closely connected to the understanding of people and social context."

"Understanding the privacy needs of individuals and the right to be forgotten can form an essential part of social consensus from the viewpoint of individuals and act as a starting point for the reduction of the cultural lag [64]."

Methods and Data Description

"Explicit legislation of the right to be forgotten does not exist, guidelines on the Right to Request Access Restriction on Personal Internet Postings provide individuals with legal rights to seek deletion of personal content regardless of the membership status of the service provider [65]."

"To investigate the reasons why people want to be forgotten online and the expected consequences of such a right, we conducted open-ended surveys with a convenience sample."

"Our research topic concerns each individual's need to be forgotten in a general online environment, our sample should consist of individuals familiar with online services and experienced in uploading content on the Internet."

"The open-coded categories in the initial phase were then aggregated through iterative discussions into broader categories of why people revise/delete the content they upload."

Motives and Interpretations

"Based on qualitative content analysis of the open-ended survey responses on the reasons people want to delete not only their personal data but also the data of others relevant to their privacy, we identified six categories of motives for wanting to be forgotten: information disclosure, content sensitivity, social reputation, control over further processing, system/process, and sociality."

"Content sensitivity Respondents' concerns can increase when the data contain potentially sensitive information, where potentially sensitive means the individuals might not understand the inherent risks of initially uploading the data because of the difficulty in estimating the potential damage due to an increase in accessibility to private content."

"When the subject is the owner of the data, respondents show the concern for sociality or content sensitivity in addition to the concern for social reputation and information disclosure."

Expected Impact of the Right to Be Forgotten on Online Content Generation

"The respondents expected the right to be forgotten to reduce privacy concerns and encourage free expression, and thus, increasing the amount of online content."

"Such respondents expressed their thoughts as: "It will reduce the risk of personal information leakage," "It will help to protect privacy," "It increases posting frequency [because] it creates safety to post," "It can lead to more freedom of expression," "It can make us share our thoughts more freely," and "Contents will be posted more if complete deletion is possible [by enforcing the right to be forgotten].""

"Such respondents expressed their concerns as: "Strong claims or content postings with the possibility of controversy will be reluctant," "The greater the awareness of the right to be forgotten, the more cautious is the content posting," "I think over time, only refined and simplified content will remain," and "It may cause long-term negative effects to the diversity of Internet content.""

Cultural Lag with the Motives Identified and Online Content Generation

"From the motives identified and related discussions, we posit that the cultural lag impacts various aspects of an individual's privacy, including information management and interaction management."

"When the right to be forgotten is appropriately implemented in addition to the fulfillment of the above-mentioned motives, individuals could enjoy several benefits of online content generation resulting from the reduced lag."

"As the motives for wanting to forgotten online that were identified in this study describe privacy concerns created or amplified by the lag, design and direction of the right to be forgotten need to consider the motives to reduce the lag."

"The motives and expectations of individuals regarding privacy and the right to be forgotten must be considered in the right's design and its implementation to truly make social adjustments and create consensus and eventually reduce the cultural lag."

Implications

"Previous privacy studies of motives for revising or deleting online content have mainly adopted a narrow context-based approach to investigate individual privacy, such as social network service [66, 67]."

"Examining the use of the Internet allows us to identify the lag between IT and the human concept of memory, generalize apprehensions about the desire to be forgotten online, and provide a direction for a better understanding of the cultural lag, individual privacy, and implementation of the right to be forgotten."

"Investigating the motives for wishing to be forgotten by considering both the cases of the subject being the data owner or not, this study presents the privacy needs of individuals using the Internet from a more general perspective [68]."

"To build a balanced system of personal information, our study provides insight into the three stakeholders in personal data management, namely, individuals, data controllers/processors, and governments."

Suggestion for Future Research

"Further studies in specific contexts, including services, personal data, and situations, can enhance the understanding of the right to be forgotten."

"The motive of social reputation and saving face determined in this study could result from a cultural factor prevalent in Korean and East Asian contexts and the high value of saving face in Korea can affect perceptions of the right to be forgotten."

"Perceived freedom of speech can be identified as a major motive for wanting or not wanting to be forgotten for individuals in the United States, which was not the case for the Koreans in this study."

"The perception of the right to be forgotten in the United States can differ from that in Korea."

"Investigations are required to identify different motives depending on different cultures and countries and how members of a society determine the legitimacy of diverse interests for implementing the right."

Conclusion

"The right to be forgotten can be a comprehensive approach to protecting individual privacy, which has become an unavoidable aspect of each individual's life."

"By revealing the motives for wanting to be forgotten, this study established a better understanding of the right to be forgotten and initiated discussions regarding this right."

"The study identified the need for and range of the right to be forgotten and estimated the positive and negative effects of its implementation."

"These findings constitute a stepping stone in the research on individuals' right to individual privacy."

Unpleasant Memories on the Web in Employment Relations: A Ricoeurian Approach [108]

This is a machine-generated summary of:

Habisch, André; Kletz, Pierre; Wack, Eva: Unpleasant Memories on the Web in Employment Relations: A Ricoeurian Approach [108]

Published in: Humanistic Management Journal (2022).

Link to original: https://doi.org/10.1007/s41463-022-00,138-0.

If you want to cite the papers, please refer to the original.

For technical reasons we could not place the page where the original quote is coming from.

Abstract-Summary

"To limit companies' access to personal information, a right to be forgotten has been introduced by the European jurisprudence."

"By discussing the notion of forgetting from the perspective of French hermeneutic philosopher Paul Ricoeur, the present article demonstrates that both, companies and employees, would be harmed if access to online information on applicants and current employees would be denied."

"Consistent with a Humanistic Management approach that promotes human dignity and flourishing in the workplace, this article proposes guidance for the responsible handling of unpleasant online memories in personnel decision-making processes, thereby following Ricoeur's notion of forgetting as "kept in reserve"."

Introduction

"Based on his book Memory, history, forgetting [69], which offers an extensive analysis of the notions of memory and forgiveness, the second aim of this paper is therefore to employ Ricœur's (69) concepts of forgetting as kept in reserve and difficult forgiveness to offer a Humanistic Management approach towards personnel decision-making."

"This article contributes to the literature on Humanistic Management in HRM by showing that processes of personnel decision-making, which acknowledge and forgive individuals' past mistakes, are better suited to foster human flourishing and personal growth than an erasure of online traces under the pressure to conform."

"The remainder of the article is structured as follows: The first part reviews the literature on the use of information from the web in personnel decisions, thereby illustrating reasons and benefits for organizations, as well as criticism relating to the ethicality of this practice."

"We will also derive practical implications for a Humanistic Management approach to the use of online information in personnel decision-making."

The Use of Information from the Web in Human Resource Management

"In recruitment and selection, cybervetting is often used as a social media background check or as a source of information in addition to the résumé [70]."

"Information from social media sites is considered useful in predicting applicants' personality [71, 72] and assessing their potential fit with an organization [73]."

"Another frequent argument for the use of social media in selection is to avoid the risk of irresponsible or even criminal employee behaviour, thus protecting the company's stakeholders from harm [74]."

"By disclosing such information to potential employers, social media sites give answers to questions that are often not allowed to be asked during a job interview [75]."

"From an employer's perspective, there exist a variety of reasons for monitoring employees' social media channels [76]."

"Monitoring employees' social media activities also raises privacy infringement issues [76], as well as concerns regarding disruption of traditional employment relations and a shift in public/private boundaries in organizational life [77]."

A Ricœurian Approach Towards Forgetting and Forgiving

"It is therefore that we turn to French hermeneutic philosopher Paul Ricœur (1913–2005) to demonstrate that human forgetting is more than the mere absence of memory."

"As Ricœur's hermeneutics approach takes an enduring interest in the analysis of human action and an understanding of the human condition [78], it has been referred to in a number of management and organization research contexts."

"His book Memory, history, forgetting [69] serves as a basis for the discussion of how unpleasant online memories can be dealt with in an HRM context."

"In Memory, history, forgetting, Ricœur [69] points to the dilemma of the duality of forgetting and remembering."

"At the same time, humans are intimidated by the thought of a memory that would never forget anything [69]."

"Concerning the latter concept, Ricœur [69] considers forgetting in relationship to the notion of latency: A memory never disappears completely, but is kept in the back of people's minds and can suddenly reappear."

A Critical Discussion of the Right to Be Forgotten

"As a means of defending users' privacy rights and information autonomy, the present right to be forgotten has received considerable praise [79], since it may allow users to erase personal information from the past that affects their present."

"Critics argue that the removal of links under the right to be forgotten reduces the information about a person available to others and thus weakens freedom of speech [80]."

"While the balancing of the right to privacy and the right to free speech is intensely debated, Hoffman and others [81], as well as Uncular [82] state that link removals under the right to be forgotten do not automatically cause unfavourable memories on the web to be truly forgotten."

"Research has shown that due to practical and technical limitations, the right to be forgotten is difficult to enforce, as personal information hardly disappears completely from the web."

A Responsible Handling of Unpleasant Memories on the Web

"Since responsible recruitment, selection, and disciplinary processes should ultimately benefit the relationship between members of an organization, it is worthwhile taking a closer look at the concept of forgiveness and the role of remembering in HRM decision-making."

"Rather, it is the act of keeping a mistake in reserve and changing the relationship with its memory [69, 83], which allows a manager to truly forgive applicants and employees by accepting them despite, or even because of past mistakes."

"To forgiveness of mistakes in existing employment relationships, a responsible hiring process characterized by remembering and forgiving would thus support the formation of such constructive organizational cultures, as well as harmonious and constructive relationships within organizations."

"Paragraphs, we will formulate implications and recommendations for practitioners aimed at establishing processes of personnel decision-making that acknowledge and forgive individuals' past mistakes and make responsible use of information from web searches."

Conclusion

"As cybervetting increasingly becomes common practice in hiring and digital tools enable employers to monitor employees' social media activities, the right to be forgotten addresses users' need to have unpleasant memories removed from the web."

"To account for a responsible handling of unpleasant memories on the web consistent with Humanistic Management, a manager's attitude of forgiveness and tolerance of mistakes can thus be regarded as more beneficial for human flourishing within an organizational culture of harmonious cooperation in the long run."

"According to Ricœur's [69] notion of forgetting as "kept in reserve", the adequate challenge is not to irretrievably silence unpleasant memories of past mistakes through a right to be forgotten."

"Employment relationships would thus benefit to a great extent if existing and prospective employees were given the possibility to position themselves on any negative information on the web in the context of their personal learning history."

3.3 International Transfer of Personal Data

Machine generated keywords: transfer, international, transfer personal, third, country, health research, exchange, transmit, german, million, datum health, controller, hold, identification, research purpose.

International transfers of personal data for health research following Schrems II: a problem in need of a solution [109]

This is a machine-generated summary of:

Hallinan, Dara; Bernier, Alexander; Cambon-Thomsen, Anne; Crawley, Francis P.; Dimitrova, Diana; Medeiros, Claudia Bauzer; Nilsonne, Gustav; Parker, Simon; Pickering, Brian; Rennes, Stéphanie: International transfers of personal data for health research following Schrems II: a problem in need of a solution [109]

Published in: European Journal of Human Genetics (2021).

Link to original: https://doi.org/10.1038/s41431-021-00,893-y.

If you want to cite the papers, please refer to the original.

For technical reasons we could not place the page where the original quote is coming from.

Abstract-Summary

"The decision may have significant effects on the legitimate transfer of personal data for health research purposes from the EU."

Introduction

"This paper considers the effects of the Court of Justice of the European Union's (CJEU) Schrems II decision—on Facebook's transfers of personal data (PD) from the EU to the US—on the international exchange of PD for health research, specifically in the context of the COVID-19 pandemic."

"We aim: (i) to outline the consequences of the case for the legitimate international exchange of PD for health research, specifically in the context of the COVID-19 pandemic; and (ii), to consider certain options to address the consequences of the case and to facilitate international data exchange moving forward."

"We then provide a more detailed overview of the Schrems II case and how the decision impacts the transfer of EU PD for health research purposes ('Schrems II: the case, the decision, and transfers of PD for health research', 'Broader consequences of Schrems II for transfers of PD for health research' and 'Limitations of Article 49 options')."

The importance of international data exchange in health research

"At least three types of arguments outlining the importance of the international exchange of PD for health research can be put forward."

"Several kinds of health research depend completely on the international exchange of PD."

"Such research includes, for example, research into questions best addressed by comparing data from multiple populations in distinct regions and research which can only be conducted with access to PD sets located in multiple different countries."

"The optimal conduct of health research requires that PD be as accessible as possible so research can be synthesised and built upon."

"In this regard: PD allows risks of bias to be assessed more thoroughly; PD permits flexible analysis, including reanalysis, through which research can be independently tested and verified with greater accuracy; and PD allows new scientific questions to be asked of existing data."

"The international exchange of PD for health research, however, is subject to legal conditions."

The approach of the GDPR to international health research transfers

"Transfers of PD from the EU to third countries are only legitimate if based on a justification outlined in the GDPR."

"Adequacy decisions can provide a framework for transfers of PD for health research."

"If an adequacy decision is not in place for a third country, transfers of PD for health research may be legitimised based on bi- or multilateral arrangements with recipients which assure the standard of protection elaborated by EU law Article 46 outlines several relevant approaches: 'a legally binding and enforceable instrument between public authorities'; 'binding corporate rules'—agreements valid within a company or group of companies; 'data protection clauses'—contractual clauses on data transfer; 'codes of conduct'; 'certification mechanisms'; and 'provisions… inserted into administrative arrangements between public authorities'."

"If neither a tier 1 or tier 2 option is available, EU researchers may still transfer PD to third countries based on 'specific situation' justifications outlined in Article 49."

[Section 4]

"The case concerns questions as to how Facebook might legitimately transfer PD from the EU to the US."

"The CJEU found the US system did not provide a level of protection 'essentially equivalent' to that provided by the EU regarding US authorities' powers to access, in bulk, EU citizens' transferred PD."

"The SCCs outlined by the Commission are, in principle, still valid and useful to legitimate health research transfers from the EU to third countries but only if EU citizens' rights in relation to the transfer(s) in question are provided with 'essentially equivalent' protection in the third country."

"SCCs can only be used to legitimate health research transfers if: (i) a third country already provides 'essentially equivalent' protection; or (ii) supplementary measures assuring 'essentially equivalent' protection—contractual, technical, organisational, or other—can be put in place [84, 85]."

[Section 5]

"Whilst Schrems II dealt specifically with privacy shield and SCCs, the logic of the decision applies to all another tier 2 transfer options for health research."

"In discussing SCCs, the CJEU was clear that to legitimately transfer PD from the EU, the law in a third country cannot function such that the standard of protection provided to EU citizens falls below the standard provided under EU law In terms of their relationship with third-country law, all other tiers 2 justifications function comparably to SCCs."

"This will require researchers to expend supplemental resources in conducting the assessments needed to ensure legitimate transfers under tier 2 justifications."

"One retort to our assertion that the consequences of the decision are significant for transfers of PD for health research might be: international transfer options outlined by non-US adequacy decisions and those in Article 49 (tier 3 options) are untouched by the decision."

Limitations of Article 49 options

"A look at the applicability conditions for Article 49 options in relation to health research reveals extensive limitations."

"The applicability of several Article 49 justifications superficially relevant for health research transfers, is in fact unclear."

"The EDPB recognises the justification may be used to legitimate certain transfers for health research in the context of the COVID-19 pandemic [86]."

"Even when a justification's applicability is clear, all Article 49 options are subject to general conditions limiting their utility for health research transfers."

"This means Article 49 justifications should not justify run-of-the-mill transfers for health research or open, ongoing, sharing arrangements."

"Even when the applicability of an Article 49 justification is clear, and the general conditions for application are met, each justification is still subject to specific criteria limiting utility for health research transfers."

Technical measures to overcome challenges

"There are technical measures that do not involve actual data transfer."

"There are perhaps three commonly used technologies that facilitate international access to health research data without transfer."

"Researchers access the data remotely to perform their analyses and may only retrieve results."

"The results of the execution of the code/script are then assessed by the data hosts at the source location before being returned to the researchers."

"This removes the need for data to be transferred or for researchers to ever actually see the data."

"Infrastructure solutions can offer health researchers greater flexibility compared to a remote execution model but without the need to see the data as required by data visitation via a thin client."

Limitations of technical measures to overcome challenges

"Approaches such as data visitation and remote execution show promise regarding analyses on single large data sets and for certain types of 'second-level' analysis."

"Data visitation and remote execution models do not permit PD to be exported and pooled in support of such ends."

"Users of data visitation or remote execution systems will require skills training and support to work in data holders' systems."

"Further, to enable reproducibility, there remains a need for guidelines and standards for reporting analyses performed under data visitation and remote execution models."

"Regarding remote execution and hybrid models, the problem appears with the definition of PD and the degree to which synthetic data or anonymous data can be returned which are both scientifically useful and evade classification as PD [87]."

"The fact technical measures may be inadequate to resolve all issues connected with current EU data protection challenges to health research transfers engenders a need to look elsewhere for solutions."

Health research transfer agreements to overcome challenges

"We propose considering the conclusion of unique bi- and multilateral agreements outlining principles specifically governing health research transfers."

"Such agreements could insulate transfers of PD for health research purposes from problematic, jurisdiction-specific, legislation—considering legal and practical limitations."

"The idea that transfers of PD for health research purposes should be considered as a unique form of international transfer should not be received as strange."

"Against this background, why should international transfers of PD for health research purposes not be treated as unique forms of transfers?"

"Should the EU and the US conclude a bilateral health research-specific transfer agreement implementing limitations on US intelligence services' activities necessary to meet EU standards, we would tentatively suggest the impact on US national security interests would be nominal."

"We do not suggest that PD transferred internationally for health research should be considered 'lower risk' in relation to non-research processing."

Conclusion

"Transfers of PD for health research from the EU to third countries must be legitimated under one of the justifications in the GDPR."

"In the recent Schrems II case, however, the CJEU made consequential pronouncements for the utility of these justifications for health research transfers."

"From a legal perspective, the assertion is applicable to many of the justifications generally relevant for the legitimation of international transfers for health research under the GDPR."

"Certain GDPR justifications for legitimating international transfers for health research remain directly unaddressed in Schrems II."

"The utility of these options for health research transfers, however, has limitations."

"Moving forward, certain technical measures provide promising approaches to allow international health research involving EU PD to proceed despite legal obstacles."

Transfers of personal data to third countries [110]

This is a machine-generated summary of:

Van den Bulck, Paul: Transfers of personal data to third countries [110]
Published in: ERA Forum (2017).
Link to original: https://doi.org/10.1007/s12027-017-0482-3.

If you want to cite the papers, please refer to the original.

For technical reasons we could not place the page where the original quote is coming from.

Abstract-Summary

"The EU's General Data Protection Regulation GDPR confirms the general principles of the existing Directive on the transfer of personal data to a third country (in particular the requirement of an "adequate level of protection")."

The stakes of personal data transfer

"Recital 6 of the General Data Protection Regulation (GDPR) provides context and emphasises the stakes involved in the transfer of personal data in these terms: "Rapid technological developments and globalisation have brought new challenges for the protection of personal data."

"Technology has transformed both the economy and social life, and should further facilitate the free flow of personal data within the Union and the transfer to third countries and international organisations, while ensuring a high level of the protection of personal data.""

"Although the transfer of data, in particular to third countries, has increased over time, the current legal issues remain the same as those prevailing at the time of the enactment of Directive 95/46 on personal data protection (the Directive) namely to ensure a high level of protection of data during and after transfers are made."

The third country transfer notion

"The most illuminating definition is the one found in a position paper of 14 July 2014 by the European Data Protection Supervisor (EDPS), which identifies the following potential elements of a "transfer of personal data": "communication, disclosure or otherwise making available of personal data, conducted with the knowledge or intention of a sender subject to the Regulation that the recipient (s) will have access to it"."

"Definition, the EDPS gave the following examples: the sending of data by post or email by an EU controller to a recipient in a third country; the internet-based transfer, known as "push" (deliberate transfer): the communication is initiated by the EU controller or central server (controller), to a recipient in a third country; the internet-based transfer, known as "pull" (permitted access): the request for the transmission of information is first submitted to the EU controller by the recipient in a third country; direct on-line collection of individual's data in the EU by a non-EU processor acting on behalf of an EU controller; publication of personal data on the internet by an EU controller."

Principles relating to transfers of personal data

"Since an adequate level of protection is guaranteed within the EU by the GDPR, Article 1(3) of the GDPR reproduces the principle set out in Article 1(2) of the Directive, namely the free flow of personal data within the EU."

"As the right to protection of personal data is a fundamental right, the GDPR further specifies in Article 44 that "All provisions in this Chapter shall be applied in order to ensure that the level of protection of natural persons guaranteed by this Regulation is not undermined"."

"In so doing, it seems that an application of the provisions of the GDPR relating to transfers to third countries which would be technically compliant, but which would not guarantee the level of protection for natural persons desired by the GDPR or the Charter of Fundamental Rights of the European Union (the Charter), would not be in conformity."

The conditional regime for transfers to third countries

"To the Directive, the GDPR allows the exporter, controller or processor, to export the data under his responsibility, without formal requirement and more specifically without any authorisation from the supervisory authority."

"As was provided for under the Directive, the European Commission (the Commission) can, pursuant to a procedure currently laid out in Article 5 of Regulation 182/2011, utilise an implementing decision to make a determination that a third country ensures an adequate level of protection."

"Once an adequate level of protection is recognised by the Commission, transfers can be made without specific authorisation."

"Until the GDPR goes into effect, third countries recognised as ensuring an adequate level of protection are recognised under the authority of the Directive."

"Article 45(9) of the GDPR states that adequacy decisions already adopted shall remain in force until amended, replaced or repealed by a Commission decision adopted in accordance with paragraph 3 or 5 of Article 45 of the GDPR."

"The same applies under Article 96 of the GDPR for any international agreement involving the transfer of personal data to a third country which was completed prior to 24 May 2016 and which complies with EU law as it was applicable prior to that date."

"This judgment shaped Article 45 of the GDPR, particularly by providing clarification on the role of the Commission and actions which it should have taken with regard to adequacy decisions (cf. infra 4.2.1, 4.2.3 to 4.2.5)."

"The Commission, when determining whether a third country, a territory, or a given sector of a third country has an adequate level of protection, must take into account: "(a) the rule of law, respect for human rights and fundamental freedoms, relevant legislation, both general and sectoral, including concerning public security, defence, national security and criminal law and the access of public authorities to personal data, as well as the implementation of such legislation, data protection rules, professional rules and security measures, including rules for the onward transfer of personal data to another third country or international organisation which are complied with in that country or international organisation, case-law, as well as effective and enforceable data subject rights and effective administrative and judicial redress for the data subjects whose personal data are being transferred; (b) the existence and effective functioning of one or more independent supervisory authorities in the third country or to which an international organisation is subject, with responsibility for ensuring and enforcing compliance with the data protection rules, including adequate enforcement powers, for assisting and advising the data subjects in exercising their rights and for cooperation with the supervisory authorities of the Member States; and (c) the international commitments the third country or international organisation concerned has entered into, or other obligations arising from legally binding conventions or instruments as well as from its participation in multilateral or regional systems, in particular in relation to the protection of personal data.""

"With regard to a third country's international commitments, recital 105 of the GDPR stipulates that the Council of Europe Convention of 28 January 1981 for the Protection of Individuals with regard to the Automatic Processing of Personal Data and its Additional Protocol will be looked upon favourably during an adequacy evaluation."

"The Commission makes adequacy decisions through the use of implementing decisions adopted in accordance with the evaluation procedure laid down in Article 93(2) of the GDPR."

"Before the committee which represents the various Member States, as detailed in Article 5 of Regulation 182/2011 and Article 93(2), votes on the draft Commission

Decision, the European Data Protection Board (EDPB) must give an opinion to the Commission on the adequacy of the level of protection offered by the third country."

"Since the adoption of the Directive, many supervisory authorities (that make up the EDPB) have become more independent and very strong advocates for personal data protection."

"The Commission is required to publish in the Official Journal of the European Union and on its website a list of third countries, territories, and specific sectors in third countries for which it has issued decisions that an adequate level of protection is ensured."

"Article 45(3) of the GDPR places an obligation on the Commission to review adequacy decisions at least every four years."

"The Advocate General, in his opinion preceding the Schrems case, did not hesitate to point out that, over a fifteen-year period, the adequacy decision for Safe Harbor, had never been subject to an actual review by the Commission."

"The Commission shall, on an ongoing basis, monitor developments in third countries and international organisations that could affect the functioning of decisions adopted pursuant to paragraph 3 of this Article and decisions adopted on the basis of Article 25(6) of Directive 95/46/EC."

"The Commission shall, where available information reveals, in particular following the review referred to in paragraph 3 of this Article, that a third country, a territory or one or more specified sectors within a third country, or an international organisation no longer ensures an adequate level of protection within the meaning of paragraph 2 of this Article, to the extent necessary, repeal, amend or suspend the decision referred to in paragraph 3 of this Article by means of implementing acts without retro-active effect."

"The Commission shall enter into consultations with the third country or international organisation with a view to remedying the situation giving rise to the decision made pursuant to paragraph 5." (emphasis added) In other words, the Commission: must permanently monitor developments in third countries which could negatively affect the application of adequacy decisions; if a third country no longer ensures an adequate level of protection, the Commission must, on a case-by-case basis, repeal, amend or suspend its adequacy decision by following the same procedure as the one used for the adequacy decision-making process."

"A decision which repeals, amends or suspends an adequacy decision does not prejudice transfers of personal data out to third countries on the basis of otherwise legal and appropriate safeguards or derogations allowed in specific situations."

"Going forward, the Commission will review adequacy decisions that are in force in accordance with all the terms and within the deadlines established by the GDPR."

"The GDPR integrates clarifications provided by the CJEU with regard to the adoption and monitoring of adequacy decisions."

"The Schrems case contains all relevant information about the supervisory duties of the CJEU and of the supervisory authorities with regard to adequacy decisions."

"The CJEU observes in the Schrems case that an adequacy decision of the Commission presents specific characteristics which preclude the CJEU, for the purposes of adjudicating an invalidation, from limiting itself to the elements of fact and law existing at the date on which the decision was issued."

"This observation will almost certainly remain valid for the GDPR, since as explained above, the Commission's obligation to permanently monitor the developments in third countries which could negatively affect the application of adequacy decisions is now explicit."

"The CJEU stated that: "the term 'adequate level of protection' must be understood as requiring the third country in fact to ensure, by reason of its domestic law or its international commitments, a level of protection of fundamental rights and freedoms that is essentially equivalent to that guaranteed within the European Union by virtue of Directive 95/46 read in the light of the Charter. (…) Even though the means to which that third country has recourse, in this connection, for the purpose of ensuring such a level of protection may differ from those employed within the European Union in order to ensure that the requirements stemming from Directive 95/46 read in the light of the Charter are complied with, those means must nevertheless prove, in practice, effective in order to ensure protection essentially equivalent to that guaranteed within the European Union.""

"The Schrems case also settled the important question of whether an adequacy decision given by the Commission has the effect of precluding a supervisory authority from making a determination as to whether the level of protection ensured by a third country is adequate and to draw any possible conclusions with regard to the individual case concerned."

"The CJEU, after a detailed analysis of the Directive, in particular in Article 28 relating to supervisory authorities and Article 8 of the Charter, concluded that adequacy decisions by no means impaired the powers of supervisory authorities outlined in the Directive."

"The supervisory authority cannot, however, revoke the decision of the Commission, as only the CJEU has this authority."

"In the absence of an adequacy decision, a controller or a processor can only transfer personal data to a third country when using appropriate safeguards or under the guise of derogations that the GDPR allows for in specific situations."

"Appropriate safeguards remain the main way to transfer personal data to a third country and derogations are an exception to be applied in very specific cases."

"Appropriate safeguards can be used to transfer personal data to a third country provided that the data subjects involved enjoy enforceable rights and effective legal remedies."

"To an adequacy decision which incorporates these criteria in the assessment of the level of protection of third country, an appropriate safeguard might not integrate these conditions."

"A data exporter, controller, or processor must ensure that the safeguard being used meets these conditions or that the third country guarantees that the aforementioned rights of the data subjects are enforceable and that effective legal remedies exist."

"The GDPR draws a distinction between appropriate safeguards that require authorisation from a supervisory authority and those that do not."

"The following safeguards do not require specific authorisation from a supervisory authority: (a) a legally binding and enforceable instrument between public authorities or bodies; (b) Binding Corporate Rules (BCRs); (c) standard data protection clauses adopted by the Commission in accordance with the examination procedure in Article 93(2) of the GDPR; (d) standard data protection clauses adopted by a supervisory authority and approved by the Commission pursuant to the examination procedure in Article 93(2) of the GDPR; (e) an approved code of conduct pursuant to Article 40 of the GDPR, together with binding and enforceable commitments of the controller or processor in the third country to apply the appropriate safeguards, including those in relation to data subjects' rights; or (f) an approved certification mechanism pursuant to Article 42 of the GDPR, together with binding and enforceable commitments of the controller or processor in the third country to apply the appropriate safeguards, including those in relation to data subjects' rights."

"The absence of a specific authorisation from a supervisory authority prior to carrying out a transfer of data abroad regarding some safeguards does not mean that the safeguard itself was drawn up without the approval of a supervisory authority."

"To ensure that appropriate safeguards that were approved prior to the GDPR remain in legal compliance, the GDPR specifies that authorisations granted by a Member State or supervisory authority, as well as decisions adopted by the Commission based upon Article 26(2) and 26(4) of the Directive remain valid until they are amended, replaced, or repealed by a supervisory authority or the Commission."

"The appropriate safeguards set out in this transitional law clause are standard clauses adopted by the Commission or a supervisory authority and the BCRs."

"The following safeguards require a special authorisation from a supervisory authority: (a) contractual clauses between the controller or processor and the controller, processor or the recipient of the personal data in the third country; or (b) provisions to be inserted into administrative arrangements between public authorities or bodies which include enforceable and effective data subject rights."

"BCRs are corporate rules which must (i) be binding upon involved companies, including their employees, and (ii) include all basic principles and enforceable rights to ensure appropriate safeguards for transfers of personal data or transfers of certain categories of personal data."

"The Directive does not expressly provide for BCRs which came into existence from discussions of the Article 29 Working Party about "sufficient guarantees" for the protection of personal data."

"The Portuguese supervisory authority found that BCRs did not provide an adequate level of protection to allow transfer."

"BCRs shall, at minimum, specify: (a) The structure and contact details of the group of undertakings, or the group of enterprises engaged in a joint economic activity and of each of its members; (b) the data transfers or set of transfers, including the categories of personal data, the type of processing and its purposes, the type of data subjects affected and the identification of the third country or countries to which data are to be transferred; (c) their extent to which they are legally binding nature, both internally and externally; (d) the application of the general data protection principles, in particular purpose limitation, data minimisation, limited storage periods, data quality, data protection by design and by default, legal basis for processing, processing of special categories of personal data, measures to ensure data security, and the requirements for onward transfers to bodies not bound by the binding corporate rules; (e) the rights of data subjects in regard to processing and the means to exercise those rights, including the right not to be subject to decisions based solely on automated processing, including profiling in accordance with Article 22 of the GDPR, the right to lodge a complaint with the competent supervisory authority and before the competent courts of the Member States in accordance with Article 79 of the GDPR and to obtain redress and, where appropriate, compensation for a breach of the binding corporate rules; (f) the acceptance by the controller or processor established on the territory of a Member State of liability for any breaches of the binding corporate rules by any member concerned not established in the EU; the controller or the processor shall be exempt from that liability, in whole or in part, only if it proves that that member is not responsible for the event giving rise to the damage; (g) how information on the BCRs, in particular the information in points (d), (e) and (f) of this paragraph and Articles 13 and 14 of the GPDR, is provided to the data subjects; (h) the tasks of any data protection officer designated in accordance with Article 37 of the GDPR or any other person or entity in charge of the monitoring compliance with the binding corporate rules within the group of undertakings, or group of enterprises engaged in a joint economic activity, as well as monitoring training and complaint-handling; (i) the complaint procedures; (j) the mechanisms within the group of undertakings, or group of enterprises engaged in a joint economic activity for ensuring the verification of compliance with the binding corporate rules."

"Results of such verification should be communicated to the person or entity referred to in point (h) and to the board of the controlling undertaking of a group of undertakings, or of the group of enterprises engaged in a joint economic activity, and should be available upon request to the competent supervisory authority; (k) the mechanisms for reporting and recording changes to the rules and reporting those changes to the supervisory authority; (l) the cooperation mechanism with the supervisory authority to ensure compliance by any member of the group of undertakings, or group of enterprises engaged in a joint economic activity, particularly by making available to the supervisory authority the results of verifications of the measures referred to in point (j); (m) the mechanisms for reporting to the competent supervisory authority any legal requirements to which any member of the group of undertakings, or group of enterprises engaged in a joint economic activity is subject in a third country which are likely to have a substantial adverse effect on the guarantees provided by the binding corporate rules; and (n) the appropriate data protection training to personnel having permanent or regular access to personal data."

"The relevant supervisory authority must approve the BCRs in accordance with the consistency control mechanism set out in Article 63 of the GDPR."

"Article 46 of the GDPR expressly draws a distinction between three types of clauses, namely: (a) Two types of standard clauses not requiring the authorisation of a supervisory authority: standard clauses adopted by the Commission; standard clauses adopted by a supervisory authority and approved by the Commission; (b) A type of ad hoc clause requiring the authorisation of the competent supervisory authority: ad hoc clauses adopted between an exporter (controller or processor) and an importer (controller, processor, or recipient)."

"The sets of standard clauses adopted by the Commission were recently modified to integrate the findings of the Schrems case, specifically, to avoid a possible invalidation by the CJEU for not appropriately recognising the powers of supervisory authorities."

"Article 40(3) of the GDPR specifies that these codes of conduct can also be of use to entities who are not subject to the GDPR, such as importers found in countries not ensuring an adequate protection."

"Articles 42(2) and 46(2)(f) of the GDPR allow importers found in countries not ensuring an adequate level of protection and not subject to the GDPR to be certified; this certification acts as an appropriate safeguard authorising the transfer."

"In the absence of an adequacy decision or appropriate safeguards, a transfer or a set of transfers of personal data to a third country can only take place on the basis of a derogation."

"the data subject has explicitly consented to the proposed transfer, after having been informed of the possible risks of such transfers for the data subject due to the absence of an adequacy decision and appropriate safeguards; 2."

"A transfer could not be based on an adequacy decision, an appropriate safeguard and none of the derogations above are applicable, a transfer to a third country may take place only if: a. this transfer is necessary for the purposes of compelling legitimate interests pursued by the controller which are not overridden by the interests or rights and freedoms of the data subject; b. the controller has assessed all the circumstances surrounding the data transfer and has on the basis of that assessment provided suitable safeguards with regard to the protection of personal data; c. the controller i. informs the supervisory authority of the transfer; ii."

"In the absence of an adequacy decision, EU or Member State law may, for important reasons of public interest, expressly set limits to the transfer of specific categories of personal data to a third country."

"The GDPR confirms the existing prohibition already confirmed by Article 29 Working Party under the terms of which any judgment of a court or any decision of an administrative authority of a third country requiring a controller or processor to transfer or disclose personal data may only be recognised or enforceable in any manner if based on an international agreement such as a mutual legal assistance treaty, in force between the requesting third country and the EU or a Member State or where the transfer took place on the grounds of transfer set out in the GDPR."

International cooperation

"The GDPR encourages international cooperation between supervisory authorities by requiring the Commission and supervisory authorities to take certain measures to that end."

"Obvious that, by definition, the degree of maturity of international cooperation depends on the will of stakeholders."

Athlete consent as a legal base for data transfers to third countries for anti-doping purposes, under EU and German law [111]

This is a machine-generated summary of:

Kornbeck, Jacob: Athlete consent as a legal base for data transfers to third countries for anti-doping purposes, under EU and German law [111]

Published in: The International Sports Law Journal (2017).

Link to original: https://doi.org/10.1007/s40318-017-0112-9.

If you want to cite the papers, please refer to the original.

For technical reasons we could not place the page where the original quote is coming from.

Abstract-Summary

"This article aims to discuss athlete consent as a legal base for data transfers to third countries for anti-doping purposes, under EU and German law, including by summarising the legal relevance of international anti-doping requirements and expectations."

Introduction

"Whenever National Anti-Doping-Organisations (NADOs) based in the European Union (EU) and the European Economic Area (EEA) are required to transfer the personal data of athletes to partners operating outside the EU/EEA, these data transfers are subjected to stricter conditions than those applying to intra-EU/EEA transfers."

"As for all data processing operations, transfers require the demonstration, inter alia, of a valid legal base, including but not limited to the consent of the data subject (in this case: the athlete concerned) or a specific legal (statutory) provision."

"The challenges involved will be discussed with reference to EU and German law The comparative analysis aims to identify the relative merits of athlete consent (the traditionally preferred legal base in the anti-doping community) as opposed to statutory provision."

"Germany recently saw the entry into force of an Anti-Doping Act (Anti-Doping-Gesetz) (AntiDopG) which is one of the most far-reaching of any EU Member State (MS): an act of parliament including specific provisions for data transfers."

Legal, political and institutional framework

"NADOs based in the EU and the EEA find themselves in a situation of double jeopardy, simultaneously facing expectations from the World Anti-Doping Agency (WADA) and sports governing bodies (SGBs) such as the International Olympic Committee (IOC) for the sharing of athletes' personal information, as opposed to strict legal requirements imposing limitations, under EU and national law, regarding the sharing of such data, in particular with partners located outside the EU."

"NADOs are expected to meet the expectations of WADA and WADA stakeholders such as sport governing bodies (SGB's), while also complying with national law In the Member States (MSs) of the European Union (EU), this poses a range of specific compliance issues, in particular because privacy and data protection are EU fundamental rights (unlike the US Constitution which extends similar protection to free speech but not to privacy), which makes compliance mandatory and subject to specific standards which must be met in accordance with a strict legality requirement: rather than being lawful by default, subject to the absence of legal impediments, the lawfulness of data processing operations, "the first named principle" of European data protection law, require the demonstration of a legal base."

Political and legal debates

"In a Swiss legal PhD thesis from before the WADC 2009, Flueckiger identified "unlawful interferences with [athletes'] personality specifically caused by data processing operations" for anti-doping purposes, as well as a general absence of awareness that established practices might pose privacy concerns."

"The retired data protection commissioner of the German state of Schleswig-Holstein (who just before retiring had co-authored a highly critical legal opinion together with his regulator colleague from the state of Rheinland-Pfalz), Weichert, published the book review already referred to, criticising the justifications used to support the "Whereabouts" requirements of the WADC and referring to the WADA-sponsored ADAMS database (Anti-Doping Administration and Management System) as "a control and surveillance system of great informational calibre, yet not particularly well thought out.""

"By early 2017, German legal scholarship had definitely recognised the acuity of privacy issues in sport and anti-doping, and the German book market was the only one boasting a data protection handbook written for the sports sector and marketed by a major national legal publisher."

Consent as a legal basis for data transfers to third countries

"EU law is more specific in enumerating the valid cases of lawfulness (Art 7 Directive 95/46/EC; Art 6(1) GDPR), including consent (Art 6(1)(a) GDPR) as well as a detailed enumeration (not found in Germany's BDSG) of the remaining valid grounds for processing: performance of a contract between the data controller and the data subject (Art 6(1)(b) GDPR); fulfilment of a legal obligation on the part of the data controller (Art 6(1)(c) GDPR); protection of vital interests of the data subject or of another natural person (Art 6(1)(d) GDPR); performance of a task carried out in the public interest or in the exercise of official authority vested in the controller (Art 6(1)(e) GDPR); protection of the legitimate interests of the controller or by a third party, as long as this does not interfere with the interests or fundamental rights of the data subject, especially if the data subject is a child (Art 6(1)(f) GDPR)."

Findings of this article

"NADOs' obligations under EU and national law are often marginal and often absent with regard to the purported obligation to share athletes' data, while their obligation to protect the privacy of athletes and ensure the protection of their personal data is enshrined in enforceable EU and national legislation with fundamental right status."

"The right to protection of personal data is guaranteed by the CFR, with EU secondary and national (including German) legislation providing specific, procedural requirements for the respect of this fundamental right, including the requirement that the lawfulness of all data processing operations must be demonstrated mandatorily."

"Whereas athlete consent has traditionally been relied upon to a very large extent in anti-doping work, it transpires from an even quite cursory assessment of EU and national German legal requirements that the consent of the data subject is highly problematic as a legal base, especially where a significant imbalance between data subject (athlete) and data controller (NADO) is found, as in anti-doping work."

Further implications

"The present article aims to have discussed athlete consent as a legal base for data transfers to third countries for anti-doping purposes, under EU and German law, including by summarising the legal relevance of international anti-doping requirements and expectations."

"It will compare and contrast the findings from that exercise with the outcome of the analysis made in the present article, with the aim to arrive at an assessment of the two options for a legal basis under EU and German data protection law The analysis provided in this article is limited to the current WADC 2015 framework."

"Based on previous experience with the WADC framework, it is highly likely that some of the proposed measures will have legal implications within the legal orders of the EU and its MS, including with regard to athletes' access to exercising their fundamental rights."

International Aspects of Radio Frequency Identification Tags: Different Approaches to Bridging the Technology/Privacy Divide [112]

This is a machine-generated summary of:

Erickson, G. Scott; Kelly, Eileen P.: International Aspects of Radio Frequency Identification Tags: Different Approaches to Bridging the Technology/Privacy Divide [112]

Published in: Philosophy & Technology (2007).

Link to original: https://doi.org/10.1007/s12130-007-9011-y.

If you want to cite the papers, please refer to the original.

For technical reasons we could not place the page where the original quote is coming from.

Introduction

"There are reportedly up to one billion tags already installed worldwide, but the potential of the technology and awakening consumer privacy concerns are only now sparking some discussion concerning the concept."

"We want to review the state of the art RFID (and potential future applications), the potential benefits of the technology, concerns about specific uses of the technology, and the role of government in sorting it all out."

"There is no doubt that the current rate of improvement in technology is well ahead of what is being done by industry or government to create standards and conditions of use."

"We will look at the privacy concerns and other potential downsides of the technology as expressed by consumer privacy organizations, and, finally, consider the role of government in supporting and/or restricting the development of the technology."

RFID: the Technology Behind the Controversy

"The mandate was for every pallet of merchandise sent to Wal-Mart to be identified with an RFID tag by the installation date [88]."

"When the RFID tags communicate with computer systems, which can hold even more data concerning the specific product (such as expiration date, transport history, end user), they enable the underlying information technology systems to create massive databases with the potential to hold incredibly detailed histories of production, distribution, and usage [89]."

"Unlike barcodes, which must be manually scanned in close proximity (a matter of inches with direct line of sight), RFID tags have ranges up to 30–100 m, depending on the estimate and the technology involved, and do not need line of sight."

"The costs of the tags themselves, the costs of new reader and data input/transfer systems, and the costs of storage (remember, information can be held on each unique item now, so the data generated by RFID could be substantial) will be considerable, at least initially."

The Case for RFID

"RFID applications are probably best illustrated by considering them at different points in the Value Chain, as well as in different types of industries."

"The most visible proposed applications to date have been in the supply chain, and we will start there before moving on to operations (including services), retail, and post-retail."

Supply Chain

"The Wal-Mart and DOD mandates obviously cover supply chain use of RFID."

"Even with its best-practice information technology-driven inventory systems, Wal-Mart estimates that it has a 7% stockout rate at any given time."

"RFID can help provide this type of information in a timely manner, as tagged pallets or items are continuously tracked and databases updated whenever they move through a scanned portal (e.g., the loading dock at a warehouse or retail store)."

"Further, to guard against stockout, most supply chains hold safety stock (not just in retail, but in manufacturing and service applications as well)."

"RFID could also help with shrinkage, both in the supply chain and in retail stores, as we will discuss shortly."

"RFID shows the potential to save companies money on labor costs, as items move more swiftly through the supply chain."

Operations

"Manufacturers with complicated processes, especially those requiring different operations performed on different individual units, will benefit from RFID tags that can identify items and link to instructions on what to do with them on the production floor."

"From ski slopes to amusement parks, RFID tags can allow service providers to track whether individuals have paid, monitor different admission packages allowing different activities, and watch over at-risk clients such as children [90]."

"Using a tag to track the movement of individuals in care facilities can allow for better design of the facilities, resulting in more attention to the types of needs most critical to quality of life [91]; so, although not as publicized as inventory and retail applications, the potential of RFID to improve both manufacturing and service operations is substantial."

Retail and post-retail

"Clothing retailers such as Zara and H&M depend on their ability to rapidly identify hot items, shifting production and distribution to accommodate current trends."

"Tesco, for example, the UK grocer, claims to be able to make 4 million distinct promotional offers to its customer base, each customized to the individual customer's purchasing patterns [92]."

"With RFID, many more retailers could take such an approach, even those without loyalty programs but who are able to identify customers by credit card."

"Further, customers could be enticed with individualized product recommendations at point-of-purchase (tie with shirt, for example) based on the individual product handled or purchased."

"In the event of another Mad Cow scare, for example, identification and recall of potentially infected products could be more targeted and more efficient [93]."

The Case Against RFID

"Those questioning RFID are chiefly concerned about consumer privacy."

"Consumer privacy is a matter of some concern that RFID adopters need to confront."

"One new issue with RFID is not just the depth of information collected, but the fact that consumers may very well be unaware of the collection."

"With a loyalty card, most consumers recognize that data collection is occurring."

"Extensive data can be collected without consumer awareness and without any indication of the depth and breadth of the underlying database."

"Business partners of the RFID tagger are one thing, but outsiders or government may have access to tag data as well, and that prospect raises even further concerns."

"When and where will the government have access to tagged data and the underlying databases?"

"What happens to customer/consumer privacy, and what happens to proprietary corporate information?"

Bridging the Divide

"The consumer privacy implications are important to broad acceptance of the technology, and existent uncertainty and the possibility of adverse reaction after a consumer backlash are the types of matters that could slow investment in the technology and the accrual of clear benefits."

"The mishmash of international attention and concern has to be an issue for corporations, considering the substantial investments required by RFID systems and the potential influence of consumer groups with privacy issues."

"Starting with the bottom left box, we see a situation where industry is moving forward, making substantial investments in RFID technology, but government has yet to set the terms of use."

"The EU, and especially Finland, have shown a willingness to more clearly delimit the privacy implications of RFID and policies for firms using the technology."

Conclusion

"The appeal to industry is obvious, and new and even more attractive applications will likely proliferate in the next few years."

"Industry has to recognize that a downside exists."

"Although self-restraint and self-regulation from industry could solve some of the problems, advocates of the technology do not seem to be headed in that direction."

"We believe government needs some involvement, at least in pushing industry toward self-restraint if a final legislative or regulatory answer is not yet obvious."

References

1. EHDS Proposal (Proposal for a regulation of the European Parliament and of the Council on the European Health Data Space COM/2022/197 final). (2022). https://eur-lex.europa.eu/legal-content/EN/TXT/?uri=CELEX%3A52022PC0197
2. Article 29 Data Protection Working Party. (2007). Opinion 4/2007 on the concept of personal data. https://ec.europa.eu/justice/article-29/documentation/opinion-recommendation/index_en.htm
3. Article 29 Data Protection Working Party. (2014). Opinion 05/2014 on anonymisation techniques. https://ec.europa.eu/justice/article-29/documentation/opinion-recommendation/files/2014/wp216_en.pdf
4. Regulation (EU) 2016/679. (2016). of the European Parliament and of the Council of 27 April 2016 on the protection of natural persons with regard to the processing of personal data and on the free movement of such data, and repealing Directive 95/46/EC (General Data Protection

Regulation) (Text with EEA relevance), Regulation (EU) 2016/679 OJ L 119. https://eur-lex.europa.eu/eli/reg/2016/679/oj
5. Convention 108+ (Convention for the protection of individuals with regard to the processing of personal data). (2018). Retrieved May 5, 2023, from https://rm.coe.int/convention-108-convention-for-the-protection-of-individuals-with-regar/16808b36f1
6. Groos, D., & van Veen, E.-B. (2020). Anonymised data and the rule of law. European Data Protection Law Review (EDPL), 6(4), 498–508.
7. Regulation (EU) 2022/868. (2022). of the European Parliament and of the Council of 30 May 2022 on European data governance and amending Regulation (EU) 2018/1724 (Data Governance Act) (Text with EEA relevance), 152 OJ L. http://data.europa.eu/eli/reg/2022/868/oj/eng
8. G Cormode, D Srivastava. Anonymized Data: Generation, Models, Usage. SIGMOD, Providence, Rhode Island. 2009.
9. Clifton C, Tassa T. On syntactic anonymity and differential privacy. Trans Data Privacy. 2013;6(2):161–83.
10. Sarathy R, Muralidhar K. Evaluating Laplace noise addition to satisfy differential privacy for numeric data. Trans Data Privacy. 2011;4(1):1–17.
11. Daries JP, Reich J, Waldo J, Young EM, Whittinghill J, Seaton DT, et al. Privacy, anonymity, and big data in the social sciences. Queue. 2014;12(7):30. 12 pages
12. Access to Classified Information, Executive Order #12968, August 4, 1995, http://www.fas.org/sgp/clinton/eo12968.html
13. Maturana HR, Varela FJ. Autopoiesis and Cognition: The Realization of the Living. Reidel: Dordrecht; 1991.
14. Hildebrandt M. 'Defining Profiling: A New Type of Knowledge and Profiling' and 'The Identity of the European Citizen'. In: Hildebrandt M, Gutwirth S (eds.). Profiling the European Citizen. A Cross-disciplinary Perspective. Springer: Dordrecht; 2008a. p. 17–30 and 303–26.
15. Fayyad UM, Piatetsky-Shapiro G, Smyth P, Uthurusamy R. Advances in Knowledge Discovery and Data Mining. AAAI Press / MIT Press Meno Park: California - Cambridge, Mass. - London England; 1996.
16. Zarsky T Z. "Mine Your Own Business!": Making the Case for the Implications of the Data Mining or Personal Information in the Forum of Public Opinion, 5 Yale Journal of Law & Technology 2002–2003, nr. 4, p.17–47.
17. Custers B. The Power of Knowledge. Ethical, Legal, and Technological Aspects of Data Mining and Group Profiling in Epidemiology. Wolf Legal Publishers: Nijmegen; 2004.
18. ISTAG. Scenarios for Ambient Intelligence in 2010. Information Society Technology Advisory Group; 2001: available at: http://www.cordis.lu/ist/istag-reports.htm .
19. Aarts E, Marzano S (eds.). The New Everyday. Views on Ambient Intelligence. 010: Rotterdam; 2003.
20. Schwartz PM. Beyond Lessig's Code for Internet Privacy: Cyberspace Filters, Privacy-Control and Fair Information Practices. Wisconsin Law Review; 2000, p. 743–788.
21. Data Protection and Privacy Commissioners, Resolution on Privacy Protection in Social Network Services, 30th International Conference of Data Protection and Privacy Commissioners Strasbourg, 17 October 2008. http://www.edps.europa.eu/EDPSWEB/webdav/shared/Documents/Cooperation/Conference_int/08-10-17_Strasbourg_social_network_EN.pdf . Accessed 10 April 2009.
22. Gross R, Acquisti A. Information Revelation and Privacy in Online Social Networks, in Proceedings of the 2005 ACM Workshop on Privacy in the Electronic Society (WPES'05), Virginia; 2005. p. 71–80.
23. Brimsted K. The JK rowling photo case–are privacy rights evolving for the online era? CLSR. 2008;24:465–8.
24. European Court of Justice , C-101/01, Bodil Lindqvist, 6 November 2003, O.J. 10 January 2004, C 7/3–7/4, available at http://curia.europa.eu .
25. Commission Nationale de l'Informatique et des Libertés, Délibération n°2005–284 du 11 novembre 2005 décidant la dispense de déclaration des sites web diffusant ou collectant des

données à caractère personnel mise en oeuvre par des particuliers dans le cadre d'une activité exclusivement personnelle (dispense n°6), J.O n° 293, 17 December 2005. http://www.cnil.fr . Accessed 16 April 2009.
26. College Bescherming Persoonsgegevens, CBP Richtsnoeren–Publicatie van persoonsgegevens op internet, December 2007. http://www.cbpweb.nl/downloads_rs/rs_20071211_persoonsgegevens_op_internet_definitief.pdf?refer=true&theme=purple . Accessed 10 April 2009.
27. Article 29 Working Party, Work Programme 2008–2009 of the Article 29 Working Party, WP146, 18 February 2008. http://ec.europa.eu/justice_home/fsj/privacy/workinggroup/wpdocs/2008_en.htm . Accessed 15 April 2009.
28. International Working Group on Data Protection in Telecommunications, Report and Guidance on Privacy in Social Network Services–"Rome Memorandum", adopted 3–4 March 2008, Rome. https://www.agpd.es/portalweb/canaldocumentacion/internacional/common/pdf/wp_social_network_services.pdf . Accessed 27 October 2008.
29. European Court of Human Rights, judgment of 16 October 2008, Taliadorou and Stylianou v. Cyprus, No. 39627/05
30. European Court of Human Rights, judgment of 22 February 1994, Burghartz v. Switzerland, No. 16213/90
31. Augsberg, I.: Article 8 of the Charter of Fundamental Rights. In: von der Groeben, H., Schwarze, J., Hatje, A. (eds.) Europäisches Unionsrecht, 7th edn. Nomos, Baden-Baden (2015)
32. Court of Justice of the European Union, judgment of 19 July 2012, Ahmed Mahamdia ./. Algeria, C-154/11. ECLI:EU:C:2012:491
33. Court of Justice of the European Union, judgment of 28 July 2016, VKI ./. Amazon EU, C-191/15, ECLI: EU:C:2016:612
34. Tolmie P, Crabtree A, Rodden T, Colley J (2016) 'This has to be the cats'—personal data legibility in networked sensing systems. CSCW 16:491–502
35. Papadopoulou E, Stobart A, Taylor NK, Howard Williams M. 2015 Enabling data subjects to remain data owners. Smart Innovation, Systems and Technologies, 239–248
36. Winter, S., Diamond, M., Green, J., Karasic, D., Reed, T., Whittle, S., et al. (2016). Transgender people: Health at the margins of society. The Lancet, 388(10042), 390–400
37. Köhler, R., Ehrt, J., & Cojocariu, C. (2017). 2nd revised legal gender recognition in Europe. Berlin: TGEU.
38. World Health Organization FAQ on Health and sexual diversity an introduction to key concepts. https://www.who.int/gender-equity-rights/news/sexual-gender-diversity-faq.pdf . Retrieved March 18 2021.
39. Reisner, S. L., Radix, A., & Deutsch, M. B. (2016). Integrated and gender-affirming transgender clinical care and research. Journal of acquired immune deficiency syndromes (1999), 72(Suppl3), S235
40. Durbach, N. (2013). Private lives, public records: Illegitimacy and the birth certificate in twentieth-century Britain. Twentieth Century British History, 25(2), 305–326
41. BBC News (2018). Portugal's parliament approves new legal gender change law. https://www.bbc.com/news/world-europe-43740417 . Accessed April 16 2019.
42. Rich, L. E., & Ashby, M. A. (2012). Today's "Sexmission": Bioethics and the quest for greater understanding of sexual and gender diversity. Journal of Bioethical Inquiry, 9(3), 229–233. https://doi.org/10.1007/s11673-012-9385-8
43. Vijfvinkel, M. M. (2016). Technology and the right to be forgotten. Master's Thesis, Radbod University, Nijmegen.
44. Şchiopu, Silviu-Dorin. Divergent views on the right to be digitally forgotten: the opinions of the general advocates vs. the decisions of the court of justice of the European Union. Law Review, vol. X, issue 2 (2020).
45. Lavelle, James J. Search query: can America accept a right to be forgotten as a publicity right? Brooklyn Law Review, vol. 83, issue 3 (2018).

46. Saifee, Seema. Penumbras, privacy, and the death of morals-based legislation: comparing U.S. constitutional law with inherent right of privacy in islamic jurisprudence. Fordham International Law Journal, vol. 27, no. 1 (2003).
47. Alsheeh, Eli. 'Do Not Enter Houses Other than Your Own': the evolution of the notion of a private domestic sphere in early Sunni Islamic thought. Islamic Law and Society, vol. 11, no. 3 (2004).
48. Junk, M.W. 2019. When Diversity Works: The Effects of Coalition Composition on the Success of Lobbying Coalitions. American Journal of Political Science 63 (3): 660–674.
49. Beyers, J., and I. De Bruycker. 2018. Lobbying Makes (Strange) Bedfellows: Explaining the Formation and Composition of Lobbying Coalitions in EU Legislative Politics. Political Studies 66 (4): 959–984.
50. Phinney, R. 2017. Strange Bedfellows: Interest Group Coalitions, Diverse Partners, and Influence in American Social Policy. Cambridge: Cambridge University Press.
51. Holyoke, T. 2009. Interest Group Competition and Coalition Formation. American Journal of Political Science 53 (2): 360–375.
52. Mahoney, C. 2007b. Networking vs. Allying: The Decision of Interest Groups to Join Coalitions in the US and the EU. Journal of European Public Policy 14 (3): 366–383.
53. Dür, A., and D. De Bièvre. 2007. The Question of Interest Group Influence. Journal of Public Policy 27 (1): 1–12.
54. Mahoney, C. 2007a. Lobbying Success in the United States and the European Union. Journal of Public Policy 27 (1): 35–56.
55. Mahoney, C. 2008. Brussels versus the Beltway: Advocacy in the United States and the European Union. Washington, D.C.: Georgetown University Press.
56. Michalowitz, I. 2007. What Determines Influence? Assessing Conditions for Decision-Making Influence of Interest Groups in the EU. Journal of European Public Policy 14 (1): 132–151.
57. Bunea, A. 2013. Issues, Preferences and Ties: Determinants of EU interest groups' preference attainment in the environmental policy area. Journal of European Public Policy 20 (4): 552–570.
58. Dür, A., P. Bernhagen, and D. Marshall. 2015. Interest Groups Success in the European Union: When (and why) does Business Lose? Comparative Politics 48 (8): 951–983.
59. Klüver, H., C. Braun, and J. Beyers. 2015. Legislative lobbying in context: towards a conceptual framework of interest group lobbying in the European Union. Journal of European Public Policy 22 (4): 447–461.
60. Klüver, H. 2013a. Lobbying as a collective enterprise: Winners and losers of policy formulation in the European Union. Journal of European Public Policy 20 (1): 59–76.
61. Klüver, H. 2013b. Lobbying in the European Union: Interest Groups, Lobbying Coalitions, and Policy Change. Oxford: Oxford University Press.
62. European Parliament. 2013. Report on the Proposal for a Regulation of the European Parliament and of the Council on the Protection of Individuals with Regard to the Processing of Personal Data and on the Free Movement of such Data (General Data Protection Regulation) (COM(2012) 0011 – C7–0025/2012 – 2012/0011(COD)). Brussels 21 (11): 2013.
63. European Parliament. (2016). Regulation (EU) 2016/679 of the European Parliament and of the Council of 27 April 2016 on the protection of natural persons with regard to the processing of personal data and on the free movement of such data, and repealing Directive 95/46. Official Journal of the European Union (OJ), 59(1–88), 294.
64. Weber, R. H. (2010). Internet of things-new security and privacy challenges. Computer Law & Security Review, 26(1), 23–30.
65. Korea Communications Commission. (2016). KCC takes measures to guarantee "right to be forgotten." Gwacheon-si: Korea Communications Commission.
66. Sleeper, M., Cranshaw, J., Kelley, P. G., Ur, B., Acquisti, A., Cranor, L. F., et al. (2013) I read my Twitter the next morning and was astonished: A conversational perspective on Twitter regrets. In Proceedings of the SIGCHI conference on human factors in computing systems, 2013 (pp. 3277–3286). ACM

67. Wang, Y., Norcie, G., Komanduri, S., Acquisti, A., Leon, P. G., & Cranor, L. F. (2011). I regretted the minute I pressed share: A qualitative study of regrets on Facebook. In Proceedings of the seventh symposium on usable privacy and security (pp. 10): ACM
68. DeNardis, L. (2014). The global war for internet governance. London: Yale University Press.
69. Ricoeur, Paul. 2004. Memory, history, forgetting. Chicago: University of Chicago Press.
70. Berkelaar, Brenda L. 2017. Different ways new information technologies influence conventional organizational practices and employment relationships: The case of cybervetting for personnel selection. Human Relations. https://doi.org/10.1177/0018726716686400 .
71. Kluemper, Donald H., and Peter A Rosen. 2009. Future employment selection methods: Evaluating social networking web sites. Journal of Managerial Psychology. https://doi.org/10.1108/02683940910974134 .
72. Rosen, Peter A., Shelby J. Solomon, Benjamin D. McLarty, Candace A. Esken, and Erik C. Taylor. 2018. The use of Twitter profiles to assess personality and hireability. Journal of Managerial Issues 30 (2): 256–272.
73. Becton, John Bret, H. Jack Walker, J. Bruce Gilstrap, and Paul H. Schwager. 2019. Social media snooping on job applicants. Personnel Review. https://doi.org/10.1108/PR-09-2017-0278 .
74. Kluemper, Donald H. 2013. Social network screening: Pitfalls, possibilities, and parallels in employment selection. Social Media in Human Resources Management (Advanced Series in Management). https://doi.org/10.1108/S1877-6361(2013)0000012005 .
75. Jones, Chris, and Susan Behling. 2010. Unchartered waters: Using social networks in hiring decisions. Issues in Information Systems. https://doi.org/10.48009/1_iis_2010_589-595 .
76. Lam, Helen. 2016. Social media dilemmas in the employment context. Employee Relations. https://doi.org/10.1108/ER-04-2015-0072 .
77. McDonald, Paula, and Paul Thompson. 2016. Social media(tion) and the reshaping of public/private boundaries in employment relations. International Journal of Management Reviews. https://doi.org/10.1111/ijmr.12061 .
78. Reagan, Charles E. E. 1998. Paul Ricoeur: His life and his work. Chicago: University of Chicago Press.
79. Lindsköld, Linnéa. 2018. Google as a political subject: The right to be forgotten debate 2014–2016. Online Information Review. https://doi.org/10.1108/OIR-06-2017-0198 .
80. Kim, Byung-Cheol, and Jin Yeub Kim. 2017. The economics of the right to be forgotten. The Journal of Law and Economics. https://doi.org/10.1086/694254 .
81. Hoffman, David, Paula Bruening, and Sophia Carter. 2016. The right to obscurity: How we can implement the Google Spain decision. North Carolina Journal of Law & Technology 17 (3): 437–482.
82. Uncular, Selen. 2019. The right to removal in the time of post-Google Spain: Myth or reality under general data protection regulation? International Review of Law Computers & Technology. https://doi.org/10.1080/13600869.2018.1533752 .
83. Faldetta, Guglielmo. 2021. Forgiving the unforgivable: The possibility of the 'unconditional' forgiveness in the workplace. Journal of Business Ethics. https://doi.org/10.1007/s10551-021-04885-2 .
84. European Data Protection Board. Frequently asked questions on the judgment of the court of justice of the European Union in case C-311/18—data protection commissioner vs Facebook Ireland Ltd and Maximillian Schrems. Policy. 2020.
85. European Data Protection Board. Recommendations 01/2020 on measures that supplement transfer tools to ensure compliance with the EU level of protection of personal data. Policy. 2020.
86. European Data Protection Board. Statement on the processing of personal data in the context of the COVID-19 outbreak. Policy. 2020.
87. Quinn P. The anonymisation of research data—a pyric victory for privacy that should not be pushed too hard by the EU data protection framework? Eur J Health Law. 2017;24:347–67.
88. Atkinson, W. (2003b). "Web-Based RFID: Hype and Glitz," in Operations & Fulfillment, (September 1).

89. Atkinson, W. (2003a). "DIS-connected," in Operations & Fulfillment, (May 1).
90. Allen, W. (2004). "Current and Anticipated Uses of RFID Technology," in FTC RFID Workshop, (June 21).
91. Fishkin, K. (2004). "RFID for Healthcare: Some Current and Anticipated Uses," in FTC RFID Workshop, (June 21).
92. Spethmann, B. (2004). "Loyalty's Royalty," in Promo, (March) 33–41, 83.
93. Siegel, J. (2003). "TraceElements," in Operations & Fulfillment, (May 1).
94. He, Zhicheng From Privacy-Enhancing to Health Data Utilisation: The Traces of Anonymisation and Pseudonymisation in EU Data Protection Law. Digital Society (2023). https://doi.org/10.1007/s44206-023-00043-5
95. Davis, John S.; Osoba, Osonde Improving privacy preservation policy in the modern information age. Health and Technology (2018). https://doi.org/10.1007/s12553-018-0250-6
96. Cimina, Veronique The data protection concepts of 'controller', 'processor' and 'joint controllership' under Regulation (EU) 2018/1725. ERA Forum (2020). https://doi.org/10.1007/s12027-020-00632-8
97. Gil González, Elena; de Hert, Paul Understanding the legal provisions that allow processing and profiling of personal data—an analysis of GDPR provisions and principles. ERA Forum (2019). https://doi.org/10.1007/s12027-018-0546-z
98. Bieresborn, Dirk The impact of the General Data Protection Regulation on Social Security. ERA Forum (2019). https://doi.org/10.1007/s12027-019-00565-x
99. Hildebrandt, Mireille Profiling and the rule of law. Identity in the Information Society (2008). https://doi.org/10.1007/s12394-008-0003-1
100. Van Alsenoy, Brendan; Ballet, Joris; Kuczerawy, Aleksandra; Dumortier, Jos Social networks and web 2.0: are users also bound by data protection regulations?. Identity in the Information Society (2009). https://doi.org/10.1007/s12394-009-0017-3
101. Wulf, Alexander J.; Seizov, Ognyan "Please understand we cannot provide further information": evaluating content and transparency of GDPR-mandated AI disclosures. AI & SOCIETY (2022). https://doi.org/10.1007/s00146-022-01424-z
102. von Hein, Jan; Bizer, Anna Social Media and the Protection of Privacy: Current Gaps and Future Directions in European Private International Law. International Journal of Data Science and Analytics (2018). https://doi.org/10.1007/s41060-018-0108-z
103. Urquhart, Lachlan; Sailaja, Neelima; McAuley, Derek Realising the right to data portability for the domestic Internet of things. Personal and Ubiquitous Computing (2017). https://doi.org/10.1007/s00779-017-1069-2
104. Correia, Mónica; Rêgo, Guilhermina; Nunes, Rui Gender Transition: Is There a Right to Be Forgotten?. Health Care Analysis (2021). https://doi.org/10.1007/s10728-021-00433-1
105. Osman, Amr The Right to be Forgotten: an Islamic Perspective. Human Rights Review (2022). https://doi.org/10.1007/s12142-022-00672-2
106. Christou, George; Rashid, Imir Interest group lobbying in the European Union: privacy, data protection and the right to be forgotten. Comparative European Politics (2021). https://doi.org/10.1057/s41295-021-00238-5
107. Kwak, Chanhee; Lee, Junyeong; Lee, Heeseok Could You Ever Forget Me? Why People Want to be Forgotten Online. Journal of Business Ethics (2021). https://doi.org/10.1007/s10551-021-04747-x
108. Habisch, André; Kletz, Pierre; Wack, Eva Unpleasant Memories on the Web in Employment Relations: A Ricoeurian Approach. Humanistic Management Journal (2022). https://doi.org/10.1007/s41463-022-00138-0
109. Hallinan, Dara; Bernier, Alexander; Cambon-Thomsen, Anne; Crawley, Francis P.; Dimitrova, Diana; Medeiros, Claudia Bauzer; Nilsonne, Gustav; Parker, Simon; Pickering, Brian; Rennes, Stéphanie International transfers of personal data for health research following Schrems II: a problem in need of a solution. European Journal of Human Genetics (2021). https://doi.org/10.1038/s41431-021-00893-y
110. Van den Bulck, Paul Transfers of personal data to third countries. ERA Forum (2017). https://doi.org/10.1007/s12027-017-0482-3

111. Kornbeck, Jacob Athlete consent as a legal base for data transfers to third countries for anti-doping purposes, under EU and German law. The International Sports Law Journal (2017). https://doi.org/10.1007/s40318-017-0112-9
112. Erickson, G. Scott; Kelly, Eileen P. International Aspects of Radio Frequency Identification Tags: Different Approaches to Bridging the Technology/Privacy Divide. Philosophy & Technology (2007). https://doi.org/10.1007/s12130-007-9011-y

Chapter 4
Data Protection—Issues and Perspectives

Introduction by the Editor

The objective is to place the data subject in focus while deciding the course of actions that protect and prevent the misuse of personal information. Often, the data controllers or those processing personal data work towards compliance within prescribed norms. These norms are set according to the legislative interventions and guidelines that jurisdictions provide. While there are common grounds in different jurisdictions, they have added their understanding in framing data protection norms. However, the operational practices following these norms are sometimes devoid of ground realities concerning the process that data subjects follow to perceive privacy and information surrounding its protection.[1] Imbibing data subjects' views and understanding are critical in practices that are meant to protect the sanctity of their personal space. For instance, in some contexts, data subjects may uphold anonymisation as a benchmark before sharing personal data.[2] This personal space is often mired by controversies when data subjects are vulnerable and require greater support from mature adults and data controllers.[3] Ethics and privacy issues exist because of the data subjects' age group and engagement with different data processing ecosystems.[4]

The extent of concerns connected with personal data processing is varied in nature. While the data subjects' rights prevail in most circumstances, the perspective of the data economy and use of data continue to ask questions about balancing data needs

[1] Annie I Antón and Julia B Earp, "A requirements taxonomy for reducing Web site privacy vulnerabilities" (2003) Requirements Engineering 9, 169–185.

[2] Eva-Maria Schomakers and others, "All of me? Users' preferences for privacy-preserving data markets and the importance of anonymity" (2020) Electronic Markets 30, 649–665.

[3] Terri Dowty, "Overlooking children: an experiment with consequences" (2008) Identity in the Information Society 1, 109–121; Jaime Fúster and others, "Analysis of security and privacy issues in wearables for minors" (2023) Wireless Networks 30, 5437–5453.

[4] Damian Murchan and Fazilat Siddiq, "A call to action: a systematic review of ethical and regulatory issues in using process data in educational assessment" (2021) Large-scale Assessments in Education 25.

I. Gupta (ed.), *Expectations vs Realities of Information Privacy and Data Protection Measures*, https://doi.org/10.1007/978-981-99-6778-0_4

with data control, including the use and reuse of data.[5] The overall idea of data protection legislation is giving back due control to an individual.[6] Several threats may hinder the transfer of such control at different stages of data processing.[7] The utility of a legislation increases that reduces such impediments and provides greater control to data subjects. Yet multiple factors are critical to its overall functioning, going much beyond disciplinary boundaries.[8] One such factor is over-dependence and over-estimation of how individuals exercise choice and decide about privacy. In this regard, the privacy design architecture adopts a discretionary approach, thereby providing options to the individuals. For multiple reasons, correctly choosing an option requires data subjects' expert knowledge and a lot of support from those involved in processing personal data.[9] At times, there is a lack of both these elements, leading to underutilisation of the perceptional autonomy that should be optimised with data protection legislation.

Aligning individuals in the ecosystem of inter-dependency, collaborative workspaces, social network sites,[10] and their overall implications on the protection of privacy rights are crucial and informed choices offered in this ecosystem are likely to have a positive impact.[11] There are issues about the role such individuals play in the context of the nature and amount of data they process and generate.[12]

The data subject's interest in personal space and personal information must be contextualised in the environments where they receive protection.[13] The settings where data subjects can exercise their choices vary greatly. It is rather difficult to

[5] Kieron O'Hara and others, "Lifelogging: Privacy and empowerment with memories for life" (2008) Identity in the Information Society 1, 155–172; Livia Iacovino and Malcolm Todd, "The long-term preservation of identifiable personal data: a comparative archival perspective on privacy regulatory models in the European Union, Australia, Canada and the United States" (2007) Archival Science 7, 107–127.

[6] Heng Xu and Sumeet Gupta, "The effects of privacy concerns and personal innovativeness on potential and experienced customers' adoption of location-based services" (2009) Electronic Markets 19, 137–149; Ann Cavoukian, "Privacy in the clouds" (2008) Identity in the Information Society 1, 89–108.

[7] I van Ooijen and Helena U Vrabec, "Does the GDPR Enhance Consumers' Control over Personal Data? An Analysis from a Behavioural Perspective" (2018) Journal of Consumer Policy 42, 91–107.

[8] J C Poindexter and others, "An experimental economics approach toward quantifying online privacy choices" (2006) Information Systems Frontiers 8, 363–374.

[9] Mauro Luis Gotsch and Marcus Schögel, "Addressing the privacy paradox on the organizational level: review and future directions" (2021) Management Review Quarterly 73, 263–296.

[10] Martin Pekárek and Stefanie Pötzsch, "A comparison of privacy issues in collaborative workspaces and social networks" (2009) Identity in the Information Society 2, 81–93.

[11] Anjuli Franz and Alexander Benlian, "Exploring interdependent privacy—Empirical insights into users' protection of others' privacy on online platforms" (2022) Electronic Markets (2022) 32, 2293–2309.

[12] Peter P Swire, "When Should "Consumers-as-Producers" Have to Comply With Consumer Protection Laws?" (2008) Journal of Consumer Policy 31, 473–487.

[13] Elin Palm, "Privacy Expectations at Work—What is Reasonable and Why?" (2008) Ethical Theory and Moral Practice 12, 201–215.

anticipate the gamut of data processing in the range of services data controllers offer.[14] The practicality of exercising choices and preferences in different complex ecosystems is overwhelming for data subjects and would, therefore, need extended support from the data controller.[15] It includes raising the bar of data security.[16] As a point of recovery, it seems raising user awareness is the preferred route added with the use of technology, which comes to the rescue of limiting data processing.[17] Privacy policies and other related documents hold prime importance in the realms of user awareness. In the context of awareness, bringing about transparency in data processing while using different digital tools can effectively increase user satisfaction.[18] The same argument of awareness extends to the issues of geoprivacy.[19] Knowledge is decisive in bringing about different approaches to exercising personal autonomy.[20] That brings the age-old debate of improving how users should be notified at the outset and further leads to the idea of optimal consent that data subjects share. In many ways, these processes are part of the circular argument, which starts with awareness and ends with the notice and consent framework. While this is an effective route and perhaps one of the better alternatives, it is far from being the complete solution to the challenges perceived before. It has to do with how the data subjects read privacy situations either by themselves or by having support from others. Termed as the privacy self-management approach, scholars have been critical of how the notice and consent framework has failed to uphold different values attached to the privacy of an individual.[21] It provides choices that are difficult for data subjects to make, and sometimes, the effects are felt beyond their lifetime.[22]

Overall, the idea of trust is at the core of all transactions in which the processing of personal data occurs. There could be different environments where the crucial

---

[14] Dag Slettemeås, "RFID—the "Next Step" in Consumer–Product Relations or Orwellian Nightmare? Challenges for Research and Policy" (2009) Journal of Consumer Policy 32, 219–244.

[15] Odnan Ref Sanchez and others, "A recommendation approach for user privacy preferences in the fitness domain" (2019) User Modeling and User-Adapted Interaction 30, 513–565.

[16] Alana Maurushat and Kathy Nguyen, "The legal obligation to provide timely security patching and automatic updates" (2022) International Cybersecurity Law Review 3, 437–465; Adrian Gregory, "Conserving customer value: Improving data security measures in business" (2008) Journal of Database Marketing and Customer Strategy Management 15, 233–238.

[17] Kounoudes Dini and others, "Enhancing user awareness on inferences obtained from fitness trackers data" (2023) User Modeling and User-Adapted Interaction 33, 967–1014.

[18] Frank Ebbers and others, "User preferences for privacy features in digital assistants" (2020) Electronic Markets 31, 411–426; Felix Beierle and others, "What data are smartphone users willing to share with researchers?" (2020) Journal of Ambient Intelligence and Humanized Computing 11, 2277–2289.

[19] Hongyu Zhang and Grant McKenzie, "Rehumanize geoprivacy: from disclosure control to human perception" (2023) GeoJournal 88, 189–208.

[20] H. Li and A. Nill, "Online Behavioral Targeting: Are Knowledgeable Consumers Willing to Sell Their Privacy?" (2020) Journal of Consumer Policy 43, 723–745.

[21] Gordon Hull, "Successful failure: what Foucault can teach us about privacy self-management in a world of Facebook and big data" (2015) Ethics and Information Technology 17, 89–101.

[22] Hiroshi Nakagawa and Akiko Orita, "Using deceased people's personal data" (2024) AI & SOCIETY 39, 1151–1169.

role of trust is validated.[23] Data subjects may feel much more comfortable sharing personal data in an environment where transparency is at the core of all services,[24] with data controllers setting security standards as the highest priority.

Significant developments are happening in several jurisdictions to balance privacy rights within and beyond the lifetime of data subjects with data needs and the overall application of data protection legislation. Of course, the EU has a long-standing approach to protecting the privacy of individuals, but other jurisdictions have followed closely to create a framework for processing personal sensitive data[25] and otherwise.[26] Whether EU data protection law impacted and inspired multiple jurisdictions is a point of research.[27] However, the foundations on which jurisdictions are either contemplating or have enacted their data protection law vary greatly.[28] It remains to be seen how multinational companies navigate their way through multiple legislation.[29] Further, jurisdictions are grappling[30] with issues like data localisation or local storage of certain categories of personal data[31] and the large-scale use of personal data as part of the public governance delivery mechanism.[32] Jurisdictions have assessed the effect of data breaches on data subjects, the conditions under which

---

[23] James Lappeman and others, "Sloane Trust and digital privacy: willingness to disclose personal information to banking chatbot services" (2023) Journal of Financial Services Marketing 28, 337–357; George Roussos and Theano Moussouri, "Consumer perceptions of privacy, security and trust in ubiquitous commerce" (2004) Personal and Ubiquitous Computing 8, 416–429.

[24] Hélène Millet, "Leveraging booking information for offer personalization and GDPR consequences" (2023) Journal of Revenue and Pricing Management 22, 152–156.

[25] Özlem Özkan and others, "Reflections on Turkish Personal Data Protection Law and Genetic Data in Focus Group Discussions" (2022) NanoEthics 16, 297–312; Laura Gianna Guntrum and others, "Using digitally mediated methods in sensitive contexts: a threat analysis and critical reflection on data security, privacy, and ethical concerns in the case of Afghanistan" (2022) Zeitschrift für Friedens- und Konfliktforschung 11, 95–128.

[26] Marilyn Prosch, "Protecting personal information using Generally Accepted Privacy Principles (GAPP) and Continuous Control Monitoring to enhance corporate governance" (2008) International Journal of Disclosure and Governance 5, 153–166; Jean Philbert Nsengimana, "Reflections upon periclitations in privacy: perspectives from Rwanda's digital transformation" (2017) Health and Technology 7, 377–388.

[27] Giulio Vittorio Cervi, "Why and How Does the EU Rule Global Digital Policy: an Empirical Analysis of EU Regulatory Influence in Data Protection Laws" (2022) Digital Society (2022) 1(18).

[28] James Castro Edwards, "Data protection: Where are we now?" (2008) Journal of Database Marketing and Customer Strategy Management 15, 285–292.

[29] Philip Andreas Weber and others, "A comparative analysis of personal data protection regulations between the EU and China" (2020) Electronic Commerce Research 20, 565–587; Vibhushinie Bentotahewa and others, "The Normative Power of the GDPR: A Case Study of Data Protection Laws of South Asian Countries" (2022) SN Computer Science 3(183).

[30] Tathagata Satpathy, "The Aadhaar: "Evil" Embodied as Law" (2017) Health and Technology 7, 469–487.

[31] Rishab Bailey and Smriti Parsheera, "Data localization in India: paradigms and processes" (2021) CSI Transactions on ICT 9, 137–150.

[32] Pam Dixon, "A Failure to "Do No Harm"—India's Aadhaar biometric ID program and its inability to protect privacy in relation to measures in Europe and the U.S." (2017) Health and Technology 7, 539–567.

they are willing to share data after a data breach[33] and how data breach incidents can be reduced by following different standardised practices internally. There has been research on institutional responses to privacy and ethical issues holistically to address the existing prejudices and protection of vulnerable groups. Some research has discussed the time during COVID-19 and the matters about processing health data vis-à-vis data protection norms.[34] During the global pandemic, jurisdictions have struggled to overcome the barriers preventing health data sharing vis-à-vis the existing data protection framework.[35] Research has suggested the need for health practices requiring identifiable data in different economies.[36] Further, the impact of advanced technology on data protection has been captured extensively with the need to introduce additional safeguards.[37] There are existing challenges pertaining to rolling out of Facial Recognition Technologies (FRT) and how greater transparency and ethical norms are expected to mitigate certain negative impacts.[38] There has been a concentrated effort to strengthen enforcement mechanisms in several jurisdictions.[39] Regardless, reaching a consensus about the basics of data protection principles is crucial in a digital space like ours.

This chapter captures all of those mentioned above and brings nuanced arguments about the ever-increasing dimensions that jurisdictions need to consider in the context of data protection measures.[40]

Machine Generated Summaries

Disclaimer: The summaries in this chapter were generated from Springer Nature publications using extractive AI auto-summarization: An extraction-based summarizer aims to identify the most important sentences of a text using an algorithm and

---

[33] Shilpa Madan and others, "Privacy please: Power distance and people's responses to data breaches across countries" (2022) Journal of International Business Studies 54, 731–754.

[34] Dana Naous and others, "Learning From the Past to Improve the Future" (2022) Business & Information Systems Engineering 64, 597–614.

[35] Alexander Bernier and Bartha Maria Knoppers, "Pandemics, privacy, and public health research" (2020) Canadian Journal of Public Health 111, 454–457.

[36] Philip AbdelMalik and others, "The perceived impact of location privacy: A web-based survey of public health perspectives and requirements in the UK and Canada" (2008) BMC Public Health 8(156).

[37] Jacob Dexe and others, "Explaining automated decision-making: a multinational study of the GDPR right to meaningful information" (2022) The Geneva Papers on Risk and Insurance—Issues and Practice 47, 669–697; Pompeu Casanovas and others, "Regulation of Big Data: Perspectives on strategy, policy, law and privacy" (2017) Health and Technology 7, 335–349.

[38] Denise Almeida and others, "The ethics of facial recognition technologies, surveillance, and accountability in an age of artificial intelligence: a comparative analysis of US, EU, and UK regulatory frameworks" (2022) AI and Ethics 2, 377–387.

[39] Ewan Nettleton and Ian Turner, "Data protection: Tougher enforcement and increased power for the Information Commissioner?" (2008) Journal of Database Marketing & Customer Strategy Management 15, 207–212.

[40] Darra Hofman and Michele A L Villagran, "Queer Privacy Protection: Challenges and the Fight within Libraries" (2023) International Journal for the Semiotics of Law—Revue internationale de Sémiotique juridique 36, 2157–2178.

uses those original sentences to create the auto-summary (unlike generative AI). As the constituted sentences are machine selected, they may not fully reflect the body of the work, so we strongly advise that the original content is read and cited. The auto generated summaries were curated by the editor to meet Springer Nature publication standards. To cite this content, please refer to the original papers.

Machine generated keywords: user, consumer, privacy, digital, customer, sector, technology, breach, identity, system, privacy protection, privacy concern, device, threat, security.

4.1 Keeping Data Subjects' Lens in Focus

Machine generated keywords: user, consumer, privacy, device, preference, customer, location, privacy concern, site, web, app, technology, adoption, mobile, willing.

Does the GDPR Enhance Consumers' Control over Personal Data? An Analysis from a Behavioural Perspective [308]

This is a machine-generated summary of:

van Ooijen, I.; Vrabec, Helena U.: Does the GDPR Enhance Consumers' Control over Personal Data? An Analysis from a Behavioural Perspective [308].

Published in: Journal of Consumer Policy (2018).

Link to original: https://doi.org/10.1007/s10603-018-9399-7.

If you want to cite the papers, please refer to the original.

For technical reasons we could not place the page where the original quote is coming from.

Abstract-Summary

"Individual control over personal data has become an important subject in European privacy law Compared to its predecessor, the General Data Protection Regulation (GDPR) addresses the need for more individual control over personal data more explicitly."

"With the introduction of several new principles that seem to empower individuals in gaining more control over their data, its changes relative to its predecessors are substantial."

"It appears, however, that, to increase individual control, data protection law relies on certain assumptions about human decision making."

"Further, we analyse the extent to which new provisions in the GDPR effectively enhance individual control through a behavioural lens."

"For each stage, we identify the pitfalls of human decision-making that typically emerge and form a threat to individual control."

"Keeping in mind the pitfalls in human decision-making, we assess how effective the new legal provisions are in enhancing individual control."

Individual Control and the GDPR

"Observing these trends, behavioural scientists have warned for a number of threats to individual control, such as information overload and data invisibility [1]."

"It has been considered desirable that individuals are able to exert some form of control over their personal data."

"The contribution of this paper is to critically examine the EU mission to enhance individual control in the data-driven economy by assessing the effectiveness of the GDPR from a behavioural perspective."

"In the approval and primary use stage, the context wherein decisions about personal data are made, as well as the manner in which data collection requests are framed, affect individual control."

"We state that each stage is associated with certain pitfalls that result from cognitive processing and human decision-making, which pose threats to individual control over personal data."

"We end with concluding on the effectiveness of the GDPR in enhancing actual individual control over personal data."

Provisions in the GDPR That are Related to Control

"Data subject rights, also referred to as subjective rights or control-rights [2], are a catalogue of eight entitlements split into three sections: (1) information and access to personal data, (2) rectification and erasure, and (3) the right to object and automated individual decision-making."

"The right to information is a key source of control, preparing a data subject for the data processing and its consequences."

"Consent facilitates control in the second stage of data processing, as it means yes or no for subsequent data processing."

"The following sections will explain how the characteristics of human cognitive processing and decision-making form a problem for individuals to gain control over their personal data in the contemporary data-economy."

Threats to Individual Control in the Information Receiving Stage

"In the information receiving stage, data collectors may provide the individual with this kind of information by means of a data use policy."

"Difficulties in cognitively processing information about data collection are an evident threat to individual control."

"Paradoxically, the more information individuals have access to about what happens to their (personal) data, the less information they are able to filter, process, and weigh to make decisions that are in line with their own privacy preferences."

"When it comes to privacy and data protection mechanisms, such as disclosures, policies, and standard forms, parties that provide the information, including regulators, assume that consumers (are able to) process information extensively."

"In order to make informed decisions, individuals should estimate the expected benefits and sacrifices that are associated with data disclosure and, subsequently, decide whether these are in line with their attitudes and preferences."

How Does the GDPR Address These Threats in the Information Receiving Stage?

"The GDPR provides three articles on information provision that aim to enhance individual control in this stage, namely the right to explanation."

"Under the GDPR, a data controller should also provide the information about "... the existence of automated decision making, including profiling [...] and at least in those cases, meaningful information about the logic involved as well as the significance and the envisaged consequence" (p. 42)."

"Ex post explanation would mean that a data subject would be informed about the logic and individual circumstances of their specific decision, the data or features that were considered in her particular case, and their weighting within the decision tree or classification structure."

"That means that by focusing too much on providing easy-to-understand information, icons could actually increase the problem of data invisibility and intangibility."

Threats to Individual Control at the Approval and Primary Use Stage

"Privacy invasive default settings can hardly result in individual control over data."

"It is likely that contextual cues that signal control, such as the mere existence of a privacy policy, implicitly signal trustworthiness, which in turn decreases privacy concerns and increases disclosure behaviour."

"These findings clearly indicate that communicating to individuals that they are in control is not enough to enhance actual control–in some cases, it may wrongfully signal trustworthiness and decrease privacy concerns in turn."

"Communicating control to individuals by means of choice architecture and design should always go together with a situation that actually enhances control, such as data protective defaults."

"The use of contextual cues and choice architecture relates to individual control in several ways."

"These factors change privacy preferences especially when people engage in intuitive, uncontrolled information processing and thus exploit the lack of individual control."

How Does the GDPR Address These Threats at the Approval and Primary Use Stage?

"What this may mean in practical terms is indicated in Recital 32, which states that unambiguous consent may be signified by ticking a box or by making a statement that clearly indicates acceptance of data processing for the specific purpose of processing at hand."

"To prohibiting pre-ticked boxes (i.e., opt-out) as consent, the GDPR embeds the more general "privacy by default" principle in Article 25, where it states: "The controller shall implement appropriate technical and organisational measures for ensuring that, by default, only personal data which are necessary for each specific purpose of the processing are processed.[...] In particular, such measures shall ensure that by default personal data are not made accessible without the individual's intervention to an indefinite number of natural persons" (p. 48)."

Control at the Data Reuse Stage

"A compelling example of data misuse is the case of Cambridge Analytica, the company that used personal information of millions of U.S. Facebook users without authorization, and used this information to profile and target users with personalized political advertisements."

"Looking at these trends, threats to individual control over personal data increasingly emerge in the third stage."

"Affordances of generated personal data in terms of its intangibility, invisibility, and scope and flow restrict individual control in the reuse stage."

"To consumers, the personal data that are being collected by parties are permanently invisible, especially when it comes to behavioural data."

"After data has been collected, it is processed and, in the third stage, it flows to third party data processors or to new controllers."

"In terms of the scope and flow of data streams, individual control is certainly challenged."

How Does the GDPR Address Threats to Control at the Data (Re)Use Stage?

"Control rights as defined in the GDPR are of particular importance for individuals, because they can ensure control over personal data also in the later stages of data processing."

"Control rights that particularly aim to increase individual control are the right to data portability, the right to access and the right to erasure."

"To some extent, the right to data access and portability could solve the problem of control in terms of increased invisibility and clarification of the scope of data streams."

"The right to data erasure, and especially the obligation of the data controller to notify other parties of data erasure responsibilities, increases individual control of the scope and flow of these data."

"Under the GDPR, individuals have more control in terms of the scope and flow of their data in situations where data processing is unlawful and the right to erasure applies."

Conclusion and Discussion

"The General Data Protection Regulation emphasises the importance of enhancing individual control in the data economy."

"We have analysed a number of relevant provisions in this regulation and have determined the extent to which they enhance individual control over consent-based data processing."

"We have examined how the provisions address threats to individual control that are increasingly present in the data economy and that appear in the three stages of the data processing timeline: (1) the information receiving stage, (2) the approval and primary use stage, and (3) the data secondary use (reuse) stage."

"This work provides an aid in exposing and addressing the problems that arise with regard to individual control in the data-driven economy."

"Although the primary aim of this work is not to provide specific solutions to the problem of declining individual control, we hope to have sparked a discussion that stimulates such cooperation and helps addressing the pressing problem of weakening individual control in the data-driven economy."

Leveraging booking information for offer personalization and GDPR consequences [309]

This is a machine-generated summary of:

Millet, Hélène: Leveraging booking information for offer personalization and GDPR consequences [309]

Published in: Journal of Revenue and Pricing Management (2022).

Link to original: https://doi.org/10.1057/s41272-022-00401-x.

If you want to cite the papers, please refer to the original.

For technical reasons we could not place the page where the original quote is coming from.

Abstract-Summary

"By creating personalized offers, airlines aim at offering their customers what they are the most likely to buy."

"Knowing customers is usually considered as a mandatory step to achieve that goal."

"We will examine Passenger Name Record (PNR) data as a source of information, and how it can be used for a better customer knowledge."

"PNR data's content is rich: not only on whom the customer is and on what he bought, but also on the history of his/her purchase (when he/she bought, canceled, changed)."

"Once cleaned and organized, PNR data provide a solid base for personalization."

"Making "personalized proposals" can come down to fitting what the customer is looking for."

"Statistical analysis of anonymous data and context usage can be combined for best results Anonymized data can be used to qualify consistent groups of customers, using very simple clustering methods."

"The content of a shopping request (= the context) can then be used to connect the customer to one of the identified groups."

PNR as source of information

"The minute the customer books a flight, a PNR (Passenger Name Record) is created, and the airline receives a track of his actions."

"The PNR is, therefore, a minimal source of data for a minimal (transactional only) but exhaustive knowledge of the customer booking behavior."

"Customer behavior and history of the PNR: changes, cancelations, disruptions Depending on the airline (and its contract with its PSS), those data can be available offline (once a day) or online (each single change generates a update message)."

"During several years, PNR data were gathered for a couple of small and medium size carriers, from any distribution channel, data cleaned and processed to analyze unique customers."

"The added value of studying PNR data (vs limiting customer information to FFP) appeared as obvious: most frequent travelers had not subscribed to any FFP program (up to seventy percents), if one was offered by the airline, which was not always the case."

How personal data usage for personalization is challenged with data privacy rules

"There are myriads of legislations across different markets and customers are increasingly aware of data privacy issues and thus more cautious when and with whom they share their data."

"To be enforced starting May 2018, the European GDPR (General Data Protection Regulation), establishes rules that should give customers more transparency regarding the use of their personal data, ensure higher level of security and give customers back control over the data they agree to share or not."

"The real challenge is to ensure that they engage their customers both in an understandable and fully transparent way at the same time about how they process personal data."

"That privacy matters need to be handled from the start: dealing with private data, one need to ensure complete accessibility for customers: they must have access to their information, for consultancy, modification and even cancelation, at any time."

Take the best out of it

"We also see that the multiplicity of sources for airline data, even for transactional information such as PNR, makes it difficult to get a clear consent from the customer for every transaction."

"Airlines need first to use the information at their disposal at best, with a few rules: 1."

"Anonymized data (considering that it is not possible to track back the customer), already grouped data: for corporates, for groups."

"Use anonymized data to create groups as explained, but never directly target customer for historical assets, use his/her live characteristics to calculate least distance from a persona (proactive personalization, rather than reactive) 3."

"Directly use information, do not store private data, whenever possible."

"In a GDPR world, to engage with customers, as an airline, you need to make sure: It is easy for them to share/correct information Whenever they share information, they see the value it brought them."

[Section 4]

"Personalized offers are now part of our daily life."

"As in many verticals, personalization has made the buzz in travel for years now: a reason to modernize distribution, to facilitate bundling or unbundling services, and to find partners to complete an offer."

"For an airline to be able to offer the customer what he wants at the minute he wants it, those are mandatory improvements."

"The first step to personalized offers is better knowing the requester."

Successful failure: what Foucault can teach us about privacy self-management in a world of Facebook and big data [310]

This is a machine-generated summary of:

Hull, Gordon: Successful failure: what Foucault can teach us about privacy self-management in a world of Facebook and big data [310]

Published in: Ethics and Information Technology (2015).

Link to original: https://doi.org/10.1007/s10676-015-9363-z.

If you want to cite the papers, please refer to the original.

For technical reasons we could not place the page where the original quote is coming from.

Abstract-Summary

"I argue that the privacy paradox masks a more important paradox: the self-management model of privacy embedded in notice-and-consent pages on websites and other, analogous practices can be readily shown to underprotect privacy, even in the economic terms favored by its advocates."

"The real question, then, is why privacy self-management occupies such a prominent position in privacy law and regulation."

"Privacy self-management isn't about protecting people's privacy; it's about inculcating the idea that privacy is an individual, commodified good that can be traded for other market goods."

"Along the way, the self-management regime forces privacy into the market, obstructs the functioning of other, more social, understandings of privacy, and occludes the various ways that individuals attempt to resist adopting the market-based view of themselves and their privacy."

Why we have a privacy paradox: the failure of privacy self-management

"Are three types of reasons why self-management can be expected to underprotect privacy: (1) users do not and cannot know what they are consenting to; (2) privacy preferences are very difficult to effectuate; and (3) declining to participate in privacy-harming websites is increasingly not a viable option for many."

"Users do not and cannot plausibly be expected to know enough—neither about the uses to which their information might be put, nor about the specific benefits and harms that might result from those uses, nor about the likelihood that such harms might result—for consent to be meaningful, especially if one makes the assumption that those users are following a risk/benefit model of economic rationality."

"According to one recent report [3], for example, Facebook now allows users to opt-out of sharing their information with data brokers."

Privacy as a social value

"It is also a social value: a society in which some level of privacy is present is better not just for the individuals who assign value to their privacy, but for everyone else too."

"Research on Facebook underscores the gap between these social practices and privacy self-management."

"Privacy self-management does not just obscure the social benefits of privacy, it actively subverts them: particularly once certain tipping points are reached, it will almost always be individually rational to forego privacy, even as the social benefits of privacy remain collectively rational."

"An important feature of the preceding analyses is epistemic: not only are individuals incapable of knowing the economic value of their privacy, either to themselves or to others (as noted in the previous section), but the social benefits of privacy are also both substantial and difficult to quantify."

Privacy and subjectification

"Notice and consent, then, fails completely as a strategy for privacy protection: it presents users with choices they cannot rationally make, and consistently fails to make legible the many social reasons why privacy is valuable."

"Its empirical success in protecting privacy or not is less relevant than the ways it presents an information environment in which individuals see themselves as functioning autonomously, treating personal information as a low-value, purely personal good they can easily trade for other desired goods and services."

"The process thus repeats the presentation of privacy as an economic decision, both occluding the actual uncertainties users face and offering the possibility of corrective strategies on the part of websites wanting to elicit "rational" behavior."

Conclusion: A note about resistance

"Not only does Facebook demand that users be their offline selves on the site, it both makes it notoriously hard to change privacy settings, and it changes the available settings routinely, a practice that dramatically increases the cost of efforts at privacy self-management."

"It is no surprise that users appear to have given up on the task of protecting themselves from Facebook, but outrage over Beacon and the emotional contagion study suggests both that users do in fact care, and that the site does a good job making it nearly impossible to achieve meaningful privacy protection against the corporation itself."

"By viewing the loss or preservation of privacy as a commercial transaction, and then treating the transaction as revealing consumer preference, the privacy self-management model obscures a social struggle, repackaging it as a well-functioning market."

"It is also part of why insisting on the evidentiary value of privacy self-management is so pernicious: doing so both actively occludes possibilities for resistance and enables a narrative of consumer preferences based on that occlusion."

[Section 5]

"How is it that the socially standard regimes of privacy protection featured in "notice and consent" policies and other examples of "privacy self-management" so completely fail to represent how individuals actually conceptualize and attempt to manage their privacy?"

"I will argue that the current reliance on privacy self-management, epitomized by notice and consent regimes, not only completely fails to protect privacy, but that it does so in a way that encourages adherence to several core neoliberal techniques of power: the belief that privacy can only be treated in terms of individual economic choices to disclose information; the occlusion of the fact that these choices are demonstrably impossible to make in the manner imagined; and the occlusion of the ways that privacy has social value outside whatever benefits or losses may accrue to individuals."

Privacy Expectations at Work—What is Reasonable and Why? [311]

This is a machine-generated summary of:

Palm, Elin: Privacy Expectations at Work—What is Reasonable and Why? [311].

Published in: Ethical Theory and Moral Practice (2008).

Link to original: https://doi.org/10.1007/s10677-008-9129-3.

If you want to cite the papers, please refer to the original.

For technical reasons we could not place the page where the original quote is coming from.

Abstract-Summary

"Privacy claims in the semi-public area of work have not been sufficiently investigated."

"The case is made that employees have reasonable expectations on privacy at work."

"A normative analysis explicates the reasons why privacy should be protected."

"The main thrust is to provide a more inclusive privacy concept and hence, a more adequate basis for privacy protection legislation and codes in the area of work."

"Contrary to prevailing workplace privacy protection, employees' need for local privacy should be accommodated as well as informational privacy."

Introduction

"Existing laws and codes typically focus on the collection and processing of personal data that should remain with the employee or within her discretionary power (cf. Council of Europe; Convention on Human Rights and Fundamental Freedoms, 1950, Convention for the Protection of Individuals with regard to Automatic Processing of Personal Data, 1981, Recommendation on the protection of personal data used for employment purposes 1989, The EU Directive 95/46/EC, The ILO code on Protection of Workers' Data, 1996)."

"In order to safeguard privacy at work, "local privacy"—employees' possibilities of withdrawing to secluded areas and to use certain areas as if private must be accommodated as well as "informational privacy"—workers' chances of controlling certain data and aspects of themselves."

"Contrary to how local privacy traditionally has been associated with interests regarding the domestic sphere, local privacy in this article includes (but is not restricted to) freedom from unwarranted intrusion or observation within certain realms at work."

"In contrast to the predominating focus on protection of unique identifiers and socially stigmatizing personal data, informational privacy concerns personal aspects that employees wish to keep to themselves."

Privacy in the Realm of Work

"In below, documented experiences of surveillance at work will be discussed alongside general intuitions regarding what employees are likely to consider privacy sensitive in the context of work."

"Although, for practical reasons, employers may not be able to offer shop or factory floor workers full seclusion, those employees may still expect to leave the pay-desk or conveyer belt in order to conduct certain work tasks in isolation."

"Privacy has a subjective element and it should also be noted that the expectations that employees have and the claims that they raise are likely to vary, not only with work culture (e.g. high or low trust environment) but with the employees' sex, cultural background, status and position within a firm and depending on the type of work that they carry out."

Why Privacy is Important at Work

"In relation to employers and managers, employees are typically in a disadvantaged position to control personal information and to advocate their privacy interests, hence to influence these particular relations."

"These aspects correlate well with some of the most common accounts of why privacy is needed—accounts that draw on the importance for individuals to be able to control personal information and other peoples' access to data about themselves [4–12]."

"When investigating employees' need for informational privacy and local privacy respectively, it becomes clear that both aspects are needed in order for workers to control and restrict other's access to them and thereby to shape their professional role and establish and influence the various relations in the context of work."

"More precisely, privacy is the spatial and intellectual sphere that enables individuals to decide over matters that concern them, to control who has access over information about themselves (both direct sensory access and stored data) and to establish and develop different types of relations."

Reasonable Expectations of Privacy

"Having established that workers' privacy ought to be protected for the reason that personal autonomy will be secured in that way, it remains to explicate how privacy should be weighed against other, conflicting interests potentially arising in the context of work."

"Employees' expectations of privacy at work are typically complicated by the fact that employers and managers, for several reasons, must require data from employees and keep a certain amount of control over their doings."

"One rather grave reason for employers to go against employees' expectations of privacy is their responsibility for the health and safety of their workers as well as of third parties."

"Few cases except those of employers' duties towards employees and third parties make up convincing cases for privacy intrusions/surveillance."

Conclusion

"Workspace privacy contains more aspects than what traditionally has been recognized in privacy protection laws and codes, and, in order to protect workers' privacy properly, a broader understanding of privacy is necessary."

"Even though there can be valid reasons for circumscribing workers' privacy (cf. [13]), privacy should be respected and protected as long as it does not violate rights and liberties of others."

"Not only single employees but also firms and society as a whole are likely to be negatively effected by an improper protection of workers' privacy."

"The main contribution of this paper however is that of establishing what privacy at work consists in and why it should be protected, thereby providing a more adequate conceptual basis for legal and other types of protection of workers' privacy."

A recommendation approach for user privacy preferences in the fitness domain [312]

This is a machine-generated summary of:

Sanchez, Odnan Ref; Torre, Ilaria; He, Yangyang; Knijnenburg, Bart P.: A recommendation approach for user privacy preferences in the fitness domain [312].

Published in: User Modeling and User-Adapted Interaction (2019).

Link to original: https://doi.org/10.1007/s11257-019-09246-3.

If you want to cite the papers, please refer to the original.

For technical reasons we could not place the page where the original quote is coming from.

Abstract-Summary

"As fitness-related data are persistently captured, stored, and processed by these devices, the need to ensure users' privacy is becoming increasingly urgent."

"We apply a data-driven approach to the development of privacy-setting recommendations for fitness devices."

"We first present a fitness data privacy model that we defined to represent users' privacy preferences in a way that is unambiguous, compliant with the European Union's General Data Protection Regulation (GDPR), and able to represent both the user and the third party preferences."

"We then examine different personal tracking data and user traits which can potentially drive the recommendation of privacy profiles to the users."

"Even though several works exist on privacy preference modeling, this paper makes a contribution in modeling privacy preferences for data sharing and processing in the IoT and fitness domain, with specific attention to GDPR compliance."

"The identification of well-identified clusters of preferences and predictors of such clusters is a relevant contribution for user profiling and for the design of interactive recommendation strategies that aim to balance users' control over their privacy permissions and the simplicity of setting these permissions."

Introduction

"Building on privacy management studies in the field of ubiquitous computing, this paper aims to fill this gap by modeling users' privacy preferences and recommending privacy settings in a fitness IoT scenario."

"The features for privacy preference modeling considered in this paper are based on the Privacy Preference for IoT ontology (PPIoT) [14] with the aim to unambiguously identify privacy preferences and data sharing permissions for both the user and TPs."

"The aim is to suggest privacy settings for tracking data that fit the user's preferences, are GDPR compliant, and reduce the effort for users to set such permissions, thereby maintaining meaningful control over their preferences without unnecessary burden."

Background and related work

"Several solutions have been proposed to improve mobile privacy protection and offer users more privacy control (e.g., [15])."

"IpShield [16] is a context-aware privacy framework for mobile systems that provides users with great control over their data and inference risks."

"Nudging allows users to be informed about both their privacy settings and how TP applications access their data [17]."

"By identifying the privacy profile that matches a new user, one can provide decision support by means of a privacy recommendation [17]."

"In a subsequent paper, the authors add new features (such as the purpose of information and app categories) for modeling user privacy profiles, as well as privacy nudges that make users more aware of unexpected data practices from TPs [17]."

Research methodology

"RQ2 Is it possible to identify well-defined privacy profiles that can represent the diversity of users' privacy preferences?"

"We conduct an unsupervised machine learning analysis (clustering) to cluster users' privacy settings into distinct profiles by recruiting a total of 310 Fitbit users."

"We collect privacy profile data by developing a fitness app installation simulator (FitPro) that captures the user privacy permission settings, followed by a questionnaire."

"We conduct a supervised machine learning analysis (tree learning) to find privacy profile items and questionnaire items (i.e., privacy attitude, negotiability, social behavior, exercise tendencies, and demographics) that best predict the user profiles from RQ2."

"Users' answers to these settings/questions subsequently allow us to provide them an accurate recommendation of the profile most suitable as a starting point for their privacy settings."

Fitness data privacy model

"The data items in the F set are by default only available in the tracker's own application, but other TPs can ask for permission to gain access to this data."

"That the F set permissions are repeated for each additional TP that requests access to this data."

"These permissions are not for the native app of the fitness tracker, but for other TP apps that the user desires to use and allow access to her/his fitness tracking data."

"Our data model, which takes a superset of permissions asked by all four fitness trackers, more granular Activity and Exercise data, and the additional G set, include 45 permissions in total."

"If users want to share their fitness data (F set) with one or more additional health or fitness tracking apps, the permissions for this must be decided upon for each additional TP individually."

Data collection

"The goal of the FitPro prototype fitness app is to collect privacy preferences of the participants in a semi-realistic environment."

"The goal was to investigate if certain user traits, some of them already studied in the literature, have relations with participants' privacy behaviors collected through FitPro."

"We aimed to measure participants' privacy-related attitudes (trust, privacy concerns, perceived surveillance and intrusion, and concerns about the secondary use of personal information), the negotiability of their privacy settings, their social behavior (social influence and sociability), exercise tendencies (a proxy for their attitude and knowledge about fitness tracking), and demographic information."

"Domain experts tend to be less concerned about their privacy than domain novices; hence, we expect an association between these questions and participants' privacy settings."

"In our questionnaire, we introduce this category to investigate if there is an association between participants' privacy settings and their demographic attributes, as resulted in previous studies (cf. [18])."

Privacy preference profiles

"Participants are more likely to give permission to share their data with their friends on social networks and to other health/fitness apps, and they are less likely to give permission to share their data with government fitness programs or publicly on social media."

"We expect that clustering F permissions should provide an unanimous deny/share for all items, while clustering G permissions should provide more nuanced clusters of different entity types receiving the data specified in the F set."

"Another modeling decision must be highlighted as well: In our dataset, the fitness data permissions (F set) are specified repeatedly for each entity type (part of the G set)."

"Unconcerned (cluster B): This subprofile shares all data requested in this dataset."

"Unconcerned (cluster A): This subprofile shares all fitness data with TPs."

"Strict (cluster B): This subprofile does not share any fitness data with TPs."

"Socially active (cluster A): This subprofile shares data with health/fitness apps and social network friends, but not with other recipients."

Profile prediction

"For each approach and for each subset of data (S, A, F, and G sets), we develop decision trees that will enable us to predict which subprofile best describes a user."

"The J48 algorithm extracts the permission items (for the direct prediction) or questionnaire items (for the indirect prediction) that classify a new user into the correct subprofile with the highest possible accuracy."

"This is expected, since unlike the direct prediction questions, the questionnaire items about user traits have no direct relationship with the permission settings in the privacy profiles."

"We also attempted to use the negotiability of users' privacy settings as input for subprofile prediction."

"For the G set, users who share their fitness app profile with other TPs if the risks decrease are classified into the "Socially active" subprofile, otherwise they are classified into the "Health-focused" subprofile."

Recommendation strategies and validation

"If users do this correctly, these manual settings should match their privacy preferences 100%."

"The next strategies exploit the results of the analysis in the previous section to provide interactive recommendations that simplify the task of privacy permission setting, with different levels and type of user intervention."

"Our first approach in this regard is to have users manually select which privacy subprofiles they prefer."

"An interactive 4-question input sequence is created to collect users' privacy-setting preference for the indirect prediction approach."

"The benefit of the indirect prediction approach is that the user does not have to answer any permission questions, not even the four needed in the direct approach to give them a subprofile recommendation."

"This approach has the highest accuracy, because only the error from the difference between the privacy profile and the users' settings is counted, omitting the errors introduced by the user classification."

Discussion, limitations, and future work

"Our privacy recommendation procedure uses the privacy calculus as a "prescriptive" model: Instead of burdening users with the task of balancing privacy and benefit, our recommendation strategy provides users with a guidance to the privacy calculus, based on an analysis of the decisions of a large number of other users (which is used to generate the profiles and the profile-assignment strategy)."

"We have conjectured that profile-based approaches reduce the hassle of making privacy settings but that the manual selection of a privacy profile might be difficult for a user."

"We are currently working on extending the use case scenarios in order to comprehensively evaluate the PDM's feasibility to model the user and TP privacy preferences in the domain of IoT personal devices."

Conclusion and contribution

"We presented a data-driven approach to the development of recommendation strategies for supporting users to set permissions regarding their personal data collected and shared by tracking devices in the fitness domain."

"These recommendation approaches are aligned with the PPIoT vocabulary: The data model and the recommendation strategies will be used by the PDM to represent the user privacy preferences and recommend privacy settings."

"Even though several works exist on privacy preference modeling, this paper makes a contribution in modeling privacy preferences for data sharing and processing of tracked data in the IoT and fitness domain, with specific attention to GDPR compliance."

"The identification of well-defined clusters of preferences and predictors of such clusters is a relevant contribution for the design of recommendation strategies and interactive user interfaces that aim to balance users' control over their privacy permissions and the simplicity of setting these permissions."

Enhancing user awareness on inferences obtained from fitness trackers data [313]

This is a machine-generated summary of:

Dini Kounoudes, Alexia; Kapitsaki, Georgia M.; Katakis, Ioannis: Enhancing user awareness on inferences obtained from fitness trackers data [313]

Published in: User Modeling and User-Adapted Interaction (2023).

Link to original: https://doi.org/10.1007/s11257-022-09353-8.

If you want to cite the papers, please refer to the original.

For technical reasons we could not place the page where the original quote is coming from.

Abstract-Summary

"In some cases these data may become available to interested third parties, who can analyse them with the intention to derive further knowledge and generate new insights about the users, that they can ultimately use for their own benefit."

"The immense increase in fitness trackers use has further increased the amount of user data generated, processed and possibly shared or sold to third parties, enabling the extraction of further insights about the users."

"We utilise the PrivacyEnhAction privacy tool, a web application we implemented in a previous work through which the users can analyse data collected from their IoT devices, to educate the users about the possible risks and to enable them to set their user privacy preferences on their fitness trackers accordingly, contributing to the personalisation of the provided services, in respect of their personal data."

Introduction

"Since transparency is key in the application of GDPR, it is essential that fitness trackers users become aware about how their personal data are processed [19]."

"These techniques are applied in the PrivacyEnhAction privacy tool introduced in our previous work, a web application through which the users can analyse data collected from their smart devices (smart water meters or motion sensors) with the objective to be informed about potential privacy vulnerabilities and possible inferences that emerge from the use of these devices, and thereupon to be able to change and set their user privacy preferences on their devices appropriately, contributing in this way to the personalisation of the provided services, in connection with their personal data."

"The main contributions of this work are: We lay out a list of possible inferences that pose a threat to user privacy when using fitness trackers, and we attempt to identify which specific inferences can be drawn from specific data collected from Fitbit, Garmin and Xiaomi fitness trackers."

Background

"We present research in the fields of privacy protection in wearables, inference extraction from fitness trackers data and also related works in the area that engage in the collection and analysis of the opinions and perceptions of fitness trackers users in relevance to the protection of their privacy."

"While Psychoula et al. [20] were occupied with user privacy awareness in the area of wearables and IoT services by presenting a framework that could be used as guidance to developers and service providers in order to integrate privacy risk user awareness in their products, no other work to the best of our knowledge has been involved with raising user awareness in relation to the inferences that can be extracted about the users from their fitness trackers data."

"A study on the privacy vulnerabilities of fitness trackers is presented by Reichherzer et al. [21], where machine learning techniques are exploited for the analysis of data from these devices in order to make meaningful inferences about user activities."

Research questions

"In order to answer this question we use the results from the literature review we performed for this work in combination with our previous research in the area, and we produce a list of possible inferences that pose a threat to user privacy when using fitness trackers."

"For providing an answer to this research question, we conduct an online questionnaire that targets fitness trackers users in order to gain an understanding of: (i) their concerns over their privacy when using their devices, (ii) their awareness of what data are collected by their fitness trackers and how these are being used and shared, (iii) their awareness on the privacy risks form fitness trackers data."

"We provide the same group of fitness trackers users with a number of datasets from the three fitness trackers brands (Fitbit, Garmin and Xiaomi) under study."

Privacy policies in fitness trackers

"Further user information is accumulated from the trackers that the user may be unaware of, like the time they wake up, the time they go to bed, their location, timezone, IP address, etc. Even though fitness trackers privacy policies usually state that no data are shared with third parties, this is not always the case as constant user tracking and data collection give fitness tracker companies the opportunity to capitalise on user data with the help of third-party sales [22]."

"We provide a review of how Fitbit, Garmin and Xiaomi fitness trackers address data sharing in their privacy policy."

"The Garmin privacy policy includes in the list of possible recipients of the users' personal data various third-party apps, platforms or service providers with whom the users ask Garmin to share their data."

Possible inferences from fitness trackers data

"An elevated heart rate could be due to a health condition, exercising at the time of the readings or heavy alcohol consumption, and consequently, the inferences that could be extracted about the user from these data are that the user may be suffering from a heart condition or could be an alcoholic [23, 24]."

"These attacks may include: (i) undesired advertising to users of products near to the user proximity, (ii) physical attacks and harassment or user profiling and tracking, when location data can be used to infer other sensitive information, such as state of health, personal habits or professional duties, (iii) political, religious, sexual persecution and discrimination, in which a person's location is used to restrict his or her freedom [25], (iv) planned break-in according to the times the user is away from home, (v) stalking."

Fitness trackers scenarios under study

"This dataset was collected by thirty eligible Fitbit users that participated in an Amazon Mechanical Turk survey, submitting physical activity, heart rate and sleep monitoring data at minute level."

"Following this procedure, we acquired a number of user datasets, containing daily physical activity data, heart rate and sleep monitoring data."

"This method led to the inclusion of four files in each dataset, containing activity data, heart rate data, sleep data and user information."

"Considering that the resting heart rate increases by 30–50% during pregnancy to match the needs of the growing baby [26, 27], we exploited the available personal user information in the Mi Band data that includes the user gender and date of birth, in order to infer the likelihood of pregnancy."

Implementation

"Inference detection analysis for Xiaomi Mi Smart Band 4C The code written in Python has been changed to include the new changes, where, depending on the selected device, the corresponding modules are called to process the files that are uploaded by the user."

"Users of these fitness trackers models can upload their data to the application through the interface, after they have exported them from their corresponding account dashboard, in order to analyse the data and view the possible inferences that could be extracted about them and be informed about the potential privacy risks that these inferences entail."

"In the illustrated example regarding the inferences that can be extracted from the user's Fitbit heart rate data, the user is informed about what information can be revealed from the heart rate in the first part of the interface."

User evaluation

"We describe the experimental process we performed in order to evaluate the impact of PrivacyEnhAction to the awareness of the users in relation to the privacy risks and the inferences that could be drawn about them from their fitness trackers data."

"In relation to the effect that the participants' gender has to answers, further analysis on this section's questions has shown that the gender is not correlated with the user awareness about the possible inferences that can be extracted from fitness trackers data."

"In the second questionnaire, the participants had to answer the same set of questions regarding their awareness on the privacy risks and the possible inferences that could be extracted from fitness trackers data, as in the first questionnaire, in our effort to seek an answer to Research Question RQ3: Does the PrivacyEnhAction application enhance the awareness of the users regarding the possible inferences that can be obtained from their fitness tracker data?".

Discussion and limitations

"These types of information were not analysed in this study at this stage; therefore, as future work, we are investigating the privacy vulnerabilities of these information types in order to raise user awareness, as it is very important that the users become aware of what information is being collected from their fitness tracker, considering that a big amount of personal information is at risk."

"The results have shown that even though a big percentage of the users are aware that their trackers may collect their personal data, they do not take any action to minimise any possible risks, such as by altering their fitness trackers privacy preferences or by reading the privacy policies of their trackers in order to get informed."

"The findings from this study provide valuable insights for the users of fitness trackers in our effort to increase their awareness; however, despite the possible privacy risks, the inferences that can be extracted from fitness trackers data can also have a positive impact to the users."

Conclusions

"We have investigated the possibility of getting insights and extracting inferences about the users from their data collected from fitness trackers."

"A list of possible inferences that pose a threat to user privacy through the use of fitness trackers and we utilise our privacy tool, PrivacyEnhAction, as a means for increasing the users' awareness about the privacy risks that emerge from the data collected by their devices, in order to enable the users to set their privacy preferences in an appropriate way, contributing to the personalisation of the provided services in connection with their personal data, while protecting their privacy at the same time."

"The results of our experimental research have showed that the interaction with the PrivacyEnhAction application can increase the user awareness on the possible inferences that can be obtained from their fitness trackers data."

Addressing the privacy paradox on the organizational level: review and future directions [314]

This is a machine-generated summary of:

Gotsch, Mauro Luis; Schögel, Marcus: Addressing the privacy paradox on the organizational level: review and future directions [314]

Published in: Management Review Quarterly (2021).

Link to original: https://doi.org/10.1007/s11301-021-00239-4.

If you want to cite the papers, please refer to the original.

For technical reasons we could not place the page where the original quote is coming from.

Abstract-Summary

"Researchers across disciplines have formulated different theories on why consumers' privacy concerns do not translate into increased protective behaviour."

"Authors generally agree that companies are in a strong position to reduce consumers' paradoxical behaviour by improving their customers' informational privacy."

"This paper aims at answering the question: How can companies address the privacy paradox to improve their customers' information privacy?"

"Reviewing a sample of improvement recommendations from 138 papers that explore 41 theories in total, we determined that companies can generally align their privacy practices more closely with customers' expectations across 4 inter-connected managerial processes: (1) strategic initiatives, (2) structural improvements, (3) human resource management, and (4) service development."

Introduction

"Martin and Murphy ([28], p. 153) also support this conclusion in their literature review on the role of privacy in marketing: "Although a coherent subset of theoretic approaches has provided a robust understanding through deep insights, in some respects this focus has limited our view of privacy too much to individual silos (...) Future research directions should embody a holistic approach, blending the many consumer, organisational, ethical, and legal concerns that feature in contemporary data privacy questions.""

"Companies are in a strong position to reduce consumers' paradoxical behaviour by improving their customers' informational privacy."

"Despite the prevalent recommendation for companies to adapt a "more proactive approach" ([29], p. 652), current literature across disciplines does not sufficiently address the privacy paradox on an organisational level."

"We review 138 papers with 41 different theoretical approaches to the privacy paradox and evaluate their recommendations on what companies could learn from their studies."

Theory

"Even though "information privacy research can be conceptualised as a multi-level concept", it "is very rarely researched as such" ([30], p. 1028) as most research either starts from or is concerned with addressing challenges arising on the personal level."

"This is not to suggest that previous privacy research offers no insights on how to address the privacy paradox on the organisational level."

"Addressing the privacy paradox on an organisational level necessitates the often demanded "holistic perspective" ([30], p. 1027; [28], p. 153), i.e. trying to unify the rational, irrational and structural approaches in order to reduce the discrepancy between expressed privacy concerns and actual privacy behaviour."

"Since no cohesive overarching framework of how to address the privacy paradox on an organisational level exists ([31], p. 180), different organisational capabilities were examined and measured across all of these studies."

Method

"We performed a literature review of publications researching the privacy paradox and offering implications or recommendations for companies on how to address it."

"Papers neither citing an explicit theoretical background nor proposing a new theory to explain the privacy paradox (or a related question) as well as papers not proposing direct or indirect implications to improve privacy orientation at an organisational level were rejected."

"We extracted ten data points for each paper—amongst them identifiers such as the author and title, classifications such as research stream and theories cited and the recommendations made to address the privacy paradox on an organisational level."

"No—the paper offers neither recommendations nor inferences on how to address the privacy paradox on an organisational level."

"We coded the recommendations using the inductive process described by Bandara and others ([32], p. 169) to capture the recommendations found in the literature to address the privacy paradox on an organisational level."

Addressing the privacy paradox

"There are few structural recommendations which go beyond compliance and trying to work with the customer to profit from data collection while addressing the privacy paradox."

"Their goal is to address the privacy paradox in two ways: On the one hand, by trying to reduce biases and increase transparency at the point of disclosure, e.g. by not asking for customer data after cognitively taxing tasks [33, 34] or by using just-in-time alerts to pro-actively warn users about the context of their disclosure before doing so [35]."

"Most recommendations in the sample revolve around improving customers' privacy at potential points of data collection, e.g., by increasing an app's number of data control settings or by improving how these features get communicated."

"These recommendations aim at addressing the privacy paradox by giving the customers both more control over and transparent information on requested data flows."

Discussion

"The purpose of this systematic review is to summarise and structure the recommendations from the past 20 years of privacy paradox research on how to address paradoxical behaviours by customers on an organisational level."

"By aggregating these processes within four dimensions, we hope to provide a common underlying structure for further research on improving customers' informational privacy."

"While there are currently no studies on the direct effect of the processes outlined in the four dimensions on the individual level, theories like the APCO model suggest that strategic initiatives and structural improvements may affect individuals' privacy concerns ([29], p. 643)."

"Despite each paper in the sample giving direct or indirect recommendations for addressing the privacy paradox on an organisational level, very few of the sampled research designs aimed at doing so from the outset."

"Company structure: Could more customer-centric company structures be used to improve organisational privacy?"

Research limitations

"Due to the limited research addressing the privacy paradox on an organisational level, this review opted to analyse recommendations and implications across different disciplines and levels."

"While the logical congruence of individual recommendations with their corresponding study results were qualitatively appraised, they are still often based on untested claims."

"The recommendation "to develop (website) privacy control features (…) to maintain the anonymity, secrecy, and confidentially of their personal information" ([36], p. 309) may be logically sound, even when extended to a company's data collection as a whole, but it still is an untested claim."

"Measuring privacy concerns at a low level of specificity while researching within a highly specific context."

"By incorporating recommendations based on such measurements, this study might exaggerate the general effectiveness of certain recommendations."

Conclusion

"The purpose of this paper was to review the recommendations made in the past 20 years of privacy paradox research on how to address it on an organisational level."

"We determined that companies can generally align their privacy practices with customers' expectations across four inter-connected managerial processes: (1) strategic initiatives, (2) structural improvements, (3) human resource management, and (4) service development."

"Should a company wish to sustainably address the privacy paradox, it would have to move beyond the reactive thinking of privacy law compliance."

"We believe that looking at the problem through a marketing management lens has the potential to both unearth new methods of addressing the privacy paradox and to gain insights into whether privacy could be leveraged by companies to obtain a strategic advantage."

All of me? Users' preferences for privacy-preserving data markets and the importance of anonymity [315]

This is a machine-generated summary of:

Schomakers, Eva-Maria; Lidynia, Chantal; Ziefle, Martina: All of me? Users' preferences for privacy-preserving data markets and the importance of anonymity [315].

Published in: Electronic Markets (2020).

Link to original: https://doi.org/10.1007/s12525-020-00404-9.

If you want to cite the papers, please refer to the original.

For technical reasons we could not place the page where the original quote is coming from.

Abstract-Summary

"Privacy-preserving data markets are one approach to restore users' online privacy and informational self-determination and to build reliable data markets for companies and research."

"We empirically analyze internet users' preferences for privacy in data sharing, combining qualitative and quantitative empirical methods."

"Study I aimed at uncovering users' mental models of privacy and preferences for data sharing."

"Users with higher privacy concerns are less willing to share data in data markets and want more privacy protection."

"The results contribute to an understanding of how privacy-preserving data markets could be designed to suit users' preferences."

Introduction

"In privacy-preserving data markets, users shall actively and voluntarily share their data."

"We do not yet have the necessary insights into what informational self-determination and privacy means to users in such data markets."

"How can users' desires for privacy be satisfied while they share data in the decontextualized online world?"

"Knowledge gap, we empirically analyze users' notions of and desires for privacy protection in a privacy-preserving data market."

"We use a threefold mixed-method approach to identify and understand users' notions of privacy in data sharing in detail as well as to quantify users wishes and observe trade-offs in data sharing decisions."

"Based on the findings of this first study, users' motives, barriers, and conditions for privacy in data sharing are then measured and quantified in a second study (survey)."

Related work: Online privacy and data sharing from the user perspective

"A number of studies have found a mismatch between users' privacy attitudes and their respective online behavior: Users are concerned about their online privacy, but they still share private information without hesitation and do not protect their privacy accordingly [37, 38]."

"This paper aims to understand which privacy protective measures and which control possibilities users desire to maintain their privacy when sharing data."

"One possibility to protect users' privacy in data markets is anonymization of the shared data."

"Prior Conjoint studies have used specific context (e.g., OSNs, medical data sharing) and different operationalization of privacy in data sharing (e.g., privacy control for visibility of posts in OSNs, binary levels of anonymized data compared to data linked to the name)."

"To understand privacy in data sharing in privacy-preserving data markets, research within this specific context considering the users' mental models of privacy protection is needed."

Material and methods: Empirical approach

"In order to understand users' notions of privacy in data sharing, we undertook a three-step empirical approach."

"In a first exploratory approach, qualitative interviews and focus groups were used to uncover mental models and attitudes regarding privacy in data sharing and conditions for a user-centered privacy-preserving data market."

"An online questionnaire was developed to quantify users' motives, barriers, and conditions as well as willingness to provide data using a larger sample of German internet users."

"In a third-step, experiments were conducted to model users' choices of data sharing in which the trade-offs between privacy protection, type of data, data receiver, and benefit of data sharing are identified."

Study I: Exploring users' understanding of privacy in data sharing

"The aim of Study I was to examine users' notions regarding privacy in data sharing and to identify motives, barriers, and conditions for the use of user-centered privacy-preserving data markets."

"A qualitative approach was chosen, as it is not yet known what users' preferences for such data markets are and especially mental models regarding privacy usually do not include the case of sharing data."

"After an introduction and warming up with discussions about privacy in the online environment, the scenario of a user-centered, privacy-preserving data market was introduced."

"In all focus groups, the participants were not only asked to talk about their notions of privacy in such a data market but also to draw visualizations and controls for privacy in data sharing."

Results of study I

"Focusing on the idea of sharing data while preserving privacy, the control over what is shared to whom for what benefit and which purpose is one central element."

"If the data would only be used for medical purposes, for improvements in science or so, then I would be more willing to provide data." (female participant, 51 years old) Perceived barriers include essentially three topics: privacy concerns and missing trust into the security of such a data market, moral concerns, and not seeing the personal benefit of the system."

"For privacy protection, the participants demanded—besides informational self-determination and transparency—anonymity and protection within the data market application (e.g., password protection, facial recognition)."

"The participants' attitudes towards user-centered privacy-preserving data markets were quite diverse."

"Some participants showed very high privacy concerns in general and regarding such data markets, while others were more laid-back."

Key Findings of Study I

"A key element of online privacy and privacy in data sharing is the protection of the identity, and thus, anonymity."

"For privacy in data sharing, it is important to control."

"Barriers against the use of privacy-preserving data markets are foremost privacy concerns, as well as moral concerns against the 'trading' with personal data."

Study II: Quantifying users' conditions for data sharing

"The second study aimed at confirming and quantifying the motives, barriers, and conditions for data sharing identified in Study I. $N = 800$ German participants completed an online questionnaire in which they were introduced to the scenario of a privacy-preserving data market."

"In Study I, varying preferences towards such data markets could be observed between people with less and more privacy concerns."

"In Study II, we also analyze differences between groups of users with lower and higher privacy concerns."

"The third and main part of the questionnaire started with a scenario of a privacy-preserving data market: A newly developed data market enables participants to share their data with interested parties and to get rewards in return."

"After the introduction to the scenario, the participants indicated their willingness to share exemplary data types, and their agreement to benefits, barriers, and conditions for the use of the data market."

Results of study II

"The willingness to provide data differs significantly between user with low privacy concern and with high privacy concern ($F(10,789) = 5.10$, $p < 0.001$, partial $\eta^2 = 0.06$): Users with lower privacy concerns are more willing to provide data than users with higher privacy concerns."

"Users with lower privacy concerns see significantly more motives to use the privacy-preserving data market than do users with higher privacy concerns ($F(10,789) = 9.24$, $p < 0.001$, partial $\eta^2 = 0.11$)."

"The trust in the data protection in such a data market is higher by those with lower privacy concerns."

"Those with lower privacy concerns do not perceive that many barriers ($F(15,784) = 12.7$, $p < 0.001$, partial $\eta^2 = 0.2$)."

"Users' with high privacy concerns put more importance on these conditions for privacy than do users with lower privacy concerns ($F(10,789) = 18.13$, $p < 0.001$, partial $\eta^2 = 0.19$)."

Key Findings of Study II

"Informational self-determination and benefits are the most important motives to use a user-centered privacy-preserving data market."

"Privacy concerns are the most relevant barriers against such a data market, followed by moral concerns."

"Users with higher privacy concerns in general are less willing to provide data in a privacy-preserving data market, agree less to motives and more to barriers, and put more emphasize on privacy conditions."

Study III: Modeling users' trade-offs in data sharing

"Choice-based conjoint experiments model these complex decision processes: Participants need to choose between n scenarios with m different attributes which vary in levels."

"The selection of the attributes for this study was based on the findings in Studies I and II, referring to which factors mostly influence users' decisions to share data."

"Here the laypersons' understanding of anonymity was considered: As most participants of Study I showed to have a notion of a 'complete' anonymization, this was included as level in this study."

"The participants were given 10 random choice tasks, as a complete design with 1024 ($4 \times 4 \times 4 \times 4 \times 4$) possible combinations of the attribute levels would not have been feasible."

"In each choice task, the participants were presented with $n = 3$ scenarios—that are random combinations of the attribute levels—plus a "none" option."

Results of Study III

"To identify which attribute has the greatest impact on users' decision to share data, relative importance scores are calculated."

"The concern groups show small but significant differences in their importance attribution ($F(4,121) = 3.14$, $p < 0.05$, partial $\eta^2 = 0.09$)."

"The relative importance of data type, data receiver, and app security does not differ."

"Users with a higher privacy concerns put more value to the level of anonymization ($M = 38\%$) than do those with lower privacy concerns ($M = 30.6\%$, $F(1,124) = 12.37$, $p < 0.01$, partial $\eta^2 = 0.09$)."

"Those with lower privacy concerns put more importance to the monetary benefit ($M = 15.2\%$) than those with higher privacy concerns ($M = 12.5\%$, $F(1,124) = 4.86$, $p < 0.05$, partial $\eta^2 = 0.02$)."

"No significant differences in the level evaluation between the groups with low and high privacy concerns can be observed ($F(15,110) = 1.71$, $p > 0.5$)."

Key Findings of Study III

"The level of anonymization has the greatest impact on the decision whether to share data, followed by the type of data."

"The most accepted scenario is sharing the social network profile with 'complete' anonymization to a public administration for 75€. The application should be protected by a double password protection."

"The least accepted scenario is sharing financial account data without anonymization to a financial institution for 5€ with the application being protected with facial recognition."

"Users with high privacy concerns attribute more importance to the anonymization level and less importance to the monetary benefit than users with lower privacy concerns."

Discussion and conclusion

"These markets are one approach to restore users' online privacy and informational self-determination as well as to build a viable data market for companies and research."

"In a choice-based conjoint experiment, Study III, five attributes (anonymization level, type of data, data receiver, monetary benefit, app security) were weighed against each other in their impact on the users' decisions to share data."

"The findings of all three studies indicate that the protection of one's identity is the core element for privacy and, correspondingly, anonymization is the most important factor that influences users' decisions to share data."

"The results are important for the design of privacy-preserving data markets as they show that good anonymization techniques are a core element to ensure users' perceived privacy."

"Our results show that when giving users informational self-determination and anonymized sharing of data, they perceive their privacy to be guarded."

Trust and digital privacy: willingness to disclose personal information to banking chatbot services [316]

This is a machine-generated summary of:

Lappeman, James; Marlie, Siddeeqah; Johnson, Tamryn; Poggenpoel, Sloane: Trust and digital privacy: willingness to disclose personal information to banking chatbot services [316]

Published in: Journal of Financial Services Marketing (2022).

Link to original: https://doi.org/10.1057/s41264-022-00154-z.

If you want to cite the papers, please refer to the original.

For technical reasons we could not place the page where the original quote is coming from.

Abstract-Summary

"This study explored digital privacy concerns in the use of chatbots as a digital banking service."

"Three dimensions of trust were tested in relation to user self-disclosure in order to better understand the consumer-chatbot experience in banking."

"Privacy concerns were found to have a significantly negative relationship with user self-disclosure in both treatment groups."

"Respondents exposed to their preferred banking brand experienced lower user self-disclosure and brand trust than those exposed to a fictitious banking brand within the South African context."

"Through this study, we concluded that, despite its power to influence decision-making, the power of a brand is not enough for consumers to considerably increase self-disclosure."

"This study addressed a theoretical gap, by adding brand trust to existing studies on cognitive and emotional trust regarding user self-disclosure."

Introduction

"To address gaps found in the literature, the following research question was proposed: Does digital-privacy concern, brand trust, emotional trust and cognitive trust influence Internet banking user's willingness to disclose personal information to chatbots used in the South African banking industry?"

"This question in turn led to four primary objectives being formed: To determine if brand trust positively influences Internet banking users' cognitive trust when engaging with chatbots used in the South African banking industry To determine if cognitive trust positively influences Internet banking users' emotional trust when engaging with chatbots used in the South African banking industry To determine if emotional trust positively influences Internet banking users' willingness to disclose personal information to chatbots used in the South African banking industry To determine if digital privacy concerns negatively influence Internet banking users' willingness to disclose personal information to chatbots used in the South African banking industry Furthermore, three secondary objectives applied: To determine if there is a difference in the level of digital privacy concerns between age categories among Internet banking users To determine if there is a difference in the level of brand trust between age categories among Internet banking users."

Literature review

"Users need to be convinced and have trust that their personal information is secured [39–41]."

"Trust and digital privacy concerns are factors that influence the user's decision to disclose private information [42, 43]."

"Their study aimed to describe the relationship between perceived personalisation and privacy in relation to cognitive and emotional trust within the technology acceptance model [44, 45]."

"This study will address this gap by exploring dimensions of trust—cognitive and emotional with the introduction brand trust as an additional dimension in relation to digital privacy concerns and user self-disclosure."

"This study is in agreement that emotional trust bridges these gaps, as it is used to account for trust evaluations that go beyond the available information in rational reasoning [44]."

Analytical model

"The work of Alashoor and others [43] is of particular prominence in this research study as it analyses digital privacy concerns and both emotional and cognitive trust as constructs that determine user self-disclosure."

"Trust encompassing brand, cognitive and emotional trust was explored as a separate linked variable influencing user self-disclosure when engaging with chatbots."

"H_1: Brand trust positively influences Internet banking users' cognitive trust when engaging with chatbots used in the South African banking industry H_2: Cognitive trust positively influences Internet banking users' emotional trust when engaging with chatbots used in the South African banking industry H_3: Emotional trust positively influences Internet banking users' willingness to disclose personal information to chatbots used in the South African banking industry H_4: Digital privacy concerns negatively influence Internet banking users' willingness to disclose personal information to chatbots used in the South African banking industry H_{5a}: There is a difference in the level of digital privacy concerns between age categories among Internet banking users H_{5b}: There is a difference in the level of brand trust between age categories among Internet banking users H_{5c}: There is a difference between age categories in the willingness of Internet banking users to disclose personal information".

Methodology

"A pre-experimental two-group one-shot case study design was selected, as respondents were observed after exposure to the treatment [46]."

"To ensure an effective measure of trust dimensions, this study utilised a familiar and fictitious banking brand in experimentation."

"To ensure that the construct of brand trust was effectively tested, display logic was utilised on the Qualtrics platform within experimental group 2, so as to tailor measurement aids to the banking brand respondents were most familiar with."

"The measurement aid depicting the simulated conversation had respondents' selected banks preferred logo to ensure familiarity with the brand respondents were engaging with."

"Three ANOVA tests were conducted comparing age and the constructs of digital privacy concerns, brand trust and user self-disclosure, thus allowing theoretical constructs to be compared across groups [46]."

Results

"It was concluded that brand trust has a positive influence on cognitive trust in Group B. H_2: Cognitive trust positively influences Cape Town Internet banking users' emotional trust when engaging with chatbots used in the South African banking industry In Group A, the null hypothesis could be rejected at the 5% level of significance, with a p value of 0.00 and a t-stat of 12.78."

"It was concluded that digital privacy concerns have a negative influence on user self-disclosure in Group B. The path coefficients in the structural model displayed the relationship between brand trust and cognitive trust, cognitive trust and emotional trust, as well as emotional trust and user self-disclosure having been more significant for Group A than Group B. While, the relationship between digital privacy concerns and user self-disclosure was more significant for Group B. This revealed that trust was more significant for respondents exposed to a fictitious brand than those exposed to their preferred brand."

Conclusions

"This research paper has explored the influence of digital privacy concerns and trust—to be brand, cognitive and emotional trust—on Cape Town Internet banking users' willingness to disclose personal information."

"Statistical analysis from PLS-SEM revealed that the relationship between brand trust and cognitive trust, cognitive trust and emotional trust, as well as the relationship between emotional trust and user self-disclosure, were more significant in the model for Group A. The relationship between digital privacy concerns and user self-disclosure proved more significant in the model for Group B. These results observed that digital privacy concerns were more significant and trust less significant in influencing user self-disclosure when respondents were exposed to their preferred banking brand as opposed to a fictitious banking brand."

"The age category of 18–35 displayed higher levels of user self-disclosure than that of 36–50 for Group A. This could be a result of the higher brand trust and lower digital privacy concerns experienced by the younger demographic."

"Brands need to develop long-term, trust-based relationships to overcome the challenge of digital privacy concerns, such that users are more willing to self-disclose personal information."

Managerial implications

"Companies using chatbots should focus on easing digital privacy concerns in order to improve engagement and lower privacy concerns."

"Rather, a bridge of trust, encompassing brand, cognitive and emotional, is needed."

"The results showed that the younger demographic, particularly 18–35, showed lower levels of privacy concerns in both treatment groups, while displaying higher levels of brand trust and user self-disclosure in Group A. This may be attributed to the younger demographic's comfort level with technology [47] in contrast to the older demographic [48]."

"The older demographic may have fewer social media touch points with a brand and thus a lower brand trust, marketers should gear research to identifying how older users engage with their brand on digital platforms [49]."

Contribution to theory

"Conducted by Awad and Krishnan [50] and Baumer and Earp [51], it was discovered that trust plays a pivotal role in the willingness of users to disclose personal information in online environments."

"Despite the abundance of literature exploring the impact of trust on privacy concerns, trust was rarely explored in a multi-faceted fashion which this research study aimed to do by including cognitive, emotional and brand trust as separately tested constructs."

"Our study acknowledged that trust, particularly within the realms of technology and digital privacy concerns, comprises of elements such as; trust in reliability of the system to perform, a feeling of security when disclosing personal information and trust in a brand to ensure data security and efficiency in use [43, 52]."

"We acknowledge that brand trust forms an integral component to understanding user willingness to disclose personal information in online environments and including both a fictitious and branded bank in the questionnaire simulated this."

Limitations and future research

"The model presented in this study may be modified in future research in order to explore the role of other factors that influence user willingness to disclose information to chatbots."

"Additional research can focus on messaging specific themes (use of language, colloquialism and personalities) in order to create a more comfortable user experience that encourages the disclosing of information as well as the possibilities of chatbot to chatbot interactions [42, 53]."

"Emergent legislation like the General Data Protection Regulation (GDPR) in Europe or the Protection of Personal Information (POPIA) in South Africa where this study was based will shape the way people feel about disclosing personal information (potentially both positively and negatively) and hence future research will need to encompass a legislative viewpoint as this develops further globally [54, 55]."

The effects of privacy concerns and personal innovativeness on potential and experienced customers' adoption of location-based services [317]

This is a machine-generated summary of:

Xu, Heng; Gupta, Sumeet: The effects of privacy concerns and personal innovativeness on potential and experienced customers' adoption of location-based services [317].

Published in: Electronic Markets (2009).

Link to original: https://doi.org/10.1007/s12525-009-0012-4.

If you want to cite the papers, please refer to the original.

For technical reasons we could not place the page where the original quote is coming from.

Abstract-Summary

"Privacy concerns associated with the use of LBS may ultimately prevent consumers from gaining the convenience of 'anytime anywhere' personalized services."

"We examine the adoption of this emerging technology through a privacy lens."

"Drawing on the privacy literature and theories of technology adoption, we use a survey approach to develop and test a conceptual model to explore the effects of privacy concerns and personal innovativeness on customers' adoption of LBS."

"The results indicate that privacy concerns significantly influence continued adoption as compared to initial adoption."

Introduction

"Although much theoretical development has occurred in regard to individual adoption behavior with new information technologies (e.g., [57]), this body of work has paid limited attention to privacy issues."

"Most of the theoretical models (e.g., TAM and TPB) address technology adoption from a positive-utility oriented perspective, while paying limited attention to potential negative consequences (e.g., the risks that consumers may experience with respect to privacy violations in the LBS context) arising from the adoption and use of new technologies."

"We attempt to develop a research model to simultaneously consider both positive and negative outcomes of adopting and using a new technology that raises a new set of concerns related to individual privacy."

"Acknowledging the difference between potential and experienced users in relation to technology acceptance [58, 59], we test our model separately for potential and experienced LBS users."

Theoretical background

"Consistently, our investigation follows the direction of technology adoption literature by specifying a model that directly captures several constructs of the UTAUT: behavioral intention (intention to use LBS), performance expectancy (instrumental value of using LBS), effort expectancy (learning cost of using LBS), and individual level factor (personal innovativeness)."

"In the field of information systems, Agarwal and Prasad [60] define personal innovativeness as willingness of an individual to try out new technology."

"Personal innovativeness has been conceptualized in terms of its operational definition, i.e., individuals are characterized as 'innovative' if they are early to adopt an innovation [60]."

"Since the nature of this study pertains to a specific technology domain (i.e. LBS), we follow Agarwal and Prasad [60] to use domain-specific conceptualization of personal innovativeness in the context of LBS."

"We define personal innovativeness as an individual's willingness to try out LBS."

Research model and hypothesis

"We adapt this definition in our research context and define performance expectancy as the degree to which an individual believes that using LBS would reduce his or her time and effort required to search or access the needed information or service."

"We hypothesize: H1: Performance expectancy is positively related to intention to use LBS."

"We hypothesize: H4: Privacy concerns are negatively related to intention to use LBS."

"Consumers are concerned about loss of privacy in using LBS whereby their whereabouts and other personal identifiable information may be tracked by service providers."

"We believe that privacy concerns, viewed as the fear of losing control over personal information, should negatively impact an individual's process expectancy—i.e., effort expectancy."

"We hypothesize: H8: Personal Innovativeness will moderate the relationship between privacy concerns and intention to use LBS."

Research methodology

"service provided by SingTel (the largest telecom operator in Singapore), can locate the nearest public library, community center, hospital, ATM, Café, cinema, horse betting outlet, Fast Food and Food Court, Petrol Station, taxi stands, Post Office, and Supermarkets."

"In Singapore, these location-based services were offered to mobile phone users via Short Messaging Service (SMS) based on the Cell-Identification (Cell-ID) technique employed by the network of telecom operators."

"One specific pull-based LBS application—*SEND-A-TAXI service, was introduced with more details in the survey."

"Performance expectancy was measured with four questions to capture the extent to which an individual would believe that using LBS would reduce his or her time and effort required to search or access the needed information or service [57]; effort expectancy was measured with questions on whether using LBS would be clear, understandable, and easy to use [57]."

"They were introduced with the usage scenario of one specific service—*SEND-A-TAXI, and then asked to complete a questionnaire regarding their behavioral intention, performance expectancy, effort expectancy, personal innovativeness, and concerns for information privacy (CFIP)."

Data analysis and results

"In PLS, three tests are used to determine the convergent validity of measured reflective constructs in a single instrument: reliability of items, composite reliability of constructs, and average variance extracted by constructs."

"After establishing the validity of the measures, we tested the structural paths in the research model using PLS."

"The explanatory power of the structural model is assessed based on the amount of variance explained in the endogenous construct (i.e., behavioral intention)."

"Our model is statistically sound for the marginal supported hypothesis H7 (for potential users) and H2 (for experienced users)."

"For the minimum acceptable statistical power (80%), the sample size in this study is able to capture effect size (lowest path coefficient) of 0.124 and above for potential users and 0.170 and above for experienced users."

Discussions and conclusions

"Personal innovativeness did not moderate the relationship between privacy concerns and behavioral intention for this sample and the specific context of LBS."

"Our findings also suggest that privacy concerns, viewed as the fear of losing control over personal information, have a larger negative impact on an individual's process expectancy (i.e., effort expectancy) and less impact on outcomes of using LBS (i.e., performance expectancy and behavioral intention)."

"An interesting finding of this study is that while for potential customers, only effort expectancy (and not privacy concerns) had a significant influence on performance expectancy; for experienced customers, only privacy concerns (and not effort expectancy) had a significant influence on performance expectancy."

"This research is one of the few studies that investigate the role of privacy concerns in the LBS adoption for both potential and experienced customers."

An experimental economics approach toward quantifying online privacy choices [318]

This is a machine-generated summary of:

Poindexter, J. C.; Earp, Julia B.; Baumer, David L.: An experimental economics approach toward quantifying online privacy choices [318].

Published in: Information Systems Frontiers (2006).

Link to original: https://doi.org/10.1007/s10796-006-9013-4.

If you want to cite the papers, please refer to the original.

For technical reasons we could not place the page where the original quote is coming from.

Abstract-Summary

"The importance of personal privacy to Internet users has been extensively researched using a variety of survey techniques."

"This paper describes our efforts at applying the techniques of experimental economics to provide a foundation for (a) estimating the values that consumers place on privacy and various forms of security (encryption, HIPAA, etc.) and for (b) quantifying user responses to changes in the Internet environment."

"Preliminary results from a series of pilot studies are consistent with optimizing behaviors, indicating that continued application of experimental economics techniques in the quantification of Internet user actions in privacy/security space will be illuminating."

"Our results show that Internet users place great value on security measures, both regulatory and technical, that make identity theft much less likely."

"Our Web-based experiments indicate that privacy- and security- enhancing protections are likely to be subject to moral hazard responses, as participants in our online experiments became more aggressive in their Internet usage with greater protection in place."

Introduction

"Dissatisfaction with the privacy and security status quo is likely to increase as computerization of personally identifying information (PII) continues to lower the costs of acquiring, storing and transferring such information [61, 62]."

"All of the following are responsible for making inroads on personal privacy: widespread use of the Internet, networked systems, radio frequency identification (RFID) tags, surveillance cameras, location-tracking wireless devices, cards that track buying patterns and electronic storage technology that allows organizations to store an abundance of personal information."

"The results of our preliminary work are reflective of our initial applications of experimental economics methods to the search for an understanding of consumer privacy demands, in particular decision-making in the context of benefits and costs of making private information available to Internet sites."

Relevant prior literature

"It is hoped that better quantification of models of user privacy choice will provide privacy/security valuations free of the idealistic notions self-reported in earlier surveys, assisting (1) computer scientists and managers in making more accurate and productive decisions regarding privacy management and security technology choices and (2) policymakers in the design of an economically efficient regulatory environment."

"A search of the experimental economics literature does not reveal prior use of this methodology to quantify Internet choice-making and security preferences, though one study used experimental economics to value privacy generally [63]."

"There are no studies of online choices that make use of experimental economics to examine the impact on user decision-making of regulatory acts and developments that change the riskiness of the Internet environment, though one study does estimate the value of protecting online information using economic incentives by presenting participants with choices in a conjoint framework [64]."

Methodology

"With choices collected and compiled manually, our initial pilot experiment was intended to provide a guide for the preparation of additional web-based experiments that will use an electronically simulated Internet environment capable of large numbers of experiment repetitions."

"Instructed that winners and losers would be tracked by the value of their dollar accounts, casual observation of the students' zealous participation in this initial experiment suggested that participants were actively working, using the information they were provided, to "optimize" their net benefits from Internet activities so as to finish in the "winner" as opposed to "loser" group."

"The payoffs provided are not linear and were judgmentally set to provide payoffs believed to be plausible to student participants in the experiments (whose net worths typically are less than $10,000), who may use Monster.com to search for a job, and who know that there are consequences, positive and negative, associated with Internet usage."

Empirical results

"This choice was not surprising since it offered the highest expected value, ignoring risks such as identity theft, etc. Of the 21 participants who selected option 2, 19 opted to pay for protection in the form of a $100 payment to a detective website to investigate whether the employer website they submitted their resume to was authentic."

"In the second round of our preliminary experiment, both Groups were presented with the same options, but the results of the first round were made visible to them, so participants from each Group began the second round knowing that: One participant was a victim of identity theft and thus had a net wealth decrease of $5,000."

"For Group B, of the 27 participants for the second experiment, 15 selected option 2 and, of those, 11 elected to pay $1,000 for detective website protection."

Web-based experiments

"Subjects clearly made use of this information, as the choice patterns revealed by the manual pilot study experiments described above are transparently rational, with Internet users attracted by positive payoffs but willing to trade off prospective expected benefits to reduce risk exposures."

"The ideal contextual setting for repetitive experiments focused on Internet usage would rely on a simulated Internet environment that allows participants to make frequent visits to the web and that would allow detailed tracking of user decisions/actions."

"In our first application of this software, focusing on general Internet use decisions, test subjects were allowed to make (and change) choices with regard to their levels of usage (e.g., hours per week of active usage) and the risk profile of their Internet usage (e.g., low risk, moderate risk, or high risk)."

Summary and future work

"The stable responses observed in our experiments indicate clearly that repetitive experiments of the kind undertaken in this preliminary study, but with a larger participant sample, with more repetitions of each experiment, and with fine gradations of shocks imposed on the system, have the potential to provide quite detailed and compelling quantification of the types of metrics needed both by policymakers and protection creators."

"These future experiments will rely on a simulated Internet environment that consists of several websites, whose privacy policies may or may not make substantive promises with regard to PII and with which the participants will interact."

"Participants will surf the simulated web environment as they try and fulfill familiar tasks, such as purchasing, emailing, online banking and other activities including file sharing."

"Similar to the actual Internet, the simulated Internet environment will expose participant to risks including vulnerability to computer viruses, identity theft, hackers, etc. As a result, participants will make decisions that will impact their privacy as well as their success in fulfilling web surfing tasks."

Consumer perceptions of privacy, security and trust in ubiquitous commerce [319]

This is a machine-generated summary of:

Roussos, George; Moussouri, Theano: Consumer perceptions of privacy, security and trust in ubiquitous commerce [319].

Published in: Personal and Ubiquitous Computing (2004).

Link to original: https://doi.org/10.1007/s00779-004-0307-6.

If you want to cite the papers, please refer to the original.

For technical reasons we could not place the page where the original quote is coming from.

Abstract-Summary

"We then proceed to report on a study on consumer perceptions of MyGrocer, a recently developed ubiquitous commerce system."

"The emphasis of the discussion is on aspects of security, privacy protection and the development of trust; we report on the findings of this study."

"We expect that this interpretation can help to guide the development of appropriate strategies for the successful deployment of ubiquitous commerce systems."

Introduction

"This trend is expected to accelerate in the coming years due to a number of different factors, including the introduction of new mobile and ubiquitous computing technologies, the wider recognition by business of the strategic advantages offered by the implementation of ubiquitous computing and communications infrastructures, the emergence of novel business models which become possible only through this technology and, last but not least, the development of new economics that can be used to understand and value ubiquitous commerce activity."

"We report on a recent study on the consumer perceptions of retail services carried out as part of the MyGrocer research project [65, 66]."

"The paper is organised as follows First, we discuss the background for the development of ubiquitous retail systems and highlight their importance from a business perspective."

"We discuss our findings and relate them to three principles for the development of effective strategies in designing ubiquitous commerce systems."

Creating consumer value with ubiquitous commerce

"Decades after the introduction of information systems in production and logistics control there are still significant inefficiencies in modern supply chains, which adversely affect the cost of retail operations."

"Upstream supply chain inefficiencies affect the relationships of all trading partners and result in high out-of-stock conditions at the point of sale, a high returns rate and long lead times."

"Ubiquitous computing technologies applied to this problem space can provide the necessary consumption data early on in the replenishment process so as to allow for greater prediction accuracy, which leads to reduced inventories and optimised supply chains both upstream and downstream."

"One of the main concerns of retailers implementing VMI has been the perception that reduced inventory will result in less product being available on the shelves at any one time and, therefore, loss of market share."

Ubiquitous retail and consumer VMI

"The next natural step for ECR is to extend the use of technology to the collection of data directly from the selves and, even more, to consider extending the supply chain to the consumer home."

"MyGrocer was primarily aimed at creating an early prototype and identifying some of the core issues in this context, while several more recent projects examine specific aspects of ubiquitous commerce systems."

"More recent work has focussed on other aspects of the system; for example, the Auto ID Centre at MIT developed some of the standards required for the facilitation of business data exchanges in the new retail environment, now taken over by EPCGlobal."

"Extending the supply chain in this way has significant repercussions for the consumer, who is now involved in the data processing pipeline."

"The argument developed in this section points to the significance of adopting ubiquitous computing technologies in FMCG retail."

The enacted view of ubiquitous commerce

"The enacted view sees technologies in the context of their use and their acceptance depending on a dynamic, unpredictable and strongly mediated process by the idiosyncrasies, needs and preferences of individuals and groups."

"The enacted view argues that, if we remove individual human agency and free choice in the actual and day-to-day use of technologies, then we reach an artificial and unhelpful understanding of their success or failure [67]."

"We need to attend to technologies-in-use, that is, the actual results of introducing technologies into particular situations, contexts, tasks and communities."

"To help conceptualise and discuss them, we have introduced three principles that may be used to guide the design of appropriate systems and help interpret behaviour when this technology is seen in its use context [68]."

The consumer viewpoint

"The three scenarios agreed upon related to the usage of the system "on-the-floor" during a supermarket visit, while on the move using a mobile device and finally at home, to monitor consumption: The consumer enters the supermarket and selects a "smart" shopping cart equipped with RFID readers and a tablet PC."

"Periodically, the consumer gives permission to her home server to upload her new shopping list to the system."

"Features that proved to be the most attractive were: The constant awareness of the total cost of the shopping cart content that offers the opportunity to accurately control spending during a shopping trip Access to complete and accurate descriptions of products, including price, size, ingredients, suitability for particular uses and so forth The ability to compare the value of similar products The provision of personalised, targeted promotions that reflect the individual consumer profile in addition to the usual generic promotions, as well as the fact that consumers could access all offers available in the specific supermarket at a single interaction point The proposed in-store navigation system, especially in the case of hypermarkets, where orientation is particularly difficult The smart checkout and the ability to bypass queues and reduce waiting time".

Research methodology

"The discussions of all groups were recorded in audio and video with the permission of the participants."

"The report delivered by Market Analysis followed the conventions of the P&G methodology, and focussed solely on the value vs. barriers to use aspects of the system."

"Since several members of the development team had observed the groups via CCTV or behind the mirrored wall, there was a strong impression that the analysis was too structured for the particular system (it is designed for products that were almost ready for market introduction) and that many significant aspects have not been raised."

"Categories that emerged ranged from the collection and processing of private data and personal security to the impact of the system on social roles, etiquette and the law In the following section, we discuss the findings of this re-analysis of the focus group transcripts."

Research findings

"Consumers perceived monitoring of home inventory levels using a network of sensors as a direct violation of their privacy, since it recorded information which they considered strictly personal and private. (W5, G2): I don't like this."

"The vast majority of the participants would resist providing their data unless they could be confident that they would be used fairly and only in the context of this particular service. (W7, G1): Personal data harm because of bad use."

"Five participants observed that it is not appropriate to make this system the only available shopping mode; they demanded that they are offered a choice between using the system and not using it. (W4, G1): If I want to shake it [the service] off, I switch off the monitor or I do not use my PIN or I take a different cart [without sensors]."

Discussion and conclusions

"In the intimate computing context of ubiquitous commerce, the development of concepts of trust on this basis is of restricted use."

"Although ubiquitous computing technology may allow busines1ess to offer new ubiquitous commerce services, consumers may choose to engage in business activities with parties for whom they have access to a comprehensible company identity and, thus, a clear set of expectations of trust."

"Following the principle of context and locality, consumers in ubiquitous commerce cannot be expected to be comfortable with a single identity profile in relation to ubiquitous and pervasive services."

"In the long term, developing a trusting relationship between ubiquitous commerce and consumers is critical for its wider acceptance."

"We expect that widening the discussion on trust in ubiquitous commerce will be essential for the development of appropriate service provision models."

Exploring interdependent privacy—Empirical insights into users' protection of others' privacy on online platforms [320]

This is a machine-generated summary of:

Franz, Anjuli; Benlian, Alexander: Exploring interdependent privacy—Empirical insights into users' protection of others' privacy on online platforms [320].

Published in: Electronic Markets (2022).

Link to original: https://doi.org/10.1007/s12525-022-00566-8.

If you want to cite the papers, please refer to the original.

For technical reasons we could not place the page where the original quote is coming from.

Abstract-Summary

"A person's own information privacy is affected by the decisions of others—a phenomenon referred to as interdependent privacy."

"Drawing on an online vignette experiment (N = 330), we investigate the efficacy of an interdependent privacy salience nudge and reveal that it can decrease the likelihood that users disclose others' personal information by 62%."

"We develop a novel measurement instrument and empirically validate that users' decision to disclose others' personal information to an online platform is formed via a serial mediation mechanism through users' realization of the data transfer, recognition of others' ownership, and respect for others' rights."

Introduction

"Among the latter, Kamleitner and Mitchell [69] have proposed the "3R Interdependent Privacy Protection Framework", which postulates a sequential chain of the underlying mechanisms realization of the data transfer (RE), recognition of others' ownership (RC), and respect for others' rights (RS) forming an individual's decision to protect others' personal data."

"While information privacy research on the individual level has investigated interventions such as transparency tools to help users make informed decisions (e.g., [70, 71]), we have little knowledge on effective countermeasures that can help to mitigate interdependent privacy violations."

"We therefore raise the following research questions: RQ1: To what extent can the 3R mechanisms underlying users' interdependent privacy decision-making be empirically validated?"

"Our results indicate a three-stage mediation of the effect of our interdependent privacy salience nudge on users' disclosure of others' information through RE, RC, and RS, which validates Kamleitner and Mitchell [69]'s theoretical model of users' interdependent privacy decision-making."

Theoretical background

"Privacy has arisen to be one of the most crucial concepts of our time: While personal information (e.g., photos, preferences or location data) is being generated and shared online at a rapid pace, recent sociopolitical movements (e.g., [72, 73]) demonstrate why privacy rights are of paramount importance for individuals' freedom and sovereignty."

"Other works have focused on empirically exploring interdependent privacy behavior, for example, by investigating the monetary value which users of online services place on their contacts' personal information [74, 75], or by analyzing the roles of information sensitivity [76] or sharers' anonymity [77]."

"While property infringements mostly arise from a failure of respect, interdependent privacy violations can be caused by failures at antecedent stages [69, 78] have derived three sequential steps that users need to take in order to protect others' personal information online: realization of the data transfer (RE), recognition of others' ownership (RC), and lastly respect for others' rights (RS)."

Research model and hypothesis development

"Building upon prior works on privacy nudging (e.g., [70, 71, 79]) and interdependent privacy protection [69], we develop a research model which suggests that users' RE, RC, and RS carry over the effect of an interdependent privacy salience nudge (IPN) to users' decision to disclose others' information (DOI)."

"Regarding the violation of interdependent privacy, we thus hypothesize the following: H1: The presence (vs. absence) of an interdependent privacy salience nudge (IPN) decreases users' disclosure of others' information (DOI)."

"We posit that the effect of the IPN on users' DOI takes place via a three-stage serial mediation through RE, RC, and RS: H3: Users' realization of the data transfer (RE), recognition of others' ownership (RC), and respect for others' rights (RS) will act as a three-stage serial mediator for the effect of the interdependent privacy salience nudge (IPN) on users' disclosure of others' information (DOI)."

Research methodology

"Drawing on prior literature on privacy nudging, we designed our interdependent privacy salience nudge (IPN) with the following ideas in mind: Building on Kamleitner and Mitchell [69]'s suggestions for interventions to improve interdependent privacy protection, we aimed to implement additional steps of decision control into the data transfer process by including an opt-in mechanism that requires users to actively confirm that they have their contacts' consent to share their personal information."

"Participants chose between the following two options on how to proceed with the Instagram prompt: (1) "Press 'Allow Access' to sync my contacts" for the control group and "Check the box 'I confirm that I have my contacts' consent to share this data' and press 'Allow Access' to sync my contacts" for the treatment group, respectively; (2) "Press 'Don't Allow Access' and not sync my contacts" for both groups."

Results

"The results of our regression analysis demonstrate a significant negative effect ($B = -0.98$; $p < 0.05$; $Exp(B) = 0.38$) of the IPN on participants' DOI."

"The model revealed a positive and statistically significant direct effect of the IPN on participants' RE ($B = 3.97$; $p < 0.01$)."

"We found a positive and statistically significant direct effect of the IPN on participants' RC ($B = 0.400$; $p < 0.05$)."

"As the direct effect of IPN on DOI ($B = -0.98$; $p < 0.01$) became insignificant after entering RE, RC and RS as mediators ($B = 0.475$; $p > 0.1$), this represents a full mediation through the 3R mechanisms ([80], 2018)."

"And most interesting to our research, 41 participants (control group: 8, treatment group: 33) stated that they chose to not allow Instagram access to their address book due to their respect for their contacts' privacy rights."

Discussion

"Whereas prior works on interdependent privacy have yielded important insights into the phenomenon of interdependent privacy, what has been missing is a deeper understanding of why and how users decide to either protect or violate others' privacy when interacting with online platforms, as well as an investigation of effective remedies."

"Our study revealed that an additional step of decision control in the form of an interdependent privacy salience nudge can decrease the likelihood that a user discloses others' information by 62%."

"This is one of the first works to explore the efficacy of an interdependent privacy salience nudge in a user study."

"We show that enabling users to make an informed decision about the disclosure of others' personal information can reduce interdependent privacy violations significantly."

"The evaluation of our interdependent privacy salience nudge can serve as a starting point for the design of user interfaces where users can make informed decisions regarding interdependent privacy."

Conclusions

"This work investigates the formation of users' interdependent privacy decisions, and how such decisions can be supported by design elements such as an interdependent privacy salience nudge."

"We empirically validate Kamleitner and Mitchell [69]'s 3R Framework of Interdependent Privacy Protection by revealing a serial mediation effect of our interdependent privacy salience nudge on users' disclosure of others' information through their realization of the data transfer (RE), recognition of other's ownership (RC) as well as respect for others' rights (RS)."

"Addressing RQ2, we show that an interdependent privacy salience nudge employing an opt-in mechanism, framing, and transparent privacy-related information can support interdependent privacy protection on online platforms, with participants in the treatment group being 62% less likely to disclose others' personal information."

User preferences for privacy features in digital assistants [321]

This is a machine-generated summary of:

Ebbers, Frank; Zibuschka, Jan; Zimmermann, Christian; Hinz, Oliver: User preferences for privacy features in digital assistants [321].

Published in: Electronic Markets (2020).

Link to original: https://doi.org/10.1007/s12525-020-00447-y.

If you want to cite the papers, please refer to the original.

For technical reasons we could not place the page where the original quote is coming from.

Abstract-Summary

"This leads to privacy concerns for users and makes privacy features an important component of digital assistants."

"This study examines user preferences for three attributes of the design of privacy features in digital assistants, namely (1) the amount of information on personal data that is shown to the user, (2) explainability of the DA's decision, and (3) the degree of gamification of the user interface (UI)."

"It estimates users' willingness to pay (WTP) for different versions of privacy features."

"The results for the full sample show that users prefer to understand the rationale behind the DA's decisions based on the personal information involved, while being given information about the potential impacts of disclosing specific data."

"Further, the results indicate that users prefer to interact with the DA's privacy features in a serious game."

"A playful design of privacy features is strongly preferred, as users are willing to pay 23.8% more compared to an option without any gamified elements."

Introduction

"This poses privacy risks, because assistants extensively track and process personal data often without the user's knowledge or control after the initial consent given by the user [82]."

"This leads to a situation in which users are "increasingly challenged with managing the complex trade-offs of technology innovation with the risks of information privacy concerns" ([81], p. 17)."

"Although user interfaces are another important feature of information systems [83] and a "fruitful avenue for human–computer interaction" ([84], p. 539), to the bets of our knowledge, there is no study that examines user preferences with respect to the design of privacy features."

"To bridge this gap, our study addresses the following research question: What are prospective end-users' preferences regarding the design of privacy features of a digital assistant, and what are they willing to pay for different variants?"

Related work

"These assistants can control and visualize the data streams from different devices and are capable of learning users' privacy preferences [85]."

"Whereas several researchers identify user preferences concerning technical and organizational privacy features of digital assistants, preferences towards the user interface design are not studied."

"This study aims to identify user preferences for the design of the DA's interface for controlling and understanding privacy settings."

"We investigate preferences concerning the amount of background information about personal data that is shown to the user, the explainability of the assistant's decisions, and the degree of gamification of the DA's interface."

"In the second "study refinement" stage (originally called "product refinement"), we conducted a representative user study in which we presented choice sets of different privacy assistants with different prices to users [86]."

Results

"The most preferred variant (1st ranking) is an assistant that explains its decisions with regard to the data basis, offering information about possible impacts from the disclosure of a specific piece of information while interacting with the user as a serious game."

"We found that participants prefer comprehensive information about the treatment of their personal data."

"With regard to costs, users are willing to pay 23.8% more compared to an option without any gamified elements (€21.39 vs. €16.29)."

"The "Best Agers" prefer an explanation based on data."

"With regard to their WTP, "Best Agers" pay on average 12.82 €/month over all feature variants."

"DA Advocates" have similar preferences, but show a significantly higher averaged WTP (15.32 €/month)."

"Participants prefer gamified elements or a serious game for interaction with the assistant."

Discussion

"To answer the research question, we examined users' preferred way to control and understand the privacy settings of a digital assistant and their willingness to pay for their favorite version."

"The study takes into account insights from previous research on information privacy and introduces different user interface elements to elicit privacy preferences."

"We find that users generally opt for a restricted set of explanations, but maximum gamification of privacy in the context of digital assistants."

"Further analysis of how these clusters differ may also lead to a more differentiated view of end-users' privacy concerns about assistant systems [82]."

"Our findings show that at least restricted privacy gamification and explainability of the underlying artificial intelligence are worth pursuing, as they have business relevance and can significantly increase users' willingness to pay for digital assistants."

Conclusions and outlook

"We analyzed users' preferences and their willingness to pay (WTP) for DAs with different designs of privacy features."

"Based on a literature review and expert discussions, we identified three promising key attributes of privacy features: (1) the amount of personal data that is shown to the user, (2) the explainability of the DA's decision, and (3) the degree of gamification of the UI."

"Our results show that an assistant that explains its decisions with regard to the data basis, and offers information about the possible impacts of the data disclosure, while interacting with the user as a serious game is most preferred."

"Our study contributes to the existing literature by investigating user preferences for the design of privacy features that can be implemented in digital assistants."

"This prototype could also investigate users' preferences for how privacy-related data can be presented in a comprehensible way."

Rehumanize geoprivacy: from disclosure control to human perception [322]

This is a machine-generated summary of:

Zhang, Hongyu; McKenzie, Grant: Rehumanize geoprivacy: from disclosure control to human perception [322].

Published in: GeoJournal (2022).

Link to original: https://doi.org/10.1007/s10708-022-10598-4.

If you want to cite the papers, please refer to the original.

For technical reasons we could not place the page where the original quote is coming from.

Abstract-Summary

"In the age of ubiquitous digital content, geoprivacy is unique because concerns in this area are constantly changing and context-dependent."

"Multiple factors influence people's location disclosure decisions, including time, culture, demographics, spatial granularity, and trust."

"Existing research primarily focuses on the computational efforts of protecting geoprivacy, while the variation of geoprivacy perceptions has yet to receive adequate attention in the data science literature."

"We explore geoprivacy from a cognate-based perspective and tackle our changing perception of the concept from multiple angles."

"It is essential that we understand the spatial variations of geoprivacy perceptions in the era of big data."

"Masking geographic coordinates can no longer fully anonymize spatial data, and targeted geoprivacy protection needs to be further investigated to improve user experience."

Introduction

"Why is the privacy of our location information unique, though?"

"Location data and privacy requirements are always changing and context-dependent."

"The objectives of this paper are to Provide an overview of the recent literature pertaining to the broad topic of privacy, framed from a big spatial data perspective, Examine and identify what makes geoprivacy unique, from a human geography perspective, and Discuss how our perceptions of geoprivacy have substantively changed nowadays and how they vary around the globe."

"Johnson's opinion aligns with our view of geoprivacy, that the privacy of one's location information is almost entirely dependent on the context in which it is collected and shared."

"When it comes to location sharing, the incentives of Quality of Service (QoS) improvements frequently outweigh privacy concerns [87]."

Dimensions of geoprivacy

"Data include various attributes ranging from spatial location information (e.g., geographic coordinates, regions, places) to social-semantic data such as social relationships and shared media (e.g., text, images or videos)."

"Temporal data is also highly indicative, containing information related to trajectories, frequency of visits, the sensitivity of places, and co-location of users."

"When we acknowledge spatial data in this conversation, do we mean a location, a place, or a space?"

"Preserving one's geoprivacy thus involves more than simply masking geographic coordinates: attention also must to be paid to non-explicitly spatial data that can be used to identify someone's location [88]."

"The difference between location privacy and geoprivacy can be referenced from the comparison between space and place, where the former is data-centric and the latter is human-centric."

Contemporary conditions behind geoprivacy

"In a recent study of location disclosure based on respondents' willingness to share, privacy-related concerns are at the top (86%)."

"The decision process of location sharing has been explained by the privacy calculus model, which calculates the perceived benefits and privacy risks on user adoption [89–92]."

"Only when benefits from service providers exceed the cost of potential privacy threats will users opt to disclose personal (location) information [90, 93]."

"Because the concept of privacy is all about the "beliefs" (rather than the actual safety of information; see Section 1) [94], service providers can manipulate user perceptions to increase their intention to share."

"We suggest that LBS developers focus on creating a safer data-exchange environment (e.g., enhancing securities and limiting third-party data transfer) to dismiss most users' privacy concerns rather than choosing an easy path that only attracts a group of people."

Cultural differences of geoprivacy

"This section first discusses the cultural and demographic variables that influence individuals' location disclosure choices and then presents the range of privacy protection laws worldwide."

"Compared to selected European countries (e.g., Italy [95] and Germany [96]), Americans showed a higher level of internet privacy concerns."

"Lin and others [97] identified some interesting differences in location privacy preferences between university students in the United States and China."

"Additional age-dependent background such as levels of education [98] and internet experience [99] also have associations with privacy concerns because experienced users are more knowledgeable about potential privacy issues."

"The individual (e.g., age and gender) and the situational factors (e.g., culture) together shape people's subjective and social norms, which in turn reflect individuals' privacy attitudes and behaviours (e.g., avoidance, opt-out, and proactive protection) [100]."

Economic implications of spatial data

"While users believe that geosocial check-in services, for example, are free to use, users' personal data are collected by service providers, which can be turned into revenues (in the "privacy information markets", for example, where location data can be purchased for research or marketing purposes) [101]."

"Grossklags and Acquisti [102] found that the average WTA was much greater than the average WTP, meaning that while people generally have less interest in paying in exchange for their privacy, they may still value their privacy and would only sell personal information at a reasonable price."

"The incentives to encourage participation are not always monetary, and researchers have studied different incentive mechanisms."

"To determine the amount of (monetary) incentives, game-theoretical, or more specifically, auction-theory-based methods were popular choices (e.g., 103, 104])."

Conclusions and recommendations

"To better understand the changing perceptions of geoprivacy, we propose a number of future research directions: Designing personalized questions: Due to the subjective nature, questionnaires on geoprivacy perceptions need to have survey questions attached to personal connections [105]."

"The more knowledgeable people are about location-based technologies, the better decisions users can make when they share location data."

"The ultimate goal of geoprivacy protection is to develop a privacy-aware system [105]."

"Researchers in this area aim to take all of the important factors of privacy into account, including transparency, controllability, user feedback, informed consent, data sensitivity, information receiver, and purpose of use [106–108]."

"With stricter information privacy laws being proposed worldwide, privacy as a service (e.g., VPN) will become ubiquitous in the future and act as a guard of people's geoprivacy."

What data are smartphone users willing to share with researchers? [323]

This is a machine-generated summary of:

Beierle, Felix; Tran, Vinh Thuy; Allemand, Mathias; Neff, Patrick; Schlee, Winfried; Probst, Thomas; Zimmermann, Johannes; Pryss, Rüdiger: What data are smartphone users willing to share with researchers? [323].

Published in: Journal of Ambient Intelligence and Humanized Computing (2019).

Link to original: https://doi.org/10.1007/s12652-019-01355-6.

If you want to cite the papers, please refer to the original.

For technical reasons we could not place the page where the original quote is coming from.

Abstract-Summary

"As filling out personality questionnaires is tedious, we propose the prediction of the user's personality from smartphone sensor and usage data."

"In order to collect data for researching the relationship between smartphone data and personality, we developed the Android app track your daily routine (TYDR), which tracks and records smartphone data and utilizes psychometric personality questionnaires."

"With TYDR, we track a larger variety of smartphone data than many other existing apps, including metadata on notifications, photos taken, and music played back by the user."

"Based on the development of TYDR, we introduce a general context data model consisting of four categories that focus on the user's different types of interactions with the smartphone: physical conditions and activity, device status and usage, core functions usage, and app usage."

"On top of this, we developed the Privacy Model for Mobile Data Collection Applications (PM-MoDaC) specifically tailored for apps that are related to the collection of mobile data, consisting of nine proposed privacy measures."

"Based on data about granting TYDR all or no Android system permissions, we find evidence that younger users tend to be less willing to share their data (average age of 30 years compared to 35 years)."

"TYDR users score higher on the personality trait openness to experience than the average of the population, which we assume to be evidence that the type of app influences the user base it attracts in terms of average personality traits."

Introduction

"Having additional data about the user's personality could improve context-aware systems from different domains, e.g., mobile health, personalization and recommendations, or mobile commerce."

"With a smartphone, we will be able to track different types of data that might reflect the user's personality: the smartphone's sensors can track the user's physical context and the operating system can track the user's interaction with the smartphone and its apps."

"In Beierle et al. [109], we highlighted some aspects about the development process of the app, relating to the implementation of sensor data collection and some privacy aspects."

"We give an overview of the implementation of the introduced privacy model in the Android app TYDR. We perform an extensive evaluation of the proposed privacy model based on data collected with TYDR and evaluate what type of users are willing to share which data."

Related work and study planning

"Most of the studies aim at use cases related to mobile health or context-aware recommender systems, e.g., recommending new apps based on the personality correlated with already installed apps [110]."

"Some studies go further than correlating data with the personality of the user."

"A dynamic study will try to find correlations between sensor/usage data and changing aspects about the user, for example, mood or stress level [111, 112]."

"Sensus [113], LiveLabs [114], and AWARE [115] aim at providing researchers with frameworks for conducting research related to collected sensor/smartphone usage data."

"In order to assess the personality traits, we use the Big Five Inventory 2 (BFI-2) questionnaire [116, 117]."

"In order to register those aspects, we utilize the personality dynamics diary (PDD) questionnaire, which captures the user's experience of daily situations and behaviors [118]."

Categorization of context data

"An additional technical category constitutes the explicit permission by the user in order to allow an app to access data from the given source."

"Physical conditions and activity deal with the physical context of the user that is not related to the interaction with the smartphone."

"As its data is related to the physical context, i.e., the light level of the environment of the user, we regard it as part of the physical category."

"Our context data categorization allows to address different specific questions based on our research question regarding the prediction of the user's personality."

"Specific questions are, for example, whether the physical context alone can predict personality, how meaningful metadata is, or how accurate the prediction can be if the user did not give any explicit permissions."

PM-MoDaC: privacy model for mobile data collection applications

"Of the reviewed related work, only one paper provides some details about the processes and measures taken to ensure user privacy [119]."

"By letting the users see the data that is being collected, they can make a more informed decision about sharing it. (C) Opt-out option The possibility of opting-out is only mentioned in Wang et al. [120]."

"The app should request a permission at the point where it is needed and explain to the user what the accessed data source is being used for. (H) Secured transfer The point of having secured data transfer between mobile device and backend is explicitly mentioned in Wang et al. [120] and Ferreira et al. [115]."

"In order to contact the study participants, contact information is needed, which might contradict PM E. In order to alleviate this concern, we developed a process for identifying individual users without linking to their collected data on the backend [109]."

Implementation of the privacy model PM-MoDaC in TYDR

"The visualization of the data that is collected about the user is TYDR's core feature (PM B)."

"The person icon in a process signifies that the user is actively doing something."

"All other processes are part of the app and do not require user interaction."

"Starting TYDR for the first time, the user has to confirm the terms and the privacy policy (cf. PM A)."

"Only then the five processes of the app are started."

"In order to collect data labeled with personality states, we conducted a study where users commit to turning this feature on for a certain period of time."

Evaluation

"We address the following research questions (RQ): RQ1 What kind of data can researchers expect when they publish an app related to mobile data collection? RQ2 Is there evidence regarding user acceptance of PM-MoDaC? RQ3 In what way are user characteristics related to granted permissions?Data sets TYDR was released on Google Play mid-November 2018."

"We have data about 778 users that confirmed terms and policy and gave the permission to access their app usage statistics."

"Privacy does not seem to be a concern for those users that want to install and use the app (RQ2). (C) Opt-out option Only one user withdrew his or her data from the study."

"All analyses were conducted using R. There are some aspects to consider when interpreting the results: Users that did not fill out the demographic or personality questionnaire because of privacy concerns are not reflected in the following statistics, simply because their data is not available."

Conclusion and future work

"Context-aware applications can potentially benefit from data relating to the user's personality."

"To be able to conduct a study on the relationship between smartphone sensor and usage data and the user's personality, we developed the Android app TYDR."

"We further developed the privacy model PM-MoDaC comprising nine proposed measures that can be taken to ensure user privacy in apps related to mobile data collection."

"Analyzing the educational background and the personality trait scores of the users, we do not observe any statistically relevant pattern with respect to sharing data."

"Comparing the average personality trait scores from TYDR users to the population average, we find some evidence that the type of app influences the user base with respect to the average user personality."

"Future work includes data analyses relating to the prediction of the user's personality from smartphone data."

Online Behavioral Targeting: Are Knowledgeable Consumers Willing to Sell Their Privacy? [324]

This is a machine-generated summary of:

Li, H.; Nill, A.: Online Behavioral Targeting: Are Knowledgeable Consumers Willing to Sell Their Privacy? [324].

Published in: Journal of Consumer Policy (2020).

Link to original: https://doi.org/10.1007/s10603-020-09469-7.

If you want to cite the papers, please refer to the original.

For technical reasons we could not place the page where the original quote is coming from.

Abstract-Summary

"Unbeknownst to many online consumers, their personal information is being traded on a flourishing and rapidly increasing market for privacy data."

"In a process often labeled online behavioral targeting (OBT), data mining companies and online brokers collect, analyze, buy, and sell consumers' personal data in an effort to enable online and offline marketers to deliver personalized, highly relevant, and ultimately profitable advertisements."

"Using survey data methodology, we find that those consumers who are more knowledgeable about OBT are willing to pay more for keeping their data private and, at the same time, are willing to sell their data for less money than less informed consumers."

What Is OBT?

"Information is collected about the individual consumer's Internet and mobile activities in order to get a broad picture of the consumer."

"OBT has the potential to truly make ads informative and only target consumers who stand to benefit most from this information [121]."

"Despite outlawing some of most pernicious tracking devices in Europe [122] and to a lesser degree in the United States of America (USA) [123], even informed consumers cannot eliminate the risk of being tracked by using publicly available technological devices such as ad blockers or privacy settings on their browsers."

"An ad network sold specific information about this consumer including her behavioral profile, medical condition, and the number of her smart phone, which via GPS calculates her physical location to the food store [124]."

"Unlike traditional information collection—via survey or focus group, etc.—which usually offers a pre-agreed reward to the consumer for divulging information, OBT does not provide any direct compensation."

How Much Do Consumers Value Their Privacy?

"75% of consumers in Turow's survey [125] incorrectly assumed that when a web site has a privacy policy, it means the site will not share personal information with other web sites or companies."

"Research shows that giving consumers more control over their data or at least the perception of more control reduces consumer privacy concerns [126, 127]."

"Wang and Wu [128] provide a model that lends support to the notion that if consumers are informed, not with a simple disclaimer, but informed of the actual usefulness of collecting information (a value-added strategy), then people are more likely to relinquish privacy controls."

"The more consumers are concerned about their privacy, the more they want to control their own data [129] and avoid OBT [130–132]."

"Even consumers, who claim to be concerned about their privacy, often provide their personal information in exchange for a small or no benefit at all."

Development of Research Questions

"When alerted by the moderator that it is potentially inconsistent to have no problem to reveal personal information to get "free" access to a website (in lieu of paying the website to not collect data) and, at the same time, not to be willing to relinquish the same information for money, no common justification emerged."

"Following the literature review and the results from our focus groups, we ask the following research questions: (1) How knowledgeable are consumers about OBT? (2) What is the impact of knowledge on consumers' willingness to accept relinquishing some of their private information in return for some payment? (3) What is the impact of knowledge on consumers' willingness to pay for keeping their private information private? (4) What is the impact of knowledge on the spread between consumers' willingness to accept and willingness to pay? (5) What are implications for marketers and regulators?"

Sample and Measures

"Since "consumer knowledge" is a key variable in this study, one might argue that a student sample provides a better test for the relationships explored because online savvy university students are likely to be more knowledgeable than a random sample of US adults."

"The survey also contained questions to measure subjects' knowledge about online-based targeting practices as well as their attitudes and beliefs towards OBT."

"We performed a series of two-sample t tests to determine that the OBT-related characteristics of non-respondents for the age variable and non-respondents for the income variable are statistically indistinguishable from respondents."

"We will first create a new variable called "knowledge," which serves to aggregate responses for the questions concerning knowledge about OBT in the survey."

"This will be a key variable in our analysis since we will see that knowledge is a driving factor behind many of the responses on this survey."

Summary Statistics

"To answer our first research question, we find that the average amount of knowledge for the subjects in our survey is about 6.4, which means that on average, people answered about half of the 11 knowledge questions correctly."

"If we look at the breakdown of these knowledge questions, we can see that there are quite a few questions for which less than half of the subjects answered correctly."

"These questions correspond to the difference between e-mail and traditional mail, knowledge about whether a website can share the information it has collected, whether one can instruct the browser to accept flash cookies and what a website privacy policy really means."

"Subjects were allowed to answer "never," meaning that they would never sell their private information at any price."

Results and Discussion

"To answer the third research question, more informed people are willing to pay more to get rid of OBT and keep their private data private (higher WTP) than less informed people."

"Sight, one would be inclined to assume that more knowledgeable consumers value their privacy more than less informed consumers since they are willing to pay more to keep their data private."

"Of the surprising result that less informed consumers are asking for more money to sell their data, and, at the same time, are willing to pay less to get rid of OBT, they have a much higher wiliness to accept (WTA) than to pay (WTP)."

"To answer our fourth research question, the more knowledgeable consumers are, the smaller is the difference between WTA and WTP."

"The most knowledgeable consumers are willing to sell their privacy data for about the same price they are willing to pay for it to keep it private."

Limitations and Conclusion

"Consumers still seem to lack knowledge about OBT and are rather reluctant to sell their data or pay for websites."

"One of our central—and somewhat counterintuitive—findings is that uninformed consumers demand more money to sell their data than informed consumers and, at the same time, are only willing to pay less money than knowledgeable consumers to keep their data private."

"Our results also show that the same less knowledgeable consumers are only willing to sell their data for a—at times absurdly—high amount of money, which would indicate that they value their privacy very much."

"In our sample, the majority of the least knowledgeable consumers were not willing to sell their data for any price."

"Our results show that the more knowledge consumers had, the more they were willing to pay keeping their data private."

[Section 8]

"While Dan's personal information has been bought and sold several times, he has not been afforded the opportunity to meaningfully participate in this market for privacy data."

"While the revenue of those advertisements allows "free" access to most of the content provided on the Internet, consumers—often without their knowledge—"pay" for it by allowing online marketers access to their private information."

"Educating and informing consumers about OBT are necessary before a market-based solution for private data."

"Consumers who lack knowledge about OBT have often unrealistic expectations about the market value of their personal information, are oblivious of the current tradeoff between free online content and collection of personal data, and would neither be willing nor capable of participating in a market for online data in a meaningful way."

A comparison of privacy issues in collaborative workspaces and social networks [325]

This is a machine-generated summary of:

Pekárek, Martin; Pötzsch, Stefanie: A comparison of privacy issues in collaborative workspaces and social networks [325].

Published in: Identity in the Information Society (2009).

Link to original: https://doi.org/10.1007/s12394-009-0016-4.

If you want to cite the papers, please refer to the original.

For technical reasons we could not place the page where the original quote is coming from.

Abstract-Summary

"Two of these types of applications—collaborative workspaces and social network sites—have a number of features in common, which are explored to provide a basis for comparative analysis."

"This basis is extended with a suitable definition of privacy, a sociological perspective and an applicable adversary model in order to facilitate an investigation of similarities and differences with regard to privacy threats."

"Analysis suggests that a combination of technical, legal, and normative solutions should be considered to counter privacy issues."

Introduction

"As part of the European Community's Seventh Framework Programme (FP7/2007–2013), one of the topics of the PrimeLife project is the investigation of privacy issues of collaborative workspaces and social network sites [133]."

"People participating in these platforms provide and adapt content, and divulge personally identifiable information in the process, thus leading to (potential) privacy issues."

"This paper investigates whether and to what extent social networks and collaborative workspaces can be treated equally when trying to solve privacy threats, and suggests a number of potential solutions that may mitigate these issues."

"Rather, the goal is to outline possible types of solutions that may be considered based on the particular features of collective workspaces and social network sites."

"The mainstay of the paper is formed by an analysis of the privacy issues arising in collective workspaces and social network sites, and it concludes with a number of suggested improvements."

Social software

"The term social software characterises infrastructures, platforms and applications that enable users to communicate, collaborate and coordinate themselves via networks, to establish and maintain relationships and thus in some way map social aspects of real life to an online environment."

"Social network sites and collaborative workspaces both aim at supporting users in online collaborations; however they follow two different approaches."

"The essential feature of a social network site is the provision of user profiles and connections between them (c.f.)"

"Collaborative workspaces can be realised only with features for creating and managing content and without any user administration."

"Vice versa an application that allows people to create profiles and indicate connections fulfils all mandatory requirements of a social network site without providing additional features for communication and producing content in addition."

"Collaborative workspaces and social network sites are both automated tools to support interactions between participants."

Investigating informational privacy in social software

"The theory behind context segregation and the risk of collapsing contexts form a powerful means to analyse the privacy issues in both social network sites and collaborative workspaces."

"The mismatch between the user's expectation of social norms and the existing practices in a particular network or workspace could be another source of arising privacy issues."

"To structurally assess potential privacy threats, we have to establish what entities are exactly threatening the privacy of the users of social network sites and collaborative workspaces."

"It is necessary to distinguish between privacy issues arising from authorised access to personal data (i.e. in compliance with the law, with social norms and/or with technical protocols) and access against these conditions [134]."

"Other users Other users in the role of an adversary have an account for the collaborative workspaces or the social network site and a similar or higher level of legitimate knowledge about the application as third parties."

Privacy issues in collaborative workspaces and social network sites

"In both collaborative workspaces and social network sites, information collected by such means may cause severe privacy issues, ranging from embarrassment to identity theft, which affects many users."

"Concerns regarding the confidentiality, correctness and integrity of personal data have already been discussed in the previous section for the example of Wikipedia, and is also appropriate for social network sites when other users are considered as potential adversaries."

"From a legal perspective, while posting information on Facebook and on Wikipedia, the user in both cases grants license for further use of the personal and content data to others as clearly stated in the Terms of Use."

"Collaborative workspaces contain much less isolated personal data items than social network sites, where it is a key feature for users to build up an own profile with much well structured personal information."

"It is therefore easier to collect users' personal data automatically from social network sites than from collaborative workspaces."

Suggested improvements

"Because of the high degree of similarity between the privacy issues originating from the use of both collaborative workspaces and social network sites, it is likely that a number of potential improvements will be applicable to both platforms."

"This section discusses a number of potential improvements that help to mitigate a number of privacy issues in collaborative workspaces and social network sites."

"In social network sites and collaborative workspaces many privacy issues have non-technical causes, i.e. adversaries disregard social norms—by accident or intentionally."

"Awareness Often users are not aware of the pervasiveness of information entrusted to social network sites and collaborative workspaces."

"Social norms As we have seen one of the main causes for privacy issues is that personal information from one context is shifted to a new context without the consent of the user."

Conclusions and future work

"Similarities and differences concerning privacy issues of both social software applications have been identified."

"The finding that social software lacks fine-grained and user-determined access control options aggravates this source of privacy issues."

"Serious privacy issues are not only the result of the breach of technical implementations, but may also be brought about through the disregard of social norms and legal provisions."

"We conclude that solutions to address privacy issues in social software can neither be only technical, nor only legal, nor only based on upholding certain social norms: it is necessary to find a comprehensive approach."

Privacy lost: the net, autonomous agents, and 'virtual information' [326]

This is a machine-generated summary of:

Gotterbarn, Donald: Privacy lost: The Net, autonomous agents, and 'virtual information' [326].

Published in: Ethics and Information Technology (1999).

Link to original: https://doi.org/10.1023/a:1010011512024.

If you want to cite the papers, please refer to the original.

For technical reasons we could not place the page where the original quote is coming from.

Abstract-Summary

"This paper describes a new and significant way in which privacy is violated."

"This examination includes a discussion of technical approaches and professional standards as ways to address this violation of 'virtual information.'"

Privacy in the clouds [327]

This is a machine-generated summary of:
Cavoukian, Ann: Privacy in the clouds [327].
Published in: Identity in the Information Society (2008).
Link to original: https://doi.org/10.1007/s12394-008-0005-z.

If you want to cite the papers, please refer to the original.

For technical reasons we could not place the page where the original quote is coming from.

Abstract-Summary

"Informational self-determination refers to the right or ability of individuals to exercise personal control over the collection, use and disclosure of their personal data by others."

"The paper advocates a "user-centric" approach to managing personal data online."

"The paper explores possible technology solutions to ensure that individuals will be able to exercise informational self-determination in an era of network grid computing, exponential data creation, ubiquitous surveillance and rampant online fraud."

"The paper describes typical "Web 2.0" use scenarios, suggests some technology building blocks to protect and promote informational privacy online, and concludes with a call to develop a privacy-respective information technology ecosystem for identity management."

Introduction

"Informational self-determination refers to the right or ability of individuals to exercise personal control over the collection, use and disclosure of their personal information by others."

"All organizations that collect and use personal data of individuals must accommodate the legitimate interests of these individuals."

"Beyond personal privacy, at stake is the confidence and trust of all individuals, consumers, and citizens in today's evolving information society."

"At the Office of the Information and Privacy Commissioner of Ontario (IPC), we have long advocated a strong role for individuals to manage their personal information, They can do this by exercising their privacy rights under Ontario law, by becoming better informed about privacy risks and choices, and by using privacy-enhancing technologies (PETs)."

"PETs empower individuals in many ways; they can minimize the disclosure and (mis)use of personal data, and help secure that data from unauthorized use by others."

The twenty-first century privacy challenge

"Examples: using Amazon's Simple Storage Service (S_3) and Elastic Computing Cloud (EC_2) to store unlimited photos on Smugmug, an online photo service; using Google Apps for Word-processing; virtual worlds such as Second Life that enable users to build 3-D environments combining Web pages and Web applications (e.g. feeding a Webcast into a virtual theatre); grid computing."

"In phase 2 of consumer computing, users usually have to establish their identity each time they use a new Internet-based application, usually by filling out an online form and providing sensitive personal information (e.g., name, home address, credit card number, phone number, etc.)."

"A truly flexible identity management system would not be limited to laptop and desktop computers; it would work on cell phones, personal digital assistants, smart cards, sensors, consumer electronics like video recorders and online game consoles—any way that a user might touch the Internet."

Case studies

"In order to facilitate this, OpenID, developed by an open community, is free "user-centric" digital identity technology that simplifies the online user experience by reducing the complexity of managing dozens, even hundreds of user names and passwords across Internet sites, and providing greater control over the personal information users are required to share with websites when they sign in."

"For online businesses, these efforts can lower password and account management costs, help reduce the overall risks of security breaches by limiting the amount of customer personal information they need to store and protect, and increase both new and return user traffic by lowering the barriers to website entry and re-entry."

"With user-centric identity management, you could establish your identity once and then be able to use the full range of services in a virtual world."

Creating a user-centric identity management infrastructure

"For a successful user-centric identity management infrastructure to emerge, it is crucial that its development be driven by a wide and open community, spanning over the different geographies and cultures, and that open source implementations of all of its infrastructural components be made available."

"To enable the federation and interoperability of the different existing and emerging identity management systems, the underlying standards and specifications need to be complete, freely accessible, and, most important, driven by the community."

"Identity management systems can support compliance with privacy laws through the use of privacy policies, enforcement mechanisms, and technologies that allow applications to use only the amount of personal information that is strictly necessary to the application."

"There is currently a great deal of diversity in identity management systems, along with a multitude of open standards that support identity federation and user-centricity for these systems."

Technology building blocks

"These different scenarios will require a number of different technology building blocks, including: Open source and proprietary identity software based on open standards which can be easily incorporated into the full range of online services and devices (similar to the open source software that is at the core of the Internet and the Web today)."

"Federated identity so that once users have authenticated themselves with one service or institution, their identity credentials will be recognized elsewhere."

"Brokering of security and authentication will eliminate the need to use a different stand-alone log-on process for each application or online service."

"Multiple and partial identities so that users can access online services, explore virtual worlds, and collaborate with others without necessarily revealing their name and true identity to everyone."

A call to action

"Different segments of the "IT ecosystem" can take steps to overcome them: Corporate and individual users can explore the evolving identity systems and demand that they have privacy protection built in, as well as implementing open standards so that different systems will be truly interoperable; Standards bodies can continue to develop and promote the fundamental standards needed for identity systems, data-centered policies, and privacy-enhancing technologies; Software vendors and website developers can embrace privacy-enhancing technologies, open standards, open identity management systems, and true interoperability; and Governments, through their procurement decisions, can support the development of open identity management systems that are designed to meet user needs for privacy, interoperability, and flexibility."

"Trust the data to behave New privacy-enhancing information technologies make it possible to attach individual privacy rights, conditions and preferences directly to their own identity data, similar to digital rights management technologies for intellectual property."

"We call upon all stakeholders and technology developers, in particular, to develop trusted mechanisms for assuring widespread and enduring user confidence in the privacy and security of their identity data in the Web 2.0 world of the future."

RFID—the "Next Step" in Consumer–Product Relations or Orwellian Nightmare? Challenges for Research and Policy [328]

This is a machine-generated summary of:

Slettemeås, Dag: RFID—the "Next Step" in Consumer–Product Relations or Orwellian Nightmare? Challenges for Research and Policy [328].

Published in: Journal of Consumer Policy (2009).

Link to original: https://doi.org/10.1007/s10603-009-9103-z.

If you want to cite the papers, please refer to the original.

For technical reasons we could not place the page where the original quote is coming from.

Abstract-Summary

"Consumer-oriented RFID research is scarce and fragmented."

"This paper seeks to chart the RFID territory, address the research that is readily available, and conclude with a call for a more encompassing research agenda on RFID and consumers."

"Research on RFID is needed to better understand the various contexts RFID appear in, to inform policy-makers not familiar with the many challenges involved, and to make industries aware of the full range of consumer-critical issues in RFID development."

Introduction

"There has been limited research on RFID and consumer-related issues, and the research conducted has mainly aimed to improve technology or address managerial strategies for coping with consumer fears."

"This article will explore the field of RFID and attempt to relate it to a consumer-oriented framework which is relevant for research on consumers and for society in general, as well as for policy-makers who need to assess future implications of RFID."

"The relative breadth and focus on technical and application areas of RFID is included in this article to capture a new audience interested in consumer-related research and policy issues, which has not (or has only partially) been exposed to RFID before."

"This interest group domination should be supported, supplemented, and moderated by comprehensive RFID research with a consumer perspective."

"The aim of this article is to propose a research agenda for studying RFID in socio-consumer contexts—which means exploring RFID technology by moving away from supply chain management and into everyday life."

RFID and Its Technological History

"The RFID tag is a tiny silicon chip (microchip) for data storage with an antenna that transmits information about the tagged product to an RFID reader."

"There are in essence three types of RFID tags in operation: active, semi-passive, and passive tags."

"The semi-passive tag uses an internal battery source to ensure data integrity, but the data transmission from tag to reader is generated by the energy of the signal itself, provided by the reader ([135], p. 89)."

"The passive tag is powered by incoming electromagnetic or inductive signals from the RFID reading device."

"The advantages of a passive tag include its long life-span, minimum maintenance, low costs, and the fact that the chip only sends information when activated by a reader [136]."

RFID and Its Societal Relevance

"The manufacturing industry and retail businesses have voiced interest in implementing RFID "closer to the consumer" in order to extract the benefits of closer interaction between products and consumers."

"According to Roussos, item-level tagging is impractical in many ways due to the high costs of RFID deployment and the low profit margin of most supermarket products."

"In terms of a consumer perspective on RFID, the most interesting aspect is retail business use of such applications, both in-store—in products, on products, on shelves, and so on—and out-of-store, as consumers take home RFID-tagged items that are live."

"Juban and Wyld [137] follow up on this, claiming that most consumers are unaware of RFID as a term, how RFID will affect how products are sold, and what benefits and risks are involved at the consumer level."

RFID and Its Potential

"For retailers and manufacturers in general, user profiling (where data about the user is collected, systematised, and stored in a profile) appears to be one of the most tempting and valuable benefits of item-level RFID deployment."

"Item-level RFID can also yield new information and insights into shopping habits, as RFID leaves an "information trail" [138] of recordings about location, how and when consumers and products have engaged, and how the products travel through the value chain."

"Jensen and others ([139], p. 44), in the same vein, assert that the benefits of RFID are well documented for industry, while consumer benefits have largely been overlooked, making it necessary to draw attention to these benefits so consumers can balance the benefits with perceived risks."

"Educating consumers about the positive aspects of RFID is not enough, according to Jensen and others; consumer evaluation of government's role in protecting privacy is imperative and a major factor in consumers' cost–benefit-(risk) analyses."

RFID in Various Applications

"The upscale clothier Prada has implemented RFID in its New York store, providing enhanced information about products to customers when they scan RFID-tagged clothes with portable readers."

"NP Collection, in a "future clothing store" in Hollola, Finland, has tagged all clothing items with RFID chips that communicate with a touch screen in the dressing rooms to supply customers with enhanced product information."

"The biggest potential of RFID is in the retail sector, with automated purchasing and service systems and product-recall opportunities."

"Wal-Mart, the world's largest retailer, actually forced the implementation of barcodes in 1984 by pushing its suppliers to implement the technology, just as the retailer recently required its top suppliers to implement RFID for inventory tracking [136]."

"Wal-Mart and Gillette also tested a Smart Shelf in 2003 where the system detected information from RFID-enabled Gillette products."

RFID-Related Privacy and Consumer Concerns

"The "Big Brother" scenario surrounding RFID has spurred some public concern by evoking the vision of a future comprehensive surveillance society, with a seamless network of millions of RFID readers and billions of RFID tags placed everywhere that constantly read, process, and evaluate information about consumer behaviour and purchases [140]."

"Roussos maintains that most consumers do not trust retailers when it comes to complying with the EU Data Protection Directive, and he believes that even though low-level mechanisms and technical tools (such as tag killers or blocker tags) can ensure some compliance with relevant EU directives, "they are unlikely to conclusively address consumer concerns because the users interact with RFID systems at a much higher conceptual level" [138]."

"The RFID Position Statement of the US-based Consumer Privacy and Civil Liberties Organizations was issued in 2003 by a range of groups and organizations concerned with individual privacy, including CASPIAN, Privacy Rights Clearinghouse, American Civil Liberties Union (ACLU), Electronic Frontier Foundation (EFF), Electronic Privacy Information Centre, and others."

Regulations and Statements Relevant for RFID

"Apart from these directives, the EU has no immediate plans for either demanding industry implementation of RFID or issuing specific rules and regulations for privacy protection in RFID applications."

"In February 2008, the EC published a preliminary draft of recommended guidelines for RFID tagging, where privacy and security issues were addressed."

"Unlike the EU, the USA does not have protection of personal privacy in its constitution, although certain aspects such as "breach of confidence" have been put into law The state of California has gone further than federal legislation by addressing RFID specifically in its state laws, due to the complexity of RFID and privacy-related issues."

"EPC Global has issued the following guidelines for RFID deployment: (1) consumer notice of the presence of RFID tags; (2) consumer choice in discarding, removing, or disabling the RFID tags from products; (3) consumer education about RFID and its applications; and (4) companies must publish information on their policies regarding retention, use, and protection of consumer-specific data."

Media Coverage of RFID-Related Issues

"The various pilots and tests of RFID products performed by retailers and technology researchers closer to the consumer have resulted in RFID reaching the public audience through consumer advocates and the mass media's critical assessments of these tests."

"Albrecht has ranked RFID risk on par with nuclear risks and consequently referred to RFID as a "spychip," a "big brother bar code," and a "tracking device" ([141], p. 222)."

"These media portrayals of RFID technology impact and form public opinion on the issue, but to what extent is not known."

"In Europe, the people who were aware of RFID attained information about the technology through printed media (37%), the Internet (29%), television (16%), and word of mouth (16%)."

"Comparative technologies to RFID, like nanotechnology, have also been discussed in terms of the effects of news coverage on public opinion and the representations of the technology drawing on Sci-Fi ([142], pp.)"

Consumer-Related Research on RFID

"This research seeks to investigate how consumers' attitudes towards RFID technology relate to their actual behaviour."

"Resatsch and others [143] study more specifically point-of-sale evaluations by consumers based on RFID-related technology called Near Field Communication to see how product-relevant information such as user-generated ratings, reviews, and opinions can be accessed at purchasing time."

"The authors claim that security and privacy concerns have dominated the consumer-centred RFID research at the expense of customer acceptance."

"Their research concludes that perceived U impacts RFID acceptance the most, followed by overall attitude toward the protection of data privacy and general attitude towards novel technologies ([144], p. 37)."

"A US study by Sill and others [145] surveys how information about RFID and personal traits of consumers affect intentions to purchase RFID-tagged products."

"The focus is on how consumers' perceptions of trust, risk likelihood, and risk harm impact their intentions to use RFID technologies."

Theoretical Platform and Research Agenda for Consumer-Related RFID Research

"A first step in the direction of establishing a theoretical framework for consumer-related RFID research, which could also be extended to future studies of the Internet of Things, would be to draw on Ulrich Beck's studies on risk and Niklas Luhmann's studies on trust."

"Thiesse states that privacy risks and trust issues related to RFID can be met technologically, i.e., with PETs or with blocker tags, which have been discussed earlier."

"We can conclude that for complex technological systems, such as RFID, risk perceptions and trust responses must be researched in an integrated manner, involving the micro (relational) and macro (social/system) levels, as well as acknowledging the plethora of possible risk categories and trust mechanisms."

"Such a framework would encompass an overarching theoretical stance grounded in sociological theories of risk and trust in modern societies characterised by pervasive technologies, complex systems, and challenges that reach beyond the single human–product relations."

Conclusion

"The suggested agenda includes a theoretical and methodological discussion, supplemented by a set of specific research questions that can guide future research on consumers and RFID."

"This research agenda strives to make RFID "available and relevant" to policy-makers and consumers, addressing not only the needs and concerns of businesses and industry sectors."

"Another aspect of policy concern that relates to RFID is the potential for a widened digital divide between those who can access RFID-communicated information and those who cannot."

"As awareness about RFID use increases, consumer recognition of potential risks will be exposed."

"It might be problematic to leave complete responsibility to the consumer for making cost–benefit-risk evaluations of RFID-related consumption."

"Policy-makers and the industry must acknowledge this, and research must address the complexity in understanding RFID's interactions with consumers (micro) and the larger societal effects (macro) of full-scale RFID implementation."

Overlooking children: an experiment with consequences [329]

This is a machine-generated summary of:

Dowty, Terri: Overlooking children: an experiment with consequences [329].
Published in: Identity in the Information Society (2008).
Link to original: https://doi.org/10.1007/s12394-009-0010-x.

If you want to cite the papers, please refer to the original.

For technical reasons we could not place the page where the original quote is coming from.

Abstract-Summary

"There are dangers that overloading an already stretched social care service with low-level concerns will damage effective child protection work, while any insecurity in the systems potentially puts all children at increased risk of harm."

"Issues around consent to data-sharing have not been adequately addressed, but the reduction in confidentiality brought about by routine inter-agency information-sharing may deter children and their families from accessing services at all."

"We risk habituating children to a very high level of surveillance, and yet the possible effect of such widespread data-gathering on their personal development and future lives has not been considered."

"This paper lists some of the systems now in use or being planned, and outlines the potential risks that they pose to children's safety and development."

Introduction

"Although other rights to life, health and development may sometimes weigh more heavily in the balance, privacy is still as fundamental to children as it is to adults."

"Developments in the use of Information Technology threaten to destroy children's privacy."

"When technology is combined with an adult-centric view of children as human beings in-the-making rather than subjects of rights in the present, it threatens to create a society in which children no longer have private lives at all."

"Their perceived 'preciousness and dangerousness' create an atmosphere in which we believe that children must be watched closely in order to curb the monsters that may prey upon them, or those that lie in wait within children themselves, and that sacrificing children's privacy rights is a fair price to pay."

"Developments in a wide range of arenas are having a cumulative effect on children's privacy and their interactions with adults."

CCTV and Webcams

"Since the camera covers the entire room, all observers see every child and adult present."

"While advertising material focuses on webcams as a way of fostering parental involvement in their children's lives, a Danish study of the use of nursery webcams concludes that they are a useful means of bridging the gap between 'traditional society' and 'radicalised modernity' and could help parents who experience: 'difficulties accepting the decreasing importance of the family in the socialization of the child.' [146] Webcams are thus seen as important for parents, and in promoting a political agenda, but children are, paradoxically, invisible."

"The assertion that they permit 'parental involvement' is simply misleading: one-sided observation cannot constitute a reciprocal relationship; it merely reinforces the idea of a child as a passive object of adult scrutiny."

Mobile location

"Some include a feature that allows parents to set a physical boundary or key in a regular route, and receive an alert if the child deviates from this."

"The advertising appeals to parental perceptions of 'stranger danger' and suggests that the product can bring parents 'peace of mind', but this is illusory: all that parents can know is the whereabouts of their child's phone but not whether it is still in the child's possession."

"Any false sense of security may increase the risk that a child who is, in reality, not yet competent to be out alone will meet the more likely danger of a traffic accident."

"Whilst it is understandable that parents would prefer to avoid the inevitable anxiety of their child's first essays in independence, it is that very anxiety that ensures they weigh up the risks and equip their children to handle the unexpected."

Biometrics

"During the last five years, increasing numbers of schools have installed library, canteen and registration systems that monitor and analyse children's habits, such as their choice of reading matter (including by age, ethnicity and gender) or their individual school meal choices."

"After undertaking a literature review of the use of children's biometrics, Dr Sandra Leaton-Gray of Homerton College, Cambridge contacted the campaign group, 'Leave Them Kids Alone' to inform them that: I have not been able to find a single piece of published research which suggests that the use of biometrics in schools promotes healthy eating or improves reading skills amongst children."

"On the issue of the potential use of biometrics by other agencies or individuals, Microsoft's Chief Identity Architect, Kim Cameron, points out: If you want to find out who owns a fingerprint, just convert the fingerprint to a template and do a search for the template in one of these databases."

Databases and assessment systems: The School Census and National Pupil Database (NPD)

"They are not able to control access to Connexions partnerships' databases, but partnerships must ensure that all processing of information about young people complies with data protection principles [147]. The Connexions service can be seen as a prototype for other childhood risk management systems—those that use assessment tools and information-sharing in an attempt to predict problems and offer early intervention."

"When these plans were scrutinised by the Parliamentary Joint Committee on Human Rights, they expressed concern that: …if the justification for information-sharing about children is that it is always proportionate where the purpose is to identify children who need welfare services, there is no meaningful content left to a child's Article 8 right to privacy and confidentiality in their personal information [148].

ContactPoint has been delayed by a series of security alerts—indeed, a government-commissioned report from Deloitte, of which only the executive summary has been published, confirms that it will not be possible to make the system fully secure [149]. The latest time estimate indicates that the population of Contactpoint will begin in January 2009."

Analysis of security and privacy issues in wearables for minors [330]

This is a machine-generated summary of:

Fúster, Jaime; Solera-Cotanilla, Sonia; Pérez, Jaime; Vega-Barbas, Mario; Palacios, Rafael; Álvarez-Campana, Manuel; Lopez, Gregorio: Analysis of security and privacy issues in wearables for minors [330].

Published in: Wireless Networks (2023).

Link to original: https://doi.org/10.1007/s11276-022-03211-6.

If you want to cite the papers, please refer to the original.

For technical reasons we could not place the page where the original quote is coming from.

Abstract-Summary

"To this, the fact that wearables have been requiring more and more personal data from the user makes them attractive devices for an attacker."

"We propose a set of tests for evaluating the security and privacy of wearables and we apply them to analyse the security and privacy of a set of commercial wearables that are targeted at minors, who represent a group with especially high requirements in this regard."

"Although the considered low-end devices are broadly speaking less secure than high-end ones, most of them present security and privacy flaws, which illustrates the necessity of regulation that ensures the fulfilment of appropriate security and privacy requirements."

Introduction

"Given this scenario and the sensitive nature of the data in question, it seems timely to study the current market situation by analysing the security and privacy of these devices."

"We propose a set of tests for evaluating security and privacy risks in wearables."

"Using these tests we analyse security and privacy issues in wearables currently being marketed to children and young people, considering that they are a specially sensitive group in terms of security and privacy that represents a remarkable part of the market."

"We aim to increase transparency and user awareness on the security and privacy of wearables, as well as to encourage manufacturers to improve their security and privacy features, paying special attention to the protection of the most vulnerable groups, like children."

"To support such tests, a toolkit for security and privacy vulnerabilities analysis in the context of wearables is also provided."

Background

"Most of the eavesdropping and traffic analysis attacks are related to inadequate implementations of the BLE publication process or the use of static device addresses."

"Information gathering attacks usually involve breaking the key exchange process during BLE pairing or gathering information about other devices, such as smartphones [150]."

"All the vulnerabilities found in the context of integrity attacks are due to weak authentication methods or the absence of encryption in communications between devices."

"This VA serves to analyse and identify security issues in three different wearable devices that communicate via BLE with a smartphone: Easy Fit by Cellular Line, Fitbit Charge and Fitbit Alta by Fitbit."

"The analysis results show that the use of weak Short-Term Key (STK) encryption generation methods (such as Just Work or Passkey Entry methods) and the lack of a pairing and binding process between the device and the smartphone are the principal vulnerabilities affecting these technologies."

Testing methodology

"The following list shows attack categories that are considered to be relevant in the context of wearables and that need to be tested. (1) Authentication: The application associated with the wearable device implements a method to authenticate the user's identity. (2) Insecure pairing method: The link between the wearable device and the smartphone uses a pairing method considered insecure or ineffective against MITM attacks or passive eavesdropping attacks and lacks privacy safeguards. (3) Unencrypted Communications: BLE communications between the wearable device and smartphone are not encrypted. (4) Encryption key capture: During the pairing process, the wearable and mobile devices exchange encryption keys in a format that the BLE sniffer can easily capture and process. (5) Static MAC address: The wearable device uses a static MAC address (i.e., it does not change when the device is turned off or restarted and does not change periodically), exposing it to tracking and user identification attacks. (6) Transmission of sensitive information to third-party servers: The application sends sensitive user information to third-party servers. (7) Sending of information and firmware updates via HTTP: The application receives firmware updates and sends requests with sensitive information using HTTP without TLS encryption."

Analysis of results

"Other higher-end wearable devices require users to use proprietary or specific applications."

"Both BIGGERFIVE devices (Fitness and Vigor) and TOOBUR devices (Smart band and Smartwatch) recommend using a third-party application, VeryFitPro, which does not require any authentication or registration, although a user account can be created."

"The TOOBUR and BIGGERFIVE wearable devices analysed paired with the smartphone directly, using the Just Works Method with no encryption, allowing the device to seamlessly connect to any other device once it has lost connectivity with the central communications hub (e.g., the user's smartphone)."

"The only wearables which can prevent MITM and eavesdropping attacks are Fitbit Ace 3 and Apple Watch Series 6 since both implement BLE Secure Connections with ECDH key exchange or secure proprietary exchange methods."

"Only the applications used by high-end wearables require user authentication, and in the case of devices specifically designed for minors, only Garmin Jr. and Fitbit apply specific measures to protect minor's data."

Discussion and recommendations

"The privacy of its users is compromised, both due to possible access to sensitive information handled by these devices or because said information is transferred through insecure connections with cloud servers and shared with third-party companies."

"The importance of protecting devices against user tracking through BLE connections is reflected by the existence of several techniques and attacks that can allow for an attacker to gain private information by passively observing the communication between smartphone and wearable device [151]."

"An attacker can pretend to be a peripheral device by looking for UUIDs or spoofing GATT profiles, and track a user whenever the smartphone application tries to connect with it."

"Another crucial factor necessary to guarantee the security and privacy of wearable devices is the implementation of up-to-date protocols that incorporate secure methods aimed to safeguard user information."

Conclusions

"We presented a set of tests for analysing security and privacy risks in wearable devices."

"Using such set of tests, we evaluated the vulnerability problems associated with prevalent wearable devices, especially those targeting children and young people."

"The results obtained during the tests show that, although devices from well-known brands tend to apply more security and privacy measures than devices from smaller companies, such as BIGGERFIVE or TOOBUR, many of them were found not to encrypt BLE communications or implement pairing methods that do not ensure the privacy of user data."

"Lower-end devices use third-party applications such as VeryFitPro that send sensitive information over HTTP and are neither secure nor private."

"One of the objectives of this research is the analysis of wearable devices targeted at minors, this vulnerability is particularly worrying."

Learning from the Past to Improve the Future [331]

This is a machine-generated summary of:

Naous, Dana; Bonner, Manus; Humbert, Mathias; Legner, Christine: Learning From the Past to Improve the Future [331].

Published in: Business & Information Systems Engineering (2022).

Link to original: https://doi.org/10.1007/s12599-022-00742-2.

If you want to cite the papers, please refer to the original.

For technical reasons we could not place the page where the original quote is coming from.

Abstract-Summary

"Contact tracing apps were considered among the first tools to control the spread of COVID-19 and ease lockdown measures."

"The public debate as well as academic research about contact tracing apps emphasizes general concerns about privacy (and the associated risks) but often disregards the value-added services, as well as benefits, that can result from a larger user base."

"While the results confirm the privacy-preserving design of most European contact tracing apps, they emphasize the role of value-added services in addressing heterogeneous user segments to drive user adoption."

"The findings thereby are of relevance for designing effective contact tracing apps, but also inform the user-oriented design of apps for health and crisis management that rely on sharing sensitive information."

Introduction

"Some of the first empirical studies that responded to this call are Trang and others [153], who analyze the impact of various app specifications (i.e., benefit appeal, privacy design, and convenience design) on app acceptance, as well as Meier and others [154] and Welrave and others [155], who examine users' intentions to use contact tracing apps in light of privacy concerns."

"Other studies have analyzed how more users can adopt contact tracing apps."

"For contact tracing apps, this implies that more attention is needed to provide services that offer benefits to users and can result in a larger user base."

"This motivates our research goal to explore the role of goal-congruent features in improving the design of contact tracing apps and as an adoption driver."

"We study the following question: What are users' preferences for contact tracing app features, and what is the impact of value-added services on users' adoption of these apps?"

"Based on a conjoint analysis (CA) study in Germany, we provide empirical insights into users' preferences for core and privacy-preserving features, as well as value-added services of contact tracing apps."

Background

"These contact tracing apps continuously track users' proximity and, in the event of possible COVID-19 exposure, notify them that they should self-isolate [156, 157]."

"Whereas most countries use BLE technology in the design of their contact tracing apps, only a few have adopted a location-tracking mechanism to cross-check paths, as Israeli app Hamagen has done."

"User adoption is a crucial factor for the success of contact tracing apps in curbing the transmission of COVID-19."

"Studies relating to contact tracing app design employ discrete choice experiments in assessing users' preferences and trade-offs for different implementation options."

"We have seen from previous studies that perceived benefits for both self and society, privacy perceptions, and usability aspects can play an important role in the adoption of contact tracing apps (e.g., [153])."

Research Approach and Design

"It aims to understand individual and group user preferences for different app designs and analyze the impact of value-added services on the adoption of contact tracing apps to improve their design."

"Applying the utility concept from economics, CA enables the estimation of a user preference structure based on his evaluation of different product attributes or features."

"In the third phase of the attributes and levels selection, we organized a focus group with five current and potential users of COVID-19 apps to identify important attributes and eliminate the unacceptable option."

"Participants are asked to build the most preferred configurations of the contact tracing app from the list of available attributes and levels."

"This allows us to estimate the user preferences for the different attributes and levels based on the choice data."

"This allows us to understand opportunities for improvement in the contact tracing app design based on feedback from users."

Results

"Our results show the contact tracing app's core services—exposure logging (19%) and test results sharing (13%)—as the two most important attributes."

"To gain insights into different user segments for contact tracing apps and varied group preferences, we performed a cluster analysis based on the individual part-worth utilities."

"With market simulations enabled by conjoint analysis, we can understand whether adding value-added services with the proposed contact tracing app can result in higher market shares and, therefore, better adoption rates."

"We observe that App 1 (simple diagnosis) and App 4 (Maps) would result in higher market shares within their categories for value-added services, with slightly better results for App 1."

"App 5 with a diagnosis service of symptoms tracking and contextual service of maps also resulted in higher market shares: 43% of users."

Discussion: The Varying Users' Preferences for Contact Tracing Apps

"With regard to individual user preferences, we find that exposure logging and test results sharing are the most important features in contact tracing apps, while exposure notification as the third core service lags far behind."

"Our findings support the dominant privacy-preserving design of most European contact tracing apps and confirm user preferences for a decentralized approach and contact tracing through proximity rather than location-based tracking via GPS."

"Despite the ongoing debate about privacy concerns and contact tracing apps, the results show that not all privacy-preserving features are valued by users."

"A platform design could help avoid that several applications functioning in silos emerge, with none of them achieving critical mass: For example, Switzerland was the first country to promote a decentralized approach that preserves user privacy, but its app design did not have goal-congruent features until 2021, when it added an event check-in option."

Contribution

"Building on the work of Wortmann and others [158], we argue that contact tracing app system design can be improved through the integration of value-added services that benefit the self as well as the public."

"Our results provide an illustrative example of the impact of goal-congruent features on system use and acceptability and help explain the uptake of contact tracing apps in Germany, France and UK."

"From a practical perspective, our results are relevant to the application developers and service providers of contact tracing apps."

"The preference model resulting from the CA study provides concrete realization options of the contact tracing app to be taken into consideration to gain sufficient critical mass and acceptability among users."

Limitations and Outlook on Future Research

"An important limitation of our study is its focus on a sample from Germany, which has specific cultural characteristics and a democratic system, as well as an a priori model of decentralized contact tracing."

"Since privacy is a dominant topic in contact tracing apps, we have to acknowledge that our study focused on understanding user preferences for app design, but we did not use a detailed questionnaire to assess our respondents' privacy awareness or concerns."

"It is important to note that our study took place while the contact tracing apps were being launched, and our selection of attributes was based on available and suggested options in June 2020."

"The suggested methodological approach could be used to study those additional goal-congruent features in order to further improve the design of contact tracing apps and expand the use of market simulations in application design."

A requirements taxonomy for reducing web site privacy vulnerabilities [332]

This is a machine-generated summary of:

Antón, Annie I.; Earp, Julia B.: A requirements taxonomy for reducing Web site privacy vulnerabilities [332].

Published in: Requirements Engineering (2003).

Link to original: https://doi.org/10.1007/s00766-003-0183-z.

If you want to cite the papers, please refer to the original.

For technical reasons we could not place the page where the original quote is coming from.

Abstract-Summary

"Consumers often have only the stated Web site policies as a guide to how their information is used, and thus on which to base their browsing and transaction decisions."

"This paper presents a taxonomy of privacy requirements for Web sites."

"Using goal-mining, the extraction of pre-requirements goals from post-requirements text artefacts, we analysed an initial set of Internet privacy policies to develop the taxonomy."

"This taxonomy was then validated during a second goal extraction exercise, involving privacy policies from a range of health care related Web sites."

"This validation effort enabled further refinement to the taxonomy, culminating in two classes of privacy requirements: protection goals and vulnerabilities."

"The identified taxonomy categories are useful for analysing implicit internal conflicts within privacy policies, the corresponding Web sites, and their manner of operation."

"These categories can be used by Web site designers to reduce Web site privacy vulnerabilities and ensure that their stated and actual policies are consistent with each other."

Introduction

"We also introduce a taxonomy of privacy protection goals and vulnerabilities, developed using a grounded theory approach (content analysis), that provides an effective mechanism for analysing and comparing privacy policies, system requirements and the functionality of the respective systems."

"We employed a content analysis technique, goal-mining (the extraction of pre-requirements goals from post-requirements text artefacts), to derive the privacy-related goals of various Internet health care Web sites."

"The approach concurrently led to the development of an integrative taxonomy and goal-mining method as well as the analysis of Internet privacy policies."

"The second privacy policy analysis effort enabled evaluation and refinement of the taxonomy and its associated goal-mining heuristics."

"The health care privacy policies we examined (and which serve as the primary focus of this paper) were in force during June of 2001, when the second goal-mining effort was conducted."

Background and related work

"P3P requires consumers to answer a set of standardised multiple-choice questions via their Web browser that address various aspects of Web site privacy policies."

"P3P cannot guarantee a company's Web site is operating according to the practices expressed in the company's privacy policy."

"A TRUSTe seal simply ensures that TRUSTe has reviewed the licensee's privacy policy for disclosure of the following uses of information by a Web site: what personal information is being gathered; how the information will be used; who the information will be shared with; the choices available regarding how collected information is used; safeguards in place to protect personal information from loss, misuse, or alteration; and how individuals can update or correct inaccuracies in information collected about them."

"Similar to TRUSTe, consumers are given a false sense of security when they encounter a BBBOnLine seal since they do not realise that a Web site can display it regardless of whether or not a privacy policy truly protects consumer privacy."

Taxonomy of privacy goals

"The goals derived from 25 Internet e-commerce privacy policies were categorised according to common characteristics that emerged and coded into the following categories: notice/awareness, choice/consent, access/participation, integrity/security, enforcement/redress, monitoring, aggregation, storage, transfer of information, collection, personalisation and solicitation."

"Some of the goals exemplified privacy protection practices while others reflected practices that introduce vulnerabilities into a site's ability to protect personal information."

"In the examined e-commerce privacy policies, few goals related to the identity of the organisation collecting the data; the examined privacy policies either did not address this issue at all or in a few cases simply noted that their sites contained links to other sites that collected PII."

"Choice and consent goals reflect ways in which a Web site ensures that consumers are provided the option to decide what personal information collected about them is to be used and whether it may be used for secondary purposes."

Case study: health care privacy policy Requirements Analysis

"This section describes the goal-mining process within the context of our health care privacy policy analysis."

"This goal, although expressed differently on different sites, appeared in 10 of the 23 analyzed health care privacy policies."

"We hypothesised that the number of protection goals in a health care privacy policy is greater than the number of vulnerabilities for that policy; this hypothesis was confirmed as true."

"In 8 of the examined privacy policies we observed the number of protection goals for a given Web site to be equal to or fewer than the number of expressed vulnerabilities in that Web site; for example, AETNA's privacy policy stated five vulnerabilities and five protection goals."

"A careful analysis of selected goals revealed that one privacy policy failed to include the goal even though the respective Web site does in fact allow cookies to be sent to third parties."

Summary and future work

"As one would expect, a site that supports e-commerce transactions will obviously require more policy statements that focus on PII related privacy."

"Goal-mining with the privacy goal taxonomy is effective for examining how Web sites claim they manage online customer data and how they convey these practices to their customers [159]."

"We have introduced a taxonomy for classifying privacy protection goals and vulnerabilities; we describe our use of a software requirements engineering content analysis technique, goal-mining, to examine privacy policies for system goals and requirements; and we codify domain specific heuristics for applying the GBRAM for goal-mining Internet privacy policies."

"Examining and comparing privacy policies using goals is an innovative and effective analysis method that provides useful guidance to software developers, policy makers and consumers."

"The taxonomy goals will be used to populate the SMaRT (Scenario Management and Requirements Tool) to support the specification of privacy related requirements by software developers, providing additional validation."

Explaining automated decision-making: A multinational study of the gdpr right to meaningful information [333]

This is a machine-generated summary of:

Dexe, Jacob; Franke, Ulrik; Söderlund, Kasia; van Berkel, Niels; Jensen, Rikke Hagensby; Lepinkäinen, Nea; Vaiste, Juho: Explaining automated decision-making: a multinational study of the GDPR right to meaningful information [333].

Published in: The Geneva Papers on Risk and Insurance—Issues and Practice (2022).

Link to original: https://doi.org/10.1057/s41288-022-00271-9.

If you want to cite the papers, please refer to the original.

For technical reasons we could not place the page where the original quote is coming from.

Abstract-Summary

"The General Data Protection Regulation (GDPR) establishes a right for individuals to get access to information about automated decision-making based on their personal data."

"By recruiting volunteering insurance customers, requests for information about how insurance premiums are set were sent to 26 insurance companies in Denmark, Finland, The Netherlands, Poland and Sweden."

"Findings illustrate the practice of responding to GDPR information requests and the paper identifies possible explanations for shortcomings and omissions in the responses."

"The paper discusses what can reasonably be expected from explanations in consumer oriented information."

Introduction

"We conducted an empirical investigation of how Swedish insurance companies responded to information requests made by consumers regarding automated decisions in the pricing of home insurance."

"In addition to repeating the same study in more countries, the present study also goes beyond the previous study in two respects: First, while Dexe et al. [160] offered a description of the history of, and recommendations for, applying the GDPR, the article did not give any further insight into why companies in practice seemed to have interpreted the requirements in such different manners."

"The following research questions are investigated: These research questions were investigated by use of actual consumer requests to 26 separate insurers in five EU countries."

"Trust plays in important role here, and insurers thus have every reason to be transparent and responsive to customer requests about how their data is used and how decisions are made (see, e.g., [161])."

Related work

"Alizadeh et al. [162] interview 13 volunteering households who requested their personal data from loyalty program providers, investigating expectations about GDPR and the right to access data."

"While the number of empirical studies on the effects of the GDPR in general are growing, the right to meaningful information about the logic involved in insurance has not, to our knowledge, been investigated before, with the exception of our previous article [160] and Sørum and Presthus [163]."

"Wachter et al. [164] are also sceptical about the right to explanation and argue that it does not exist in the GDPR."

"Of particular relevance to the topic of insurance is prior work by Binns and others, who aim to "find explanation styles which could plausibly meet or exceed the regulatory requirements regarding transparency of automated decisions, in particular the requirement that organisations provide 'meaningful information about the logic involved' in an automated decision (GDPR art 15(h))" ([165], p. 4)."

Method

"The approach can be summarised in the following manner: (1) Identify the relevant type of insurance—extant insurance offerings will differ between countries and companies. (2) Get an overview of the market—identify main actors and rough market shares (to know which ones are most relevant to include). (3) Translate the request. (4) Recruit volunteers—making sure that there is only one per company. (5) Send requests—volunteers are asked to note date and means of contacting their insurer. (6) Gather responses from the volunteers and analyse these."

"Home insurance was initially chosen for several reasons: (i) multiple market actors offering very similar products, (ii) a straightforward product which is comparable across companies, (iii) it is nearly ubiquitous on the Swedish market (nearly 97% of Swedes have home insurance [166], and (iv) it is based on non-sensitive data which makes it easier for volunteers to share responses (if the study been done on, e.g., health insurance it would probably have been more difficult to recruit volunteers)."

Results

"Three companies (DK2, DK4, and DK5) provided generic examples of the data categories used to calculate the home insurance without referring to the specific categories used to calculate the pricing of the enquirer."

"One company gave three category examples (address, area of living, and price of insurance)."

"The other company (DK1) gave a general description (resembling "input influence-based explanation") of the company's algorithmic calculation stating that calculating insurance price is "based on data and associated statistical models."

"One company provided examples of the data categories that were relevant in the calculation of the price, including both data collected through public information (the living area, year of construction, and type of home) as well as information provided by the customer (year of birth, family composition)."

"Four companies provided examples of data categories taken into account in calculating the insurance premiums."

Discussion

"We suggest two types of interpretations identified in our discussions: The narrow interpretation of Article 15(1)(h) implies that the phrase 'at least in those cases' relates to the 'automated decision-making, including profiling, referred to in Article 22(1) and (4)' as a whole phrase."

"The broad interpretation suggest that 'at least in those cases' could refer to 'automated decision-making', meaning that profiling in scope of Article 22 GDPR would be regarded as a non-exhaustive example of processing covered by this provision, and that meaningful information is equally applicable to automated decision-making in general."

"It suggests an explanation for some of the differences in that if companies can assume that customers have prior knowledge and are aware of the context in which the requests for information were sent, then the companies have less reason to explain every single data point that is included in the pricing algorithm."

Conclusions

"With respect to RQ3, we have shown that one complicating factor for companies and other actors is the fact that the requirement for meaningful information about algorithmic decisions is subject to differences in interpretation—both in terms of how the paragraph is phrased (in any language) and the fact that the different language versions seem to point to different interpretations."

"For the most part, the responses only give partial explanations for what goes into the pricing algorithm and not one seems to contain an exhaustive explanation for all data points that are reasonably part of an automated decision regarding pricing in home insurance."

"That does not mean that such an explanation is acceptable in the eyes of the law, or that it is right or just for a company to assume such prior knowledge when responding to requests for meaningful information."

Lifelogging: Privacy and empowerment with memories for life [334]

This is a machine-generated summary of:

O'Hara, Kieron; Tuffield, Mischa M.; Shadbolt, Nigel: Lifelogging: Privacy and empowerment with memories for life [334].

Published in: Identity in the Information Society (2008).

Link to original: https://doi.org/10.1007/s12394-009-0008-4.

If you want to cite the papers, please refer to the original.

For technical reasons we could not place the page where the original quote is coming from.

Abstract-Summary

"The growth of information acquisition, storage and retrieval capacity has led to the development of the practice of lifelogging, the undiscriminating collection of information concerning one's life and behaviour."

"We argue that some of the privacy concerns are overblown, and that much research and commentary on lifelogging has made the unrealistic assumption that the information gathered is for private use, whereas, in a more socially-networked online world, much of it will have public functions and will be voluntarily released into the public domain."

Introduction

"The logical end-point of the drive to store digital information is to store indiscriminately, a practice called lifelogging."

"Lifelogging can be passive—one stores the by-products of the life one would have lived anyway—or active—one surrounds oneself with sensors and information capture tools to create as rich a picture of one's life as possible."

"The purpose of storage of such information can vary, and may not be clear even to the lifelogger at the point of storage, but it is relatively straightforward to gather and likely to be of interest either in itself (to the individual who generated it), or to help with associative searching."

"In an information-intensive age where the surrender of digital identity is a commonplace, for purposes ranging from commerce, marketing, social networking, government, receipt of services, travel or security, lifelogging has the potential to reaffirm the individual's control of his or her own identity."

Lifelogging

"There are services which allow people to upload medical records, such as Google Health or Microsoft's HealthVault, while social networking methods have been used to support Web 2.0-style information sharing about symptoms and treatments of various illnesses, such as PatientsLikeMe, which hosts a number of MySpace-like communities discussing conditions such as Motor Neurone Disease or Parkinson's."

"Some social networkers will use lifelogging techniques to generate large quantities of information for their own use and to share with friends or like-minded people, while also incorporating social information to add further contextual information to their own lifelog."

"Personal items, such as Web browsing habits, geodata or emails, can be stored in a personal knowledge base, while public domain information, including personal information that the user has deliberately exposed on social networking sites, can be stored separately."

"Social lifelogging extends the solipsistic model by adding in shared information, and sharing information one generates oneself."

Privacy, identity and accountability

"It is clear, particularly on the reasonable conceptualisation of privacy as a condition of restricted access to information about a person [11], that the Durkheimian trust that is exhibited in social networking generally, and social lifelogging in particular, will threaten privacy."

"It is arguable that Jennifer Ringley was not properly aware of the nature of the threat to her privacy posed by the JenniCam [167], but a different model of lifelogging would focus on information both public and private, with two-way traffic between the individual and his or her social network(s)."

"In a situation where the lifelogger is accountable for something, he or she is empowered by the lifelog to the extent that it allows him or her to provide information to satisfy the expectation, or provide evidence that competing accounts are false."

Discussion

"As lifelogging tools are currently a matter for research rather than commercial application, there is a window for reflection about privacy and regulation."

"Tools to gather data will have obvious applications for surveillance, and this is a powerful worry, as evinced by the fate of the DARPA LifeLog project."

"Many presume that the data are necessarily personal, though they may be public domain, but why think that information gathered should be kept private?"

"In this more nuanced environment, lifelogging tools can provide the user with somewhat more control, gathering together his or her appearances in the public domain."

"Knowing what can be seen is an important first step in the preservation of privacy, and it is important that restrictions or distortions are not brought into lifelogging tools that prevent this empowerment of the user."

"Paradoxically, one could use lifelogging techniques to decide which information to delete (although deletion goes against the grain of the lifelogging ethos)."

Conclusion

"The dangers exist, but the discussion so far is framed on possibly false assumptions that lifelogs will (a) consist of personal information, (b) be universal in scope, (c) include information that has traditionally been held private by owners, and (d) become a mainstream activity, possibly via social pressure."

"We have also argued that lifelogging can be empowering for the logger, allowing him or her information which can be used in the construction of a personal online identity, or identities, which is not under the control of authorities."

"A final point: lifelogging sounds somewhat recondite, but personal knowledge management is an issue for anyone with a significant Web presence, or who uses digital technologies."

"To that extent, lifelogging tools are tools for everyone to exert more control over their personal data, their public presence online and their digital identity."

4.2 A Roundup of International Data Protection Developments Outside EU

Machine generated keywords: digital, breach, sector, user, global, datum breach, threat, detail, protection law, jurisdiction, system, confidentiality, loss, identity, million.

Data localization in India: Paradigms and processes [335]

This is a machine-generated summary of:

Bailey, Rishab; Parsheera, Smriti: Data localization in India: paradigms and processes [335].

Published in: CSI Transactions on ICT (2021).
Link to original: https://doi.org/10.1007/s40012-021-00337-4.

If you want to cite the papers, please refer to the original.

For technical reasons we could not place the page where the original quote is coming from.

Abstract-Summary

"This paper examines the debates around data localization in India, using the term to mean any kind of mandatory requirements of local storage or processing of data within the country."

"We focus on three policy initiatives—the Personal Data Protection Bill, the Reserve Bank of India's payments directive, and the localization conditions in telecommunication licenses."

"We find each of these processes to be lacking in terms of robustness, transparency, and deliberative policy making with the data protection discussions faring better than the other two initiatives."

Introduction

"Requiring the local presence of business assets, personnel, and the processing of data, are some of the ways in which these demands are being manifested."

"We focus on the debates around data localization in India, using the term to mean any kind of mandatory requirements of local storage and/or processing of data within the territory of the country."

"This paper is aimed at understanding the justifications and challenges of data localization and the robustness of the policy processes through which such norms are being introduced in India."

"We examine the extent to which the policy processes surrounding data localization have adhered to the desirable principles of good governance in policy formulation."

"We examine how India's policy processes around data localization fare in terms of adherence with the desirable practices of problem identification, evaluation of alternatives, and consultations."

Understanding localization and its policy context

"The phrase 'data localization' is commonly used to refer to a requirement for local storage of data."

"Under this view localization measures may include for instance, prohibitions on foreign data transfers, requirements to obtain consent prior to cross-border transfers, or even the imposition of taxes on data exports."

"We use the term in its commonly understood sense, implying a mandatory requirement for local data storage."

"While some, such as Russia, China, Vietnam and Indonesia, have opted for relatively broad based localization requirements, most others tend to apply differential standards based on the nature of the data and the sector to which it pertains."

"To take a few examples, sectoral localization norms are found in Australia (health data), France (data relating to judicial proceedings) and Germany (telecommunications metadata and tax accounting data) [168]."

"The Information Technology Act, 2000 (IT Act) implements horizontal conditional localization requirements in the context of sensitive personal data."

Policy motivations for localization measures

"To the core issue of privacy and security of data, it also articulates concerns regarding the broader strategic and security interests of India, issues concerning access to data for law enforcement and government functions, as well as the need to develop a local AI ecosystem [169]."

"Surveillance: localization is often justified on the grounds that this may limit the ability of foreign states to access Indian data."

"Localization may not actually limit the ability of foreign intelligence agencies to access data stored in India, given the pervasive and sophisticated nature of intelligence gathering tactics used today [170–172]."

"Looking at the digital ecosystem in India, it appears that the costs imposed by broad localization measures would be non-trivial, given the underdeveloped state of India's data center infrastructure."

An assessment of the regulatory processes

"Sections we examine three data localization initiatives in India to assess how they fared in terms of the effectiveness of the regulation making process."

"The selected initiatives are, the localization proposals under the PDP Bill, the RBI's directive on localization of payments data, and the provision contained in the licenses issued to TSPs."

"The selected localization initiatives represent a mix of different types of policy instruments, ranging from a proposed legislation (data protection) to sector specific requirements in the form of directives and licensing conditions."

"Like in the case of the earlier version, the phrase critical personal data continues to be undefined in the draft law The RBI issued a directive on 6 April, 2018, requiring all payment system providers to store all their data related to payment transactions only in India [173]."

Conclusion

"We examined the policy motivations and challenges associated with the implementation of data localization norms in India and the robustness of the policy processes through which such norms are being introduced."

"Following an assessment of each of these perspectives we find that the costs of introducing broad and sweeping data localization norms are likely to outweigh its benefits, from a rights-based perspective as well as an economic one."

"We therefore examine the extent to which the policy processes in India surrounding data localization have adhered to the desirable principles of good governance in policy formulation."

"We examine whether the introduction of the localization norms was accompanied by processes such as a clear articulation of the problem, examination of alternatives and their costs and benefits, and a consultative decision making process."

"We find that India's move towards data localization is not accompanied by a robust and deliberative policy making process."

A Failure to "Do no Harm"—India's Aadhaar biometric ID program and its inability to protect privacy in relation to measures in Europe and the U.S. [336]

This is a machine-generated summary of:

Dixon, Pam: A Failure to "Do No Harm"—India's Aadhaar biometric ID program and its inability to protect privacy in relation to measures in Europe and the U.S. [336].

Published in: Health and Technology (2017).

Link to original: https://doi.org/10.1007/s12553-017-0202-6.

If you want to cite the papers, please refer to the original.

For technical reasons we could not place the page where the original quote is coming from.

Abstract-Summary

"It is important that digital biometric identity systems be used by governments with a Do no Harm mandate, and the establishment of regulatory, enforcement and restorative frameworks ensuring data protection and privacy needs to transpire prior to the implementation of technological programs and services."

"Digital identity policy development, as well as technical deployment of biometric technologies and enrollment processes, may all differ markedly, and could depend in some part at least, on the overall economic development of the country in question, or political jurisdiction, among other factors."

"This article focuses on the Republic of India's national digital biometric identity system, the Aadhaar, for its development, data protection and privacy policies, and impact."

"Two additional political jurisdictions, the European Union, and the United States are also situationally analyzed as they may be germane to data protection and privacy policies originated to safeguard biometric identities."

"Since biometrics are foundational elements in modern digital identity systems, expression of data protection policies that orient and direct how biometrics are to be utilized as unique identifiers are the focus of this analysis."

"Both vigorous and descriptive means of data protection should be well situated within any jurisdictionally relevant deployment area, prior to in-field deployment of digital identity technologies."

Introduction

"Of the digital biometric identity systems in existence today, the most notable information exploitation case is that of the Republic of India."

"India's Aadhaar provides a unique and prominent case study for how risks that are endemic to identity systems have developed, and have since been welded-in to their digital biometric system."

"Joseph Atick describes digital identity ecosystems as "a platform consisting of a collection of technologies, processes and policies that are integrated together to enable unique natural persons to prove, unambiguously and securely, who they are to an information system and to empower them to assert their legal rights in a digital context." [174] The ability to merge inexpensive computer storage, and substantial computing power has increased both the capacity for building these large identity systems to accommodate digitized biometric elements from populations, and the appetite for doing so."

"The use of digital identity systems and biometrics for identifying or authenticating individuals need not be onerous, if appropriate protections are in place."

India's national digital biometric ID system and policies

"Despite the near-ubiquity of the Aadhaar, and its increasing use in everyday life, India's government has still not passed national data protection and privacy legislation for the Aadhaar identity system, even though suitable proposals have been drafted that would provide a version of globally accepted and widely implemented data protection standards."

"Given the broad access of the government to the Aadhaar database, including for law enforcement purposes, and the ability of the Indian government to link DNA data to the card at a future data, combined with the lack of privacy protections in The Aadhaar Act, it is regrettable that the National Identification Authority of India Bill 2010—which contained privacy provisions—was quietly withdrawn from Parliament after the passage of The Aadhaar Act, thus making it even more difficult for the Indian government to debate and pass a privacy bill for Aadhaar."

Europe's general data protection regulation and biometrics

"Countries and economic or political jurisdictions outside of the EU that permit the widespread use of biometrics in their respective societies will either have to have in force regulations that meet EU standards, or they will need to have a dual system that imposes definitive protections and standards for the biometric data of EU residents, and then a separate system implementing the standards of their own jurisdiction."

"Any member state in Europe seeking to use an individual's biometric data as defined in the GDPR, with few exceptions, will have Special Processing obligations under European law in regards to the collection, processing, and use of the biometric data in addition to ensuring that all other rights, such as notification of data breach, among other activities, are conducted."

"The European approach to biometric data differs markedly from the approach to biometric data in the Republic of India, in that India has not passed baseline data protection regulation for its digital biometric identity system, Aadhaar."

US data protection and privacy regulatory framework and biometrics

"A few states have specific biometric laws, including Illinois, which has passed the Biometric Information Privacy Act."

"The Privacy Act remains a law of substantial consequence for federal agency privacy practices, including the use of biometrics."

"Beyond REAL ID, much of existing federal regulation relates to government use of biometric data in federal and state law enforcement activities."

"The most important state level law is the Illinois Biometric Information Protection Act (BIPA), which is the strongest state biometric privacy law to date."

"While the Privacy Act of 1974 should theoretically provide protection in the case of Federal government uses of biometric technology, the US Government Accountability Office (GAO) report on the implementation of Federal privacy was not encouraging regarding compliance and transparency in uses by law enforcement."

Discussion: biometrics policy

"In the Republic of India, the Aadhaar Act and other legislation does not provide comprehensive data protections and privacy for the Aadhaar program and its use of biometric data."

"The U.S. has a patchwork of focused, sector-based regulation that applies unevenly in regards to data protection and privacy for biometrics, including a lack of broadly applicable data protection and privacy legislation on the use specifically of digital biometric identity systems."

"In non-EU jurisdictions, much progress is possible if serious attempts at legislation aimed at improving data protections and privacy specifically for biometrics use, including digital biometric identity data, are undertaken."

"When the technology for the Aadhaar system—including the collection of biometrics—was discussed as a potential program, legislation regulating the targeted and limited use of the Aadhaar identity and data should have been put forward as a mandatory step prior to any widespread technical deployment or biometric enrollment of residents."

Conclusion: what are the stakes for a failure to act?

"In considering India's Aadhaar program and its lack of adequate protections of privacy and autonomy, what stands out the most is the continuum of choices that have to be made to protect privacy rights, freedom of choice, and how the timing of making the right choices appears to matter a great deal."

"With 97% of adults enrolled in the Aadhaar biometric scheme in India, India's policy around its government-issued national biometric identity card may have garnered benefits, but it is also riddled with highly problematic human rights and other challenges."

"Given the increasing deployment of patient biometric authentication in health care settings, and the high potential of a national digital biometric identity system in the future, the US will need to pay close attention and take focused action in order to address the forthcoming and significant security and privacy challenges."

The Aadhaar: "Evil" Embodied as Law [337]

This is a machine-generated summary of:

Satpathy, Tathagata: The Aadhaar: "Evil" Embodied as Law [337].
Published in: Health and Technology (2017).
Link to original: https://doi.org/10.1007/s12553-017-0203-5.

If you want to cite the papers, please refer to the original.

For technical reasons we could not place the page where the original quote is coming from.

Abstract-Summary

"However, India, and the distressing privacy annulling actions on the sub-continent have become the chief discussion point among those in the world who are alarmed over the ways in which personal privacy is being encroached upon by moneyed interests of all variety, and by those governments who are willing to collude with them, under false pretenses."

"In a nation with nearly 1.4 billion people, one readily identified as the world's largest democracy, forces are now at work, arbitrarily, autocratically, undemocratically, and unconstitutionally deploying, what has come to be known as the world's largest biometric national ID program (scans and captures iris, fingerprints and facial inputs into a government database), called The Aadhaar."

"As a member of the national parliament, and as the only opposing voice to the implementation of the Aadhaar, in a chamber of 545 members, the author aims to candidly introduce the reader to the leading issues that are enabling the Central government and moneyed interests to collude, and to suppress democratic processes to ramrod the Aadhaar legislation through the parliament as a 'money bill,' and what this Indian program and experience could herald."

Introduction

"The debate around Privacy in India has been mainstreamed primarily due to the introduction and implementation of a national biometric identification program called "the Aadhaar" (termed 'basis', when loosely translated to English)."

"The agency mandate is to issue the Aadhaar, or a set of unique identifying numbers (12 digit) to each individual Indian citizen, which will be foundationally reinforced by highly sensitive citizen biometric information."

"Politicians of all hues have been coaxed and convinced to accord extreme importance to the implementation of Aadhaar, while the Supreme Court of India seems totally out of touch, being indecisive, and unable to deliver any judgement—in any matter coming up before it regarding the Aadhaar, and all too importantly, matters related to privacy issues."

"In all, this article shall seek to present certain important multidimensional perspectives, as to why, the largest national biometric identification program in the world, called the Aadhaar, now being illegitimately deployed, should, and in no uncertain terms—be terminated."

[Section 2]

"In order to better explain why a reference to the term "evil" is being made, in relation to the Aadhaar, and to better represent the author's views regarding parliamentary responsibilities—especially given the social implications of the way the Aadhaar program is being deployed and used, a proper examination of the term "evil" must first be briefly made."

"The answer is fairly simple; in that, if the meaning of the term "evil" as it was being referred to in this article could be provided, and made clear, then the reader could suitably grasp how the author, a Citizen of India, and a parliamentarian would opposed the Aadhaar scheme from the very beginning."

"The very advancement of the Aadhaar program is invigorated in fact, by manufactured realities, packaged neatly with prosaicism, a certain banality, if you must, by the program's inceptors."

A brief sanity check

"To underscore the aforementioned point relating to the Aadhaar program being deployed without generating any great loathing at all by most Indian citizens, the reader's attention is drawn to a statement recently made by the Attorney General of India, before the Supreme Court on 2 May 2017, where the Attorney General presented before the esteemed court that 'Citizens (of India) should not expect to have, and in fact do not have, absolute right over their own bodies'."

"These words were presented in relation to the Supreme Court's query to whether there exists a requirement to compel Indian Citizens to avail themselves before government, or government sanctioned entities, to provide biometric information such as, Iris scans, Fingerprints and other type of scans as may be provisioned for, through incremental legislation—including DNA data—which could be made mandatory for obtaining the Aadhaar card."

Programmatic background

"Through Mission-Creep, the Government of India now claims that it has enrolled 99% of the national population into the Aadhaar program, which de facto would make the Aadhaar, the largest biometric identification system in the world."

"This KYC allows private companies to acquire Indian government issued licenses that would subsequently permit them to access the Aadhaar server for drawing-out information on registered citizens."

"The Delhi State government recently pushed the use of the Aadhaar program even further, by making the unique identification number a compulsory requirement for children who desire to seek admission into any school."

"Within the article's scope and other limitations, these examinations, at the root, aim to introduce the reader to reasons for which the general population has, more or less proposed that they are blameless in the Aadhaar's launch, and remain largely unperturbed at present by the manic Aadhaar push undertaken by the Government of India."

Cultural background—A contributor to privacy insensitivity

"There must be a harmonization of Privacy related personal viewpoints, social attitudes, and national frameworks for privacy protections, defence and repatriation mechanisms for Indian citizens, despite the vastly differentiated political and socio-economic nature of Indian States, and respective administrative frameworks, if India is to be a strong economic member of the international community of goods and services producers."

"Adjoiningly, one should take note that Social Media platforms like Facebook and Twitter are more recent advents, and in terms of India's past, cultural history, technology adoptions, and technocratic evolution, contiguously considering also, that previous generations possess neither knowledge nor skills, transmittable through modalities of cultural transmission, that would inherently benefit India's technologically savvy middle-class with respect to privacy."

"That question being: what exactly did it mean for India—to accede to the International Conventions above, if the Country's founders and her subsequent leaders/administrators did not (or do not) believe in granting Indian Citizens, those human rights 'most fundamental to all' in the world irrespective of colour, creed, faith, sex, education, beliefs, culture and socio-economic standing etc. ?"

Social corruption

"The harms that such a violation of personal privacy would entail are only just surfacing in the past few months."

"Various news and views published in the recent past throw immense light on the many irregularities associated with the biometric data collection program undertaken in India."

"It may sound far-fetched now, but, so much of the data collected, could possibly lead to the realization of outcomes and themes, now vividly portrayed in the highly popular 'Black Mirror' TV series."

System & Security Failures

"With respect to the Aadhaar program in India, there are reports almost every day about data leaks originating from Government websites where sensitive biometric information of citizens, including their caste, religion, bank account details, address, next of kin and more, is being made publicly available for anyone to access through a simple Google search."

"The government has been making Aadhaar mandatory for availing numerous schemes like State rationed items, State guaranteed employment, old age pension, crop insurance, disability welfare, meals for school children and even compensation for victims of the Bhopal Gas Tragedy."

"To add insult to injury, in the span of 1 week—during late February and early March of this year—the government of India made the use of the Aadhaar obligatory to access 11 more government programs, such as mid-day meals for school children, scholarships for disabled citizens and job training help for women rescued from sexual trafficking."

[Section 8]

"How could the AG, in a Republic, and without taking the fundamental steps to ensure 'due-process of law,' unilaterally determine that his decision should assert ultimate validity for 1.4 billion people and that it was to carry legal sufficiency to make a corrosive program such as the Aadhaar, mandatory in Indian society?"

"And importantly, in asking the court to make something that is unconstitutional—mandatory, nowhere was it clearly qualified how the Aadhaar program, which then had 440,000 fake students in just 3 Indian States officially enrolled [175], was going to be an adequate security barrier for all the fraud that AG Rohatgi was claiming to exist, and that which was supposedly funneling state-funds to shell companies."

"Although the Indian Supreme Court has been feeble on all matters related to the Aadhaar program, especially on matters related to privacy, it remains the last bastion for hope and sanity to prevail in India."

[Section 9]

"In an Harvard Business Review article called "Values in Tension: Ethics Away from Home [176]," Tom Donaldson of the Wharton School of Business painted a tale of how, in the 1980, European tanneries and pharmaceutical companies shopped around the world, for cheap waste-dumping sites; countries where those companies could unload highly toxic wastes."

"By the presentation, Donaldson wanted to have the reader adequately contemplate the application of the requisite template for ethical business conduct, both domestically and internationally."

"The speed and rates of modern business data and other forms of information flows have curtailed such practices as Donaldson describes, but have not altogether eliminated them, as the Aadhaar demonstrates in India today."

"Strangely, Donaldson's story is useful to analyze what lies ahead for India in terms of privacy."

Privacy context specificity to India's economic future

"It is this author's assessment that the reliability, robustness, stability, safety, and security as it is related to the shaping of India's privacy framework will decidedly shape the nation's economic future also."

"Personal Privacy in the Digital Century is translatable synonymously to the safety and security of information technology assets and the critical information that it holds, anywhere."

"India's ability to offer vibrant Information Technology Enabled Services competitively, and in a changing word, is inextricably linked to the country's ability to recognize the critical value of privacy, and the established need for the security and safety of information, and information systems."

"And structurally, in this regard, India must instrument privacy, information safety and security policies and practices."

Privacy related contextualization example: India-EU economic relations

"The EU and, or other states cannot be expected to "rubber-stamp" such declarations, when the awareness of privacy and data protection constructs by the Indian workforce is generally weak, and is usually coupled to the existence of weak institutional policies, practices and frameworks related to information safety and security practices."

"It is this author's read that while the government has attempted to convey to the outside world through the existence of "Reasonable Security Practices and Procedures and Sensitive Personal Data or Information Rules, 2011", India has met key data and privacy protection "adequacies", the government will only be fooling itself."

"Greenleaf qualified it best saying, "[a] bizarre aspect of the Bill, for a country seeking an EU adequacy finding, is that it limits its protection to Indian businesses… India is therefore one of the few countries to have enacted data privacy laws for its private sector, but not for its public sector."

Conclusion

"Vs State of Andhra Pradesh And Ors. (1993) have recognized one's enumerated right to personal privacy."

"Being a member of the national legislature, this author can attest to the fact that no such compelling State interest test has ever been discussed, conceived, deliberated, consented to, or developed—with the cooperation and, or consent of the Parliament, in relation to the Aadhaar program."

"And again as a member of the national legislature, this parliamentarian can further attest that, no compelling state interest in relation to the need to have the Aadhaar program deployed has ever been introduced, and, or formally brought forward for deliberation before the national legislature; either supported by results of any compelling national interest test, or separately by the widely acknowledged presence of any compelling state interest, which has been incontestably surfaced, and through incontrovertible means."

Privacy please: Power distance and people's responses to data breaches across countries [338]

This is a machine-generated summary of:

Madan, Shilpa; Savani, Krishna; Katsikeas, Constantine S.: Privacy please: Power distance and people's responses to data breaches across countries [338].

Published in: Journal of International Business Studies (2022).
Link to original: https://doi.org/10.1057/s41267-022-00519-5.

If you want to cite the papers, please refer to the original.

For technical reasons we could not place the page where the original quote is coming from.

Abstract-Summary

"Although data breaches can cause significant harm to users, businesses, and society, there is significant individual and national variation in people's responses to data breaches across markets."

"Eight studies using archival, correlational, and experimental methods find that high power distance makes users more willing to continue patronizing a business after a data breach (Studies 1–3)."

"This is because they are more likely to believe that the business, not they themselves, owns the compromised data (Studies 4–5A) and, hence, do not reduce their transactions with the business."

"Making people believe that they (not the business) own the shared data attenuates this effect (Study 5B)."

"Study 7 shows that high uncertainty avoidance acts as a moderator that mitigates the effect of power distance on willingness to continue patronizing a business after a data breach."

Introduction

"We examine people's willingness to continue patronizing a business after a data breach, which has significant implications for businesses, users, and public policy."

"We propose that power distance affects people's attributions of data ownership, and this, in turn, shapes their willingness to continue patronizing a business after a data breach."

"We expect uncertainty avoidance to moderate the relationship between power distance beliefs and users' willingness to patronize a business after a data breach."

"We argue that even users high in power distance beliefs will be more concerned about the potential risks of sharing new data with a firm following a data breach if they are also high in uncertainty avoidance."

"Our focus on data breaches adds a novel perspective to prior research in the international business literature, that has primarily focused on people's privacy concerns when they are deciding whether to share their data [61, 177]."

Conceptual background and hypotheses

"As high uncertainty avoidance individuals feel more threatened by uncertainty, they might be less willing to attribute data ownership to the business after a data breach as they are more concerned about the ways in which the data may get compromised again in the future, thereby weakening the effect of power distance beliefs on ownership attribution to the company and, consequently, willingness to continue patronizing the business after a data breach."

"As users who are low in uncertainty avoidance are not particularly threatened by uncertainty, those with high power distance beliefs would continue to assign greater data ownership to the company, and be willing to patronize a business after a data breach, but not those with low power distance beliefs."

Methods

"We conducted eight studies to test these hypotheses."

"Study 1 tested Hypothesis 1 at the country level using archival data."

"Study 2 tested Hypothesis 1 with victims of actual data breaches."

"Study 3 tested Hypothesis 1 in the context of the Facebook–Cambridge Analytica data breach in the US."

"Study 6 tested the underlying mechanism using an interaction design in the context of a recent data breach at Twitter."

Study 1: Country-level power distance

"Using archival data, we test whether users in high power distance countries will continue to patronize a business even after a potential breach of user data is widely covered in the news media."

"We operationalized our dependent variable, the continued popularity of FaceApp, as the number of days FaceApp was the highest-grossing app in the Apple App Store in that country after the data breach issues were first reported in the media on July 17, 2019."

"The effect of power distance on the continued popularity of FaceApp remained significant in each case (see the Online Appendix for additional analysis and the complete list of countries in the dataset).[6] Study 1 provides preliminary support for Hypothesis 1: people in high power distance countries are more willing to patronize a business after a data breach."

"A limitation of this study is that we do not control for the amount of news coverage about FaceApp's data breach in each country."

Study 2: Experience of data breach

"Participants were first asked to specify the organization with which they had experienced the data breach."

"Participants named a heterogeneous set of organizations with which they had experienced data breaches (e.g., Equifax, Facebook, Google, PayPal)."

"We averaged the four items assessing participants' willingness to continue patronizing the business after the data breach ($\alpha = 0.79$), which we used as the dependent measure."

"People's willingness to stop patronizing a business after a data breach may be affected by the number of alternatives available for the services provided by these businesses."

"Further, heterogeneity in the severity of the data breach across businesses in this study may have also affected users' willingness to continue patronizing the business."

Study 3: Facebook–Cambridge analytica data breach

"We investigated whether users with high power distance beliefs would be more likely to continue patronizing Facebook even after this data breach was revealed."

"We reasoned that users high in power distance beliefs would blame Facebook less, because they would attribute greater ownership of the compromised data to Facebook than to themselves."

"Participants were told that Facebook's CEO, Mark Zuckerberg, was recently called to the US Senate to testify about a personal data breach for 87 million users."

"We found that participants with high power distance beliefs were less likely to reduce their Facebook use ($b = -0.201$, $SE = 0.091$, $\beta = -0.201$, $t_{(117)} = 2.16$, $p = 0.033$), in support of Hypothesis 1."

"Studies 1–3 show that power distance strongly predicts users' reactions to data breaches at both global and individual levels."

Study 4: Causal evidence manipulating organizational power distance

"In the high power distance condition, participants were told that VIF Corporation believed in a hierarchical culture."

"A t test found that participants in the high power distance condition ($M = 6.82$, $SD = 0.69$) were more likely to believe that VIF was a hierarchical organization than participants in the low power distance condition ($M = 1.44$, $SD = 0.96$; $t_{(198)} = 45.33$, $p < 0.001$, Cohen's $d = 6.41$), thus indicating that the manipulation was successful."

"As hypothesized, a t test indicated that participants in the high power distance condition ($M = 15.64$, $SD = 24.56$) attributed lower data ownership to the employee than participants in the low power distance condition ($M = 30.049$, $SD = 27.55$; $t_{(199)} = 3.91$, $p < 0.001$, Cohen's $d = 0.55$)."

Study 5a: Causal evidence manipulating national power distance

"In the high power distance condition, participants were told that the people of the Solomon Islands believed in hierarchy."

"Participants in the low power distance condition read that the people of the Solomon Islands believed in equality and egalitarianism."

"A t test indicated that participants in the high power distance condition ($M = 5.49$, $SD = 1.43$) were more likely to agree that citizens of the Solomon Island accept hierarchy in their society than participants in the low power distance condition ($M = 3.073$, $SD = 1.66$; $t_{(198)} = 11.042$, $p < 0.001$, Cohen's $d = 1.56$), thus indicating that the manipulation was successful."

"As hypothesized, a t test revealed that participants in the high power distance condition ($M = 4.09$, $SD = 1.83$) were more likely to indicate that they would continue using the app than those in the low power distance condition ($M = 2.59$, $SD = 1.66$; $t_{(198)} = 6.071$, $p < 0.001$, Cohen's $d = 0.86$), thus providing causal evidence for Hypothesis 1."

Study 5b: Manipulating ownership attribution

"Study 5B provides another test of the role of ownership attribution as the underlying mechanism (Hypothesis 2) by using the experimental causal chain paradigm (Spencer, Zanna, and Fong 2005)."

"Study 5A tested the first half of the causal chain, that is, experimentally manipulating power distance to influence people's ownership attribution of their data."

"Study 5B tests the second half of the causal chain, that is, whether experimentally manipulating the ownership attributions would influence people's willingness to continue patronizing a service after a data breach."

"In the ownership-attribution-to-self condition, participants were told that "When users install the app, the terms of use state that all personal and location data collected by the app belongs to the user.""

"In the ownership-attribution-to-provider condition, participants read that, "When users install the app, the terms of use state that all personal and location data collected by the app belongs to the government.""

Study 6: Process by interaction

"In accordance with Hypothesis 2, we expected participants' power distance to be positively correlated with their willingness to continue using Twitter, despite a data breach, when they were told that the app/website owns the data (i.e., in the ownership-attribution-to-business condition)."

"We expected the relationship between power distance and willingness to continue using Twitter, despite a data breach, to be attenuated when participants were told that they (the users) own the data (i.e., ownership-attribution-to-self condition)."

"To test this, we regressed participants' willingness to continue using Twitter on their power distance beliefs (mean-centered), ownership attribution (business = –1, self = 1), and their interaction (mean-centered)."

"We found that, in the ownership-attribution-to-business condition, power distance beliefs predicted willingness to use Twitter after the data breach ($b = 0.83$, $SE = 0.24$, $\beta = 0.38$, $t_{(69)} = 3.39$, $p = 0.001$), in support of Hypothesis 2."

Study 7: Moderating role of uncertainty avoidance

"We found a significant indirect effect of power distance beliefs on greater willingness to use the FacePlay app through lower ownership attribution of personal data to the user ($b = 0.071$, $SE = 0.022$, 95% CI (0.0302, 0.12)]."

"The positive indirect effect of power distance beliefs on willingness to use the FacePlay app through reduced attribution of data ownership to the user (vs. the app) was significant for low uncertainty avoidance participants (–1 SD: $b = 0.10$, $SE = 0.04$, 95% CI (0.04, 0.18)]."

"For low uncertainty avoidance participants, power distance beliefs were a significant predictor of willingness to use an app with data breach issues because they attributed low ownership of their data to themselves."

"For high uncertainty avoidance participants, the link between power distance beliefs and willingness to use the app was attenuated because they attributed greater data ownership to themselves (rather than the app)."

General Discussion

"Across eight studies using archival, correlational, and experimental methods, we find converging evidence that power distance influences people's willingness to continue patronizing a business after a data breach."

"We provide converging causal evidence for the role of power distance in determining people's willingness to continue patronizing a business after a data breach."

"This finding has important consequences, because believing that shared data are owned by businesses (vs. the user) leads individuals who are high in power distance beliefs to be less bothered by data breaches that affect their information, and thus the data breach does not affect their willingness to continue patronizing the business after a data breach."

"Explicitly clarifying data ownership (as legally relevant) may help attenuate the relationship between power distance beliefs and willingness to continue patronizing businesses after a data breach if legal terms attribute data ownership to the user."

Notes

"Participants responded to the power distance beliefs scale (e.g., "People in higher positions should make most decisions without consulting people in lower positions"; $\alpha = 0.84$; [178]) and six items measuring their perceived authority of businesses in society (adapted from Rigby [179, 180]; e.g., "Businesses are authority figures in our society," "It is difficult to challenge the authority of businesses in our society"; $\alpha = 0.73$), all on 7-point scales ranging from completely disagree to completely agree."

"Regressing participants' perceived authority of businesses in society on their power distance beliefs revealed a significant effect of power distance beliefs ($b = 0.32$, $SE = 0.078$, $\beta = 0.38$, $t_{(98)} = 4.12$, $p < 0.001$)."

"13 A pilot study ($n = 196$ US participants) found that clarification of legal ownership rights (that attributed ownership to the business) led to significant reduction in blame assigned to the company for the data breach ($M = 4.91$, $SD = 2.23$ vs. $M = 6.49$, $SD = 0.98$, $t(194) = 6.43$, $p < 0.0001$)."

Queer Privacy Protection: Challenges and the Fight within Libraries [339]

This is a machine-generated summary of:

Hofman, Darra; Villagran, Michele A. L.: Queer Privacy Protection: Challenges and the Fight within Libraries [339].

Published in: International Journal for the Semiotics of Law—Revue internationale de Sémiotique juridique (2023).

Link to original: https://doi.org/10.1007/s11196-023-09994-x.

If you want to cite the papers, please refer to the original.

For technical reasons we could not place the page where the original quote is coming from.

Abstract-Summary

"Libraries–particularly public and school libraries–have once again become sites of conflict and resistance, with queer people and materials increasingly coming under attack both physically and legally."

"Librarians, as professionals, proclaim a commitment to privacy embedded in such documents as the American Library Association's Library Bill of Rights and the International Federation of Library Associations and Institution's Statement on Privacy in the Library Environment."

"This article examines the challenges of queer digital privacy within libraries in the United States, focusing on the polysemous, boundary-crossing nature of queerness, the digital and the material, privacy, and libraries (as both concepts and institutions)."

"This article demonstrates how binary-bound, individual-rights-oriented legal approaches to privacy have arisen, and been mediated, by cis-heteronormative patriarchal values and how the sociotechnical materialities in which they occurred (such as paper-based recordkeeping) are fundamentally incompatible with queer privacy needs."

Introduction

"The public library in Jamestown, Michigan, in the United States, was defunded after refusing to ban books by queer authors [181]."

"The 2022 winner of the American Library Association (ALA)'s Lemony Snicket Prize for Noble Librarians Faced with Adversity, Martha Hickson, is a high school media specialist who refused to pull Lesbian, Gay, Bi, Transgender, and Queer + (LGBTQ+) titles, with those attacking the books labeling her as a pornographer and pedophile, subjecting her to personal attacks, hate mail, threats, and vandalism."

"Both 'privacy' and 'libraries' have deep histories in the oppressive structures of the state; the most committed library (or librarian) must confront the individualistic orientation of privacy rights and policy, as well as the cis-heteronormative, patriarchal structures in which both privacy and libraries are embedded if they are to protect queer privacy and intellectual freedom truly."

(The Challenge of) Defining Privacy in Libraries

"Library and information science scholar Hoffmann, in her article on privacy, intellectual freedom, and self-respect in libraries, describes the normative understanding of libraries' roles in American society: It is the "grand tradition" of libraries that ready, and equitable access to information is integral to a functioning democracy [...]. [L]ibraries support a robust public sphere [...] enacting "the principle of critique and rational argumentation through the commitment to balanced collection, preserving them over time, and furthering inclusion through active attempts to make collections and resources reflect historical and current intellectual diversity" [...as well as playing] a role in the development and maintenance of communities. [Libraries] "function as places where communities get constructed" [...] [and] the place of the library in society is informed, in part, by the moral and political claims of both individual citizens and the communities within which they are embedded [182]."

Queering Privacy

"A long line of feminist scholarship exposes how privacy perpetuates patriarchy and gender norms and critiques the ability of an individualistic rights approach to protecting the "majesty of human experience.""

"A library that cannot protect the sexual privacy of its community may cause active harm, "entrenching the subordination of women and marginalized communities" [183], including queer people, who typically "shoulder the abuse [of invasions of sexual privacy]" [184]."

"As a relatively recent example, Hazeldean points to the characterization of any attempt to "ban sexual orientation and gender identity discrimination in employment, housing, or public accommodations by labeling them 'bathroom bills'" that will allow "men" (a purposeful mischaracterization of transgender women) to enter women's washrooms in violation of women's privacy [185]."

"While individual librarians operate from an ethics of care perspective in serving their queer patrons, libraries as institutions operate primarily in the space of rights and policies when it comes to privacy, typically treating all patrons the same (fairly, as opposed to necessarily justly)."

[Section 4]

"Hildebrandt states that the inscribing of law into Information and Communications Infrastructures (ICI) such as the printing press, and now digital technologies, brings material changes in the law itself and its operation, arguing that "the era of the printing press made possible the rise of the bureaucratic state [... which in turn] ushered in the era of the rule by law" [186]."

"Molecular HIV surveillance vividly illustrates the risks LGBTQ + people face, with no legal recourse to protect their privacy as surveillance technology develops apace."

"How equipped are librarians whose commitment is only to an ill-defined, neoliberal idea of "privacy," with little engagement with, or even exposure to, the broader privacy risks faced by queer patrons and stakeholders, to protect such patrons' digital privacy."

"Even for librarians who are committed to both privacy and equity, adequately identifying information risks for users who engage queer information, materials, and programming is immensely difficult, given the opaque nature and third-party control of many libraries' digital infrastructures."

Conclusion: Privacy Beyond Checklists

"For the queer library user, all ideas of privacy are relevant, though perhaps of different weight in different contexts."

"In choosing to keep transphobic materials in its collection, a library may claim that it is upholding intellectual freedom and the intellectual privacy of those users who choose to read such literature."

"Privacy, and its library companion, intellectual freedom, is not value-neutral, nor is it a value in and of itself."

"We encourage libraries and librarians to interrogate the privacy and intellectual freedom issues set out within this article."

"For "privacy" to be meaningful for queer persons in a library context, it must be understood that privacy itself is not ethical but merely serves to enable or constrain any number of ethical goods."

A call to action: a systematic review of ethical and regulatory issues in using process data in educational assessment [340]

This is a machine-generated summary of:

Murchan, Damian; Siddiq, Fazilat: A call to action: a systematic review of ethical and regulatory issues in using process data in educational assessment [340].

Published in: Large-scale Assessments in Education (2021).

Link to original: https://doi.org/10.1186/s40536-021-00115-3.

If you want to cite the papers, please refer to the original.

For technical reasons we could not place the page where the original quote is coming from.

Abstract-Summary

"Even though the use of process data in assessment holds significant promise, the ethical and regulatory implications associated with it have not been sufficiently considered."

"And to provide an overview of how ethical and regulatory requirements interface with process data from assessments in primary and secondary education (K-12), we conducted a systematic literature review."

"The studies identified as relevant interpret issues of privacy largely in terms of informed consent and the research pays little attention to ethical and privacy issues in the use of process data in assessment."

"The implications for the field of educational assessment and the use of process data are discussed."

"This includes the need to develop a specific code of ethics to govern the use of process- and logfile data in educational assessment."

Introduction

"The present study addresses this gap through a systematic review of the extent to which studies recognise and address ethical and legal concerns in contexts where process data are used in educational assessment research in K-12 education (this term includes Kindergarten to Grade 12, that is, primary and lower- and upper secondary school in many educational systems)."

"To what extent are ethical, privacy and regulatory considerations reflected in recent research that draws on process data in K-12 assessment?"

"To what extent are ethical, privacy and regulatory considerations reflected in recent research that draws on process data in educational assessment more broadly?"

"What elements associated with ethics, privacy and regulations are evident in recent research drawing on process data in educational assessment?"

"The study intends to capture research practice in relation to ethics, privacy and regulations existing at the time in the context of the use of process data."

Background

"The use of process data in assessment is underpinned by two developments recently: the increase in digital assessment and the subsequent availability of complex log files containing rich information about how students engage with tasks and systems [187]."

"This combination allows for the generation of process data that enable test developers and researchers to go beyond the student response data (number of items answered correctly, quality of response) to analyse and interpret other granulated data captured in the digital log files, such as records of the actions that students take."

"Within this space of digital assessment and process data, the present study explores a specific under-investigated related challenge, namely ethical and privacy issues associated with the use of examinee data for purposes that may not have been adequately explained to the students."

Theoretical framework

"In framing our investigation of the implications of process data use in assessment we explore the concepts of fairness and validity in assessment, their links with ethics and privacy and how these concepts are expressed in established standards and guidelines for educational and psychological assessment."

"Our study explores how ethical and privacy issues are addressed in relation to process data in the context of K-12 assessment."

"Frameworks that address ethical and privacy issues related to process data and assessment are limited."

"Other studies have reflected on the ethical and privacy issues in the field of LA and process data use more generally [188] and some authors have proposed generic frameworks and/or guidelines to deal with such issues."

"We conclude that there is considerable, though not complete overlap between GDPR and the categories used in the present study to evaluate the extent to which the studies selected addressed issues of ethics, privacy and regulation in relation to process data."

Method

"We applied the systematic review methodology, and followed guidelines for conducting systematic reviews proposed by Boland and others [189] and Gough and others [190] which include the following steps: predefining research questions, developing search strategy, defining eligibility criteria, screening of studies according to inclusion/exclusion criteria, data extraction, appraisal of studies, and synthesis."

"The initial screening following the eligibility criteria revealed a somewhat unexpected finding: only three of the studies met the key inclusion criteria, namely, dealt with ethical and/or regulatory issues associated with the use of process data in K-12 assessment."

"We reduced the focus on including studies related to assessment in K-12, and widened our primary inclusion conditions applying a second order of eligibility criteria."

"Qualitative description was used to report the extent to which studies met various inclusion criteria and how the sample of studies was sub-divided into groups depending on the extent to which they addressed issues of assessment, ethics and regulations."

Results

"This study sought to explore the extent to which issues of ethics, privacy and regulation are reflected in studies that investigate the use of logfile data in educational assessment contexts."

"One inclusion variable of interest focused on ethics/privacy issues in the context of process data from student assessments, and the results indicated a relative lack of studies meeting these criteria."

"The picture overall represents a mix of empirical and theoretical studies drawn from different continents and countries, significantly focused on university level, with a varied emphasis on assessment, and issues of ethics and/or privacy."

"The data indicate higher averages overall for theoretical studies, reflecting not only greater emphasis on categories that are addressed in both type of study designs, but also the inclusion of a broader range of ethics dimensions, for example issues of Responsibility, Validity, Enabling interventions and Minimising adverse impact."

Discussion and implications

"This research revealed a dearth of studies focusing on process data use in assessment that also incorporates consideration of ethics and privacy."

"We could find only three studies where there was a significant and clearly articulated focus on assessment in K-12 education, a finding in line with previous research which showed that the ethical and regulatory issues related to use of process data has mostly been dealt with in higher education studies [191]."

"The relative absence of attention to ethical issues in assessment-related process data studies in K-12 is noteworthy also in terms of two foundation elements of assessment—validity and fairness."

"Only 5 of the 22 studies could be identified having a significant assessment and ethics/privacy focus in the context of process data."

"We encourage future research to include additional search terms that focus more specifically on areas within ethical and regulatory issues in use of process data in assessment."

Conclusion

"This systematic review explored the extent to which ethical and regulatory norms informed the use of process data in studies incorporating assessment of students."

"Set in the context of a burgeoning harvesting of granular process data in digital assessments, the study contrasted the attention paid to technical refinement of procedures with the incorporation of ethical safeguards for participants in the process."

"Our study indicates that there is a need to develop a specific code of ethics to govern the use of process- and logfile data in assessment."

"Such standards need to inform a perception amongst some test-developers and researchers that privacy and data protection are interchangeable with ethics."

"Our review suggests that some test developers and researchers may feel that by addressing privacy and data protection, ethical requirements are met."

"We also identify the need for more empirical studies that focus on the twin issues of process data in assessment and ethical implications of using these data."

Protecting personal information using Generally Accepted Privacy Principles (GAPP) and Continuous Control Monitoring to enhance corporate governance [341]

This is a machine-generated summary of:

Prosch, Marilyn: Protecting personal information using Generally Accepted Privacy Principles (GAPP) and Continuous Control Monitoring to enhance corporate governance [341].

Published in: International Journal of Disclosure and Governance (2008).

Link to original: https://doi.org/10.1057/jdg.2008.7.

If you want to cite the papers, please refer to the original.

For technical reasons we could not place the page where the original quote is coming from.

Abstract-Summary

"In a recent speech given by the Federal Trade Commission's (FTC) Chief Privacy Officer, Marc Groman, he emphasised to the National Association of Secretaries of States to prepare well because 'you will have a data breach' (Washington, DC, 18th February, 2008)."

"This article discusses the present privacy crisis and offers a solution to managing and reducing privacy risk through the use of the AICPA/CICA's Generally Accepted Privacy Principles (GAPP)."

"When the court of public opinion or the FTC is grilling an organisation regarding a data breach, the best defence for the organisation is to provide evidence, such as adherence to GAPP criteria or an actual GAPP audit, that they were diligent and serious about data protection policies."

"Top management, including CEOs, CFOs, CISOs, and CPOs all need to be aware of privacy risk management issues and techniques to reduce privacy risk."

Introduction

"A plague of data breaches has affected organisations of all types."

"According to the Privacy Rights Clearinghouse data, the types of data breaches publicly reported for 2007 in the US affected the following types of organisations: 95 government and non-profit agencies; 87 educational institutions; 51 healthcare organisations; 93 for-profit organisations."

"In the UK, data breaches have also plagued organisations."

"Data breaches occurred before this 'privacy storm' began; however, new data breach notification laws, such as California's Civil Code Sections 1798.29 and 1798.82, require that organisations notify the victims."

"Managing this data becomes an important corporate governance issue because all organisations that collect, use, or store PI face privacy risk."

"The second objective is to introduce the AICPA/CICA's GAPP and to discuss how these principles can be used in enhancing corporate governance over PI by developing and implementing Continuous Control Monitoring (CCM) environments that contain PI."

Technology, privacy, and cultural lag

"[192] A fundamental difference between the European Commission's widely recognised and pioneering EU Data Protection Directive and the US privacy regime regarding employee PI is succinctly stated as employer data collection requirements (US) vs employee privacy rights (EU)."

"Since consumers in the US do not have the same sort of data protection mandated in the EU and Canada, among other countries, the question to businesses is whether this privacy is valued."

"As members of society begin to realise that the data are being collected, shared, generated, and sold among companies, and that the data are not well protected against theft or loss, cries for greater protection begin to be made at time b. During the period between points b and c, the awareness of privacy issues is raised and the increased privacy expectations cause the community to seek enhanced material culture in the form of privacy-enhancing technologies and controls."

Data touchpoints and associated technologies

"Personal data may be collected in digital format directly from the individual when he/she is on the internet."

"Once personal data are collected, it may be stored in an organisation's operational databases and further copied into data warehouses where it can be accessed and downloaded by employees for various purposes."

"Organisations that collect personal data should disclose with whom and why personal data will be shared."

"Some privacy advocates are concerned about these devices being placed in items which can be read very easily from a distance and personal items being used to track both the whereabouts of individuals and personal use patterns of such devices."

"Prosch [193] discusses the need to protect PI from 'cradle to grave' as well as the types of technologies currently available for each phase of the data lifecycle."

"The current marketplace offers various piecemeal privacy-enhancing technologies, that when implemented, can dramatically reduce the risk of data breaches or loss of PI."

GAPP and accountants as providers of assurance services

"The AICPA formed a Privacy Task Force to develop a Privacy Framework and Criteria that resulted in the formulation of GAPP that can be used by accounting firms of all sizes to provide privacy services and assurances to their clients."

"Accountants have the ability to: comprehensively understand both current and future statutory regulations that may be applicable to a firm; assess the risk faced by a firm for inadequate privacy policy and practices; align systems and infrastructure with developed policies and close any privacy gaps; design both high-level plan and detailed working documents in order to achieve privacy compliance; identify the various privacy components, as well as build and implement any necessary changes to them to close any privacy gaps; assess the adequacy of privacy policies and practices of relevant business partners and service providers; monitor the implemented system's compliance with its stated policies and practices."

CCM Environments and GAPP

"Developing privacy-enhancing technologies and methodologies with a continuous monitoring/assurance perspective makes good business sense given the plethora of recent privacy breaches, and it will also strengthen governance over an organisation's operations."

"With traditional financial audits, 'the major obstacle to adopting a continuous process remains the lack of commitment by organizations to invest in the technology required to develop and implement a continuous auditing process'."

"Great opportunity exists for companies developing/enhancing privacy regimes to design continuous monitoring and reporting components into the system since we are still in the early development stages."

"GAPP does not specifically have continuous audit requirements, but it does have many process-related criteria that lend themselves well to continuous monitoring and reporting processes."

A privacy lifecycle maturity model

"The CMM model is a five-level evolutionary path of increasingly organised and systematically more mature processes, and its terminology has been adapted to reflect data protection processes."

"At the managed level, an organisation monitors and controls its own processes through data collection and analysis."

"Organisations that outsource any of their data processes need to have an assurance that each of their third-party processors have data protection policies at least equivalent to their own."

"The implication is that, in order for businesses to progress along the privacy maturity model, the third-party data processors must provide their customers with assurances regarding their own data protection policies and procedures."

"Monitoring techniques may indicate that a piece of data that is currently being collected is no longer used in any process, and management may, as a result, determine that the risk of collecting and protecting such data is greater than the benefit of possessing it."

Discussion

"The accounting profession has recognised that protection of PI is an important aspect of controlling an organisation's systems and business environment."

"Many of the tasks involved in protecting PI build upon characteristics and skills that members of the accounting profession already possess."

"[194] Some of the complications and potential problems associated with the use of traditional assurance methods for businesses with digital operating environments have been previously discussed with a call for members of the accounting profession and researchers to think more holistically about continuous controls."

"Further, the accounting profession has tangibly responded to the cultural lag by developing a comprehensive set of auditable privacy criteria, GAPP, that fosters the enhancement of corporate governance over PI."

[Section 8]

"'If you haven't heard of the Generally Accepted Privacy Principles (GAPP), take stock: They're likely to become the most important new source of requirements for your IT projects since Y2k and Sarbanes–Oxley."

"The accounting industry has closed ranks around the idea that the GAPP is the best international framework for assessing the privacy health of an organization…."

"I have to agree with the auditors on this one."

"It's the best attempt so far to address the main point of pain for global chief privacy officers: the growing complexity of privacy regulations around the world.' (Cline [195])"

Private Parts: A Global Analysis of Privacy Protection Schemes and a Proposed Innovation for Their Comparative Evaluation [342]

This is a machine-generated summary of:

Pincus, Laura B.; Johns, Roger: Private Parts: A Global Analysis of Privacy Protection Schemes and a Proposed Innovation for Their Comparative Evaluation [342].

Published in: Journal of Business Ethics (1997).

Link to original: https://doi.org/10.1023/a:1005746000135.

If you want to cite the papers, please refer to the original.

For technical reasons we could not place the page where the original quote is coming from.

Abstract-Summary

"The challenge in the business community is to make the most of the opportunities presented by the growth in communication technology while, at the same time, protecting what remains of individual privacy."

"The conflict between technological advances and privacy concerns is not new, but it has grown exponentially in recent years, and the development of a data protection scheme in the European Union lends a certain economic urgency to meeting the challenge."

"Pursuant to a Directive adopted by the European Union, privacy protections throughout the EU must become more stringent and consistent throughout EU member countries."

"One area of great concern to the United States is the Directive's requirement for certain minimum standards of privacy protection in countries receiving information from member countries."

When Should "Consumers-as-Producers" Have to Comply with Consumer Protection Laws? [343]

This is a machine-generated summary of:

Swire, Peter P.: When Should "Consumers-as-Producers" Have to Comply With Consumer Protection Laws? [343].

Published in: Journal of Consumer Policy (2008).

Link to original: https://doi.org/10.1007/s10603-008-9082-5.

If you want to cite the papers, please refer to the original.

For technical reasons we could not place the page where the original quote is coming from.

Abstract-Summary

"Consumer protection law has been built on the contrast between large "producers" and small "consumers.""

"As unregulated consumers become regulated producers, a major legal question is whether and when the individual should have to comply with consumer protection laws."

"The discussion here selects four examples of US legal rules that might apply to consumers-as-producers, with recommendations spanning the range of possibilities: (1) consumer privacy legislation: creating a threshold, with no compliance required for databases of fewer than 5,000 names, is recommended; (2) advertising substantiation: concerning the requirement that advertisers have a "reasonable basis" for their claims, applying current law to small advertisers is recommended; (3) spam: current law does not create a threshold for those who send a few commercial emails, but such a threshold is worth considering; (4) political blogging: the author agrees with the US Federal Election Commission decision to create a major exemption from campaign finance laws for online political advocacy, even for large blogs or websites."

"The sorts of harm are likely to be created by consumers-as-producers, the analysis tilts towards requiring compliance."

"The sorts of harms are unlikely to be caused by consumers-as-producers, then the case for an exception is stronger."

Consumer Privacy Legislation

"The author testified in a hearing on the subject in 2006, and major industry players have now called for a national law A group of companies has specifically stated: In principle, such legislation would address businesses collecting personal information from consumers in a transparent manner with appropriate notice; providing consumers with meaningful choice regarding the use and disclosure of that information; allowing consumers reasonable access to personal information they have provided; and protecting such information from misuse or unauthorized access."

"The problem is illustrated by the example of the author's teenage son, who made money one recent summer by cutting the lawns for perhaps eight neighbors."

"In the terms of this article, when the author's son acts as a "producer"—a lawn-cutting enterprise—should he have to comply with this consumer protection law?"

"The teenager, based on the text of the law, is supposed to comply with the full set of data protection rules."

Advertising by Small Producers

"The discussion here of substantiation law, moreover, illustrates the analysis that might apply to consumers-as-producers for other advertising laws, such as truth-in-labeling with respect to third-party endorsements, or guidance concerning word-of-mouth advertising."

"The question here is the extent to which CAN-SPAM should also apply to consumers-as-producers."

"For ordinary businesses, the main requirements of CAN-SPAM are that the business be easy to contact and that it stop sending commercial emails once the consumer asks."

"CAN-SPAM has a much more significant effect on how legitimate businesses send email to consumers."

"How should CAN-SPAM apply to consumers-as-producers?"

"In the face of these compliance requirements on millions of consumers-as-producers, it may be possible to craft a threshold, such as a dozen or 50 commercial emails per day."

"The CAN-SPAM prohibitions on false email headers could apply even to consumers-as-producers."

The Federal Election Commission and Blogging: When Do Favorable Comments Become "Campaign Contributions"?

"Under the McCain–Feingold reforms of 2002, to take another example, political advertisements require a disclaimer along the lines of "this advertisement was paid by the campaign to re-elect Senator X." An important question, which erupted into a major debate in 2005 and 2006, is how these campaign finance rules apply to that vocal species of consumer-as-producer, the political blogger."

"The political blogging community suddenly became aware that it might become subject to complex campaign finance regulations, as well as FEC fines and other enforcement actions."

"The final rules "make plain that the vast majority of Internet communications are, and will remain, free from campaign finance regulation.""

"Exempting Internet speech from campaign finance laws thus furthers the goals of such laws."

"If political statements on a blog require compliance with complex campaign finance rules, then the costs of blogging rise substantially."

Lessons for When "Consumers" Should be Regulated as "Producers"

"The concept of consumer-as-producer proves helpful in analyzing the extent to which diverse regulatory regimes should apply to individuals who are equipped with personal mainframes."

"For fraud in advertising, the harm can be created even by individual producers, and the existing regime should apply to small producers."

"For consumer privacy and anti-spam legislation, the harms are likely to be significantly greater for large producers."

"Regulatory regimes more appropriately apply to consumers-as-producers the more that they track the standards of ordinary people acting in good faith."

"One should consider the likelihood that enforcement will be used to harass ordinary consumers-as-producers."

"The different risk of over-enforcement thus supports having the advertising rule but not the campaign finance rules apply to consumers-as-producers."

"The methodology here is to assess the costs and benefits of consumer protection regulation, as applied both to large producers and producers-as-consumers."

What is New About Today's "Consumers-as-Producers"?

"The objection points out, correctly, that production from home is nothing new."

"Many small businesses in the physical world today are based at home, including landscape services, solo plumbers, or accountants, and so on."

"For consumer protection law, however, there are important distinctions between this history and today for reasons of scale, geographic scope, and the visibility of possible violations."

"Those enterprises are historically subject to enforcement under consumer protection laws, but in the Web 2.0 world, much of the "production" is done by individuals working at home in a setting not historically regulated in the same way."

Conclusions

"The personal mainframe means that an individual at home can produce at a scale that quite possibly outstrips local demand."

"The producer has to consider whether enforcers in any relevant jurisdiction will seek to enforce consumer protection or other laws."

"Consumers-as-producers confront the visibility of much online activity to go along with this multiplicity of possible enforcers."

"The consumer protection rules discussed in Part II all tend to produce clear evidence of violations by consumers-as-producers: 1."

"There is thus considerably greater urgency to clarify consumer protection law for online consumers-as-producers than for traditional home producers who usually sell locally, on a small scale, and with less evidence of violations."

"Online consumers-as-producers are much more able to produce at a large scale, they have their activities visible in multiple jurisdictions, and they often produce clear evidence of violations as a by-product of online technology."

[Section 7]

"Consumer protection law has been built on the contrast between large "producers" and small "consumers.""

"As unregulated consumers become regulated producers, a major legal question is whether and when the individual should have to comply with consumer protection laws."

"The individual might now perpetrate the sort of harm that led to the consumer protection law in the first instance."

"Consumer protection laws should thus apply in full measure to consumers-as-producers."

"Individuals also face potentially significant barriers to compliance with consumer protection laws."

"Consumers-as-producers should thus be exempt from having to comply with consumer protection laws."

"The sorts of harm are likely to be created by consumers-as-producers, the analysis tilts towards requiring compliance."

"The sorts of harms are unlikely to be caused by consumers-as-producers, then the case for an exception is stronger."

The long-term preservation of identifiable personal data: a comparative archival perspective on privacy regulatory models in the European Union, Australia, Canada and the United States [344].

This is a machine-generated summary of:

Iacovino, Livia; Todd, Malcolm: The long-term preservation of identifiable personal data: a comparative archival perspective on privacy regulatory models in the European Union, Australia, Canada and the United States [344].

Published in: Archival Science (2007).

Link to original: https://doi.org/10.1007/s10502-007-9055-5.

If you want to cite the papers, please refer to the original.

For technical reasons we could not place the page where the original quote is coming from.

Abstract-Summary

"The authors recommend that the further processing of personal data under data protection law be given a wider scope of interpretation for archival preservation purposes in both the public and private sector, coupled with the use of researcher and archival codes in relation to access to personal data."

Introduction

"Despite global terrorism that has begun to erode privacy rights in a number of countries, there is a move not only to protect identifiable personal information, but also to ensure further legal protection on the basis of perceived sensitivity to certain classes of personal information."

"These trends include limiting the collection, use and disclosure of sensitive information such as that of racial, ethnic and political persuasion, and placing even higher protection on very sensitive personal data such as health-related information."

"Government policies world-wide are encouraging the re-use of personal data in shared networked environments by using unique identifiers to link personal data across jurisdictions and institutions, thus increasing potential privacy abuses in relation to secondary uses [196]."

"It analyses some specific archival exemptions in privacy legislation in representative European Union member countries as at 2003, and comparable provisions of Australian, American and Canadian federal law, as they impact on long-term preservation of personal data in electronic records."

Privacy and the preservation of personal data: legal, archival and ethical perspectives

"The legal obligations are found in freedom of information, privacy and record-keeping legislation, and legal duties of confidentiality, contractual and other special relationships, balanced with the correlative rights of access to information by the record subject or a third party."

"The record subject's informed consent to the collection, use and disclosure of his/her personal information is an essential element of privacy protection and must be obtained by all recordkeeping participants."

"Privacy and personal data protection legislation conforms with international privacy principles that personal information should only be collected, used or disclosed for its primary or original purpose, and that its use and disclosure for secondary or other purposes is subject to strict limitations."

"Privacy regimes focus on consent to the collection, use and disclosure of personal information in its immediate context, while archival regimes focus on preservation of authentic records for general public disclosure, which may include personal information once it has lost its sensitivity."

European Union, privacy rights and archival exemptions

"Italy has been the first country in the EU to incorporate a code of conduct and professional practice for the processing of personal data for historical purposes into a national privacy law, in part, in response to a Recommendation of the Committee of Ministers to Member States on European Policy on Access to Archives [197]."

"Section 102 deals with the issuance of a deontological and good conduct code to be promoted by the Privacy Commissioner for both public and private persons, including scientific and professional associations interested in using personal data for historical purposes, with particular reference to the collection, use and disclosure of health, sexual or other private matters to be handled with care, and the researcher informing the data subject who may be affected before disclosure."

General observations on the EU framework and personal data preservation

"Many member states could have made broader use of the Directive's derogation powers or could have made them more specific in their application to archives or archival processes (such as preservation activity) that may only involve incidental processing of the personal data."

"The extent to which record creators and preservers can rely on article 6 (1)(b) of Directive 95/46/EC for archival preservation of identifiable personal data is questionable if "historical" and other research or archival exemptions have been enacted narrowly to constitute "further processing" of personal data that is not allowable."

"Although one example of a sector code relates to the press—freedom of press expression is also subject to a specific exemption under the Directive—only the Dutch scientific research code relies on the exemption in Dutch law for historical, statistical and scientific purposes (see Article 10[2] of the Personal Data Protection Act 2000)."

Privacy regimes and the preservation of personal data: Canada, Australia and the United States

"In Australia, at the federal level and in most of its states, archival and freedom of information legislation has included provisions on disclosure and protection of personal information that predate privacy acts."

"Although on the surface Australian privacy laws do not have exemptions to the principle of "further processing" or "secondary uses" on the basis of research, history and statistics, the deference to archival and freedom of information legislation taking precedence (at least in the Commonwealth and Victoria), as well as the use of public interest determinations, and a broad reading of secondary purposes, provide a balanced approach to the long-term preservation of personal data and its protection in the public sector, but are far less clear in the private sector."

Conclusion

"From the research to date it can be concluded that the EU Data Protection Directive 95/46/EC introduced not total harmonisation, but an extremely full and detailed regime that required many member states to revamp substantially their privacy provisions, but there are important differences in how the archival perspective has been addressed."

"Cross-jurisdictional sharing of personal data in the web environment threatens the efficacy of the broader privacy regulatory framework, which in some countries has included freedom of information and archival legislation and the protection from disclosure by professionals and archivists well before the passing of data protection legislation."

"The recommendations arising from this study include promoting the addition of archival and researcher ethics codes into the privacy legislative framework, as has been done in Italy."

"Privacy needs to be integrated with freedom of information and archival regimes, and more needs to be done to promote the preservation of archives by private sector organisations."

Pandemics, privacy, and public health research [345]

This is a machine-generated summary of:

Bernier, Alexander; Knoppers, Bartha Maria: Pandemics, privacy, and public health research [345].

Published in: Canadian Journal of Public Health (2020).

Link to original: https://doi.org/10.17269/s41997-020-00368-5.

If you want to cite the papers, please refer to the original.

For technical reasons we could not place the page where the original quote is coming from.

Abstract-Summary

"Sharing data expediently for pandemic response purposes exposes healthcare providers in Canada to significant regulatory uncertainty."

"Institutions should foster robust data stewardship and standardize their practices to those recognized among the international health informatics community."

"Reform is critical to ensuring Canadian healthcare providers can deliver efficient health responses that are integrated with dispersed and disparate national and international approaches."

Introduction

"Worldwide, the COVID-19 pandemic is inspiring clinicians and researchers to pool data and design data repositories for long-term public health research."

"Ensuring compliance with Canadian privacy regulations creates significant challenges for such efforts."

"Coordinated and collaborative data stewardship strategies are needed."

Challenges for collaboration

"For a pandemic response, this can mean that private-sector clinicians attempting to share data with larger institutional actors such as hospitals and universities will face challenges in ensuring that the privacy law obligations imposed on the clinic and those imposed on the hospital are compatible [198]."

"Canada's private sector privacy laws differ considerably from one province to the next, and the applicable law varies depending on whether data are used within a province's borders or interprovincially."

"In the case of internal use, the local provincial laws will apply unless the province has not enacted such laws, in which case Canada's federal Personal Information Protection and Electronic Documents Act (PIPEDA) applies instead."

"In some provinces such as Quebec, provincial and federal authorities disagree as to whether the local provincial law, the federal PIPEDA, or both are applicable to interprovincial data transfers [198]."

Challenges for public health and research

"Certain provincial laws, such as Alberta's Health Information Act (HIA), impose special requirements on the contents of consent forms that could create difficulties in standardizing pan-Canadian consent materials across the clinical and research settings [199]."

"Healthcare practitioners face more onerous requirements in obtaining research consent than clinical consent."

"Canada's privacy laws create special powers for national public health authorities to requisition and share data liberally across healthcare actors and across provincial borders during a public health emergency without obtaining individual consent [200]."

"If clinicians and researchers wish to reuse the data generated for future epidemiological and healthcare research, they may be required to satisfy the more onerous burden of obtaining either individual participant consent or research ethics waivers of consent, coupled with institutional custodians' approvals, as well as the necessary, project-specific research ethics approvals [201, 202]."

Responsive and responsible data stewardship

"Faced with ethical and regulatory barriers to efficient data pooling, we impel clinicians and researchers to turn to the governmental actors most capable of orchestrating a coordinated, data-driven response to a health crisis."

"Some Canadian privacy laws allow the broad transfer of health data to combat a health risk to known persons, or to provide healthcare to the person that the data concerns [203]."

"For a coordinated long-term response to epidemic threats, including COVID-19, Canadian researchers should coordinate their data collection and data governance practices with those of established international consortia to facilitate efficient data deposit [202]."

"Fundamental reforms will be necessary to enable efficient responses to future health emergencies and to ensure that Canadian healthcare providers are not unduly precluded from granting Canadians high-quality medical care as data-driven approaches continue to reshape the health sector [204]."

Proposed regulatory reforms

"Requirements to restrict data reuse to the initial purposes of collection, or to retain data within prescribed territorial boundaries, are difficult to reconcile with the international nature of a collaborative pandemic response, and the potential to use data for flexible secondary purposes [202]."

"Government and industry stakeholders, as well as academic commentators, have recognized that procedural mechanisms for regulating privacy are difficult to reconcile with continuous technological disruption and innovation in the digital age; technology-neutral, principle-driven protections are more adaptable to the ever-shifting context of data use than are formalistic rules [205]."

"The recommendations favour decreased procedural burdens for data use generally, counterbalanced by increased substantive protections for individuals against unconscionable uses of their data [205–207]."

Harmonizing data stewardship and regulatory oversight

"To implement such a method without derogating from the shared federal and provincial oversight of health data, privacy regulators can subject government-created data repositories to delegated oversight performed by a designated Privacy Commissioner in accordance with a commonly chosen privacy law, irrespective of the provenance of the data stored in the repository."

"Such rights must be supported by broad audit and enforcement powers apportioned to Privacy Commissioners, directly actionable rights in data for patients and research participants, and obligations for data custodians to demonstrate that privacy has been accounted for in organizational structure and technological design [207]."

"In empowering individuals and Privacy Commissioners to hold entities accountable for their uses of health data, legislators can ensure that the interests of vulnerable groups are not disregarded, as they might be in the present model that relies principally on self-regulation."

"Using destruction obligations and preclusions on data reuse to safeguard health data can further stigmatize marginalized populations, and deny vulnerable individuals access to research benefits [203]."

Lessons for the future

"Challenges faced in sharing health data to respond to the COVID-19 epidemic demonstrate how Canada's duplicative privacy legislation can create considerable barriers for health data sharing in Canada with minimal gains in personal privacy."

"The failure to standardize Canada's health data regime imposes considerable institutional costs and legal risks on the healthcare sector and causes clinicians and researchers to eschew data centralization and secondary data use [208]."

"Reform is the only viable solution for the future, and the future is now!"

The legal obligation to provide timely security patching and automatic updates [346]

This is a machine-generated summary of:

Maurushat, Alana; Nguyen, Kathy: The legal obligation to provide timely security patching and automatic updates [346].

Published in: International Cybersecurity Law Review (2022).

Link to original: https://doi.org/10.1365/s43439-022-00059-6.

If you want to cite the papers, please refer to the original.

For technical reasons we could not place the page where the original quote is coming from.

Abstract-Summary

"Would it surprise you to learn that some of these products no longer receive security support or automatic updates?"

"If so, you may be surprised to hear that you are being exposed to security risks, as many cyber incidences are the direct result of an absence of security patching and automatic updates."

"Most companies provide security patches, but they are not always timely and many are not automated, requiring manual effort (often unbeknownst to consumers and businesses)."

"Timely security patching is, upon discovery or notification of a security flaw in a system or product, the release of a security update within a reasonable time that patches and updates the security of a system—sometimes this is automatic, sometimes the security patch is merely a notification that you can and should patch your own system."

"This means that there is no legal requirement to patch known security vulnerabilities and bugs or issue automatic updates."

"This paper asks whether or not Australia should have a legal obligation to ensure timely security patching and require automatic updates by default in all consumer systems."

Introduction

"One contributing factor why security vulnerabilities remain unpatched is that there is no legal requirement to continue security support of products or to issue timely security updates and patching."

"This article has a simpler mission of convincing the reader that the era of no legal compliance for fixing software bugs and providing timely security patching needs to comes to an end."

"The European Union is moving towards the legal obligation of timely security patching, and Australia should follow suit."

"The second part looks more deeply why timely security support/patching coupled with the automatic updates by default are essential tools to combat cybersecurity threats."

"Part three and four examine the legal frameworks for security patching in Australia and the European Union."

Why security patching?

"Before we can get to the many facets of this question, we need to know more about security vulnerabilities and bugs, why they are so prevalent and how they can be fixed along with the role that security patches and updates play in protecting our systems."

"While there are many forms of threat vectors, the most serious arise from exploiting security vulnerabilities as this can lead to system exploitation of every device and system globally that uses the technology."

"A security vulnerability is a software vulnerability or weakness in a computer system that can be exploited by an attacker."

"Responsible organisations upon learning of a security vulnerability in their systems, issue what is known as a security patch and send out updates for users to install the latest security update."

"Security patching and automatic updates are not easily navigable by users."

Australia

"ASIC alleges that RI contravened Sections 912A(1)(a), (b), (c), (d), and (h) of the Corporations Act due to its failure to have and to have implemented (including by its annual reports, policies, plans, procedures, strategies, standards, guidelines, frameworks, systems, resources and controls which were reasonably appropriate to adequately manage risk in respect of cybersecurity and cyber resilience (Minimum Cybersecurity Requirements)."

"The Code, which forms part of Australia's 2020 Cyber Security Strategy, is intended to be a voluntary set of measures the Australian Government recommends for manufacturers, service provides, mobile application developers, and vendors as the minimum standards for Internet of Things (IoT) devices."

European Union (EU)

"The DCA's allegations in this case were that Samsung breached its duty of care as expected of it in society or were in breach of its obligations arising from: (a) Article 7:17 of the Dutch Civil Code and/or Article 13 of the Dutch Personal Data Protection Act by failing to provide its smartphones with updates in a timely manner, at least for a period of four years from the device's introduction to the market or at least within a month of becoming aware of a vulnerability and the availability of Google's patch for that vulnerability."

"The DCA requested that the Court ordered Samsung to: (a) provide all Samsung smartphones in the Netherlands with updates that fix software vulnerabilities for a period of four years after a device enters the market and/or two years after a device's purchase, (b) fix security vulnerabilities in the Android operating system within one month after Google makes the update available, and (c) inform consumers in a clear and unambiguous manner before each device's sale about Samsung's update and upgrade policy."

Reform proposals

"While these developments are heading in the right direction towards strengthening cyber resilience generally, they still fall short of imposing a legal obligation on vendors and companies to correct known vulnerabilities promptly; to move to automatic security updates by default; and to require companies to support their products."

"Any security patching framework requires several essential elements: A mandatory unambiguous specified time for product support (Eg.)"

"Any timely legal requirement to issue a security patch will need to be fully flexible to take into account the complexities involved along with the context but this is not impossible."

"Governments and larger organisations could commence the practice of requiring companies to fully support their products for a certain amount of years and with regular security patches."

"As Europe moves towards more formal legal frameworks to security systems, Australia should follow suit starting with requirements to implement timely security patches, and by requiring systems to be automatically updated by default."

Regulation of Big Data: Perspectives on strategy, policy, law and privacy [347]

This is a machine-generated summary of:

Casanovas, Pompeu; De Koker, Louis; Mendelson, Danuta; Watts, David: Regulation of Big Data: Perspectives on strategy, policy, law and privacy [347].

Published in: Health and Technology (2017).

Link to original: https://doi.org/10.1007/s12553-017-0190-6.

If you want to cite the papers, please refer to the original.

For technical reasons we could not place the page where the original quote is coming from.

Abstract-Summary

"This article encapsulates selected themes from the Australian Data to Decisions Cooperative Research Centre's Law and Policy program."

"It is the result of a discussion on the regulation of Big Data, especially focusing on privacy and data protection strategies."

"Big, Linked, and Open Data constitute complex phenomena whose economic and political dimensions require a plurality of instruments to enhance and protect citizens' rights."

Perspectives on Big Data

"To frame the topic we will commence with a few brief remarks about the term 'Big Data' before discussing views expressed by participants in a recently-concluded study undertaken by the Law and Policy Program of the Data to Decisions Cooperative Research Centre."

"Research participants in the UK reported similar types of barriers to the use of Big Data; for example, they operated under similar resource constraints."

"Australian research participants raised privacy (12/38), misuse of data (10/38), misplaced trust in technology and assumptions behind analytics (10/38) and data security (9/38) as the most significant risks of using Big Data."

"The views expressed by the research participants and reflected above provide some insight into the range of views held in relation to Big Data."

Regulating Big Data

"These include: The benefits of Big Data and Open Data The associated data protection risks The ways that the risks can be managed/mitigated The focus in this section is on one aspect of risk mitigation, privacy enhancing technology [PET]."

"Further, what if the data in each block contains personal information—such as your health records, or your recent bankruptcy, or your change of gender?"

"The distributed ledger technology model operates in a way that challenges one of the main information privacy assumptions—that an organisation, whether public or private, collects, uses and discloses personal information, and thus is both accountable and responsible for those activities."

"It is important, when considering privacy issues as they pertain to Big Data and Open Data, to avoid simplistic analysis and to take account of all of the risks and benefits of each in formulating a response that respects privacy rights."

An improved legal framework

"The statute (s 2(4)) identifies the following additional factors that, depending on context, must be considered: "(a) the interests of national security or of the economic well-being of the United Kingdom, (b) the public interest in preventing or detecting serious crime, (c) other considerations which are relevant to: (i) whether the conduct authorised or required by the warrant, authorisation or notice is proportionate, or (ii) whether it [conduct] is necessary to act for a purpose provided for by this Act, (d) the requirements of the Human Rights Act 1998 (UK), and (e) other requirements of public law.""

"In Chapter 1 of Part 8, the legislation creates the Office of the Investigatory Powers Commissioner whose duties include review (by way of audit, inspection and investigation) the agencies' functions relating to the interception of communications; the acquisition or retention of communications data; the acquisition of secondary data [i.e. metadata] or related systems data; equipment interference (whether under warrants and authorisations or otherwise)."

Defining digital regulations

"Collecting data and producing metadata—i.e., machine understandable information—constitutes the new stage."

"When such protections are embedded into formal systems, and represented through the languages of the Web of Data we face the layer of the meta-rule of law That is, with the development of the web, the rule of law needs to evolve to a meta-rule of law, based on legal knowledge, and incorporating tools to regulate and monitor the semantic and algorithmic layer of the web [209]."

"Legal scholars, computer and social scientists have identified the main elements of this new regulatory framework—including the results of fifteen years of research on legal XML, Legal RuleML, and legal ontologies; linked data publication and consumption, copyright related terms, licensing, patents, privacy risks, and the emergence of data and metadata markets [210, 211]."

Conclusions

"Digital legal pluralism (how to deal with different legal cultures, jurisdictions, and the binding power of national states)."

"Legal forum-shopping (how to deal with markets, companies and corporate power)."

"Balancing citizens' rights (privacy) and security (how to deal with individual rights and collective needs)."

"Creating a global, digital, legal culture (how to deal with the general human–machine framework in which agency, risk scenarios and social ecosystems emerge)."

"Improving National and International legal frameworks to include new digital terminologies and concepts, harmonising them and allowing legal interoperability across jurisdictions and legal cultures."

"Building privacy means accepting that there is no absolute privacy."

"It deals rather with building communities and generating trust in citizens and consumers—within national and transnational markets."

"Identity means building, controlling and monitoring knowledge about safety, security, and personage (data and metadata)."

[Section 6]

"This article contends that the effective regulation of Big Data requires a combination of legal tools and other instuments of a semantic and algorithmic nature."

"It commences with a brief discussion of the concept of Big Data and views expressed by Australian and UK participants in a study of Big Data use in a law enforcement and national security perspective."

"Reflecting on these developments, the paper closes with an exploration of the complex relationship between law and Big Data and the implications for regulation and governance of Big Data."

The perceived impact of location privacy: A web-based survey of public health perspectives and requirements in the UK and Canada [348]

This is a machine-generated summary of:

AbdelMalik, Philip; Boulos, Maged N Kamel; Jones, Ray: The perceived impact of location privacy: A web-based survey of public health perspectives and requirements in the UK and Canada [348].

Published in: BMC Public Health (2008).

Link to original: https://doi.org/10.1186/1471-2458-8-156.

If you want to cite the papers, please refer to the original.

For technical reasons we could not place the page where the original quote is coming from.

Abstract-Summary

"The "place-consciousness" of public health professionals is on the rise as spatial analyses and Geographic Information Systems (GIS) are rapidly becoming key components of their toolbox."

"This paper describes the views and requirements of public health professionals in Canada and the UK on privacy issues and spatial data, as collected through a web-based survey."

"Perceptions on the impact of privacy were collected through a web-based survey administered between November 2006 and January 2007."

"The survey targeted government, non-government and academic GIS labs and research groups involved in public health, as well as public health units (Canada), ministries, and observatories (UK)."

"Of 112 invitees in Canada and 75 in the UK, 66 and 28 participated in the survey, respectively."

"Ninety three percent of participants indicated a requirement for personally identifiable data (PID) in their public health activities, including geographic information."

"Privacy was identified as an obstacle to public health practice by 71% of respondents."

"The clash between PID requirements—including granular geography—and limitations imposed by privacy and its associated bureaucracy require immediate attention and solutions, particularly given the increasing utilisation of GIS in public health."

"Solutions include harmonization of privacy legislation with public health requirements, bureaucratic simplification, increased multidisciplinary discourse, education, and development of toolsets, algorithms and guidelines for using and reporting on disaggregate data."

Background

"While an analysis of privacy legislation as it pertains to health data and the concept of "place" is beyond the scope of this paper, suffice it to say that such clauses are often ambiguous and subjective, particularly when combined with vague definitions of "sensitive personal information" and the scale at which geographic data becomes "identifiable"."

"It is no surprise, therefore, that the increasing popularity of "place" in public health has further exacerbated the public health research-privacy debate."

"From a health research perspective, both Canada and the UK place strong emphasis on evidence-based public health policies and services [212], yet in both countries, this seems to be hampered by privacy issues."

"This paper describes the views and requirements of public health professionals in Canada and the UK on privacy issues and spatial data, as collected through a web-based survey."

Methods

"The survey was first developed on paper in the summer of 2006, and piloted with select public health individuals in Canada and the UK."

"The consent screen outlined the voluntary and anonymous nature of the survey, indicated the approximate time it would take to complete the survey, the risks and benefits to the participants, the intellectual property and ownership of all data collected, and the protection of any personal data provided under Canadian and UK law Failure to successfully complete either of these two requirements resulted in termination of the survey."

"Progress through the survey required the selection of a response for each question, and participants could terminate the survey at any time or complete it over multiple sessions, at their convenience."

Results

"When asked "Are you or have you been restricted in your use of GIS for any public health activity because of privacy concerns (i.e. map or data might identify an individual or community)?""

"Of 83 participants who responded to the question "In your opinion, do current restrictions to PID pose an obstacle to any aspects of public health practice?""

"Fifty seven percent of participants felt that under 10% of the public population is aware of the impact of restricted access to PID on public health practice; 74% felt it to be under 20%, and 84% felt the proportion to be less than 30% (cumulative frequencies)."

"When then asked what proportion of the public they felt would allow the use of their PID if they were educated on the usefulness of such data to public health practice, 67% said 50% or higher."

Discussion

"This survey and user-needs assessment on privacy and public health shows a definite requirement by public health professionals—in various fields and positions in both Canada and the UK—for personally identifiable data, including spatial data."

"Not only did public health professionals in both countries generally rate themselves high on knowledge of privacy legislation and related issues, but those with the highest self-rated scores also tended to rate privacy as more of an obstacle."

"Although the findings of this paper may be generalisable to public health professionals in Canada and the UK, issues of privacy and public health are not unique to these countries."

"Since knowledge of privacy legislation and policies was based on self-rated scores, a thorough review and assessment of privacy legislation as it pertains to public health practice is required in both Canada and the UK to validate the findings of this survey."

Conclusion

"These provide a more holistic portrayal of public health community viewpoints, and can be used to educate the public, and as evidence for decision makers to implement changes in policies and legislation."

"The clash between a requirement for personally identifiable data—including exact, individual location—by public health professionals, and the limitations imposed by privacy and its associated bureaucracy, must be addressed and appropriate solutions developed, particularly given the increasing utilisation of geographic information systems in public health and the imminent completion of comprehensive electronic health systems."

"Since health is not limited by political boundaries, this must be pursued at an international level, and solutions must address these perceptions in the public health community, simplify the bureaucratic process, promote multidisciplinary discussions between legislators, bureaucrats and the public health community, educate communities, and develop and provide public health professionals with toolsets, algorithms and guidelines for using and reporting on disaggregate data."

Conserving *Customer Value: Improving Data Security Measures in Business* [349]

This is a machine-generated summary of:

Gregory, Adrian: Conserving customer value: Improving data security measures in business [349].

Published in: Journal of Database Marketing & Customer Strategy Management (2008).

Link to original: https://doi.org/10.1057/dbm.2008.20.

If you want to cite the papers, please refer to the original.

For technical reasons we could not place the page where the original quote is coming from.

Abstract-Summary

"The causes of security breaches are varied and prevention has, until now, been largely unregulated."

"This paper looks at the security and accessibility of data sets, investigates where threats to security come from, what can be done to prevent security breaches and what can be done if a breach occurs."

Identity fraud—the background

"According to Cifas, the police investigate less than 1 per cent of identity fraud cases, and figures from KPMG reveal that just 200 fraud cases came to court in 2007."

"According to CardWatch, card-not-present (CNP) is the most popular form of identity fraud in the UK, while figures from APACS show a 74 per cent growth in CNP since 2003."

"Year, to give just a few examples, 15,000 Standard Life customers were put at risk of fraud after a courier lost a computer disk containing personal information; Newcastle City Council admitted that personal data and payment card details of up to 54,000 local residents had been downloaded from an insecure server to an IP address outside the country, nine NHS trusts admitted to losing the personal details of more than 160,000 patients and Nationwide was fined nearly £1 million by the Financial Service Authority after a laptop containing customer account details was stolen from an employee's home."

Tackling the problem

"Data security breaches are concerned the issue is, perhaps most importantly, being addressed almost exclusively from the point of view of stopping data leaving the organisation through, or to, an unauthorised party."

"As has already been mentioned, the security and accessibility of data sets is frequently viewed as a purely internal issue."

"If an organisation were to admit that it had experienced a breach of its data security, that might open it up to potential legal liability as well as exposure of its reputation."

"Even the requirement of the Data Protection Act 1998 to keep personal data secure has tended to be viewed as an entirely internal process."

"It is an uncomfortable fact that most breaches of data security are carried out by an organisation's own staff, including its director and senior managers."

Track and trace

"Yell invests in data tracking to spot any abuse of its intellectual property rights, and its directories and website clearly state that the Yell products and services should not be used for activities such as the compilation of databases, directories and lists or for telemarketing or direct mailing purposes."

"When consultants are hired to investigate the risks and exposures to a company's database they check that it complies with the Data Protection Act, the European Directive on Privacy and Electronic Communications, The Direct Marketing Association Best Practice and Information Security Best Practice (based on ISO 27001 standards)."

"The database might then be subjected to a data hygiene and quality assessment that analyses standards in terms of factors such as people and businesses that have moved location, people who have died, firms that have ceased trading, duplicate records, out-of-date or inaccurate records, incomplete records and market coverage."

Notification requirements

"Notification of data security breaches is likely to become a legal requirement."

"In the US, in 2002, California became the first state to pass a Notice of Security Breach law requiring any organisation that suffers a breach of its data security and the loss of personal data to disclose this fact and to offer assistance to the data subjects affected."

"These require the notification, to the national regulator, of any security breaches involving the loss of personal data."

"Notification to the customer is also required where any breach of data security leads to the loss, modification or destruction of, or unauthorised access to, personal data."

Conclusion

"UK public and private sector organisations are holding an increasing volume of data on customers and citizens."

"If such organisations are to continue to be allowed to use this information to improve customer service, they also have to take on the responsibility of keeping it safe and secure."

"The exposure of 25 million personal records every year [213] to potential theft and fraud is already unacceptable."

"Individuals themselves must become more responsible about the way they keep and dispose of their personal records."

Best practice for safeguarding personal details and maintaining customer confidence [350]

This is a machine-generated summary of:

Ozimek, John: Best practice for safeguarding personal details and maintaining customer confidence [350].

Published in: Journal of Direct, Data and Digital Marketing Practice (2008).

Link to original: https://doi.org/10.1057/dddmp.2008.33.

If you want to cite the papers, please refer to the original.

For technical reasons we could not place the page where the original quote is coming from.

Background

"Figures from APACS [214]—the UK Trade Association for Organisations delivering payment services—suggest that this move has had some limited success in reducing the incidence of credit card fraud."

"It is harder to commit fraud by one means, however, fraudsters quickly turn to other approaches—some highly sophisticated—in order to hoax individuals into giving away personal details."

"CIFAS—the UK Fraud Prevention Service—reports an increase of c. 8 per cent year on year between 2006 and 2007 in cases of reported fraud."

"This ranges from the simple 'vanilla control'—where companies claim to delegate greater control to their customers—while in practice carrying on much as before, to genuine delegation of control: for instance, giving customers the ability to edit their own personal information, as well as determine the time and manner in which they may be contacted."

Issues

"Key issues are as follows:—ID theft, leading to a rethink by customers as to how and when they interact with organisations—resistance to existing supplier-oriented business practices—growing pressure from customers for far greater control in their interactions with suppliers—long-term risk that outbound calling may cease to be a viable channel of communication."

"For companies, this presents a stark challenge."

"Their current approach to doing business is increasingly out of touch with how customers wish to do business: it can appear insensitive and, in the worst cases, to actually aid criminal behaviour."

"Although this stems from a negative, it is also opportunity for those organisations that manage to resolve this demand in ways that appeal to customers."

Approach

"Rather, it draws upon a long and growing list of case studies in this area maintained by the author, in order to illustrate poor practice."

"Each of the instances documented below occurred in the last 12–18 months."

"Although each instance is a one-off report, it is reasonable to assume that they are examples of a campaign that has been approved by the organisation in question, and—if the organisation was who it claimed to be—that it is typical of their behaviours and attitudes towards their customer base."

"Steps have been taken—by contacting the organisation concerned—to confirm whether the behaviours identified are or may have been carried out by the organisation."

Case studies

"ID fraudsters could obtain credit card details quickly and easily just by raiding an individual's rubbish: credit card numbers were printed in full on discarded slips and many individuals just threw slips away with little or no attempt to obliterate the numbers."

"Since this has changed, a fairly obvious extension of fraudulent activity is to use what information remains (last four digits) as a means to extract the full card number."

"It is a phone company calling out, unable to verify its own identity, but expecting the call recipient to provide information to prove that they are who they say they are."

"It is indicative of an attitude that appears to have completely ignored the public statements of government and industry bodies in respect of giving out personal information."

Discussion

"There is ID fraud, consumer concern about that issue and the advice from a range of official bodies, including government, that you should NEVER divulge personal or banking details to anyone who phones unexpectedly."

"It would not be too far-fetched to imagine a consumer who unwittingly gave away banking details over the phone—and was subsequently refused compensation—suing their bank and calling BT in witness to what is considered 'acceptable practice'."

"The court would then be required to determine whether consumers should listen to government and official advice, or simply respond to what is current accepted practice in the UK telemarketing industry."

"Continuing failure to address this issue suggests a preference to put consumer livelihoods at risk, rather than to adopt practices that are designed to support customers."

Towards best practice

"If a customer is worried that they may have been targeted for ID fraud, they can have their minds set at rest—or their anxiety levels increased—by a simple confirmation—or denial—that they have just been targeted by the organisation in question."

"When an individual first becomes a customer, best practice already suggests that an organisation should obtain consent to contact that individual—and agree on what channels can be used for contact."

"In such cases, however, it should be fairly easy for the individual to argue that they had not been negligent—and if banks wished to pursue anyone for negligence, it would have to be the organisation whose security was compromised."

"Online tokens or ID systems to verify that an organisation is who it says it is are definitely one way forward—and only time will tell whether there is sufficient consumer demand to put them in place."

Conclusion

"The purpose of this paper has been to highlight a growing problem for organisations that need to elicit information from individuals as part of their outbound (tele)marketing."

"Customer resistance to traditional ways of doing business is growing—and good secure systems for safeguarding customer data are not yet in place."

"This paper does not provide neatly packaged solutions; it does, however, begin the debate on what such solutions might look like—and suggests that the time has come for organisations to refocus their trading practices in a more customer-centric way."

"I hope that this issue will strike a chord in some marketing departments—and may form the basis for an emergent best practice in this area."

Don't bin your Data Protection responsibilities [351]

This is a machine-generated summary of:

Nettleton, Ewan; Jensen, William: Don't bin your Data Protection responsibilities [351]

Published in: Journal of Database Marketing & Customer Strategy Management (2007).

Link to original: https://doi.org/10.1057/palgrave.dbm.3250060.

If you want to cite the papers, please refer to the original.

For technical reasons we could not place the page where the original quote is coming from.

Abstract-Summary

"Database marketers being 'data controllers' within the meaning of the Data Protection Act ought to be aware of the UK's data protection legislation and may even have had dealings with the UK regulator, the Information Commissioner's Office (ICO)."

Introduction

"They should be aware of the UK's central piece of legislation governing the protection of data, the Data Protection Act 1998 and will no doubt be familiar with dealing with the UK's regulatory body, the Information Commissioner's Office (ICO)."

"Recent actions taken by the ICO would, however, indicate that not all those who process personal information are as familiar with the Act as perhaps they should be."

"In March 2007, several banks were found to be in breach of their data protection responsibilities following customer complaints about sensitive information found in rubbish bins outside their premises."

"This paper describes some of the ICO's recent enforcement actions and considers how the ICO is seeking to achieve compliance with the UK's data protection requirements."

The data protection act 1998—An overview

"[215] To ensure information is handled correctly, the Act sets out eight data protection principles, which data controllers must comply with."

"The principles, which are also referred to as the principles of 'good information handling' require that data controllers ensure that information is: [216] 1 fairly and lawfully processed; 2 processed for limited purposes; 3 adequate, relevant and not excessive; 4 accurate and up to date; 5 not kept for longer than is necessary; 6 processed in line with an individual's rights; 7 secure and 8 not transferred to other countries without adequate protection."

"The Act sets out what each of these principles requires, and the ICO publishes guidelines to help data controllers to comply with the principles."

"The guidelines published by the ICO to assist data controllers to comply include helpful good practice notes and compliance checklists."

Enforcement by the ICO

"While the undertakings did not impose any punitive measures as such and sought instead to ensure that the Act would be complied with in the future, they provide the ICO with a right to audit the banks' data protection procedures in the future, which the ICO would not otherwise have had."

"The undertakings required that: 1 the personal data of the customer in question be suppressed from all company databases thereby ensuring that she would not receive any future marketing material from Littlewoods and 2 Littlewoods would review procedures currently in place to ensure that customers' rights under Section 11 of the Act (which gives individuals the right by written notice to require a data comptroller to cease or not to begin to process their personal data for the purpose of direct marketing) are upheld."

Conclusions

"The recent actions of the ICO against a range of businesses demonstrate that it is seeking to ensure data protection compliance, and they underline the need for data controllers such as database marketers to be acquainted with and adhere to the data protection rules in addition to the requirements that apply to electronic marketing communications."

"The ICO has tended to require undertakings as to future compliance from businesses it has found to be in breach of the data protection rules, including breaches in relation to the sending of unsolicited marketing communications."

"While unlike sanctions imposed for data protection violations by other bodies such as the Financial Services Authority, [217] such undertakings do not in themselves impose any punitive measures, they allow the ICO to scrutinise these businesses' future activities in a way it would not otherwise be able to under its statutory powers under the Data Protection Act."

Data protection: Tougher enforcement and increased power for the Information Commissioner? [352]

This is a machine-generated summary of:

Nettleton, Ewan; Turner, Ian: Data protection: Tougher enforcement and increased power for the Information Commissioner? [352].

Published in: Journal of Database Marketing & Customer Strategy Management (2008).

Link to original: https://doi.org/10.1057/dbm.2008.9

If you want to cite the papers, please refer to the original.

For technical reasons we could not place the page where the original quote is coming from.

Abstract-Summary

"Database marketers, more than most, should be aware of their obligations as regards protection of data and the role of the Information Commissioner's Office."

"The perception of those who hold personal data, so-called 'data controllers' under the Data Protection Act, has been affected by headline news in the past year or so."

Driving force for change

"This has not been an isolated incident, with many cases of loss of data and breaches of the data protection requirements being uncovered in the private and public sectors since that time."

"The announcement of the HMRC loss was closely followed by a finding from the Information Commissioner's Office (ICO) that the Foreign Office had breached the Data Protection Act, [218] and since the HMRC announcement, the ICO has been notified of almost 100 data breaches (62 in the public sector and 28 in the private sector)."

"[219] This has led to Marks & Spencer being required to encrypt all such data by April 2008."

"Unsurprisingly in these circumstances, there have been calls from numerous sources for a tightening up of data protection law and for more powers for the ICO."

Data Protection and the ICO's Powers—A Brief Reminder

"The Data Protection 1998 Act, which implemented EU-wide legislation on the protection of personal data, gives individuals rights, such as the right to ascertain what information is held about them, and imposes obligations on data controllers to ensure that personal data is processed properly."

"[220] These data protection rules, coupled with the rules relating to unsolicited marketing communications set out in the UK's Privacy and Electronic Communications Regulations, [221] are the principal rules with which the ICO will expect database marketers to comply."

Summary of the Ongoing Reviews

"This 'Data Sharing Review' was to provide recommendations on changes to the law and how policy should be developed to ensure accountability."

"Even before the review began, the ICO had been widely reported as interested in broadening its powers in relation to the UK's data protection regime."

"Mr Hannigan is considering the procedures in the government's departments and agencies for the protection of data, their consistency with current government-wide policies and standards, and the arrangements for ensuring that procedures are being fully and properly implemented."

"Both Mr Poynter and Mr Hannigan have issued interim reports in relation to their reviews, [222, 223] and their full reports are expected later this year."

"That said, more concrete recommendations for changes to the law are proposed in the Cabinet Office's interim review (as discussed further below), suggesting that Robert Hannigan's review, when issued in full, may contain more substantive recommendations."

Proposed changes to the law

"[224] The ICO's principal recommendations are that the following powers and penalties should be created:—a penalty for knowingly or recklessly failing to comply with the data protection principles so as to create a substantial risk that damage or distress will be caused to any person;—a power for the Information Commissioner to inspect personal data and the circumstances surrounding its processing in order to assess whether or not any processing of the data is carried out in compliance with the Act;—a power for the Information Commissioner to require a data controller to provide him with a report by a skilled person;—enhanced enforcement powers to enable the Information Commissioner to bring seriously unlawful processing to an immediate halt, to place formal undertakings on a statutory basis and to enable the Information Commissioner to take enforcement action to prevent breaches of the Data Protection Act that are likely to occur; and—information notices that can be served on any person rather than just a data controller."

Conclusions

"Recent high profile losses of large quantities of personal data by public sector institutions, such as HMRC in particular, have affected the perception of data controllers in general and have led to calls from many quarters for an enhancement of the UK's data protection requirements and enforcement provisions."

"Calls for enhanced powers for the ICO from both the Information Commissioner himself and from a number of heavyweight committees are likely to be acted upon, and this would tighten up the UK's data protection regime, with rights of audit to identify data breaches and requirements for such breaches to be communicated to affected individuals being one possible outcome."

"While the conduct of data controllers within the public sector may have triggered the calls for these changes, individuals such as database marketers, working in the private sector, need to be aware of the changes made and being proposed, because they could apply to them."

Data protection: where are we now? [353]

This is a machine-generated summary of:

Castro Edwards, James: Data protection: Where are we now? [353].

Published in: Journal of Database Marketing & Customer Strategy Management (2008).

Link to original: https://doi.org/10.1057/dbm.2008.18.

If you want to cite the papers, please refer to the original.

For technical reasons we could not place the page where the original quote is coming from.

Introduction

"Notwithstanding some shortcomings, EU data protection law has served to lead the way for the rest of the world, with other jurisdictions following suit by introducing legislation aimed to protect individuals' personal data."

"This paper first looks at the UK, and considers the extent to which the Data Protection Act (DPA) achieves its original aims, turning to its recognised weaknesses and proposed developments."

"It then considers developments in data protection throughout the EU, as the data protection directive from which the DPA originated comes of age."

"Data protection is not just a UK- or EU-centric issue: globally, jurisdictions are recognising the importance of creating an environment that protects individuals' personal information."

"This paper also considers the extent to which global and multinational organisations can take a 'one-size-fits-all' approach to data protection compliance."

The UK

"The last two years have seen so many high-profile data protection breaches that data protection is at last taken seriously by members of the public, who, not so long ago, may not have given the issue a second thought."

"The government's piece de resistance was the loss, last November, of the entire child benefit database, containing the personal details of 25 million people, including their bank details, National Insurance numbers and information about their children."

"With such high-profile data security breaches, public awareness of data protection issues has never been greater."

"To illustrate public consciousness of the issue, a survey conducted by the Information Commissioner's Office (ICO) in March of this year revealed that eight out of ten people are taking greater care to protect personal information following recent data loss blunders, and seven out of ten individuals feel powerless about how their personal information is looked after."

Does the DPA do what it says on the box?

"The DPA came into force on 1st March, 2000, implementing the Data Protection Directive 95/46/EC (Directive) as law in the UK."

"One issue is the definition of sensitive personal data."

"Sensitive personal data is a special category of data defined by the DPA, which requires a higher standard of care by data controllers."

"The DPA definition of sensitive personal data does not, however, include financial information."

"In an age where 'cybercrime' is rife, financial information is a form of personal data that could potentially cause an individual considerable damage, and there is a persuasive argument that this should be included in the definition of sensitive personal data."

"Another issue where the DPA attracts criticism relates to the definitions of data controller and data processor."

"A further criticism of the DPA is that the Directive aimed to harmonise the approach taken by EU member states to data protection."

Proposed developments in the data protection act

"A further development is the Consultation on the Information Commissioner's inspection powers and funding arrangements under the DPA 1998 (Consultation), which was launched on 16th July, 2008."

"[225] Both developments potentially strengthened the DPA, first by increasing the potential fines and penalties for serious breaches and secondly by giving the Information Commissioner a statutory power to gain entry to premises to audit their data-processing facilities."

"The first recommendation is that the ICO should have a statutory power to gain entry to relevant premises to carry out an inspection, with a corresponding duty of the organisation to cooperate and supply any necessary information."

"The second recommendation relates to changes to the notification fee, payable by data controllers to the ICO, through the introduction of a tiered system to ensure that the ICO receives a significantly higher level of funding to carry out its enhanced duties."

Developments in the EU

"The EU Commission has approved three sets of approved model clauses by which organisations located in an EU member state may transfer personal data outside the EEA."

"The EU-approved Model Clauses were developed for use throughout the EU, to enable businesses to transfer personal data to organisations based outside the EEA."

"Some EU regulators operate a 'permission-driven' regime, where the regulator's permission must be granted in order for personal data to be transferred outside the EEA."

"Some EU data protection regulators, however, simply will not permit the transfer of sensitive personal data outside the EEA despite the fact that a whistleblower scheme is mandatory under US law This makes it extremely difficult for a US-headquartered multinational to simultaneously comply with its SOX obligations and those of the applicable EU data protection legislation."

The Asia–Pacific economic cooperation countries

"There is no harmonised data protection approach among APEC members: some have no regulation at all, whereas others have adopted comprehensive laws."

"APEC members' governments, however, recognise, that a trusted data protection regime assists economic growth and is therefore an important issue."

"Japan's data protection regime is, however, some way ahead of China's data protection legislation: to date, China has no comprehensive privacy law The Chinese Academy of Social Science is currently drafting legislation that is likely to be moved forward by the State Council Information Office."

"In Taiwan, the Computer Processed Personal Data Protection Law applies only to computer-processed personal data and to a limited range of businesses."

"Developments in the APEC countries suggest that each country is seeking its own way to protect personal data."

"Only time will tell how such diverse data protection legislative regimes will deal with multinational organisations, given that frequently data protection laws are in an embryonic state."

The US: Doing its own thing?

"What is noticeable by its absence is a comprehensive data protection law Perhaps a closer parallel to the legislation adopted by the EU and much of APAC is the recently enacted 'security breach notification' legislation, which operates at a state level."

"The large number of separate state data privacy laws, each with its own scope and requirements as to the level of breach that must be notified, means that data controllers operating in the US should consider each separate law to ensure compliance."

"US state security breach notification laws define 'personal information' much more narrowly than EU data protection legislation."

"Some of the US state data breach laws address two issues that have been suggested as deficiencies in EU data protection law: (i) the recognition of financial data as a category of personal data that requires special treatment and (ii) the requirement that data controllers notify data subjects of breaches."

Conclusion

"In the EU, data protection laws have been in force for many years, and the legislation and regulators are tried and tested."

"The EU data protection legislative regime is well established, sophisticated, subject to ongoing refinement and fundamentally driven by the protection of human rights laws."

"EU data protection legislation is primarily aimed at protecting individuals' rights."

"[226] Accordingly, in some APEC countries, data protection laws have been passed recently, or are in the process of being enacted, to encourage trade with the rest of the world rather than to protect human rights."

"The examples of data protection legislative regimes we have considered demonstrate the very different approaches taken by different jurisdictions."

"This is even the case in the EU, where the Directive was intended to harmonise individual member states' data protection laws."

The ethics of facial recognition technologies, surveillance, and accountability in an age of artificial intelligence: a comparative analysis of US, EU, and UK regulatory frameworks [354].

This is a machine-generated summary of:

Almeida, Denise; Shmarko, Konstantin; Lomas, Elizabeth: The ethics of facial recognition technologies, surveillance, and accountability in an age of artificial intelligence: a comparative analysis of US, EU, and UK regulatory frameworks [354].

Published in: AI and Ethics (2021).

Link to original: https://doi.org/10.1007/s43681-021-00077-w.

Copyright of the summarized publication:

The Author(s) 2021

If you want to cite the papers, please refer to the original.

For technical reasons we could not place the page where the original quote is coming from.

Abstract-Summary

"The rapid development of facial recognition technologies (FRT) has led to complex ethical choices in terms of balancing individual privacy rights versus delivering societal safety."

"This analysis focuses on the regulatory contexts and recent case law in the United States (USA), United Kingdom (UK), and European Union (EU) in terms of the use and misuse of FRT by law enforcement agencies."

"Within the context of the EU and the UK, there has been a critical focus on the development of accountability requirements particularly when considered in the context of the EU's General Data Protection Regulation (GDPR) and the legal focus on Privacy by Design (PbD)."

"Globally, there is no standardised human rights framework and regulatory requirements that can be easily applied to FRT rollout."

"This article contains a discursive discussion considering the complexity of the ethical and regulatory dimensions at play in these spaces including considering data protection and human rights frameworks."

"It concludes that data protection impact assessments (DPIA) and human rights impact assessments together with greater transparency, regulation, audit and explanation of FRT use, and application in individual contexts would improve FRT deployments."

Introduction

"Each of these technologies changes the relationships between law enforcement operatives and citizens and requires the negotiation of new boundaries and revised accountability requirements."

"This latter question of accountability and acceptable privacy limits is critical in terms of balancing rights and responsibilities for FRT."

"FRT usage by law enforcement agencies provides a strong case study for considering aspects of FRT and AI ethics more generally."

"Ten critical ethical questions are framed that need to be considered for the ethical development, procurement, rollout, and use of FRT for law enforcement purposes."

"It should be established that while law enforcement agencies are at the forefront of FRT adoption, others can learn valuable ethical lessons from the frameworks put in place to safeguard citizens' rights and ensure accountability through time."

Facial recognition technologies (FRT)

"The context of the situation of FRT rollout and data gathering is potentially all important in terms of how it aligns with citizens' security versus privacy concerns in differing situations."

"In each citizen space, and dependent upon the form of law enforcement, there will be different perspectives and concerns from individuals and groups of stakeholders."

"A significant example of this, is the recent use of FRT by human rights activists and protesters as a way to identify, and hold accountable, law enforcement officers who might be abusing their power [227]."

"While a group of people who typically do not hold power may in limited circumstances use FRT to hold law enforcement accountable, that does not make the technology ethically viable."

The EU and UK legislative landscape for FRT in a law enforcement context

"Another important GDPR requirement is the need to have a Data Protection Officer (DPO) within any public authority or private entities where the core activities require large scale, regular, and systematic monitoring of individuals or large-scale processing of special category data or data relating to criminal convictions or offences."

"There are some limitations to these publications and the recent publications by the EU have been criticized for not bringing in a moratorium on biometric technologies such as FRT [228] In an EU context, in addition to the GDPR which dictates rules around managing personal data, privacy is further legislated for through the European Convention on Human Rights."

"Bridges, a civil rights campaigner, argued that the active FRT deployed by the police at public gatherings infringed on the right to respect for human life under the Human Rights Act 1998 and his privacy rights under the Data Protection Act 2018 (DPA 2018), the UK implementation of the GDPR."

The USA ethical and legislative landscape for FRT in a law enforcement context

"The second issue is centered around state-by-state differences; it occurs thanks to an absence of a general federal privacy legislation, with state law often providing only very basic rights for holding FRT operators accountable."

"The extent of privacy laws in most states is limited to notifying an individual if their data have been stolen in a security breach [229]—hardly a consolation for someone who has been affected by unintentionally biased or malicious use of FRT."

"In a USA context, the law on privacy and use of FRT for localised law enforcement operates very much at a state-by-state level."

"Clearview AI, one of the leading providers of FRT to the USA law enforcement agencies, supplies the technologies to more than 600 police departments across USA [230]; the ACLU filed a lawsuit against the company in the state of Illinois, arguing that it collected faceprints without consent, as required by the state's Biometric Information Privacy Act [231]."

Conclusion and recommendations

"Critical in the success of FRT is the transparency and accountability in each stage of its development and usage and the ability to audit and challenge as required."

"This discussions in this article makes the case that at all stages of the FRT process in all aspects of design and use including specific contexts, there is a requirement to document and account for the usage ensuring mechanisms for transparency and challenge."

"In conjunction with these steps, global regulators are needed with powers to actively investigate each aspect of the development and deployment processes of FRT in case contexts, and with powers to step in, stop and fine inappropriate FRT development and deployment."

"There should be more normal audit processes required for FRT deployment just as there are for financial oversights."

"Who should control the development, purchase, and testing of FRT systems ensuring the proper management and processes to challenge bias?"

Reflections on Turkish Personal Data Protection Law and Genetic Data in Focus Group Discussions [355]

This is a machine-generated summary of:

Özkan, Özlem; Şahinol, Melike; Aydinoglu, Arsev Umur; Aydin Son, Yesim: Reflections on Turkish Personal Data Protection Law and Genetic Data in Focus Group Discussions [355].

Published in: NanoEthics (2022).

Link to original: https://doi.org/10.1007/s11569-022-00431-0.

If you want to cite the papers, please refer to the original.

For technical reasons we could not place the page where the original quote is coming from.

Abstract-Summary

"And more rigorously since the 1990s, many countries have regulated data protection and privacy laws in order to ensure the safety and privacy of personal data."

"This study aims to explore the experts' opinions on the regulations of the health information systems, data security, privacy, and confidentiality in Turkey, with a particular focus on genetic data, which is more sensitive than other health data as it is a permanent identifier that is inherited to next of kin and shared with other family members."

"Two focus groups with 18 experts and stakeholders were conducted, discussing topics such as central data collection, legalized data sharing, and the management of genetic information in health information systems."

"Specific laws or articles dedicated to genetic data that also overlook the dimension of discrimination based on genetic differences in Turkey should be taken into consideration."

Background

"Unlike them, the Convention for the Protection of Individuals concerning Automatic Processing of Personal Data ("Convention 108"), which the Council of Europe adopted in 1981, obliged signatory countries to apply the principles of processing personal information."

"The Directive regulated privacy protection, and the transfer to a third country of personal data shall be provided by the member states to enforce other countries who conduct business within Europe [232, 233]."

"Health data is considered a particular category of data and, therefore, has stricter privacy regulations, so much that processing it is prohibited—except if explicit consent to processing personal data is given and used for medical purposes (GDPR Article 9(3))."

"The requirement for different treatment of genetic data was specified and guaranteed by law Hence, the study aims to explore the experts' opinions on the regulations of the health information systems, data security, privacy, and confidentiality in Turkey."

Methodology

"We conducted two focus group meetings to explore the experts' opinions on the regulations of the health information systems, data security, privacy, and confidentiality in Turkey, with a particular focus on genetic data."

"In the second focus group lasting 6 h, eleven participants (three females, seven males) discussed the collection of genetic data; discussions covered a wide range of topics on health information systems, ranging from electronic health records, comparisons of differences in the implementation of law in other countries to genomic data itself."

"As the main aim of the meetings was to explore the experts' opinions on the regulations of the health information systems, data security, privacy, and confidentiality in Turkey, we examined important issues of the regulations of the health information systems from legal, ethical, and security perspectives, where implementation of the new personal data protection (PDP) law in Turkey on genetic data was investigated in further detail."

Findings

"Another NGO lawyer, P10, pointed out that the PDP law gives this right to the Ministry of Health as well: "Within the same law, the Ministry of Health is also assigned the task related to collecting this data for certain purposes, like protecting public health, etc." According to P10's (NGO Lawyer) claim, this part is added to the PDP law to legalize data collection of the e-Nabız system since it was suspended previously by the council's decision (November 2015)."

"It says that when someone comes to you, you have to send me all of the data you have obtained from him." (P1, NGO Lawyer) And two, starting the data collection prior to the establishment of the personal data protection board (PDPB), which is regulated in article 47 (4) as the "Ministry shall take the necessary measures to ensure the safety of personal health data obtained under this law For this purpose, it establishes a security system that allows controlling for what purpose the registered information is used by which officer.""

Discussion and Conclusion

"Data collection and processing without consent, especially for sex life data exceptions, are the most criticized part of the PDP law Contrary to the situation in Turkey, the GDPR strengthens the issue of explicit consent."

"Considering these constraints, instead of ignoring the risks by accepting genetic/genomic data as relatively anonymized, accepting it as pseudoanonymous from the start can help us be aware of the risks, control the shared persons/institutions, and inform them about their responsibilities for undesirable events."

"On the administrative and regulatory level, collecting health and genetic information is not a problem unless it is not centrally collected, managed, processed, and accessible- without the data owner's consent."

"Even though the new PDP law separated genetic data from health data, it was collected as health data by the government agency, SGK."

Using digitally mediated methods in sensitive contexts: a threat analysis and critical reflection on data security, privacy, and ethical concerns in the case of Afghanistan [356].

This is a machine-generated summary of:

Guntrum, Laura Gianna; Güldenring, Benjamin; Kuntke, Franz; Reuter, Christian: Using digitally mediated methods in sensitive contexts: a threat analysis and critical reflection on data security, privacy, and ethical concerns in the case of Afghanistan [356].

Published in: Zeitschrift für Friedens- und Konfliktforschung (2022).
Link to original: https://doi.org/10.1007/s42597-022-00088-2.

If you want to cite the papers, please refer to the original.

For technical reasons we could not place the page where the original quote is coming from.

Abstract-Summary

"Both technical and non-technical research-related risks are analyzed, paying attention to research ethics, data security, and privacy."

"Our analysis reveals that in certain research processes, human security does not necessarily coincide with data security and that an ambivalence exists between privacy and usability."

"To the concrete threat analysis, the paper identifies some general technical solutions (e.g., encryption methods, communication software) for different research steps to foster secure and ethically justifiable research."

Communication software as an ethnographic tool

"In any research via digital means, the question should be asked whether risks (e.g., digital surveillance) are likely and whether more data secure alternative applications could be used during the entire research process, including data analysis and storage, since multiple programs (e.g., for transcription) raise privacy concerns."

"Inspired by these research gaps, we formulated the following research question (RQ): How can a threat model generally be elaborated in order to identify potential risks and adversaries in sensitive context, having privacy, security, and ethical concerns in mind?"

"We present a literature review of previous studies related to ethics in digital research and information technology (IT) security (section "Related work and theoretical embedding") and explain our research design in detail ("Methodology" section)."

"Based on existing categories from IT security for secure and privacy-compliant communication, we examine which communication tools appear suitable for conducting secure digitally mediated research, having the empirical study of Afghanistan in mind (section "What to consider when working with digitally mediated methods")."

Related work and theoretical embedding

"This status quo is changing slowly, partly because technical achievements and digitally mediated methods facilitate participation of historically marginalized groups and provide access to research contexts, which were previously difficult to reach because of security issues or travel restrictions [236]."

"According to Dobrick et al. [237], to date little research has been done on privacy violations and data security in social sciences, including peace and conflict studies."

"In such cases, it might be helpful to study existing security guides for digital research [238, 239] or to seek online support from IT security experts, who are specialized in technical support for human rights activists and journalists in conflict-related contexts [240]."

"Sometimes less data-secure tools that violate privacy principles may be an appropriate option when conducting research in sensitive contexts—for example, if such tools, unlike their more privacy-friendly alternatives, do not raise suspicion during identity checks, physical harm can be better prevented."

Methodology

"For our threat model for Afghanistan, we refer to two sources: The first is the Digital Care Guide for Afghan journalists, produced by the non-governmental organization Reporters Without Borders (RSF) in 2021 to provide Afghan journalists with digital security advice [241]."

"An in-depth literature search, including IT security blog posts and websites of organizations and foundations dealing with our research interest was conducted."

"Access to interviewees, coming from countries such as Iraq, Ukraine, and Afghanistan was facilitated by in-depth internet research (e.g., civil peace organizations, well-known IT security blogs) and recommendations (snowball sampling) [242]."

"The interviews are intended to complement given external recommendations on Afghanistan and personal experiences by suggesting other communication tools and strategies to conduct research in vulnerable contexts, highlighting that recommendations may vary depending on the specific circumstances."

Hypothetical case study: Afghanistan (2021)

"Before we discuss the data generated by our study, we need to briefly summarize our hypothetical study design: First, we would obtain contact to study participants

(Afghan journalists) through contacts recommended by RSF and subsequently use snowball-sampling to find further interlocutors."

"With regard to the first part of threat modeling (data mapping), this produces several data artefacts like audio recordings and textual notes, including contact data."

"Western intelligence services, such as the NSA (National Security Agency) or GCHQ (Government Communications Headquarters), utilize mass surveillance in the form of bulk metadata collection in order to identify terrorist threats and criminal organizations."

"We would physically destroy (shred) the USB pen drive containing the VeraCrypt keyfile, as the recorded data encrypted by this keyfile are of high protection value."

"Even if the Taliban were to detect the communication with our interlocutors, we can minimize the harm by using an Afghan phone number and contact name to avoid attention."

What to consider when working with digitally mediated methods

"As already illustrated (see sections "Security Plan and Threat Modeling" and "Hypothetical case study: Afghanistan (2021)"), prior to initial contact with potential respondents, researchers should consider whether data misuse has previously occurred in the country or context of interest, what risks might be associated with the research, which people are potentially at risk, and who might be potential adversaries (IP1)."

"It should be considered that people also use less data secure applications for their own private communication (IP8)."

"Appropriate data storage locations: Generally, three options of data storage locations may be considered: (1) encrypted data storage on a personal computer, (2) data storage on portable devices such as external hard drives (encrypted by tools such as VeraCrypt [IP6]), and (3) data storage on a cloud service that offers hosting data on a remote server accessible at any time via the internet."

Discussion and conclusion: secure research with digitally mediated methods

"There has been little debate about research ethics, data security, and privacy violations during empirical studies in the social sciences."

"Some scientists (particularly in the social sciences) seem to have little IT security knowledge [243] and the fast pace and complexity of technology development complicate many researchers' ability to maintain an overview of existing applications and their associated risks."

"This implies, amongst others, that human security does not necessarily go hand in hand with data security (for example in our case study), that different applications are used around the globe, and that different risks are associated with each application and research procedure."

"Our specific case study clearly shows that, in the example of Afghanistan, a less data secure and non-GDPR-compliant instant messenger (WhatsApp) seems the most appropriate for contacting potential interlocutors and for conducting research, for the reasons explained in the sections "Hypothetical case study: Afghanistan (2021)" and "What to consider when working with digitally mediated methods"."

Limitations

"Our threat model has been developed with utmost conscience of the circumstances in Afghanistan since summer 2021."

"At any moment, the socio-political situation may change, new security bugs may be discovered, or new applications developed, meaning that this paper can only reflect on the status-quo and on one specific country."

"Further studies may complement these insights, capturing more voices from multiple contexts and further highlighting how circumstances may change over time."

Reflections upon periclitations in privacy: Perspectives from Rwanda's digital transformation [357].

This is a machine-generated summary of:

Nsengimana, Jean Philbert: Reflections upon periclitations in privacy: perspectives from Rwanda's digital transformation [357].

Published in: Health and Technology (2017).

Link to original: https://doi.org/10.1007/s12553-017-0196-0.

If you want to cite the papers, please refer to the original.

For technical reasons we could not place the page where the original quote is coming from.

Abstract-Summary

"In part, this declaration is foundationally germane to Rwanda's digital transformation, and our society's approach to erecting Privacy protections, harnessing the deep value that Privacy had once held in Rwanda's pre-colonial tradition, and has again, in her most recent history."

"In view of a rapid transformation that is presently turning the country into a digital hub for the Continent of Africa, Rwanda is charting a path forward to achieve a double purpose; on one hand the prospect of leveraging information technology for the benefits of her citizens, and on the other hand, securing and protecting the privacy of her citizens, as a key component of Rwanda's societal identity."

"This article further presents that given the leadership that Rwanda has demonstrated in Information and Communication Technology (ICT) for Development in Africa, the nation's course of action in relation to data and personal privacy protection will have a positive influence on the rest of the African continent that is now bracing to form a single digital market through the Continent-wide Smart Africa initiative."

Introduction

"The first industrial revolution was marked by the introduction of mechanical production, enabled by the steam engine during the second half of the eighteenth century."

"The second industrial revolution was ushered by the advent of electricity and it enabled further improvements in mass production."

"The third revolution often referred to as the digital revolution was born as a product of the 1960s, and it was enabled by the development of semiconductors, mainframe computing, parallel processing, personal computing and Internet."

"The world is at the brink of yet another revolution, the 4th, driven by a rapid convergence of technologies across the biological, physical and multiple digital domains [244]."

"As part of her vision 2020 [245], Rwanda is undergoing a massive digital transformation, aimed at moving the economy and society from a subsistence agriculture base, to an information and technology hub for the region and the continent."

Privacy: the Rwandan context

"Several social relations required even higher levels of privacy."

"Within in-law relations, parties had to observe strict behaviors such that none would invade the privacy of parents-in-law, even accidentally."

"No information about the former's personal activities, health conditions, or social relations was to be shared."

"Privacy, as understood by Rwandans was dealt a great blow by many aspects of colonialism."

"Pre-existing 'social taboos,' which included a great deal of personal privacy protection dimensions were destroyed by the spread of Christianity."

"Agents of the Catholic Church gave blanket absolutions for those Rwandan people who violated their 'traditions' such as Privacy Protections."

"One of the taboo's that was widely disregarded or violated was that people's cows could now be counted publically—to levy tax, and, or categorize people into social classes and ethnic origins; an act—for which, the Church delivered absolution."

The global dash to a hyper connected society: a ministerial view

"For the sake of encouraging the build-out of energetic, creative, healthful, productive, stable and a thriving "virtual commons" in our Digital Century, humanity must address the great insufficiency before us in terms of Privacy, that is to identify and disentangle the natural truth from the "Two Chief Systems of the World" as presented; and to construct a 'manifest' life and living habitat, where none will ever have to make a trade-off between Privacy, Safety and Security, on account of some circumstance that would impose an unnecessary and definitely harmful requirement to people and communities."

Rwanda's digital transformation journey—a review

"With increased penetration of mobile devices and Internet use, the need to provide a stronger framework for protecting personal Privacy, while ensuring that information technology utility does not turn into a threat for national economic security—has become very pressing, and to the forefront of national leadership considerations."

"In anticipation of the increased production and use of personal data by the different applications that the technology companies are going to produce for the pan-African market, Rwanda is currently investing time, energy and financial resources to have in place, the necessary technical, legal and regulatory guideposts and protections, to position the country as a competitive, trustworthy, safe and secure place to host globally produced data."

"In line with this commitment, Rwanda has invested in a Digital Forensic Laboratory under the Rwanda National Police, a CSIRT and PKI under the Rwanda Information Technology Authority as well as a Regional Internet Exchange (IX), all of which have the vision of enabling the provision of secure digital services."

Privacy protection and digital transformation—essential national components

"The bill has other provisions related to protecting users' personal information and privacy by private and public entities while performing operations such as billing, customer registration, issuing of digital signatures, etc. The regulatory authority which has the mandate to protect the customers of electronic communications and services, has availed several avenues for initiating and escalating complaints related to the services including those complaints related specifically to privacy."

"The Smart Rwanda Master Plan's implementation is expected to define Privacy as a central concern, even as we continue to learn how best to leverage: continually evolving best practices in the world, the best available Privacy Protection institutional and infrastructural frameworks from those in the world, that have been actively engaged in making their environment and services safer for their users, and the availability of tools and techniques across the multiple domains of digital transformation."

Privacy, data security and safety—a top priority

"From TRACnet implementation in 2005 to "Rwanda Online" in 2015, the need for a strong framework for Privacy protection has been clear to all key stakeholders involved in the country's digital transformation; ranging from citizens, to service providers, and to policy makers."

"This trust rose from the painful yet, highly fruitful unity and reconciliation journey that all Rwandans had to collectively undertake, post the 1994 genocide against the Tutsi, in order to mold Rwanda the nation-state it has become today; a stable, thriving and forward looking country with pronounced prospects."

"While being actively engaged in Rwanda's digital transformation, governance instruments to exercise and to ensure Government's responsibility to all Citizens had to be put into place, including those that direct and institute the preservation and protection of critical societal values—such as personal privacy."

Conclusion

"The challenges related to the protection of personal information in our digital world, and indeed, this century, are complex, and requires the aggressive and focused matching of immense will, with a fiery commitment by all who are involved, toward the stewardship and the custodianship of people's Personal Privacy, to achieve civilized society's desired Privacy oriented goals."

"As Rwanda continues to embrace the many processes of digital transformation for her society, the commitment from the central government will be to continue to ensure that the use of ICT, and associated societal rewards are not realized at the expense of sacrificing people's privacy."

"Given Rwanda's ambition of becoming an ICT Connecting-Point for the Continent of Africa, it is important that the country leads by example for all those on the Continent of Africa, and beyond, by upholding a sustained commitment to the protection of personal privacy, and to the preservation of deeply rooted Rwandan traditions related to Privacy."

A comparative analysis of personal data protection regulations between the EU and China [358].

This is a machine-generated summary of:

Weber, Philip Andreas; Zhang, Nan; Wu, Haiming: A comparative analysis of personal data protection regulations between the EU and China [358].

Published in: Electronic Commerce Research (2020).

Link to original: https://doi.org/10.1007/s10660-020-09422-3.

If you want to cite the papers, please refer to the original.

For technical reasons we could not place the page where the original quote is coming from.

Abstract-Summary

"With the EU's implementation of a new data protection regulation in 2018, the development of personal data protection globally has reached an important turning point and has sparked the interest of scholars and businesses."

"Text coding was employed to compare the current personal data protection regulation landscape in the EU and in China to discover the differences between the General Data Protection Regulation and the personal data protection regulations of the fastest-growing economy in e-commerce."

"The results show that while there are several similarities in regard to general requirements, such as principles of data processing and basic rights for data subjects, China's personal data protection regulations tend to lack specific operational requirements and strong legal enforcement."

Introduction

"With regard to the EU e-commerce market, a study conducted by the European Union in 2016 revealed that two-thirds of the participants engaged in e-commerce, but among all the internet users questioned in the survey, 71% reported having given away their personal data online [246]."

"Another survey conducted by the European Union revealed that only 26% of social network users and less than 20% of people engaging in e-commerce activities feel that they have total control of their own personal data [247]."

"For governments and businesses in China and other Asian countries, on the one hand, personal data protection has not only been one of the major concerns in e-commerce transactions [248, 249] but has also been of great importance to cross-border communication and cooperation at the strategic level [250], especially in the context of data-driven business models [251]."

"To better understand the differences between the current personal data protection policy systems of China and the EU, a comparative study of the text content of policies, laws, and regulations was conducted, and the results and implications are reported in this paper."

Literature review

"There are certain laws and policies that have no digital industry or information security focus yet still contain relevant content that relates to the protection of personal data."

"The GB/T 35273–2017 became effective in May 2018, in line with the majority of personal data protection laws, including the GDPR."

"With the EU's personal data protection system, especially the GDPR, China is generally rated as having insufficient legal protection of the right to privacy [250, 252]."

"While this judgment is largely correct, it is worth noting that in the absence of a GDPR-compliant personal data protection regulation in China, previous comparative studies have mainly discussed the general logic and main features of the relevant legislation at the macro level without making detailed comparisons of the specific rules and often without even specifying the texts to which the relevant Chinese regulations refer [250]."

Methodology

"Since the code book is based on the GDPR structure, the chosen Chinese regulations had to include relevant content that is comparable to the GDPR document."

"For the practical method of conducting the coding and documenting relevant text content, the Coding Manual for Qualitative Researchers by Saldana [253] was used as a guideline during the process."

"During the second step, the text content from the selected documents on Chinese regulations was categorized and organized into the initial version of the code book."

"The presence of four levels gave enough details to compare the regulation texts on more than just a surface level but kept the codes sufficiently comprehensive to find matching content despite comparing documents from different political entities."

"After the final version of the code book was finalized and the respective text content of all the documents was distributed."

Results and discussion

"By employing the method of text coding, a comparative analysis of the GDPR and six Chinese documents was conducted to understand the different and similar focal points of the personal data protection schemes and requirements in both the EU and China to help answer the research question."

"The second example is the code OR-coc, which is codes of conduct set up as additional requirements by the GDPR and has the lowest possible match rate of 0%, which indicates that none of the Chinese documents contain relevant content."

"For future research, it would be very interesting to code and analyze the match rates for the Chinese personal data protection law after it has been published."

"The overall amount of matched content for Chinese documents in this category is very low, which was anticipated since China currently has no specific authority dedicated to personal data protection."

Conclusion

"At the practical level, (1) keeping adequate records of activities pertaining to data processing was identified as an important practical step to ensure personal data protection, which leads to the recommendation to make this into a legal requirement for China instead of merely a recommendation. (2) With regard to personal data transfers, China should consider finalizing the relevant rules that are currently being developed and provide sufficient legal enforcement or including stipulations in the upcoming personal data protection law (3) An analysis of rights for data subjects suggests that Chinese regulations should expand the overall protection of data subjects and consider whether the rights that were added in GB/T 35273–2017 are receiving adequate enforcement or to include them into a new law Moreover, Chinese processing entities should be aided with more information to help them conduct their processing activities with adequate security measures as well as more guidance to identify what laws and regulations apply to certain industries and operations."

The Normative Power of the GDPR: A Case Study of Data Protection Laws of South Asian Countries [359].

This is a machine-generated summary of:

Bentotahewa, Vibhushinie; Hewage, Chaminda; Williams, Jason: The Normative Power of the GDPR: A Case Study of Data Protection Laws of South Asian Countries [359].

Published in: SN Computer Science (2022).

Link to original: https://doi.org/10.1007/s42979-022-01079-z.

If you want to cite the papers, please refer to the original.

For technical reasons we could not place the page where the original quote is coming from.

Abstract-Summary

"The failure to securely protect data would potentially give rise to far-reaching catastrophic consequences."

"It is crucially important to have national, regional, and global data protection policies and regulations to penalise those engaged in unethical use of technology and abuse the system vulnerabilities of technology."

"This research paper aims to analyse GDPR inspired Bills in the South Asian Region and to identify their appropriateness for developing a global level data protection mechanism, given that Asian nations are far more diverse than those of the European nations."

"Against that background, the objectives of this paper are to identify GDPR inspired Bills in the South Asian Region, identify the similarities and disparities, and the barriers to developing a regional level data protection mechanism, thereby fulfilling the need for developing a global level mechanism."

"The key findings of this research indicate that many countries in the South Asian region are in the process of reviewing their current data protection mechanisms, in line with GDPR."

"In concluding, the researcher emphasised the need to develop adequate data protection mechanisms and believed that going forward it would be the appropriate and practical way to develop a consensus-based regional mechanism that would ultimately enable to develop a lasting global level data protection mechanism."

Introduction

"In the South Asian Association for Regional Cooperation (SAARC) region, most of the countries are in the process of developing data privacy laws [254], but no visible progress is being made towards producing a region relevant law/s resulting from the SAARC agreements [255]."

"To that end, this research paper aims to analyse the GDPR inspired Bills in the South Asian Region and identify the course of action they would take towards the development of a global level data protection mechanism."

"In pursuing that aim, the objectives of this paper are to identify GDPR inspired Bills in the South Asian Region, identify the similarities and dissimilarities, and identify the barriers to developing a regional level data protection mechanism that would meet the requirement for developing a global level mechanism."

"In addressing the research aim, the researcher reviewed available literature surrounding GDPR and GDPR inspired bills in the South Asian region to analyse the contribution of national level data protection mechanisms towards developing a global level data protection mechanism."

General Data Protection Regulations (GDPR)

"The European Union (EU) enacted the GDPR [256] governing personal data protection to promote the establishment of a regional strategy for information security based on fundamental rights underpinned by democracy."

"Amongst other factors, personal data protection is one important element of the rights of the people, and it offers sanctuary to the individuals and makes them feel secure from unethical intrusions to their personal data."

"The European Union (EU) prohibits data transfer from an enterprise in the EU region to countries that do not match the same level of EU Regulations on data protection [257]."

"The enterprises that undertake activities relevant to the processing of personal data are required to employ a data protection officer, and [257] reporting of data security breaches is mandatory as recommended by the commission."

"European Union by endorsing the GDPR, has taken the lead in instituting data privacy regulations."

Actions Taken by Countries in the South Asian Region

"This is the first time the Supreme Court has pronounced the right of individuals to their personal data, and privacy and data protection has been placed high on the national agenda as data use is considered a key element in the growth and economic development."

"The Information Technology Act 2000 governs the protection of personal information, specifically electronic data and transactions [258]."

"In that respect, the basic framework for data protection and privacy sets out the rights of privacy granted under the Constitution of Bangladesh, alongside the Information Communication Technology Act 2006 and the newly enacted Digital Security Act 2018 [259]."

"This Act explicitly requires obtaining consent or authorisation from data subjects before collecting, storing, and processing personal information [260]."

"The reports suggest that seven of the ten second-generation principles in the 1995 EU Data Protection Directive have been included in Bhutan's Information, Communications and Media Act, which came into force in 2018, but with limited coverage on privacy and privacy law [254]."

Discussion

"The South Asian region has no regional level data protection mechanism in place yet, but having one will encourage them to develop a global level data protection mechanism."

"The understanding of discrepancies would help to resolve implicit issues between the countries and bring them together to establish a common stance on developing a unified regional level data protection mechanism that would eventually level up towards developing a global level data protection mechanism."

"A country's overall economic strength influences internet diffusion and the resources and capital required to expand technology [261], and there is also a demand for capital investment for developing data protection mechanisms."

"These social differences and beliefs form the most significant barriers to developing a global level data protection mechanism."

"These social differences and beliefs are the biggest barriers facing developing a global level data protection mechanism."

"To meet that requirement, most countries revisited and developed data protection mechanisms at the national level lining with GDPR."

Conclusion

"In this research, the researcher selected South Asian region and sought to explore existing data protection mechanisms and identified the challenges and barriers they faced in developing data protection mechanisms."

"The available literature suggests that most countries in the South Asian region possess at least a draft data protection mechanism in place at the national level."

"The researcher believes that it is essential to have unified purposeful data protection mechanisms developed collectively by the nations to overcome security and privacy challenges and prevent data breach perpetrators escape with impunity."

"To overcome this, respective countries need to revisit and develop their national-level data protection mechanisms."

"The literature readings indicated that whilst some countries in the South Asian region were in the process of drafting their national-level data protection mechanisms in parallel to the GDPR, some countries failed to keep up."

Recommendations

"The researcher recommends that the SAARC should support South Asian countries to take the initiative to develop national and regional level data protection mechanisms."

"To make it happen, the SAARC should take the initiative to make awareness of the importance of protecting personal data and should provide appropriate guidance and support."

"The recommendation is to set up a project team consisting of like-minded professionals with expertise in negotiating and influencing those at the highest level in governments, the judiciary, and policy development, preferably facilitated by the SAARC."

Why and How Does the EU Rule Global Digital Policy: An Empirical Analysis of EU Regulatory Influence in Data Protection Laws [360].

This is a machine-generated summary of:

Cervi, Giulio Vittorio: Why and How Does the EU Rule Global Digital Policy: an Empirical Analysis of EU Regulatory Influence in Data Protection Laws [360].

Published in: Digital Society (2022).

Link to original: https://doi.org/10.1007/s44206-022-00005-3.

If you want to cite the papers, please refer to the original.

For technical reasons we could not place the page where the original quote is coming from.

Abstract-Summary

"Why is the EU capable of regulating the digital environment and how is it able to exert regulatory influence outside its boundaries?"

"As European policymakers appear increasingly keen on setting global standards in new areas of the digital domain, this article provides insight into why and how the EU has been able to exert global influence via its General Data Protection Regulation (GDPR)."

"Empirical evidence collected in this study points to three main drivers of the GDPR's global success: the EU's internal market appeal, its credibility as a regulator and enforcer, and the timing of its regulatory actions in line with evolving policy needs."

"This has enabled the EU to exert regulatory influence in a unilateral and indirect way—via market forces or independent import by third states—as well as in a more direct way—via adequacy decisions."

Introduction

"Supposedly, by virtue of its market size, regulatory capacity and stringent regulation on specific markets and targets, the EU exerts its influence both de jure—foreign countries emulate EU regulations domestically—and de facto—multinational companies abide by EU rules even when subject to other jurisdictions."

"One area that is said to be especially prone to EU regulatory influence is that of the digital economy."

"Why is the EU capable of regulating the digital environment, which is global by nature, and how is it able to exert regulatory influence outside its boundaries?"

"There are many factors favouring the EU in exerting global regulatory influence."

"The ways in which EU regulatory influence plays out are contingent on the foreign jurisdiction's situation."

"It applies this framework to a policy field that is becoming increasingly relevant in view of technological progress, examining why and how the EU exerts regulatory influence."

Policy Review

"Only three European countries adopted national data protection legislation."

"The EU has been pushing for higher standards across its Member States, as it adopted the so-called data protection directive in 1995."

"As of April 1, 2022, data protection legislation is in force in more than half of the world, and more than one in six countries either have adopted or are currently discussing regulations."

"Considering countries in the EU and the European Economic Area to which it applies, the GDPR accounts for one-third of all data protection laws now in force globally."

"In Oceania, both Australia and New Zealand recently amended their data protection laws."

"While the United States lack a federal data protection law, in recent years a few bills were adopted at state level and many others a have been introduced in Congress and State Assemblies."

Literature Review

"The theory of the "Brussels effect" [262–264] holds that there are five key elements to explain the EU's unilateral regulatory power: (1) its market size; (2) regulatory capacity to adopt and enforce regulations effectively; (3) political will to generate stringent rules; (4) regulation of inelastic targets, such as consumer markets; (5) regulation of target's conduct or production which is non-divisible across global markets."

"The Brussels Effect is not the only way in which the EU is able to exercise global regulatory influence."

"The Brussels Effect is an example of market-driven influence: it is unilateral and indirect since it is completely driven by the private actors—and only thereafter by foreign governments—subscribing to EU rules, while the EU is not actively seeking regulatory influence."

"To unilateral export or import of rules, the EU can be directly or indirectly involved in regulatory influence via its cooperation with international partners [265]."

Methodology

"Target participants in group A were government officials, members of parliament working on data protection laws, regulators from DPAs, experts and activists."

"In this group, 140 individuals from 35 countries were contacted, leading to 19 interviews."

"Differentiating factors within this group were whether companies are headquartered in the EU or not, whether individuals worked in the policy function, thus more concerned with influencing regulations, or data protection, thus more concerned with enforcement."

"Interviews started with factual questions on the state of the art of data protection policy in the interviewee's country (or area of reference) and their organisation's role within that framework."

"All interviews were conducted directly by the author, as was the transcription and coding of empirical data."

"As participation in the interviews was voluntary, the final sample composition rested upon their willingness to contribute to a research project on EU regulatory influence."

Findings

"Some participants were more granular in their responses, citing specific examples of regulatory capacity: 25% of industry stakeholders referred to thorough policy process, 21% of third-country interviewees and another 25% of industry stakeholders attached significant value to the longstanding European experience in legislating on data protection; finally, 10.5% of third-country interviewees defined the GDPR as a very good piece of legislation, an obvious advantage on the global marketplace for policies."

"Industry stakeholders were unanimous in identifying the market-driven mechanism as the primary way through which EU data protection rules exert global influence; they also referred to the extraterritoriality element, while giving less importance to state-driven or treaty-driven influence."

"While 21% of interviewees assigned no role to domestic companies in the policy debate around data protection in their country, a greater number of participants (31.6%) stated that national players are indeed vocal in expressing their support for GDPR-style provisions."

Conclusions and Further Research

"This article has examined the question of why the EU is capable of regulating the global digital environment via its data protection policy and how it exerts international regulatory influence."

"It reviewed global trends in data protection and presented a comprehensive framework of the possible mechanisms through which the EU impacts those trends."

"EU rules in data protection were found to have a sizeable international impact."

"Among the several potential enablers of EU regulatory influence, the main drivers of the GDPR's global success are the EU's internal market appeal, its credibility as a regulator, and the timing of its regulatory actions in line with evolving policy needs."

"Other mechanisms through which EU regulatory influence works."

"The extent of EU regulatory influence depends heavily on global compliance by private companies and international recognition of EU rules as gold standards."

The Japanese sense of information privacy [361].

This is a machine-generated summary of:

Adams, Andrew A.; Murata, Kiyoshi; Orito, Yohko: The Japanese sense of information privacy [361].

Published in: AI & SOCIETY (2009).

Link to original: https://doi.org/10.1007/s00146-009-0228-z.

If you want to cite the papers, please refer to the original.

For technical reasons we could not place the page where the original quote is coming from.

Abstract-Summary

"As we show, there is good evidence for social norms of limits on the sharing and use of personal information (i.e. information privacy) from traditional interactions in Japanese society, as well as constitutional evidence from the late nineteenth century (in the Meiji Constitution of 1889)."

"The growing awareness of the Japanese public about problems with networked information processing by public sector and commercial organisations from the 1980s (when a law governing public sector use of personal information was first passed) to recent years (when that law was updated and a first law governing commercial use of personal information was adopted) are not the imposition or adoption of foreign practices nor solely an attempt to lead Japanese society into coherence with the rest of the OECD."

"They are drawing on the experience of the rest of the developed world in developing legal responses to the breakdown of social norms governing interchange and use of personal information, stressed by the architectural changes wrought by networked information processing capabilities."

Introduction

"The myth of a lack of Japanese conception of privacy seems to be, at least partly, grounded in the Japanese tradition of importing external concepts into law and social appearances while either using them as a mask for continued traditional practices or misunderstanding their original conception and producing a new Japanese custom with the same name as the external concept but a different meaning [266]."

"Unlike the physical conceptions they describe where, for example, the puritan influence on most Western societies strongly militates against nudity, and where a different word in English may be better aligned with the social reality of Japanese expectations than 'physical privacy' or 'bodily privacy', our contention is that the same structures are present in information privacy in Japan and the West, but that for certain people and certain types of information the barriers to information flow are placed at different distances from the subject."

Traditions of information sharing

"Just as with the British Upper Class whose servants long had access to a great deal of information on the personal habits, issues and relationships, this lack of an architectural [267] limit on the sharing of information was solved by the development of social norms requiring at least the illusion of privacy and also a restriction on the spread of information beyond the household."

"This is reflected particularly in the as-if tradition [268], which still has a strong influence on social practices of information sharing in Japan today, particularly within the business organisations that until recently replaced the older household as the focus of loyalty within male Japanese society."

"While Japan may be at or near an extreme of information revelation, the concept of information privacy does exist, and is a key part of Japanese psychology and social structures."

Psychology and linguistics

"We present a broader definition of information privacy which includes but is not limited to this concept, allowing a fuzzy boundary to the self and the definition of 'private' information (and thus of privacy), but nevertheless allowing consistent social guidelines for the availability of and 'right to process' private information common between Japan and other societies."

"The connection between concepts of physical privacy and informational privacy is clearly signified by the Japanese social use of terms denoting places (inside and outside) to refer to social order (us and them)."

"Doi's analysis includes a reinforcement of the idea that the Japanese have always had a strong, though different to the West, concept of the role of the individual in society ([269], p. 56): Just as honne exists behind tatemae, the individual, in principle, exists in and under the shelter of the group."

The insular collective society

"The strong identification of the individual with the group in Japanese society, expressed principally in terms of vertical relationships up and down, suppressing horizontal links (at least partly by how much information about the self is revealed, as we discuss below) leads to a fragmentation of society into competing islands."

"The speech level becomes more formal, and when Akira and Bunta refer to Chiaki they now use humble forms (as they do of themselves and the rest of the technical department)."

"Since Eiko is from a separate organisation, Akira, Bunta and Daisuke now form an uchi group with Eiko in the soto zone."

"Akira, Bunta and Daisuke refer to all members of Company Fuji using humble forms, even their company president, due to the presence of the soto member Eiko."

"Akira, Bunta and Daisuke use honorific terms to refer to Eiko and all members of Company Bando."

Networked processing of personal data

"Mainichi Daily News 20 July 2007 MPD to sack police officer over leak of files through file-sharing program Winny Daily Yomiuri 13 July 2007 Govt eyes punishment for data leaks Asahi Shimbun 15 July 2007 Data leaks included sex crime details The Japan Times 24 February 2006 Confidential MSDF info leaked on Internet via Winny Until there was evidence that organisations with whom Japanese people did not have regular direct personal contact were misusing or being careless with data which could present them with difficulties, Japanese people tended to regard the data collected as 'just data' and not actually connected to the individual."

"The creation of data doubles, and the ease with which information may now be transmitted between soto entities (and the requirement to deal with such entities) has led to a rapid awakening amongst the Japanese that social norms are no longer sufficient to regulate the protection of information privacy."

"The speed with which Japanese society has moved from reliance on social norms to the development of legal protection for information privacy demonstrates just how strong the Japanese sense of information privacy is, despite an acceptance of a broader, but still limited, distribution of some personal information."

The Japanese constitutional right to privacy

"Although the current Japanese constitution, originally adopted in 1946 under US occupation, does not include an explicit general right to privacy, it does include a balance of respect for individual freedom with the needs of society in Article 13, ([270], p. 118–119) All of the people shall be respected as individuals."

"Starting in 1964, various elements of the information privacy rights recognised elsewhere have been adjudicated as deriving from Article 13 of the Japanese Constitution."

"In the 1969 final judgement on the case, the Grand Bench of the Supreme Court ruled against the claims of the protesters to an invasion of privacy because of the lack of a warrant or permission."

The Japanese sense of information privacy

"According to Yamagishi [271] and Yamagishi and Yamagishi [272] the insular collective model of Japanese society reduces trusting behaviour because the privileging of the interests of the insular mean that in a prisoner's dilemma type of situation, expectations of betrayal by members of the soto group ensure mutual defection."

"Uchi group interactions are more complex, but still provide evidence of a strong sense of information privacy in the Japanese social structure."

"The pressure not to disrupt the harmony of the uchi group by revealing opinions contrary to agreed policy is very strong in Japanese society, though not unique to it."

"Even allowing for the early and continuing socialisation of Japanese children into sacrificing individual desires for those of the group ([273], Chaps. 3 and5), these pressures require an outlet."

"There are three ways in which information revealed during such sessions in liminal zones may be treated by the rest of the group: The As-if tradition."

Conclusion

"Despite claims from Benedict [274], Itoh [275] and others, the Japanese sense of information privacy is a solid part of traditional Japanese culture, gradually updating itself, as do all parts of culture, due to the pressures of technological change and other factors."

"Although the exact boundaries of what information may be passed to whom under which circumstances differ from other cultures, the Japanese are not uniquely possessed of a lack of a sense of information privacy."

"The Japanese have had and retain as strong a sense of information privacy as citizens in other industrialized nations, while its expression in social norms and laws remain similar in scope but significantly different in detail, to those elsewhere, as shown by the cases involving factional novels such as After the Banquet and The Fish Swimming in the Stone."

Using deceased people's personal data [362].

This is a machine-generated summary of:

Nakagawa, Hiroshi; Orita, Akiko: Using deceased people's personal data [362]

Published in: AI & SOCIETY (2022).

Link to original: https://doi.org/10.1007/s00146-022-01549-1.

If you want to cite the papers, please refer to the original.

For technical reasons we could not place the page where the original quote is coming from.

Abstract-Summary

"We identify the problems with the commercialization of deceased users' images and personal data, which becomes postmortem entertainment."

"Considering these problems, we seek out the ideal form of deceased users' personal data for commercialization."

"By analyzing our survey results approximately 20% of respondents would allow the commercial use of their personal data, such as browsing their social network service logs, if they could receive compensation during their lifetime."

Introduction

"After platform users die, their data forms their digital heritage."

"Users do not necessarily have a consensual understanding about how their digital heritage will be managed and by whom."

"Some users may wish their digital heritage to be deleted after their death, while others want their family members to manage their digital heritage."

"Internet platforms and service providers cannot only manage deceased users' digital heritage but also profit from its commercialization."

"Society is considering how to best manage deceased users' digital heritage with respect for their wishes, especially by Internet platform business operators."

Background

"These data are often referred to as the deceased users' digital heritage."

"This study explores how deceased users' digital heritage, i.e., their lifetime personal data, should be managed."

"Before their death, users may give their consent to allow their personal data to become informationally immortal if many people can use or interact with the deceased users' data."

"We consider informationally immortal data as the deceased users' immortal digital personalities, which the bereaved can use as a commercial memorial service."

"According to Harbinja [276], the legislature regarding the deceased's personal data, such as the 2018 UK Data Protection Act, defined personal data as "data relating to a living individual" and denies any rights to access deceased users' personal data after their death."

"This law allows the deceased individual's personal representative to have default access to a "catalog" of the deceased users' electronic communications and other digital assets that are not protected by federal privacy laws."

Immortal digital personality

"Immortal digital personality creators collect individuals' intentional and unintentional behaviors over a long period of time before their deaths, including tracking deceased individuals' online history data, such as emails, voice mails, blog posts, SNS posts, global positioning system location data, bank account transactions."

"By paying for these services, individuals can continue to live after death through their immortal digital personalities."

"Immortal digital personalities do not necessarily have to be conscious because commercial digital heritage services can maintain the illusion of the immortal digital personalities as a model consistent with the way the deceased individuals behaved while they were alive."

"Eternime (http://eterni.me/), started in 2014, was designed for individuals to train their own immortal digital personalities through daily interactions with the system before their deaths."

"Interactivity: Digital technologies combine data and machine learning to create virtual avatars that provide immortal digital personalities that can be interacted with in both directions."

Current unfavorable problems and future directions for digital heritage management

"In considering the informational body after death, Öhman and Floridi [277] criticized the current situation where the use of the deceased users' personal data has become too commercialized."

"The following commercial services are currently being considered to deal with deceased users' informational body, i.e., their digital heritage."

"When the deceased users' personal data are used to generate commercial value, the deceased becomes an immortal digital personality who performs life activities (i.e., labor) to earn money from living Internet users."

"DAI should at least weigh their ethical constraints against economic rationality in managing the deceased users' informational body."

"While Öhman and Floridi's [277] idea of preserving the dignity of the dead is quite legitimate, the deplorable reality of the use of the deceased users' personal data as described above is due in large part to human nature."

Research questions and method

"Google's Inactive Account Manager (IAM, an account deactivation management tool), Facebook's Legacy Contact (a memorial account manager) [278], and Apple's Digital Legacy [279] were recently launched in Japan as services that include designating a person to handle deceased users' data after their death."

"To determine the potential of commercial services using postmortem personal data in Japan in the near future, it is important to generally understand Japanese people's honest feelings about how they want their personal data, such as their SNS logs, handled after their death."

"Based on these surveys and earlier studies, this study conducted an online questionnaire in an exploratory quantitative survey of the Japanese general public to clarify the following research questions: RQ1: How do people want their personal data to be managed after their death?"

"RQ2: How do people allow their personal data, such as SNS message logs, to be used after their death?"

Distribution of respondents' attributes

"We asked the following questions, which we labeled Q-rel, about how the respondents dealt with religion in their daily lives."

"Q-rel: The respondents were asked to choose from the following options to describe how they dealt with religion: (1) I believe in a particular religion. (2) I have no religious faith, but I follow the appropriate customs from different religions for weddings, funerals, and other ceremonies. (3) I have no religious beliefs, and my weddings, funerals, and ceremonies are conducted without religion. (4) I do not want to answer."

"The respondents were asked about their individual religious beliefs in the following question, labeled as Q-religion."

"This seems to be because many respondents answered very superficially to the Q-religion question."

"No respondents said they followed a specific religion in response to Q-rel and no religion in Q-religion."

How personal data are managed after death

"Q-real: We asked the respondents if there was anyone, they could recommend to manage their personal data, such as posts made to their current social network accounts on Twitter, Facebook, or Instagram after their death."

"The absolute difference in the number of respondents who can recommend someone to manage their data after death between the real-name and anonymous accounts is not large."

"Q-wh: The respondents were asked to choose from the following options who they would like to read the data (e.g., posts, photos) left in their personal accounts and Internet services, such as SNS, after their death, in cases where the data are linked to their real-name account and where the data are not linked to them because the account is anonymous: (1) spouse, (2) children, (3) siblings, (4) SNS friends (e.g., followers), (5) others, (6) everyone (public), and (7) no one."

Acceptable uses of personal data after death

"The significantly large number of respondents who allowed the use of their own personal data after death in Q-self indicates that many respondents valued the economic benefit to themselves, even if the use of their own personal data after death is a psychologically critical issue."

"The number of respondents aged 18–29 years who allowed all eight use types was significantly higher, which suggests that they were less resistant to these uses."

"The number of respondents who allow all eight use types at the significance level of 1% or less was higher than the mean of 13.9%."

"The number of respondents people who do not allow the eight use types was lower than the mean of 86.1%."

"The respondents who followed a religion only for weddings, funerals, and other ceremonial occasions were significantly more likely to accept the use of browsing and the use of the AI dialog system in response to the question in Q-self."

Conclusions

"In the second half of our study, we empirically analyzed the acceptable ways for personal data to be managed after death."

"The use of the personal data after death, such as interacting with an AI dialog system built using personal information or adding something to a photo, was much less acceptable."

"The limitations of our study are that respondents may not have a sufficiently concrete understanding of the use of their personal data after death and may have answered without being able to imagine it."

"Future studies could conduct quantitative and qualitative surveys to review the use of practical services to manage personal data after death to explore the use of data and monetary returns by considering individuals' perspectives of life and death, in addition to their willingness to "resurrect the deceased" virtually."

Regional Differences in Information Privacy Concerns After the Facebook-Cambridge Analytica Data Scandal [363].

This is a machine-generated summary of:

González-Pizarro, Felipe; Figueroa, Andrea; López, Claudia; Aragon, Cecilia: Regional Differences in Information Privacy Concerns After the Facebook-Cambridge Analytica Data Scandal [363].

Published in: Computer Supported Cooperative Work (CSCW) (2022).

Link to original: https://doi.org/10.1007/s10606-021-09422-3.

If you want to cite the papers, please refer to the original.

For technical reasons we could not place the page where the original quote is coming from.

Abstract-Summary

"Prior work argues that people's cultural backgrounds might shape their privacy concerns; thus, we could expect people from different world regions to conceptualize them in diverse ways."

"We employed word embeddings and qualitative analysis to identify which information privacy concerns are present and characterize language and regional differences in emphasis on these concerns."

"We also observe a greater emphasis on data collection in English than in Spanish."

"Our results call for more diverse sources of data and nuanced analysis of data privacy concerns around the globe."

Introduction

"The study of information privacy has advanced our understanding of individuals' concerns regarding organizational practices associated with collecting and using their personal information [281]."

"The review's authors hypothesized that individuals from different world regions have diverse cultures, values, and laws, which can, in turn, result in different conceptualizations of information privacy and its impacts [282, 283]."

"Inspired by Rho and others [284], where text analysis was used to analyze answers about individuals' privacy concerns, we analyze the semantic context in which privacy-related terms were used in tweets written by different groups of people."

"We contrast our results with one of the most used information privacy concerns framework to find terms and tweets matching different concerns."

"We test a null hypothesis that there is no difference in emphasis on information privacy terms across languages and world regions."

Information Privacy Concerns

"A multidimensional approach allows identifying to what extent users are concerned about different aspects of information privacy [285, 286]."

"Different authors have proposed alternative conceptualizations to measure information privacy concerns."

"Two questionnaires have been widely used to evaluate individuals' information privacy concerns [282, 287, 288]: Concerns for Information Privacy (CFIP) and Internet Users' Information Privacy Concerns (IUIPC)."

"As the Internet enabled new ways to collect and process data, it was expected that new concerns about information privacy might emerge [126], and a new framework was proposed: the IUIPC."

"Malhotra and others [126] introduced the IUIPC framework and conceptualized Internet users' concerns about information privacy from a perspective of fairness."

"Prior studies suggest that older Internet users are more concerned about online information privacy than younger ones [100]."

Research Questions

"When information privacy concerns questionnaires are delivered in English to speakers of other languages, key differences among countries may be obscured, as has happened with other cross-national research [289, 290]."

"We seek to assess the feasibility of using social media data to identify information privacy concerns and characterize language and regional differences."

"Following this trend of research, we aim to investigate whether Twitter data can reveal people's information privacy concerns."

"If information privacy concerns are present in a Twitter dataset, we could explore how they differ across people who live in different parts of the world and those who speak different languages."

"There are no differences in information privacy concerns by language H0b."

"There are no differences in information privacy concerns by world region."

Data & Methods

"As the Cambridge Analytica scandal triggered Twitter users from different world regions (who speak diverse languages) to spontaneously share their opinions, experiences, and perspectives about data privacy, we decided to use a sample of these tweets to answer our research question and test our hypotheses."

"Our datasets contain 36,056 human-owned accounts that created 74,644 tweets in Spanish and 410,180 accounts that created 975,678 tweets in English."

"To compare information privacy concerns by geographical regions, we divided the Spanish Twitter dataset in two sets: tweets written by users from (1) Latin America and (2) Europe."

"We built word embeddings for the Spanish and English datasets (containing both geolocated and non-geolocated tweets)."

"The closest terms retrieved to the keyword information in the English word embedding were info, data, details, and personal, in that order."

"Two of the authors conducted open coding of the 320 terms retrieved from the Spanish and English word embeddings."

Results

"The semantic context is operationalized as the 40 closest terms of each keyword: information, privacy, company, and users."

"Examples of tweets that include terms that fit this category are: '@ hidden_username @ hidden_username This is bigger than facebook because all social media outlets collect and store this data on every user."

"Three of our categories are associated with this dimension: (1) privacy and security terms that include words associated with data privacy and security such as "confidentiality", "transparency" and "safety"; (2) security mechanisms that refers to tools and techniques that implement security services (e.g., "password", "encryption"); and, (3) privacy & security risks that denote entities or bad practices that can compromise sensitive data, for example: "troyano" (trojan), "databreach", "grooming" and "ciberdelincuente" (cybercriminal)."

"Force-directed graphs of the categories and terms associated with the other keywords (information, company, and users) are available online."

"We observe that the semantic contexts of four privacy-related keywords include terms corresponding to information privacy concerns."

Discussion

"This work explores the potential of social media as a data source to study cross-language and cross-regional differences in information privacy concerns."

"We conduct an analysis of Twitter data related to a particular data breach news to deepen our understanding of how people from different world regions and who speak different languages frame privacy concerns."

"Collecting and analyzing the semantic contexts of these privacy-related keywords allows us to observe the presence of terms related to information privacy concerns in the collected tweets."

"We carefully analyzed more than a thousand terms of the semantic contexts, conducted open coding to formulate a data-grounded categorization, and then contrasted our categorization with IUIPC [126], one of the well-accepted theoretical conceptualizations of information privacy concerns."

"While we find similar rates of terms related to information privacy concerns across languages and regions, we observe significant differences in emphasis on collection and awareness."

Conclusion

"We conducted a cross-language and cross-regional study on social media content about a major data privacy leakage: the Cambridge Analytica scandal."

"Our results shed light on language and regional differences on information privacy concerns by (1) creating word embeddings by language and world regions to leverage social media data about a data breach scandal, (2) conducting open coding and content analysis of the semantic contexts (generated by the embeddings) of privacy-related keywords, (3) mapping the results to a well-known information privacy framework, and (4) conducting a comparative analysis across two languages and four world regions."

"We discuss how our findings extend current conceptualizations of information privacy concerns, and how they might relate to regulations about personal data usage in the regions we analyzed."

"Future studies might build upon our work to examine privacy concerns considering more languages, more geographical locations or different information privacy frameworks."

Guidelines adopted by agile teams in privacy requirements elicitation after the Brazilian General Data Protection Law (LGPD) Implementation [364].

This is a machine-generated summary of:

Canedo, Edna Dias; Calazans, Angelica Toffano Seidel; Bandeira, Ian Nery; Costa, Pedro Henrique Teixeira; Masson, Eloisa Toffano Seidel: Guidelines adopted by agile teams in privacy requirements elicitation after the Brazilian general data protection law (LGPD) implementation [364]

Published in: Requirements Engineering (2022).

Link to original: https://doi.org/10.1007/s00766-022-00391-7.

If you want to cite the papers, please refer to the original.

For technical reasons we could not place the page where the original quote is coming from.

Abstract-Summary

"Extending our previous work, we have investigated the actions taken by organizations regarding the LGPD, specifically in software development, considering the perception of agile development teams after two years of the LGPD implementation."

"We also investigated the perception of an agile team regarding the practices, techniques, and tools previously cited by practitioners as potential solutions for use in this context, along with techniques already in use in the current context."

"The LGPD principles are known by most agile teams and are being implemented by the organizations, although the existing tools to support privacy requirements elicitation are still underused by agile teams."

"Agile teams consider that software requirements and software construction are the most impacted areas of knowledge by the LGPD, and most of them use user stories in privacy requirements elicitation."

"Our findings reveal that agile teams and Brazilian organizations are more concerned with user data privacy issues after the LGPD became effective."

"Agile teams still face challenges in privacy requirements elicitation."

Introduction

"[291], we carried out a systematic literature review (SLR) to identify the techniques, methodologies, and tools used in the literature to perform privacy requirements elicitation in the context of Agile Software Development (ASD)."

"We also conducted an online survey to investigate the perception of agile teams regarding the impact that the Brazilian General Data Protection Law (LGPD) had had on their activities during software development."

"We investigated whether agile teams correctly interpret privacy principles and implement these concepts and principles during software development, along with what actions could have been taken to reduce the impact of needing to implement systems according to data privacy laws by agile development teams."

"The main contribution of our work is to understand the perception of agile teams regarding the actions and changes adopted by the organizations after the LGPD became effective, as well as to identify which organizational procedures were adopted concerning user data privacy and the changes made in how the teams work to ensure LGPD compliance."

Background and related works

"Aiming to gather the perception of agile software development team members from different organizations regarding the impact that LGPD will have on the activities of the software development process, Canedo and others [291] investigated said context and identified, among other findings, that agile teams know the concepts related to data privacy legislation."

"Among the authors' findings, they identified that participants lacked resources and online materials for reference and guidance when implementing data privacy, and the implementation of GDPR, when developing privacy-preserving software systems, is affected by organizations or the customer."

"A small amount of work focused on privacy techniques—which relate to procedures, prescribed language or notation, for performing privacy engineering tasks or activities—and requirements, especially when these techniques are related to agile methodologies and specifically to user story and use case."

Method

"We performed data analysis through triangulation using 3 sources: (1) systematic literature review, (2) survey and (3) semi-structured interviews."

"We have defined the following research questions to conduct this research: RQ.1 What practices/techniques, and tools are used by agile teams to elicit privacy requirements that comply with the Brazilian General Data Protection Law?"

"In the protocol were defined the research questions, the search strategies adopted to identify the relevant primary studies, the search string to use in the databases, the exclusion/inclusion criteria, and the quality assessment criteria."

"The study should propose or use/evaluate existing methods, methodologies, techniques or tools to perform privacy requirements elicitation in the context of agile software development."

"The works selected in step 3 were read in full and the volume of studies resulting in this step (36 primary studies) was used to compose the SLR and support the answers to the research questions."

Results

"Regarding the procedures adopted by software development teams in privacy requirements elicitation (Q.11 of Supplementary Material Section 2), the interviewees mentioned: (1) definition of data privacy and security needs to be automated or modified; (2) holding weekly meetings with stakeholders and users during the specification of all software privacy and security requirements; (3) modeling of all business processes, containing the improvements identified and discussed among stakeholders and users; (4) implementation of privacy and security requirements, as well as their respective business rules, use cases, and user stories; and (5) execution of user tests of the implemented privacy and security requirements."

"The following challenges for privacy requirements elicitation were mentioned (Q.13 of Supplementary Material Section 2): (1) It is difficult to get all stakeholders involved; (2) Most organizations do not have their business processes mapped; (3) The organization does not have professionals qualified in terms of the LGPD; (4) Organizations are not investing in software solutions to ensure LGPD compliance; (5) It is not easy to align business needs with the LGPD without creating new processes or work demands; (6) Agility in defining requirements and their implementation."

Limitations and threats to validity

"We cannot guarantee that all primary studies related to privacy requirements elicitation in the context of agile software development have been selected in the execution of the systematic literature review, or even if journals or conferences with a good impact factor or studies with a good number of citations were prioritized."

"This survey's findings cannot claim to cover the perception of the whole population of practitioners working in agile software development (ASD)."

"The personal interest in the LGPD, agile teams, and privacy requirements elicitation topics related to our survey, might have influenced the decision of the practitioners to participate."

"We also included surveys questions about the experience with agile software development."

"Regarding the experience in the privacy requirements elicitation activities, our sample is very balanced toward people that work more frequently in agile teams software development with data privacy concerns."

Conclusions

"This paper investigated the perception of agile teams regarding the actions Brazilian organizations are implementing to develop LGPD compliant software."

"We updated the systematic literature review conducted in our previous study [291] and conducted a survey with agile teams' practitioners from several Brazilian organizations to understand their perception regarding the organizational actions adopted due to LGPD."

"Agile teams still face some challenges in requirement elicitation, such as difficulty aligning business needs with the LGPD without creating new processes or work demands, lack of specialized qualification of practitioners, and lack of software solutions to support LGPD compliance."

"An important finding is that agile teams stated that most organizations are making changes to their organizational processes and software development process to ensure compliance with LGPD."

The fundamental right of Confidentiality and integrity of IT systems in Germany: A call for "IT Privacy" Right in Brazil? [365]

This is a machine-generated summary of:

Alberton Coutinho Silva, Cecília: The fundamental right of confidentiality and integrity of IT systems in Germany: a call for "IT Privacy" right in Brazil? [365].

Published in: International Cybersecurity Law Review (2021).

Link to original: https://doi.org/10.1365/s43439-021-00037-4.

If you want to cite the papers, please refer to the original.

For technical reasons we could not place the page where the original quote is coming from.

Abstract-Summary

"In the said scenario, this paper will seek to answer the question: Starting from the premises set by the BVerfG in the ruling rendered on February 27th, 2008, are there similar grounds to sustain the existence of an IT Privacy right in Brazil, regarding the Brazilian juridical scenario, mainly as to data protection?"

"To that end, the paper is divided into four main parts to: (i) assess the fundamentals of the decision rendered by the BVerfG in the case mentioned; (ii) present the privacy and data protection legal scenario in Brazil; (iii) point out how information security is provided for in Brazilian legislation; and (iv) validate whether the premises adopted by the BVerfG are also coherent in Brazil, considering the legal landscape presented."

"The research concludes in the sense that the Brazilian and German Constitutional Legal Orders are different, not only relating to the way in which new fundamental rights are acknowledged, but also in regard to the privacy and data protection legal culture, which directly impacts the feasibility of a fundamental right to confidentiality and integrity of IT systems."

Introduction

"The North-Rhine Westphalian Act provisions allowed the government to conduct online surveillance of personal computers, which, according to the BVerfG, were unconstitutional due to the fact that they did not sufficiently respect the individual's right to confidentiality of data stored on information technology systems and the integrity of information technology systems themselves."

"That said, this brief paper aims to answer the following question: Starting from the premises set by the BVerfG in the ruling rendered on February 27th, 2008, are there similar grounds to sustain the existence of an IT Privacy right in Brazil, regarding the Brazilian juridical scenario, mainly in terms of data protection?".

"The research concludes in the sense that the Brazilian and German Constitutional Legal Orders are different, not only relating to the way in which new fundamental rights are acknowledged, but also in regard to the privacy and data protection legal culture, which directly impacts the feasibility of a fundamental right to confidentiality and integrity of IT systems."

The rise of the fundamental right of confidentiality and integrity of IT systems

"Following said trend, on February 27th, 2008, the German Federal Constitutional Court (Bundesverfassungsgericht) annulled provisions of the North-Rhine Westphalian Act on the Protection of the Constitution (NRW-VSG), which allowed the government to conduct online surveillance and secret investigations of personal computers of people that have allegedly committed criminal offenses (Online-Durchsuchung) and also to monitor all the online behaviors of suspected individuals in the internet, giving rise to a new fundamental right, in a way, decurrent of the mentioned informational self-determination right, the fundamental right to guarantee the confidentiality and integrity of technical and informational systems (Grundrecht auf Gewährleistung der Vertraulichkeit und Integrität informationstechnischer Systeme), which were exemplified by the BVerfG as being "Personalcomputer", "Telekommunikationsgeräte" or "elektronische Geräte, die in Wohnungen oder Kraftfahrzeugen enthalten sind" [292, p. 365]."

Privacy and data protection regulations in Brazil

"With no exemption, privacy and data protection in Brazil are also provided for in other laws and regulations, which follow the general protection of the private life of the individual as a fundamental right foreseen in the Brazilian Constitution, as is the case in the following examples."

"The LGPD does not apply, however, when the processing of personal data is done: (a) by a natural person exclusively for private and non-economic purposes; (b) exclusively for journalistic, artistic, or academic purposes; (c) exclusively for purposes of public safety, national defense, state security, activities of investigation, and prosecution of criminal offenses; or when the personal data (d) has its origin outside the national territory and is not the object of communication, shared use of data with Brazilian processing agents, or the object of international transfer of data with another country that is not the country of origin, since the country of origin provides a level of personal data protection adequate to that established in the LGPD (as per art."

Information security in Brazil

"Companies will be required to implement, for example, an information security policy, which should include the set of standards, definitions, practices, security measures, responsibilities, and penalties applicable to all employees of the company who may use the company's information technology infrastructure in order to minimize the risk of loss or breach of any IT asset; that is, companies will have to make sure that all security measures are being taken to protect corporate information against possible threats of data leaks and security incidents, especially considering that, under the LGPD, the systems used for processing personal data shall be structured to meet the security requirements, standards of good practices and governance, general principles provided in the LGPD, and other regulatory rules (as per art."

"IT Privacy" right in Brazil?

"With that, if on the one hand the fundamental right to confidentiality and to IT integrity systems, as seen, was construed by the BVerfG based on elements that have been long developed and discussed in the country, at least since the 1980s, as is the case with the right to informational self-determination, on the other, in spite of the fact that Brazil already has a regulatory scenario on privacy and data protection, the country still lacks enforceability and general culture on the matter, which makes it difficult at this point to recognize an IT Privacy right in Brazil."

"The Brazilian Constitutional Order does not provide for the same unity as does the German, which is important for the full protection of the person and the most diverse aspects of personality based on human dignity [293, p. 220] and impacts the way a fundamental right such as of the confidentiality and integrity of technical and informational systems would be interpreted and recognized in Brazil."

Closing remarks

"Beyond that, even though the main goal of the research was not to investigate precisely and in depth the right to personality (in Germany, the general right to personality or "Allgemeines Persönlichkeitsrecht") and the requirements that must be fulfilled in order to recognize the existence of a fundamental right, one can also infer that the fundamental right to the confidentiality and integrity of technical and informational systems was deduced from the general personality clause in Germany, but it is also linked to the dignity of the human person in the context of the right to information and the protection of personal data."

Sabiá: An authentication, authorization, and user data delivery architecture based on user consent for health information systems in Brazil [366].

This is a machine-generated summary of:

de Paiva Marques Carvalho, Túlio; de Paiva, Jailton Carlos; de Medeiros Valentim, Ricardo Alexsandro; Silva, Carlos Breno Pereira; de Lima, Diêgo Ferreira; Silva, Emerson Costa: Sabiá: an authentication, authorization, and user data delivery architecture based on user consent for health information systems in Brazil [366]

Published in: Research on Biomedical Engineering (2020).

Link to original: https://doi.org/10.1007/s42600-020-00058-8.

If you want to cite the papers, please refer to the original.

For technical reasons we could not place the page where the original quote is coming from.

Abstract-Summary

"Health information systems in Brazil have been designed and developed in a heterogeneous manner based on local regional characteristics, resulting in a lack of health information integrity."

"This paper presents Sabiá, a platform for authentication, authorization, and data delivery based on user consent for health information systems in Brazil."

"Sabiá's architecture is designed to achieve the following requirements: (R1) Provide a Federated Identity; (R2) Be a Federated Resource Manager; (R3) Collect user data from different information systems; and (R4) Deliver user data to systems based on user consent."

"Sabiá consists of three main components: (1) Sabiá Authorization Server, responsible for implementing Open Authentication; (2) Sabiá Collector, responsible for collecting data from different information systems; and (3) Sabiá Resource Server, responsible for delivering data previously authorized by the user to the systems."

Introduction

"Given this situation, there are initiatives from the MoH that point to the need for solutions that allow the interoperability of health information systems in Brazil and the implementation of mechanisms that interoperate with the national identification databases such as the SUS National Card (CNS) and the National Health Facilities Registry (CNES), as well as other systems external to the MoH. Besides that, solutions must be in accordance with Brazilian law, taking into account aspects such as privacy, agility, security, authentication, and users' consent to store and use their data [294–297]."

"Given the above, the purpose of this paper is presenting Sabiá, which is an FRM architecture that addresses some issues raised by MoH, offering SSO authentication through federated identity and delivering user's data according to their express consent, respecting the current Brazilian legislation."

Methods

"In this app, the user can manage their profile and change their data; clearly view their information and which systems can access each one of them; and can include or revoke the access permission of each feature for each application, as well as cancel their account in the entire ecosystem, respecting, therefore, the Brazilian law The Sabiá Authorization Server component is responsible for R1 and R2, communicating with the apps and implementing the OAuth protocol, which has been adopted because it presents benefits over SAML such as interoperability with a wider range of device types (Web, Mobile, Desktop, etc.), lower traffic for being a lightweight standard, less customer deployment effort, and finally because it requires user consent before sharing its resources [298]."

"The Sabiá Collector component is responsible for R3 and has the aim of collecting and storing user information from different data sources, which may be Sabiá client applications or other external systems."

Results

"Performance testing is a technique in which simulated workloads are submitted to a system within a controlled environment for verifying whether a system will meet its performance objective when deployed in a real-world situation [299]."

"The objective of the experiments was to analyze the average response times of the functionality that most influence the performance (R4) in the heaviest workload scenario based on historical data."

"The tool used in the experiments was Apache JMeter, which is an Open-Source Tool used by testing engineers for the load testing, performance testing, and functional behavior of the web applications [300]."

"JMeter was configured to execute with a constant workload of 20 requests per second during 3600 s, which is a value mentioned by Kao and others [301] for performance testing RESTFul web applications."

"After 1 h of experiments, we got 70,510 requests with 61-ms average response time and 56-ms median response time."

Discussion

"In the tests, the average, median, 90th and 95th response times were below 100 ms considering the heaviest historical scenario, which is a behavior that the user perceives the response instantaneously."

"The TA results show that the median response time of 56 ms is close to the average of 61 ms, which denotes a symmetrical distribution and points to stability and consistency in the system behavior."

"Besides that, Sabiá has been deployed in a production environment since 2016, meeting 4 architectural requirements and integrating over 100 applications and 700,000 users."

Perspectives regarding privacy in clinical research among research professionals from the Arab region: An exploratory qualitative study [367]

This is a machine-generated summary of:

Adarmouch, Latifa; Felaefel, Marwan; Wachbroit, Robert; Silverman, Henry: Perspectives regarding privacy in clinical research among research professionals from the Arab region: an exploratory qualitative study [367].

Published in: BMC Medical Ethics (2020).

Link to original: https://doi.org/10.1186/s12910-020-0456-9.

If you want to cite the papers, please refer to the original.

For technical reasons we could not place the page where the original quote is coming from.

Abstract-Summary

"Protecting the privacy of research participants is widely recognized as one of the standard ethical requirements for clinical research."

"It is unknown, however, how research professionals regard concepts of privacy as well as the situations in the research setting that require privacy protections."

"The aim of this study was to explore the views of research professionals from Arab countries regarding concepts and scope of privacy that occur in clinical research."

"Respondents discussed several privacy issues that occurred before the research began (e.g., recruitment practices); during research (e.g., data collection and physical exams), and after the research (e.g., secondary use of data and data sharing)."

"Respondents also recommended training regarding data protections for individuals involved in research."

"Our study shows that research professionals discussed a range of privacy issues that are present during the different stages of research."

"We recommend (1) development of standards regarding privacy protections during recruitment efforts; (2) additional training for individuals involved in research regarding best practices with data security in secondary research; (3) a quantitative study involving investigators and REC members to determine their knowledge, attitudes and practices regarding privacy issues that occur in research; and (4) a quantitative study involving patients to elicit their views regarding their privacy concerns in research."

Background

"While the authors recognized the different concepts of privacy, they stated: "in a clinical-research context, the focus is typically on informational privacy" [302]."

"The Guideline for Good Clinical Practice focus only on informational privacy as it states: "The confidentiality of records that could identify subjects should be protected, respecting the privacy and confidentiality rules…" [303]."

"The Declaration of Helsinki states: "Every precaution must be taken to protect the privacy of research subjects and the confidentiality of their personal information" [304]."

"The document states later that "Privacy risks in research relate to the identifiability of participants…" thus emphasizing protections only in the realm of informational privacy, i.e., data security [305]."

"Morocco recently passed legislation, the Protection of Persons Participating in biomedical research, which states "The privacy of the participant and the confidentiality of the data concerning him must be respected…." [306]."

Methods

"We adopted an exploratory qualitative approach by the use of focus group discussions (FGDs)."

"The guide included the following questions: What are conceptions of privacy in daily life and why is it important? What are expectations of privacy in healthcare? What are expectations of privacy in clinical research? Please give examples of privacy situations before the research begins, during research, and after research."

"The interview guide was continually revised to explore the additional concepts that emerged from the focus groups."

"Participants gave verbal informed consent, which was recorded in the beginning of the focus group discussion."

"The focus groups were conducted either in English, French or spoken Arabic and were audio recorded."

"Individuals who agreed to participate were invited to a focus group discussion."

"We analyzed focus group data via a constant comparison approach [307]."

Results

"One respondent brought up an issue when a third party becomes involved with translation of participants' responses and hence, has access to the data:" I had two experiences, the first was a questionnaire we did in a village where most of the people were Amazigh [berber] and I don't speak the Amazigh, we were three doctors who don't speak it, so they brought someone to translate for us, that person was from the Douar [local community], not someone we brought with us, we had a 14 years old girl who speaks Arabic and Amazigh and she had access to all the information of the women of that Douar" "The second involved when we were working on a questionnaire with the prisoners and for sure it was impossible to interview them without the guard standing beside you the whole time even when the patient answers or says something, the guard starts to look at him or judge him by his look, I think these conditions didn't respect any privacy.""'

Discussion

"Our major findings include: (1) respondents discussed a broad range of privacy issues that occur during the three major stages of the research process: before, during, and after the completion of research; (2) research professionals revealed their perceptions of patients' concerns regarding assurances of privacy in the research setting compared with clinical care; and (3) respondents promoted the need for clarification of existing data protection standards as well as the need for training in practices involving the use of data in secondary research."

"Investigators need to explain to potential research participants (and also included in informed consent forms) that the benefits of research is to advance the health of society, which could serve as a justifiable reason for patients to waive their privacy rights coupled with assurances that their privacy will be protected."

References

1. Kamleitner, B., & Mitchell, V. W. (2018). Can consumers experience ownership for their personal data? From issues of scope and invisibility to agents handling our digital blueprints. In J. Peck & S. Shu (Eds.), Psychological ownership and consumer behavior (pp. 91–118). Cham: Springer.
2. Lynskey, O. (2015). The foundations of EU data protection law. Oxford: Oxford University Press.
3. Albergotti, R. (2014, June 12). Facebook to target ads based on web browsing. Wall Street Journal.
4. Warren, S., Brandeis, L.: The right to privacy. Harvard Law Rev. 4(5), 193–220 (1890). Available at https://www.jstor.org/stable/1321160?seq=1#metadata_info_tab_contents.
5. Westin A (1967) Privacy and freedom. Anthema, New York.
6. Rachels J (1975) Why privacy is important. Philos Public Aff 4:323–333.
7. Gavison R (1980) Privacy and the limits of law. Yale Law J 89:421–471.
8. Parent WA (1983) A new definition of privacy for the law. Law Philos 2:305–338 (December).
9. Fried C (1984) Privacy (a moral analysis). In: Schoeman F (eds) Philosophical dimensions of privacy: an anthology. Cambridge University Press, Cambridge.
10. Schoeman F (1984) Privacy and intimate information. In: Schoeman, F (ed) Philosophical dimensions of privacy: an anthology. Cambridge University Press, Cambridge.
11. Allen A (1988) Uneasy access: privacy for women in a free society. N.J Rowman and Littlefield, Totowa.
12. Elgesem D (1999) Privacy, respect for persons and risk. In: Ess C (eds) Philosophical perspectives on computer mediated communication. University of New York Press, New York State.

13. Persson AJ (2005) Applying ethical criteria for privacy. In: Hansson SO, Palm E (eds) The ethics of workplace privacy. P.I.E. Lang Publisher, Brussels.
14. Sanchez, O., Torre, I., Knijnenburg, B.: Semantic-based privacy settings negotiation and management. In: Future Generation Computer Systems (2019). (Under publication).
15. Beresford, A.R., Rice, A., Skehin, N., Sohan, R.: Mockdroid: trading privacy for application functionality on smartphones. In: Proceedings of the 12th Workshop on Mobile Computing Systems and Applications, pp. 49–54. ACM (2011).
16. Chakraborty, S., Shen, C., Raghavan, K.R., Shoukry, Y., Millar, M., Srivastava, M.B.: ipshield: A framework for enforcing context-aware privacy. In: NSDI, pp. 143–156 (2014).
17. Liu, B., Andersen, M.S., Schaub, F., Almuhimedi, H., Zhang, S., Sadeh, N., Acquisti, A., Agarwal, Y.: Follow my recommendations: A personalized privacy assistant for mobile app permissions. In: Twelfth Symposium on Usable Privacy and Security, pp. 26–41 (2016).
18. Knijnenburg, B.P., Kobsa, A.: Helping users with information disclosure decisions: potential for adaptation. In: Proceedings of the 2013 International Conference on Intelligent User Interfaces, pp. 407–416. ACM (2013).
19. Becher, S., Gerl, A., Meier, B.: Don't forget the user: from user preferences to personal privacy policies. In: 2020 10th International Conference on Advanced Computer Information Technologies (ACIT), pp. 774–778. IEEE (2020).
20. Psychoula, I., Chen, L., Amft, O.: Privacy risk awareness in wearables and the internet of things. IEEE Pervasive Comput. 19(3), 60–66 (2020).
21. Reichherzer, T., Timm, M., Earley, N., Reyes, N., Kumar, V.: Using machine learning techniques to track individuals & their fitness activities. In: CATA 2017, pp. 119–124. ISCA (2017).
22. Challa, N., Yu, S., Kunchakarra, S.: Wary about wearables: potential for the exploitation of wearable health technology through employee discrimination and sales to third parties. Intersect Stanford J. Sci. Technol. Soc. 10(3) (2017).
23. Cooney, M.T., Vartiainen, E., Laakitainen, T., Juolevi, A., Dudina, A., Graham, I.M.: Elevated resting heart rate is an independent risk factor for cardiovascular disease in healthy men and women. Am. Heart J. 159(4), 612–619 (2010).
24. Alhalabi, L., Singleton, M.J., Oseni, A.O., Shah, A.J., Zhang, Z.-M., Soliman, E.Z.: Relation of higher resting heart rate to risk of cardiovascular versus noncardiovascular death. Am. J. Cardiol. 119(7), 1003–1007 (2017).
25. Cremonini, M., Braghin, C., Ardagna, C.A.: Chapter 42—privacy on the internet. In: Vacca, J.R. (ed.), Computer and Information Security Handbook, 2 edn, pp. 739–753. Morgan Kaufmann, Boston (2013). ISBN: 978-0-12-394397-2. https://doi.org/10.1016/B978-0-12-394397-2.00042-8.
26. Maganti, K., Rigolin, V.H., Sarano, M.E., Bonow, R.O.: Valvular heart disease: diagnosis and management. In: Mayo Clinic Proceedings, vol. 85, pp. 483–500. Elsevier (2010).
27. Hunter, S., Robson, S.C.: Adaptation of the maternal heart in pregnancy. Br. Heart J. 68(6), 540 (1992).
28. Martin KD, Murphy PE (2017) The role of data privacy in marketing. J Acad Mark Sci 45(2):135–155. https://doi.org/10.1007/s11747-016-0495-4.
29. Dinev T, Mcconnell AR, Smith HJ (2015) Informing privacy research through information systems, psychology, and behavioral economics: thinking outside the "APCO" box. Inf Syst Res 26(4):639–655. https://doi.org/10.1287/isre.2015.0600.
30. Bélanger F, Crossler RE (2011) Privacy in the digital age: a review of information privacy research in information systems. In: MIS quarterly: management information systems, vol 35(4). University of Minnesota, pp 1017–1041. https://doi.org/10.2307/41409971.
31. Palmatier RW, Martin KD (2019) The intelligent marketer's guide to data privacy. Palgrave Macmillan.
32. Bandara W, Furtmueller E, Gorbacheva E, Miskon S, Beekhuyzen J (2015) Achieving rigor in literature reviews: insights from qualitative data analysis and tool-support. Commun Assoc Inf Syst 34:154–204.

33. Alashoor T, Baskerville R (2015) The privacy paradox: the role of cognitive absorption in the social networking activity. In: Proceedings of the 22nd Americas conference on information systems, pp 1–20.
34. Barth S, de Jong MDT, Junger M, Hartel PH, Roppelt JC (2019) Putting the privacy paradox to the test: online privacy and security behaviors among users with technical knowledge, privacy awareness, and financial resources. Telematics Inform 41:55–69. https://doi.org/10.1016/j.tele.2019.03.003.
35. Sundar SS, Kim J, Rosson MB, Molina MD (2020) Online privacy heuristics that predict information disclosure. In: Conference on human factors in computing systems-proceedings, pp 1–12. https://doi.org/10.1145/3313831.3376854.
36. Dinev T, Xu H, Smith JH, Hart P (2012) Information privacy and correlates: an empirical attempt to bridge and distinguish privacy-related concepts. Eur J Inf Syst 22:295–316. https://doi.org/10.1057/ejis.2012.23.
37. Gerber, N., Gerber, P., & Volkamer, M. (2018). Explaining the privacy paradox: A systematic review of literature investigating privacy attitude and behavior. Computers & Security, 77, 226–261. https://doi.org/10.1016/j.cose.2018.04.002.
38. Kokolakis S (2017) Privacy attitudes and privacy behaviour: a review of current research on the privacy paradox phenomenon. Comput Secur 64:122–134. https://doi.org/10.1016/j.cose.2015.07.002.
39. Benbasat, I., and S.Y. Komiak. 2004. Understanding customer trust in agent-mediated electronic commerce, web-mediated electronic commerce, and traditional commerce. Information Technology and Management 5 (1–2): 181–207.
40. Chattaraman, V., W. Kwon, J.E. Gilbert, and K. Ross. 2019. Should AI-Based, conversational digital assistants employ social- or task-oriented interaction style? A task-competency and reciprocity perspective for older adults. Computers in Human Behavior. 90: 315–330.
41. Vance, A., C. Elie-dit-cosaque, and D.W. Straub. 2008. Examining trust in information technology artifacts: The effects of system quality and culture. Journal of Management Information Systems 24 (4): 73–100.
42. de Cosmo, L.M., L. Piper, and A. Di Vittorio. 2021. The role of attitude toward chatbots and privacy concern on the relationship between attitude toward mobile advertising and behavioral intent to use chatbots. Italian Journal of Marketing 2021 (1): 83–102.
43. Alashoor, T., M. Boodraj, and K. Saffarizadeh. 2017. Conversational Assistants: Investigating Privacy Concerns, Trust, and Self-Disclosure. In Proceedings from the Thirty Eighth International Conference on Information Systems. 10–13 December 2017. Seoul, South Korea.
44. Benbasat, I., and S.Y. Komiak. 2006. The effects of personalization and familiarity on trust and adoption of recommendation agents. MIS Quarterly 30 (4): 941–960.
45. Gong, X., K.Z.K. Zhang, S.J. Zhao, and M.K.O. Lee. 2016. The Effects of cognitive and emotional trust on mobile payment adoption: A trust transfer perspective. AISeL Proceedings. 350: 1.
46. Malhotra, N.K. 2015. Essentials of Marketing Research: A Hands-On Orientation, 1st ed. New York: Pearson.
47. Landrum, S. 2017. Millennials, Trust And Internet Security (Online). https://www.forbes.com/sites/sarahlandrum/2017/06/28/millennials-trust-and-internet-security/#1bfb8735555e. 2018, August 30.
48. Zeissig, E., C. Lidynia, L. Vervier, A. Gadeib, and M. Ziefle. Online Privacy Perceptions of Older Adults. In Proceedings from Third International Conference Human Aspects of IT for the Aged Population. 9–14 July 2017. Vancouver, Canada.181–200.
49. Harding, S. 2017. Trends 2018. (Report 0018A). London, United Kingdom: Mindshare Futures.
50. Awad, N.F., and M.S. Krishnan. 2006. The personalization privacy paradox: An empirical evaluation of information transparency and the willingness to be profiled online for personalization. MIS Quarterly 30 (1): 13–28.

51. Baumer, D., and J.B. Earp. 2003. Innovative web use to learn about consumer behavior and online privacy. Communications of the ACM 46 (4): 81–83.
52. Hong-Youl, H. 2004. Factors influencing consumer perceptions of brand trust online. Journal of Product & Brand Management. 13 (5): 329–342. https://doi.org/10.1108/10610420410554412.
53. Hasal, M., J. Nowaková, K. Ahmed Saghair, H. Abdulla, V. Snášel, and L. Ogiela. 2021. Chatbots: Security, privacy, data protection, and social aspects. Concurrency and Computation: Practice and Experience 33 (19): 6426.
54. Coopamootoo, K.P., M. Mehrnezhad and E. Toreini. 2022. "I feel invaded, annoyed, anxious and I may protect myself": Individuals' Feelings about Online Tracking and their Protective Behaviour across Gender and Country. arXiv:2202.04682.
55. Netshakhuma, N.S. 2020. Assessment of a South Africa national consultative workshop on the Protection of Personal Information Act (POPIA). Global Knowledge, Memory and Communication 69 (1/2): 58–74.
56. Pedersen, E. P. (2005). Adoption of mobile internet services: An exploratory study of mobile commerce early adopters. Journal of Organizational Computing and Electronic Commerce, 15(2), 203–222.
57. Venkatesh, V., Morris, G. M., Davis, B. G., & Davis, F. D. (2003). User acceptance of information technology: Toward a unified view. MIS Quarterly, 27(3), 425–478.
58. Gefen, D., Karahanna, E., & Straub, D. W. (2003). Inexperience and experience with online stores: The importance of TAM and trust. IEEE Transactions on Engineering Management, 50(3), 307–321.
59. Taylor, S., & Todd, P. (1995). Assessing IT usage: The role of prior experience. MIS Quarterly, 19(4), 561–570.
60. Agarwal, R., & Prasad, J. (1998). A conceptual and operational definition of personal innovativeness in the domain of information technology. Information Systems Research, 9(2), 204–215.
61. Milberg, S. J., Smith, H. J., & Burke, S. J. (2000). Information privacy: Corporate management and national regulation. Organization Science, 11(1), 35–57.
62. Raul, A.P. (2002). Privacy and the Digital State: Balancing Public Information and Personal Privacy, Boston, MA: Kluwer.
63. Hui, K., Teo, H., & Lee, S. (2006). The value of privace assurance. MIS Quarterly (in press).
64. Horn, I., Hui, K., Lee, T., & Png, I. P. L. (2002). Online information privacy: Measuring the cost-benefit trade-off paper presented at the twenty-third international conference on information systems.
65. Kourouthanassis P, Roussos G (2003) Developing consumer-friendly pervasive retail systems. IEEE Pervasive Comput 2(2):32–39.
66. Roussos G, Kourouthanasis P, Gryazin E, Pryzbliski M, Kalpogiannis G, Giaglis G (2003) Systems architecture for pervasive retail. In: Proceedings of the 18th annual ACM symposium on applied computing (SAC 2003), Melbourne, Florida, March 2003. pp 631–636
67. Orlikowski W, Iacono CS (2001) The truth is not out there: an enacted view of the digital economy. In: Brynjolfsson E, Khim B (eds) Understanding the digital economy: data, tools and research. MIT Press, Cambridge, Massachusetts.
68. Roussos G, Peterson D, Patel U (2003) Mobile identity management: an enacted view. Int J E-Commerce 8(1):81–100.
69. Kamleitner, B., & Mitchell, V. (2019). Your data is my data: a framework for addressing interdependent privacy infringements. Journal of Public Policy & Marketing, 38(4), 433–450. https://doi.org/10.1177/0743915619858924.
70. Almuhimedi, H., Schaub, F., Sadeh, N., Adjerid, I., Acquisti, A., Gluck, J., Cranor, L. F., & Agarwal, Y. (2015). Your location has been shared 5,398 times! A field study on mobile app privacy nudging. Proceedings of the 33rd annual ACM conference on human factors in computing systems, Seoul. https://doi.org/10.1145/2702123.2702210.
71. Wang, Y., Leon, P. G., Acquisti, A., Cranor, L. F., Forget, A., & Sadeh, N. (2014). A field trial of privacy nudges for facebook. In Proceedings of the SIGCHI conference on human factors in computing systems, Toronto.

72. Harwell, D., & Harris, S. (2021). White House has spoken to Israeli officials about spyware concerns following Pegasus Project revelations.The Washington Post. Retrieved 06.08.2021 from https://www.washingtonpost.com/technology/2021/07/29/pegasus-white-house-israel-concerns/.
73. Isaac, J., and M.J. Hanna. 2018. User Data Privacy: Facebook, Cambridge Analytica, and Privacy Protection. Computer 51 (8): 56–59.
74. Marsch, M., Grossklags, J., & Patil, S. (2021). Won't you think of others?: Interdependent privacy in smartphone app permissions. Proceedings of the ACM on Human-Computer Interaction, 5(CSCW2), 1–35.
75. Pu, Y., & Grossklags, J. (2015). Using conjoint analysis to investigate the value of interdependent privacy in social app adoption scenarios. Proceedings of the International Conference on Information Systems, Fort Worth, United States.
76. Wirth, J., Maier, C., Laumer, S., & Weitzel, T. (2019). Perceived information sensitivity and interdependent privacy protection: A quantitative study. Electronic Markets, 29, 359–378. https://doi.org/10.1007/s12525-019-00335-0.
77. Pu, Y., & Grossklags, J. (2017). Valuating friends' privacy: Does anonymity of sharing personal data matter? Thirteenth symposium on usable privacy and security (SOUPS 2017), Santa Clara.
78. Kamleitner, B., & Sotoudeh, M. (2019). Information sharing and privacy as a socio-technical phenomenon: Interview with Bernadette Kamleitner. TATuP-Zeitschrift für Technikfolgenabschätzung in Theorie und Praxis, 28(3), 68–71. https://doi.org/10.14512/tatup.28.3.68.
79. Zhang, B., & Xu, H. (2016). Privacy nudges for mobile applications: Effects on the creepiness emotion and privacy attitudes. In Proceedings of the 19th ACM conference on computer-supported cooperative work & social computing, San Francisco.
80. Hayes, A. F. (2018). Introduction to mediation, moderation, and conditional process analysis: A regression-based approach. Guilford Publications.
81. Brill, T. M., Munoz, L., & Miller, R. J. (2019). Siri, Alexa, and other digital assistants: A study of customer satisfaction with artificial intelligence applications. Journal of Marketing Management, 35, 1401–1436. https://doi.org/10.1080/0267257X.2019.1687571.
82. Knote, R., Janson, A., Söllner, M., & Leimeister, J, M. (2019). Classifying smart personal assistants: An empirical cluster analysis. In Hawaii International Conference on System Sciences (HICSS). Maui, Hawaii, USA.
83. Acquisti, A., Adjerid, I., Balebako, R., Brandimarte, L., Cranor, L. F., Komanduri, S., et al. (2017). Nudges for privacy and security: Understanding and assisting users' choices online. ACM Computing Surveys (CSUR), 50(3), 1–41.
84. Maedche, A., Legner, C., Benlian, A., Berger, B., Gimpel, H., Hess, T., Hinz, O., Morana, S., & Söllner, M. (2019). AI-based digital assistants. Business & Information Systems Engineering, 61, 535–544. https://doi.org/10.1007/s12599-019-00600-8.
85. Liu, B., Andersen, M. S., Schaub, F., Almuhimedi, H., Zhang, S., Sadeh, N., et al. (2016). Follow my recommendations: A personalized privacy assistant for Mobile app permissions. In Proceedings of the twelfth USENIX conference on usable privacy and security (SOUPS '16) (pp. 27–41). Berkeley, CA: USENIX Association.
86. Gensler, S., Hinz, O., Skiera, B., & Theysohn, S. (2012). Willingness-to-pay estimation with choice-based conjoint analysis: Addressing extreme response behavior with individually adapted designs. European Journal of Operational Research, 219, 368–378. https://doi.org/10.1016/j.ejor.2012.01.002.
87. Keßler, C., & McKenzie, G. (2018). A geoprivacy manifesto. Transactions in GIS, 22(1), 3–19. https://doi.org/10.1111/tgis.12305.
88. McKenzie, G., Janowicz, K., & Seidl, D. (2016). Geo-privacy beyond coordinates. Geospatial data in a changing world (pp. 157–175). Springer. https://doi.org/10.1007/978-3-319-33783-8_10.
89. Culnan, M. J., & Bies, R. J. (2003). Consumer privacy: Balancing economic and justice considerations. Journal of Social Issues, 59(2), 323–342. https://doi.org/10.1111/1540-4560.00067.

90. Hassandoust, F., Akhlaghpour, S., & Johnston, A. C. (2021). Individuals' privacy concerns and adoption of contact tracing mobile applications in a pandemic: A situational privacy calculus perspective. Journal of the American Medical Informatics Association, 28(3), 463–471. https://doi.org/10.1093/jamia/ocaa240.
91. Naous, D., Kulkarni, V., Legner, C., & Garbinato, B. (2019). Information disclosure in location-based services: An extended privacy calculus model. Fortieth International Conference on Information Systems.
92. Xu H, Luo X, Carroll JM, Rosson MB (2011) The personalization privacy paradox: an exploratory study of decision making process for location-aware marketing. Decision Support Syst 51(1):42–52. https://doi.org/10.1016/j.dss.2010.11.017.
93. Culnan, M., & Armstrong, P. (1999). Information privacy concerns, procedural fairness, and impersonal trust: An empirical investigation. Organization Science, 10, 104–115.
94. Wacks, R. (2015). Privacy: A very short introduction. OUP Oxford.
95. Dinev, T., Bellotto, M., Hart, P., Russo, V., & Serra, I. (2006). Internet users' privacy concerns and beliefs about government surveillance: An exploratory study of differences between Italy and the United States. Journal of Global Information Management, 14(4), 57–93.
96. Krasnova, H., & Veltri, N. F. (2010). Privacy Calculus on social networking sites: Explorative evidence from Germany and USA. Proceedings of the annual Hawaii international conference on system sciences, 1–10. https://doi.org/10.1109/HICSS.2010.307.
97. Lin, J., Benisch, M., Sadeh, N., Niu, J., Hong, J., Lu, B., & Guo, S. (2013). A comparative study of location-sharing privacy preferences in the united states and china. Personal and Ubiquitous Computing, 17(4), 697–711. https://doi.org/10.1007/s00779-012-0610-6.
98. Zhang, M., Zhao, P., & Qiao, S. (2020). Smartness-induced transport inequality: Privacy concern, lacking knowledge of smartphone use and unequal access to transport information. Transport Policy, 99, 175–185. https://doi.org/10.1016/j.tranpol.2020.08.016.
99. Hong, W., Chan, F. K., & Thong, J. Y. (2019). Drivers and inhibitors of internet privacy concern: A multidimensional development theory perspective. Journal of Business Ethics, 168(3), 539–564. https://doi.org/10.1007/s10551-019-04237-1.
100. Cho, H., Rivera-Sánchez, M., & Lim, S. S. (2009). A multinational study on online privacy: Global concerns and local responses. New Media & Society, 11(3), 395–416.
101. Kummer, T.-F., Ryschka, S., & Bick, M. (2018). Why do we share where we are? The influence of situational factors on the conditional value of check-in services. Decision Support Systems, 115, 1–12. https://doi.org/10.1016/j.dss.2018.08.012.
102. Grossklags, J., & Acquisti, A. (2007). When 25 cents is too much: An experiment on willingness-to-sell and willingness-to-protect personal information. WEIS.
103. Cvrcek, D., Kumpost, M., Matyas, V., & Danezis, G. (2006). A study on the value of location privacy. Proceedings of the 5th ACM workshop on Privacy in electronic society (pp. 109–118). https://doi.org/10.1145/1179601.1179621.
104. Danezis, G., Lewis, S., & Anderson, R. J. (2005). How much is location privacy worth? WEIS, 5, 56.
105. Alrayes, F., Abdelmoty, A., El-Geresy, W., & Theodorakopoulos, G. (2020). Modelling perceived risks to personal privacy from location disclosure on online social networks. International Journal of Geographical Information Science, 34(1), 150–176. https://doi.org/10.1080/13658816.2019.1654109.
106. Adams, A., & Sasse, M. A. (1999). Privacy issues in ubiquitous multimedia environments: Wake sleeping dogs, or let them lie? Interact, 25, 214–221.
107. Friedman, B., Lin, P., & Miller, J. K. (2005). Informed consent by design. Security and Usability, 52, 503–530.
108. Langheinrich, M. (2001). Privacy by design-principles of privacy-aware ubiquitous systems. International Conference on Ubiquitous Computing (pp. 273–291). https://doi.org/10.1007/3-540-45427-6_23.
109. Beierle F, Tran VT, Allemand M, Neff P, Schlee W, Probst T, Pryss R, Zimmermann J (2018b) TYDR—track your daily routine. Android app for tracking smartphone sensor and usage data. In: 2018 ACM/IEEE 5th international conference on mobile software engineering and systems (MOBILESoft '18). ACM, New York, pp 72–75. https://doi.org/10.1145/3197231.3197235.

110. Xu R, Frey RM, Fleisch E, Ilic A (2016) Understanding the impact of personality traits on mobile app adoption–insights from a large-scale field study. Comput Human Behav 62(Supplement C):244–256. https://doi.org/10.1016/j.chb.2016.04.011.
111. LiKamWa R, Liu Y, Lane ND, Zhong L (2013) MoodScope: building a mood sensor from smartphone usage patterns. In: Proc. 11th annual international conference on mobile systems, applications, and services. ACM, MobiSys '13, pp 389–402. https://doi.org/10.1145/2462456.2464449.
112. Bogomolov A, Lepri B, Ferron M, Pianesi F, Pentland AS (2014) Daily stress recognition from mobile phone data, weather conditions and individual traits. In: Proc. 22nd ACM international conference on multimedia, MM '14. ACM, New York, pp 477–486. https://doi.org/10.1145/2647868.2654933.
113. Xiong H, Huang Y, Barnes LE, Gerber MS (2016) Sensus: a cross-platform, general-purpose system for mobile crowdsensing in human-subject studies. In: Proc. 2016 ACM international joint conference on pervasive and ubiquitous computing (UbiComp). ACM, UbiComp '16, pp 415–426. https://doi.org/10.1145/2971648.2971711.
114. Jayarajah , Balan RK, Radhakrishnan M, Misra A, Lee Y (2016) LiveLabs: building in-situ mobile sensing & behavioural experimentation TestBeds. In: Proc. 14th annual international conference on mobile systems, applications, and services. ACM, MobiSys '16, pp 1–15. https://doi.org/10.1145/2906388.2906400.
115. Ferreira D, Kostakos V, Dey AK (2015) AWARE: mobile context instrumentation framework. Front ICT. https://doi.org/10.3389/fict.2015.00006.
116. Danner D, Rammstedt B, Bluemke M, Treiber L, Berres S, Soto C, John O (2016) Die deutsche Version des Big Five Inventory 2 (BFI-2). In: Zusammenstellung Sozialwissenschaftlicher Items und Skalen. https://doi.org/10.6102/zis247.
117. Soto CJ, John OP (2017) The next big five inventory (BFI-2): developing and assessing a ierarchical model with 15 facets to enhance bandwidth, fidelity, and predictive power. J Personal Soc Psychol 113(1):117–143. https://doi.org/10.1037/pspp0000096.
118. Zimmermann J, Woods WC, Ritter S, Happel M, Masuhr O, Jaeger U, Spitzer C, Wright AGC (2019) Integrating structure and dynamics in personality assessment: first steps toward the development and validation of a personality dynamics diary. Psychol Assess 31(4):516–531. https://doi.org/10.1037/pas0000625.
119. Kiukkonen N, Blom J, Dousse O, Gatica-Perez D, Laurila J (2010) Towards rich mobile phone datasets: Lausanne data collection campaign. In: Proceedings of the ACM International Conference on Pervasive Services (ICPS).
120. Wang R, Chen F, Chen Z, Li T, Harari G, Tignor S, Zhou X, Ben-Zeev D, Campbell AT (2014) StudentLife: assessing mental health, academic performance and behavioral trends of college students using smartphones. In: Proc. 2014 ACM international joint conference on pervasive and ubiquitous computing (UbiComp). ACM, UbiComp '14, pp 3–14. https://doi.org/10.1145/2632048.2632054.
121. Tucker, C. (2012). The economics of advertising and privacy. International Journal of Industrial Organization, 30(3), 326–329.
122. Luzak, J. A. (2014). Privacy notice for dummies? Towards European guidelines on how to give "clear and comprehensive information" on the cookies' use in order to protect the Internet users' right to online privacy. Journal of Consumer Policy, 37(4), 547–559.
123. Aalberts, R., Nill, A., & Poon, P. S. (2016). Online behavioral targeting: What does the law say? Journal of Current Issues and Research in Advertising, 37(2), 95–112.
124. King, N., & Jessen, P. W. (2010). Profiling the mobile customer - Privacy concerns when behavioral advertisers target mobile phones - Part I. Computer Law and Security Review, 26(5), 455–478.
125. Turow, J., Hennessy, M., & Bleakley, A. (2008). Consumers' understanding of privacy rules in the marketplace. The Journal of Consumer Affairs, 42(3), 411–424.

126. Malhotra, N., Kim, S., & Agarwal, J. (2004). Internet users' information privacy concerns (IUIPC): The construct, the scale, and a causal model. Information Systems Research, 15(4), 336–355.
127. Tucker, C. (2014). Social networks, personalized advertising, and privacy controls. Journal of Marketing Research, 51(5), 546–562.
128. Wang, S.-C., & Wu, J.-H. (2014). Proactive privacy practices in transition: Toward ubiquitous services. Information & Management, 51(1), 93–103.
129. Smit, E., Noort, G., & Voorveld, H. (2014). Understanding online behavioural advertising: User knowledge, privacy concerns and online coping behaviour in Europe. Computers in Human Behavior, 32(C), 15–22.
130. Sipior, J., Warda, B., & Mendoza, R. (2011). Online privacy concerns associated with cookies, flash cookies, and web beacons. Journal of Internet Commerce, 10(1), 1–16.
131. Baek, T., & Morimoto, M. (2012). Stay away from me. Journal of Advertising, 41(1), 59–76.
132. Hoy, M. G., & Milne, G. (2010). Gender differences in privacy-related measures for young adult Facebook users. Journal of Interactive Advertising, 10(2), 28–45.
133. PrimeLife (2008), http://www.primelife.eu (last access 30 October 2008).
134. Schultz EE. A framework for understanding and predicting insider attacks. In: Computers & Security. 21(6), 2002, pp. 526–531.
135. Curtin, J., Kauffman, R. J., & Riggins, F. J. (2007). Making the 'MOST' out of RFID technology: A research agenda for the study of adoption, usage and impact of RFID. Information Technology and Management, 8, 87–110.
136. Gragg, J. (2003). The emergence of RFID technology in modern society. ECE 339H: Exploratory paper, Oregon State University, 12. November 2003.
137. Juban, R. L., & Wyld, D. C. (2004). Would you like chips with that?: Consumer perspectives of RFID. Management research News, 27(11/12), 29–44.
138. Roussos, G. (2006). Enabling RFID in retail. Computer, 39(3), 25–30.
139. Jensen, A. S., Cazier, J. A., & Dave, D. S. (2008). Mitigating consumer perceptions of privacy and security risks with the use of residual RFID technologies through governmental trust. Journal of Information System Security, 4, 41–66.
140. Langheinrich, M. (2006). RFID and privacy. In M. Petkovic & W. Jonker (Eds.), Security, privacy and trust in modern data management (pp. 214–231). Berlin: Springer.
141. Thiesse, F. (2007). RFID, privacy and the perception of risk: A strategic framework. Journal of Strategic Information Systems, 16(2), 214–232.
142. Anderson, A., Petersen, A., Wilkinson, C., & Allan, S. (2009). Nanotechnology, risk and communication. Basingstoke: Macmillan.
143. Resatsch, F., Sandner, U., Leimeister, J. M., & Krcmar, H. (2008). Do point of sale RFID-based information services make a difference? Analyzing consumer perceptions for designing smart product information services in retail business. Electronic Markets, 18, 216–231.
144. Müller-Seitz, G., Dautzenberg, K., Creusen, U., & Stromereder, C. (2009). Customer acceptance of RFID technology: evidence from the German electronic retail sector. Journal of Retailing and Consumer Services, 16, 31–39.
145. Sill, H. E., Fisher, S. L., & Wasserman, M. E. (2008). Consumer reactions to potential intrusiveness and benefits of RFID. International Journal of Information Technology and Management, 7, 76–97.
146. Jørgensen V, The apple of the eye: parents' use of webcams in a Danish Day Nursery. Surveillance & Society, CCTV Special (eds. Norris, McCahill and Wood), 2004; 2(2/3): pp. 446–463.
147. Timms S. HM Government Minister for Schools, Hansard: Written Answer 20th March 2002, Hansard Column 341W.
148. Joint Select Committee on Human Rights (JCHR). 19th Report, 8th September 2004.
149. Deloitte. Contactpoint Data Security Review: executive Summary. Deloitte; 2008.
150. Seneviratne, S., Hu, Y., Nguyen, T., Lan, G., Khalifa, S., Thilakarathna, K., et al. (2017). A survey of wearable devices and challenges. IEEE Communications Surveys & Tutorials, 19(4), 2573–2620.

151. Das, A.K., Pathak, P.H., Chuah, C.-N., & Mohapatra, P. (2016) Uncovering privacy leakage in ble network traffic of wearable fitness trackers. In Proceedings of the 17th International Workshop on Mobile Computing Systems and Applications, HotMobile '16, pp. 99–104, New York, NY, USA, 2016. Association for Computing Machinery.
152. Cho H, Ippolito D, Yu YW (2020) Contact tracing mobile apps for COVID-19: privacy considerations and related trade-offs. arXiv preprint. https://arxiv.org/abs/2003.11511.
153. Trang S, Trenz M, Weiger WH, Tarafdar M, Cheung CM (2020) One app to trace them all? Examining app specifications for mass acceptance of contact-tracing apps. Eur J Inf Syst 29(4):415–428.
154. Meier Y, Meinert J, Krämer N (2021) Investigating factors that affect the adoption of Covid-19 contact-tracing apps: a privacy calculus perspective. Technol Mind Behav. https://doi.org/10.1037/tmb0000040.
155. Walrave M, Waeterloos C, Ponnet K (2020) Adoption of a contact tracing app for containing COVID-19: a health belief model approach. JMIR Public Health Surv 6(3):e20572.
156. Ferretti, L., et al.: Quantifying SARS-CoV-2 transmission suggests epidemic control with digital contact tracing. Science 368(6491), eabb6936 (2020). https://science.sciencemag.org/content/368/6491/eabb6936.
157. Legendre F, Humbert M, Mermoud A, Lenders V (2020) Contact tracing: an overview of technologies and cyber risks. http://arxiv.org/abs/2007.02806.
158. Wortmann F, Thiesse F, Fleisch E (2019) The impact of goal-congruent feature additions on core IS feature use: when more is less and less is more. J Assoc Inf Syst 20(7):13.
159. Antón AI, Earp JB, Potts C, Alspaugh TA (2001) The role of policy and privacy values in requirements engineering. IEEE 5th International Symposium on Requirements Engineering (RE'01), Toronto, Canada, pp 138–145, 27–31 August 2001.
160. Dexe, J., J. Ledendal, and U. Franke. 2020. An empirical investigation of the right to explanation under GDPR in insurance. In Trust, privacy and security in digital business. The 17th international conference on trust, privacy and security in digital business—TrustBus 2020. Springer. https://doi.org/10.1007/978-3-030-58986-8_9.
161. Scott, J. 2004. Ethics, governance, trust, transparency and customer relations. The Geneva Papers on Risk and Insurance: Issues and Practice 29 (1): 45–51.
162. Alizadeh, F., T. Jakobi, A. Boden, G. Stevens, and J. Boldt. 2020. GDPR reality check—Claiming and investigating personally identifiable data from companies. In 2020 IEEE European symposium on security and privacy workshops (EuroS PW), 120–129. https://doi.org/10.1109/EuroSPW51379.2020.00025.
163. Sørum, H., and W. Presthus. 2020. Dude, where's my data? The GDPR in practice, from a consumer's point of view. Information Technology and People 34 (3): 912–929. https://doi.org/10.1108/ITP-08-2019-0433.
164. Wachter, S., Mittelstadt, B., & Floridi, L. (2017). Why a Right to Explanation of Automated Decision-Making Does Not Exist in the General Data Protection Regulation. International Data Privacy Law, 7(2), 76–99. https://doi.org/10.1093/idpl/ipx005.
165. Binns, R., M. Van Kleek, M. Veale, U. Lyngs, J. Zhao, and N. Shadbolt. 2018. 'It's reducing a human being to a percentage': Perceptions of justice in algorithmic decisions. In Proceedings of the 2018 CHI conference on human factors in computing systems, ACM, CHI '18, 1–14. https://doi.org/10.1145/3173574.3173951 .
166. Insurance Sweden. n.d. Insurance in Sweden 2010–2019. Technical report. Insurance Sweden. https://www.svenskforsakring.se/globalassets/engelska/statistics/insurance-in-sweden-2010-2019.pdf. Accessed 14 June 2021.
167. Bailey J, Kerr I. Seizing control? The experience capture experiments of Ringley & Mann. Ethics Inf Technol 2007;9:129–39.
168. Cory N (2017) Cross border data ows: where are the barriers and what do they cost? Information Technology and Innovation Foundation. https://itif.org/publications/2017/05/01/cross-border-data-ows-where-are-barriers-and-what-do-they-cost.
169. Srikrishna Committee (2018) A free and fair digital economy: protecting privacy, empowering indians. Report of the Committee of Experts under the Chairmanship of Justice B.N. Srikrishna. http://meity.gov.in/writereaddata/files/Data_Protection_Committee_Report.pdf.

170. Gallagher R, Greenwald G (2014) How the nsa plans to infect 'millions' of computers with malware. The Intercept. https://theintercept.com/2014/03/12/nsa-plans-infect-millions-computers-malware/.
171. Greenwald G (2014) How the nsa tampers with us-made internet routers. The Guardian. https://www.theguardian.com/books/2014/may/12/glenn-greenwald-nsa-tampers-us-internet-routers-snowden.
172. Press Trust of India (2014) Vodafone secretly sharing data with british intelligence: home ministry. The Hindu. https://timesofindia.indiatimes.com/india/Vodafone-secretly-sharing-data-with-British-intelligence-Home-ministry/articleshow/31797862.cms/.
173. RBI (2018c) Storage of payment system data. https://www.rbi.org.in/scripts/NoticationUser.aspx?Id=11244.
174. Atick J.: Digital identity: the essential guide, ID4Africa Identity Forum. 2016:1–3. Available at: http://www.id4africa.com/prev/img/Digital_Identity_The_Essential_Guide.pdf.
175. Worstall T. Aadhaar scheme uncovers 440,000 fake students in just three states. FORBES, 27 March 2017 https://www.forbes.com/sites/timworstall/2017/03/27/aadhaar-scheme-uncovers-440000-fake-students-in-just-three-states/#1b3dd3d95bc9.
176. Donaldson T. Values in tension: ethics away from home. Harvard Business Review, September–October 1996 https://hbr.org/1996/09/values-in-tension-ethics-away-from-home.
177. Bellman, S., Johnson, E. J., Kobrin, S. J., & Lohse, G. L. (2004). International differences in information privacy concerns: A global survey of consumers. The Information Society, 20(5), 313–324. https://doi.org/10.1080/01972240490507956.
178. Yoo, B., Donthu, N., & Lenartowicz, T. 2011. Measuring Hofstede's five dimensions of cultural values at the individual level: Development and validation of CVSCALE. Journal of International Consumer Marketing, 23(3–4): 193–210.
179. Rigby, K. 1982. A concise scale for the assessment of attitudes towards institutional authority. Australian Journal of Psychology, 34(2): 195–204.
180. Rigby, K. 1987. An authority behavior inventory. Journal of Personality Assessment, 51(4): 615–625.
181. Cantor, M. "US library defunded after refusing to censor LGBTQ authors: 'We will not ban the books,'" The Guardian, Aug. 05, 2022. Accessed: Aug. 14, 2022. [Online]. Available: https://www.theguardian.com/books/2022/aug/05/michigan-library-book-bans-lgbtq-authors.
182. Hoffmann, A. L. 2016. "Privacy, intellectual freedom, and self-respect: Technological and philosophical lessons for libraries". Emerald Group Publishing Limited 41: 49–69. https://doi.org/10.1108/S0065-283020160000041003.
183. Citron, D. K. 2019. "Why sexual privacy matters for most". Washington University Law Review 96 (6): 1189.
184. Citron, D. K. 2019. "Sexual privacy". The Yale Law Journal 128 (7): 1870.
185. Hazeldean, S. 2019. "Privacy as pretext". Cornell Law Review 104 (7): 1719–1774.
186. Hildebrandt, M., Smart technologies and the end(s) of LawNovel entanglements of law and technology. Cheltenham, UK: Edward Elgar Publishing. 2015. [Online]. Available: https://www.elgaronline.com/view/9781849808767.xml.
187. Siddiq, F., Gochyyev, P., & Wilson, M. (2017). Learning in digital networks–ICT literacy: A novel assessment of students' 21st century skills. Computers & Education, 109, 11–37. https://doi.org/10.1016/j.compedu.2017.01.014.
188. Kay, D., Korn, N., & Oppenheim, C. (2012). Legal, risk and ethical aspects of analytics in Higher Education. Jisc Cetis Analytics Series, 1(6). Retrieved from http://publications.cetis.ac.uk/2012/500.
189. Boland, A., Cherry, G & Dickson, G. (2017). Doing a Systematic Review: A Student's Guide. 2nd edition. Sage Publications Ltd.
190. Gough, D., Oliver, S., & Thomas, J. (2012). An introduction to systematic reviews. SAGE Publications Ltd.
191. *Rodríguez-Triana, M. J., Martínez-Monés, A., & Villagrá-Sobrino, S. (2016). Learning Analytics in Small-scale Teacher-led Innovations: Ethical and Data Privacy Issues. Journal of Learning Analytics, 3(1), 43–65. https://doi.org/10.18608/jla.2016.31.4.

192. McQuay, T. (2006) ‘Privacy is changing outsourcing in Canada’, Globeandmail.com, 27 July.
193. Prosch, M. (2008) ‘Preventing identify theft throughout the entire data life cycle’, Working Paper, Arizona State University.
194. Greenstein, M. and Ray, A. (2002) ‘Holistic, continuous assurance integration: E-business opportunities and challenges’, Journal of Information Systems, 16, 1–20.
195. Cline, J. (2007) ‘Mind the GAPP: Accountants bring GAAP-like principles to the privacy sphere’, Computerworld Online, 6th December, 2007.
196. Paterson M, Iacovino L (2004) Health privacy: the draft Australian national health privacy code and the shared longitudinal electronic health record. Health Inf Manag J 33:5–11.
197. Council of Europe (2000) Committee of ministers. Recommendation No. R (2000) 13 on European Policy on Access to Archives: https://www.wcm.coe.int/ViewDoc.jsp?id=366245&Lang=en Consulted July 2007.
198. Power, M. (2017). The law of privacy, second edition. Markham: LexisNexis Canada Inc.
199. Inions, N. J., Tran, L. E., & Rozovsky, L. E. (2018). Canadian health information: A practical legal and risk management guide, fourth edition. Toronto: LexisNexis Canada Ltd., 347 p.
200. Office of the Privacy Commissioner of Canada. (2020). Privacy and the COVID-19 Outbreak. Government of Canada. Retrieved from: https://priv.gc.ca/en/privacy-topics/health-genetic-and-other-body-information/health-emergencies/gd_covid_202003/.
201. Dove, E. S., & Phillips, M. (2015). Privacy law, data sharing policies, and medical data: a comparative perspective. In A. Gkoulalas-Divanis & G. Loukides (Eds.), Medical data privacy handbook (pp. 639–678). Springer Link: Switzerland.
202. Bjerring, A. K., Brownell, M., El Emam, K., Fortier, I., Henry, D., Knoppers, B. M., Laurie, G., Lemmens, T., Morgan, M., Noseworthy, T. W., Saunders, S., Wolfson, M., & Zelmer, J. (Council of Canadian Academies). (2015). Accessing health and health-related data in Canada: The expert panel on timely access to health and social data for health research and health system innovation. Council of Canadian Academies.
203. Canadian Institutes of Health Research, Natural Sciences and Engineering Research Council of Canada, and Social Sciences and Humanities Research Council (Canada Tri-Council). (2018). Tri-Council policy statement 2. Canada: Government of Canada.
204. Beauvais, M., & Knoppers, B. M. (2020). COVID-19: who’s afraid of data sharing? Canadian Science Policy Centre. Retrieved from: https://sciencepolicy.ca/news/covid-19-whos-afraid-data-sharing.
205. Department of Justice. (2019). Modernizing Canada’s privacy act: stakeholder engagement discussion papers 1–5. Government of Canada.
206. Innovation, Science, and Economic Development Canada (Department of Industry). (2019). Strengthening privacy for the digital age: proposals to modernize the Personal Information Protection and Electronic Documents Act.
207. Office of the Privacy Commissioner of Canada. (2019). 2018–2019 Annual report to parliament on the privacy act and the personal information protection and electronic documents act. Gatineau, Quebec.
208. El Emam, K., Mercer, J., Moreau, K., Grava-Gubing, I., Buckridge, D., & Joncker, E. (2011). Physician privacy concerns when disclosing patient data for public health purposes during a pandemic influenza outbreak. BMC Public Health, 11, 454. Retrieved from: https://bmcpublichealth.biomedcentral.com/articles/10.1186/1471-2458-11-454.
209. Casanovas P. Conceptualisation of Rights and Meta-rule of Law for the Web of Data, Democracia Digital e Governo Eletrônico (Santa Caterina, Brazil) vol.12. 2015: 18–41; repr. Journal of Governance and Regulation / Volume 4, Issue 4, 2015; p. 118–129. Available at: http://buscalegis.ufsc.br/revistas/index.php/observatoriodoegov/article/viewFile/34399/33229 . Cited 5 October 2016.
210. Casanovas P, Palmirani M, Peroni S, van Engers T, Vitali, F. Special Issue on the Semantic Web for the Legal Domain, Guest Editors Editorial: The Next Step. Semantic Web Journal. 2016, 7 (2): 1–13. Available at: available at SSRN: http://ssrn.Com/abstract=2765912. Cited 5 October 2016.

211. Rodríguez-Doncel V, Santos C, Casanovas P, Gómez-Pérez A. Legal aspects of linked data – The European framework, Computer Law & Security Review: The International Journal of Technology Law and Practice, 2016;32(6): 799–813, https://doi.org/10.1016/j.clsr.2016.07.005.
212. Naylor D, Basrur SH, Bergeron MG, Brunham RC, Butler-Jones D, Dafoe G, Ferguson-Paré M, Lussing F, McGeer A, Neufeld KR, Plummer F: Learning from SARS: renewal of public health in Canada. A report of the National Advisory Committee on SARS and Public Health. 2003, Ottawa, Ontario, Canada, Health Canada, Publication Number: 1210.
213. DQM Group (2006) 'Fraud fright'.
214. APACS: Card Fraud Facts and Figures, http://www.apacs.org.uk/resources_publications/card_fraud_facts_and_figures.html.
215. Guide entitled 'Notification Exemptions. A Self Assessment Guide' (available at http://www.ico.gov.uk/upload/documents/library/data_protection/forms/notification_exemptions_-_self-assessment_guide.pdf).
216. Data Protection Act 1998, Schedule 1, Part I.
217. On 14th February, 2007, the Nationwide was fined £980,000 by the Financial Services Authority as described in the article entitled 'Regulators Get Their Teeth Into Data Breaches' by Mark Watts dated 9th March, 2007 (see http://www.bristows.co.uk/articles/detail.asp?frmAreaID=3&frmarticleid=950&frmpdtid=2).
218. ICO press release 'Foreign office in breach of the Data Protection Act', 13th November, 2007.
219. ICO press release 'ICO takes enforcement action against Marks and Spencer', 25th January, 2008.
220. For an overview of the data protection requirements on data controllers, see 'Don't bin your data protection responsibilities', Journal of Database Marketing and Consumer Strategy Management, Vol. 14, No. 4, pp. 311–314.
221. These rules have also been summarised in previous articles. See for example the following articles published in the Journal of Database Marketing and Consumer Strategy Management: 'Electronic marketing and the new anti-spam regulations', Vol. 11, No. 3, pp. 235–240; 'Getting Tough on Spam', Vol. 12, No. 4, pp. 357–361; and, 'Telephone Marketing Out in the Cold?', Vol. 12, No. 2, pp. 172–176.
222. 'Data Handling Procedures in Government: Interim Progress Report', Cabinet Office publication, December 2007 (available at http://www.cabinetoffice.gov.uk/reports/~/media/assets/www.cabinetoffice.gov.uk/publications/reports/data/data_handling%20pdf.ashx).
223. Letter from Kieran Poynter to Rt Hon Alistair Darling MP, 14th December, 2007 (available at http://www.hm-treasury.gov.uk/media/E/E/poynter_review171207.pdf).
224. 'Data Protection Powers and Penalties', ICO publication of December 2007 (available at http://www.ico.gov.uk/upload/documents/library/corporate/detailed_specialist_guides/data_protection_powers_penalties_v1_dec07.pdf).
225. Ministry of Justice Consultation Paper 'The Information Commissioners' inspection powers and funding arrangements under the Data Protection Act 1998, published 16th July, 2008.
226. See for example Brown, A . (2008) 'Pakistan IT guidelines to precede data protection law', Data Protection Law and Policy, Vol. 5, No. 9, front page.
227. Hill, K. (2020) Activists Turn Facial Recognition Tools Against the Police, The New York Times, 21 October. https://www.nytimes.com/2020/10/21/technology/facial-recognition-police.html. Accessed: 23 Feb 2021.
228. Wiewiórowski, W. (2021) 'Artificial Intelligence Act: a welcomed initiative, but ban on remote biometric identification in public space is necessary', Press Release, 23 April 2021. Brussels: European Commission. Available at: https://edps.europa.eu/system/files/2021-04/EDPS-2021-09-Artificial-Intelligence_EN.pdf. Accessed: 23 Apr 2021.
229. Privacy Rights Clearinghouse (2018) Data Breach Notification in the United States and Territories: The International Association of Privacy Professionals. https://iapp.org/media/pdf/resource_center/Data_Breach_Notification_United_States_Territories.pdf. Accessed: 13 Feb 2021.

230. Hill, K. (2020) The Secretive Company That Might End Privacy as We Know It: The New York Times. https://www.nytimes.com/2020/01/18/technology/clearview-privacy-facial-recognition.html. Accessed: 13 Feb 2021.
231. ACLU (2020) ACLU v. Clearview AI. Available at: https://www.aclu.org/cases/aclu-v-clearview-ai. Accessed: 13 Feb 2021.
232. Greenleaf G, Laws GDP (2021) despite COVID delays, 145 laws show GDPR dominance (February 11th, 2021). (2021) 169 Privacy Laws & Business International Report, 1, 3–5. UNSW Law Research Paper No 21–60:6–19. https://doi.org/10.2139/ssrn.3836261. Accessed March13th 2022.
233. Fromholz JM (2000) The European Union Data Privacy Directive. Berkeley Technology Law Journal 15(418):461–484. https://doi.org/10.15779/Z383D48.
234. Moss, Sigrun Marie, Özden Melis Uluğ, and Yasemin Gülsüm Acar. 2019. Doing research in conflict contexts: practical and ethical challenges for researchers when conducting fieldwork. Peace and Conflict: Journal of Peace Psychology 25(1):86–99.
235. Mwambari, David, Andrea Purdeková, and Aymar Nyenyezi Bisoka. 2021. Covid-19 and research in conflict-affected contexts: distanced methods and the digitalisation of suffering. Qualitative Research https://doi.org/10.1177/1468794121999014.
236. Chiumento, Anna, Laura Machin, Atif Rahman, and Lucy Frith. 2018. Online interviewing with interpreters in humanitarian contexts. International Journal of Qualitative Studies on Health and Well-being https://doi.org/10.1080/17482631.2018.1444887.
237. Dobrick, Farina Madita, Jana Fischer, and Lutz M. Hagen. 2018. Research ethics in the digital age. Ethics for the social sciences and humanities in times of mediatization and digitization. Wiesbaden: Springer VS.
238. AccessNow. 2022. AccessNow helpline guides. https://guides.accessnow.org/. Accessed 21 Mar 2022.
239. Front Line Defenders. 2021. What do you need to protect? https://securityinabox.org/en/. Accessed 22 Mar 2022.
240. Digital First Aid Kid. 2022. Find support. https://digitalfirstaid.org/en/support/ . Accessed 22 Mar 2022.
241. Reporters Without Borders. 2022. Afghanistan digital care guide. https://helpdesk.rsf.org/digital-security-guide/afghanistan-digital-care-guide/. Accessed 8 Apr 2022.
242. Moser, Albine, and Irene Korstjens. 2018. Series: practical guidelines to qualitative research. Part 3: sampling, data collection and analysis. European Journal of General Practice https://doi.org/10.1080/13814788.2017.1375091.
243. Moßbrucker, Daniel. 2020. Threat modeling guide. How to identify digital risks in international development projects. Bonn: Deutsche Welle.
244. Schwab K. The fourth industrial revolution. Geneva: World Economic Forum; 2015.
245. Ministry of Finance and Economic Planning. Vision 2020. Ministry of finance and economic planning, government of Rwanda, Kigali: Rwanda; 2017. http://www.minecofin.gov.rw/fileadmin/templates/documents/NDPR/Vision_2020_.pdf.
246. European Union. (2018). Digital economy and society. Retrieved from https://europa.eu/european-union/topics/digital-economy-society_en.
247. Ryz, L., & Grest, L. (2016). A new era in data protection. Computer Fraud and Security,2016(3), 18–20.
248. Ren, Y., Cheng, F., Peng, Z., Huang, X., & Song, W. (2011). A privacy policy conflict detection method for multi-owner privacy data protection. Electronic Commerce Research,11(1), 103–121.
249. Xu, F., Michael, K., & Chen, X. (2013). Factors affecting privacy disclosure on social network sites: an integrated model. Electronic Commerce Research, 13(2), 151–168.
250. Van Deursen, S., & Kummeling, H. (2019). The New Silk Road: a bumpy ride for Sino-European collaborative research under the GDPR? Higher Education, 78(5), 911–930.
251. Krämer, J., & Wohlfarth, M. (2018). Market power, regulatory convergence, and the role of data in digital markets. Telecommunications Policy, 42(2), 154–171.

252. Ermakova, E. P., & Frolova, E. E. (2019). Legal regulation of digital banking in Russia and foreign countries (European Union, USA, PRC). Vestnik Permskogo Universiteta-Juridicheskie Nauki, 4, 606–625.
253. Saldana, J. (2009). The coding manual for qualitative researchers. New York: Sage.
254. Greenleaf, G. Advances in South Asian data privacy laws: Sri Lanka, Pakistan and Nepal. Privacy Laws and Business International Report. 2019. file:///C:/Users/sm77809/Downloads/SSRN-id3549055.pdf. Accessed 2 Jan 2020.
255. Greenleaf, G W. Asian Data Privacy Laws: trade and human rights perspectives. First edition. Google book. 2014. https://books.google.co.uk/books?id=3yfSBAAAQBAJ&pg=PA438&lpg=PA438&dq=data+privacy+and+security+policies+in+nepal&source=bl&ots=HJCoGnm1xK&sig=ACfU3U2iMecNAbiaFx4aGdgMXh2OxAVk0Q&hl=en&sa=X&ved=2ahUKEwjQrqKKzr7qAhXUnFwKHVq_DSk4ChDoATAFegQIChAB#v=onepage&q=data%20privacy%20and%20security%20policies%20in%20nepal&f=false . Accessed 15 Apr 2018.
256. Wolford B. What is GDPR, the EU's new data protection law? 2022. https://gdpr.eu/what-is-gdpr/. Accessed 22 Nov 2020.
257. ICO. Guide to the General Data Protection Regulation (GDPR). 2018. https://assets.publishing.service.gov.uk/government/uploads/system/uploads/attachment_data/file/711097/guide-to-the-general-data-protection-regulation-gdpr-1-0.pdf. Accessed: 14 Feb 2019.
258. Deloitte. The Asia Pacific Privacy Guide. 2019. https://www2.deloitte.com/content/dam/Deloitte/in/Documents/risk/in-ra-Deloitte_AP-PrivacyGuide_Interactive-noexp.pdf. Accessed 12 Dec 2019.
259. Doulah N. Bangladesh-data protection overview. 2020. https://www.dataguidance.com/notes/bangladesh-data-protection-overvie. Accessed 3 Oct 2020.
260. Mishbah ABMH. Bangladesh steps into the data protection regime. 2019. https://www.thedailystar.net/opinion/human-rights/news/bangladesh-steps-the-data-protection-regime-1726351. Accessed 22 May 2018.
261. Hargittai E. Weaving the western web explaining differences in internet connectivity among OECD countries. Telecommunications Policy. 1991. https://citeseerx.ist.psu.edu/viewdoc/download?doi=10.1.1.79.7637&rep=rep1&type=pdf. Accessed: 3 Mar 2020.
262. Bradford, A. (2012). The Brussels Effect. Northwestern University Law Review, 107(1), 1–67. https://scholarlycommons.law.northwestern.edu/nulr/vol107/iss1/1.
263. Bradford, A. (2015). Exporting standards: the externalization of the EU's regulatory power via markets. International Review of Law and Economics, 42, 158–173. https://doi.org/10.1016/j.irle.2014.09.004.
264. Bradford, A. (2020). The Brussels Effect: How the European Union Rules the World. Oxford University Press.
265. Schwartz, P. M. (2019). Global data privacy: The EU way. New York University Law Review, 94(4), 771–818. https://www.nyulawreview.org/issues/volume-94-number-4/global-data-privacy-the-eu-way/.
266. Masuhara Y (1984) /Tatemae to Honne (Superficiality and truth). Kodansha, Tokyo (in Japanese).
267. Lessig L (1999) Code and other laws of cyberspace. Basic Books, New York.
268. Mizutani M, Dorsey J, Moor JH (2004) The internet and Japanese conception of privacy. Ethics Inform Technol 6:121–128.
269. Doi T (1985) The anatomy of self. Kodansha International, Tokyo.
270. Hata H, Nakagawa G (1997) Constitutional law of Japan. Kluwer Law International, The Hague.
271. Yamagishi T (2003) Trust and social intelligence in Japan. In: Schwartz FJ, Pharr SJ (eds) The state of civil society in Japan, Chap 13. Cambridge University Press, Cambridge.
272. Yamagishi T, Yamagishi M (1994) Trust and commitment in the United States and Japan. Motiv Emot 18(2):129–166.
273. Hendry J (2003) Understanding Japanese society. Routledge, Oxford.
274. Benedict R (1946) The chrysanthemum and the sword: patterns of Japanese culture. Houghton Mifflin, Boston.

275. Itoh M (1964) Issues in the 'After the Banquet' decision. Jurist 309:47–51 [translated as Bradshaw (1967)].
276. Burden D (2020) Building a digital immortal. In: Savin-Baden M, Mason-Robbie V (eds) Digital afterlife: death matters in a digital age. Taylor & Francis, New York, pp 143–160. https://doi.org/10.1201/9780429322198.
277. Öhman C, Floridi L (2017) The political economy of death in the age of information: a critical approach to the digital afterlife industry. Minds Mach 27:639–662. https://doi.org/10.1007/s11023-017-9445-2.
278. Facebook (2022) What is a legacy contact and what can they do with my Facebook account? Facebook. https://www.facebook.com/help/1568013990080948. Accessed 1 Jan 2022.
279. Apple (2022) Digital legacy. Apple. https://digital-legacy.apple.com. Accessed 16 Feb 2022.
280. Mikolov, Tomas; Sutskever, Ilya; Chen, Kai; Corrado, Greg S; and Dean, Jeff (2013). Distributed representations of words and phrases and their compositionality. In NIPS 2013. Advances in neural information processing systems, Lake Tahoe, 5–10 December 2013, pp. 3111–3119.
281. Smith, H Jeff; Milberg, Sandra J.; and Burke, Sandra J. (1996). Information privacy: measuring individuals' concerns about organizational practices. Management Information Systems Quarterly, MIS Quarterly, pp. 167–196.
282. Bélanger, France; and Crossler, Robert E. (2011). Privacy in the digital age: a review of information privacy research in information systems. Management Information Systems Quarterly, MIS Quarterly, vol. 35, no. 4, pp. 1017–1042.
283. Mohammed, Zareef A.; and Tejay, Gurvirender P. (2017). Examining Privacy Concerns and Ecommerce Adoption in Developing Countries. Computers & Security, 67(C), 254–265.
284. Rho, Eugenia Ha Rim; Mark, Gloria; and Mazmanian, Melissa (2018). Fostering Civil Discourse Online: Linguistic Behavior in Comments of #MeToo Articles across Political Perspectives. In CSCW 2018. Proceedings of the ACM on Human-Computer Interaction. Association for Computing Machinery, 2, article 147, 28.
285. Yun, Haejung; Lee, Gwanhoo; and Kim, Dan J. (2019). A chronological review of empirical research on personal information privacy concerns: an analysis of contexts and research constructs. Information & Management, vol. 56, no. 4, pp. 570–601.
286. Hong, Weiyin; and Thong, James Y. L. (2013). Internet privacy concerns: an integrated conceptualization and four empirical studies. Management Information Systems Quarterly. MIS Quarterly, vol. 37, no. 1, pp. 275–298.
287. Cockcroft, Sophie; and Rekker, Saphira (2016). The relationship between culture and information privacy policy. Electronic Markets, vol. 26, no. 1, pp. 55–72.
288. Morton, Anthony; and Angela Sasse, M. (2014). Desperately seeking assurances: Segmenting users by their information-seeking preferences. In PST 2014. Proceedings of the 12th International Conference on Privacy, Security and Trust, Toronto, Canada, 23–24 July, 2014. IEEE, pp. 102–111.
289. Harzing, Anne-Wil; and Maznevski, Martha (2002). The interaction between language and culture: a test of the cultural accommodation hypothesis in seven countries. Language and Intercultural Communication, vol. 2, no. 2, pp. 120–139.
290. Harzing, Anne-Wil (2006). Response styles in cross-national survey research: a 26-country study. International Journal of Cross Cultural Management, vol. 6, no. 2, pp. 243–266.
291. Canedo ED, Calazans ATS, Cerqueira AJ, Costa PHT, Masson ETS (2021) Agile teams' perception in privacy requirements elicitation: Lgpd's compliance in brazil. In: 29th IEEE international requirements engineering conference, RE 2021, Notre Dame, IN, USA, September 20–24, 2021, pp 58–69. IEEE. https://doi.org/10.1109/RE51729.2021.00013.
292. Hoeren T (2008) Was ist das Grundrecht auf Integrität und Vertraulichkeit informationstechnischer Systeme? Multimed Recht 6:365–366 (https://www.itm.nrw/wp-content/uploads/grundrecht_auf_integritaet.pdf).
293. Cunha AS (2021) A privacidade como direito guarda-chuva na Alemanha e nos Estados unidos e seu reflexo nos modelos de proteção de dados. In: Ody LFW (ed) Direito Comparado Brasil-Alemanha: temas de direito privado em estudos originais e traduzidos. Faculdade de Direito da UFRGS, Porto Alegre, pp 213–240.

294. Brasil. Presidência da República. Lei n° 12.965 de 23 de abril de 2014. 2014. http://www.planalto.gov.br/ccivil_03/_ato2011-2014/2014/lei/l12965.htm. Accessed 1 Sep 2019.
295. Brasil. Ministério da Saúde. Sistemas de Informação da Atenção à Saúde: Contextos Históricos, Avanços e Perspectivas no SUS; 2015. http://www.escoladesaude.pr.gov.br/arquivos/File/sistemas_informacao_atencao_saude_contextos_historicos.pdf. Acessed 1 Feb 2020.
296. Brasil. Ministério da Saúde. Estratégia e-Saúde para o Brasil; 2017. http://portalarquivos.saude.gov.br/images/pdf/2017/julho/12/Estrategia-e-saude-para-o-Brasil.pdf. Accessed 1 Feb 2020.
297. Brasil. Presidência da República. Lei n° 13.709 de 14 de agosto de 2018. 2018. http://www.planalto.gov.br/ccivil_03/_ato2015-2018/2018/lei/L13709.htm . Accessed 1 Sep 2019.
298. Naik N, Jenkins P. Securing digital identities in the cloud by selecting an apposite Federated Identity Management from SAML, OAuth and OpenID Connect. 2017 11th International Conference on Research Challenges in Information Science (RCIS), IEEE; 2017. https://doi.org/10.1109/rcis.2017.7956534.
299. Shams M, Krishnamurthy D, Far B. A model-based approach for testing the performance of web applications. Proceedings of the 3rd international workshop on Software quality assurance—SOQUA '06, ACM Press; 2006. https://doi.org/10.1145/1188895.1188909.
300. Pradeep S, Sharma YK. A pragmatic evaluation of stress and performance testing technologies for web based applications. 2019 Amity International Conference on Artificial Intelligence (AICAI), IEEE; 2019. https://doi.org/10.1109/aicai.2019.8701327.
301. Kao CH, Lin CC, Chen J-N. Performance testing framework for REST-based web applications. 2013 13th International Conference on Quality Software, IEEE; 2013. https://doi.org/10.1109/qsic.2013.32.
302. Nurmi SM, Kangasniemi M, Halkoaho A, Pietila AM. Privacy of clinical research subjects: an integrative literature review. J Empir Res Hum Res Ethics. 2019;14(1):33–48. https://doi.org/10.1177/1556264618805643.
303. U.S. Department of Health and Human Services Food and Drug Administration. E6(R2) good clinical practice: integrated addendum to ICH E6(R1) guidance for industry. 2018; Available at: https://www.fda.gov/media/93884/download.
304. World Medical Association. Declaration of Helsinki. Available at: http://www.wma.net/en/30publications/10policies/b3/; 2013.
305. Canadian Institutes of Health Research. Tri-council policy statement: Ethical Conduct for Research Involving Humans. 2014; Available at: https://www.toronto.ca/wp-content/uploads/2018/03/97a8-tri-council-policy-statement-2014.pdf.
306. Ministre de la Sante. loi n° 28–13 relative à la protection des personnes participant aux recherches biomédicales promulguée par le dahir n° 1–15-110 du 4 août 2015. 2015; Retrieved at: https://www.sante.gov.ma/Reglementation/Pages/REGLEMENTATION-DES-PRATIQUES-MEDICALES.aspx.
307. Onwuegbuzie AJ, Dickinson WB, Leech NL. A Qualitative Framework for Collecting and Analyzing Data in Focus Group Research. Int J Qual Methods. 2009; 8:1–21.
308. van Ooijen, I.; Vrabec, Helena U. Does the GDPR Enhance Consumers' Control over Personal Data? An Analysis from a Behavioural Perspective. Journal of Consumer Policy (2018). https://doi.org/10.1007/s10603-018-9399-7.
309. Millet, Hélène Leveraging booking information for offer personalization and GDPR consequences. Journal of Revenue and Pricing Management (2022). https://doi.org/10.1057/s41272-022-00401-x.
310. Hull, Gordon Successful failure: what Foucault can teach us about privacy self-management in a world of Facebook and big data. Ethics and Information Technology (2015). https://doi.org/10.1007/s10676-015-9363-z.
311. Palm, Elin Privacy Expectations at Work—What is Reasonable and Why?. Ethical Theory and Moral Practice (2008). https://doi.org/10.1007/s10677-008-9129-3.
312. Sanchez, Odnan Ref; Torre, Ilaria; He, Yangyang; Knijnenburg, Bart P. A recommendation approach for user privacy preferences in the fitness domain. User Modeling and User-Adapted Interaction (2019). https://doi.org/10.1007/s11257-019-09246-3.

313. Dini Kounoudes, Alexia; Kapitsaki, Georgia M.; Katakis, Ioannis Enhancing user awareness on inferences obtained from fitness trackers data. User Modeling and User-Adapted Interaction (2023). https://doi.org/10.1007/s11257-022-09353-8.
314. Gotsch, Mauro Luis; Schögel, Marcus Addressing the privacy paradox on the organizational level: review and future directions. Management Review Quarterly (2021). https://doi.org/10.1007/s11301-021-00239-4.
315. Schomakers, Eva-Maria; Lidynia, Chantal; Ziefle, Martina All of me? Users' preferences for privacy-preserving data markets and the importance of anonymity. Electronic Markets (2020). https://doi.org/10.1007/s12525-020-00404-9.
316. Lappeman, James; Marlie, Siddeeqah; Johnson, Tamryn; Poggenpoel, Sloane Trust and digital privacy: willingness to disclose personal information to banking chatbot services. Journal of Financial Services Marketing (2022). https://doi.org/10.1057/s41264-022-00154-z.
317. Xu, Heng; Gupta, Sumeet The effects of privacy concerns and personal innovativeness on potential and experienced customers' adoption of location-based services. Electronic Markets (2009). https://doi.org/10.1007/s12525-009-0012-4.
318. Poindexter, J. C.; Earp, Julia B.; Baumer, David L. An experimental economics approach toward quantifying online privacy choices. Information Systems Frontiers (2006). https://doi.org/10.1007/s10796-006-9013-4.
319. Roussos, George; Moussouri, Theano Consumer perceptions of privacy, security and trust in ubiquitous commerce. Personal and Ubiquitous Computing (2004). https://doi.org/10.1007/s00779-004-0307-6.
320. Franz, Anjuli; Benlian, Alexander Exploring interdependent privacy—Empirical insights into users' protection of others' privacy on online platforms. Electronic Markets (2022). https://doi.org/10.1007/s12525-022-00566-8.
321. Ebbers, Frank; Zibuschka, Jan; Zimmermann, Christian; Hinz, Oliver User preferences for privacy features in digital assistants. Electronic Markets (2020). https://doi.org/10.1007/s12525-020-00447-y.
322. Zhang, Hongyu; McKenzie, Grant Rehumanize geoprivacy: from disclosure control to human perception. GeoJournal (2022). https://doi.org/10.1007/s10708-022-10598-4.
323. Beierle, Felix; Tran, Vinh Thuy; Allemand, Mathias; Neff, Patrick; Schlee, Winfried; Probst, Thomas; Zimmermann, Johannes; Pryss, Rüdiger What data are smartphone users willing to share with researchers?. Journal of Ambient Intelligence and Humanized Computing (2019). https://doi.org/10.1007/s12652-019-01355-6.
324. Li, H.; Nill, A. Online Behavioral Targeting: Are Knowledgeable Consumers Willing to Sell Their Privacy?. Journal of Consumer Policy (2020). https://doi.org/10.1007/s10603-020-09469-7.
325. Pekárek, Martin; Pötzsch, Stefanie A comparison of privacy issues in collaborative workspaces and social networks. Identity in the Information Society (2009). https://doi.org/10.1007/s12394-009-0016-4.
326. Gotterbarn, Donald Privacy lost: The Net, autonomous agents, and 'virtual information'. Ethics and Information Technology (1999). https://doi.org/10.1023/a:1010011512024.
327. Cavoukian, Ann Privacy in the clouds. Identity in the Information Society (2008). https://doi.org/10.1007/s12394-008-0005-z.
328. Slettemeås, Dag RFID—the "Next Step" in Consumer–Product Relations or Orwellian Nightmare? Challenges for Research and Policy. Journal of Consumer Policy (2009). https://doi.org/10.1007/s10603-009-9103-z.
329. Dowty, Terri Overlooking children: an experiment with consequences. Identity in the Information Society (2008). https://doi.org/10.1007/s12394-009-0010-x.
330. Fúster, Jaime; Solera-Cotanilla, Sonia; Pérez, Jaime; Vega-Barbas, Mario; Palacios, Rafael; Álvarez-Campana, Manuel; Lopez, Gregorio Analysis of security and privacy issues in wearables for minors. Wireless Networks (2023). https://doi.org/10.1007/s11276-022-03211-6.
331. Naous, Dana; Bonner, Manus; Humbert, Mathias; Legner, Christine Learning From the Past to Improve the Future. Business & Information Systems Engineering (2022). https://doi.org/10.1007/s12599-022-00742-2.

332. Antón, Annie I.; Earp, Julia B. A requirements taxonomy for reducing Web site privacy vulnerabilities. Requirements Engineering (2003). https://doi.org/10.1007/s00766-003-0183-z.
333. Dexe, Jacob; Franke, Ulrik; Söderlund, Kasia; van Berkel, Niels; Jensen, Rikke Hagensby; Lepinkäinen, Nea; Vaiste, Juho Explaining automated decision-making: a multinational study of the GDPR right to meaningful information. The Geneva Papers on Risk and Insurance—Issues and Practice (2022). https://doi.org/10.1057/s41288-022-00271-9.
334. O'Hara, Kieron; Tuffield, Mischa M.; Shadbolt, Nigel Lifelogging: Privacy and empowerment with memories for life. Identity in the Information Society (2008). https://doi.org/10.1007/s12394-009-0008-4.
335. Bailey, Rishab; Parsheera, Smriti Data localization in India: paradigms and processes. CSI Transactions on ICT (2021). https://doi.org/10.1007/s40012-021-00337-4.
336. Dixon, Pam A Failure to "Do No Harm"—India's Aadhaar biometric ID program and its inability to protect privacy in relation to measures in Europe and the U.S.. Health and Technology (2017). https://doi.org/10.1007/s12553-017-0202-6.
337. Satpathy, Tathagata The Aadhaar: "Evil" Embodied as Law. Health and Technology (2017). https://doi.org/10.1007/s12553-017-0203-5.
338. Madan, Shilpa; Savani, Krishna; Katsikeas, Constantine S. Privacy please: Power distance and people's responses to data breaches across countries. Journal of International Business Studies (2022). https://doi.org/10.1057/s41267-022-00519-5.
339. Hofman, Darra; Villagran, Michele A. L. Queer Privacy Protection: Challenges and the Fight within Libraries. International Journal for the Semiotics of Law—Revue internationale de Sémiotique juridique (2023). https://doi.org/10.1007/s11196-023-09994-x.
340. Murchan, Damian; Siddiq, Fazilat A call to action: a systematic review of ethical and regulatory issues in using process data in educational assessment. Large-scale Assessments in Education (2021). https://doi.org/10.1186/s40536-021-00115-3.
341. Prosch, Marilyn Protecting personal information using Generally Accepted Privacy Principles (GAPP) and Continuous Control Monitoring to enhance corporate governance. International Journal of Disclosure and Governance (2008). https://doi.org/10.1057/jdg.2008.7.
342. Pincus, Laura B.; Johns, Roger Private Parts: A Global Analysis of Privacy Protection Schemes and a Proposed Innovation for Their Comparative Evaluation. Journal of Business Ethics (1997). https://doi.org/10.1023/a:1005746000135.
343. Swire, Peter P. When Should "Consumers-as-Producers" Have to Comply With Consumer Protection Laws?. Journal of Consumer Policy (2008). https://doi.org/10.1007/s10603-008-9082-5.
344. Iacovino, Livia; Todd, Malcolm The long-term preservation of identifiable personal data: a comparative archival perspective on privacy regulatory models in the European Union, Australia, Canada and the United States. Archival Science (2007). https://doi.org/10.1007/s10502-007-9055-5.
345. Bernier, Alexander; Knoppers, Bartha Maria Pandemics, privacy, and public health research. Canadian Journal of Public Health (2020). https://doi.org/10.17269/s41997-020-00368-5.
346. Maurushat, Alana; Nguyen, Kathy The legal obligation to provide timely security patching and automatic updates. International Cybersecurity Law Review (2022). https://doi.org/10.1365/s43439-022-00059-6.
347. Casanovas, Pompeu; De Koker, Louis; Mendelson, Danuta; Watts, David Regulation of Big Data: Perspectives on strategy, policy, law and privacy. Health and Technology (2017). https://doi.org/10.1007/s12553-017-0190-6.
348. AbdelMalik, Philip; Boulos, Maged N Kamel; Jones, Ray The perceived impact of location privacy: A web-based survey of public health perspectives and requirements in the UK and Canada. BMC Public Health (2008). https://doi.org/10.1186/1471-2458-8-156.
349. Gregory, Adrian Conserving customer value: Improving data security measures in business. Journal of Database Marketing & Customer Strategy Management (2008). https://doi.org/10.1057/dbm.2008.20.
350. Ozimek, John Best practice for safeguarding personal details and maintaining customer confidence. Journal of Direct, Data and Digital Marketing Practice (2008). https://doi.org/10.1057/dddmp.2008.33.

351. Nettleton, Ewan; Jensen, William Don't bin your Data Protection responsibilities. Journal of Database Marketing & Customer Strategy Management (2007). https://doi.org/10.1057/palgrave.dbm.3250060.
352. Nettleton, Ewan; Turner, Ian Data protection: Tougher enforcement and increased power for the Information Commissioner?. Journal of Database Marketing & Customer Strategy Management (2008). https://doi.org/10.1057/dbm.2008.9.
353. Castro Edwards, James Data protection: Where are we now?. Journal of Database Marketing & Customer Strategy Management (2008). https://doi.org/10.1057/dbm.2008.18.
354. Almeida, Denise; Shmarko, Konstantin; Lomas, Elizabeth The ethics of facial recognition technologies, surveillance, and accountability in an age of artificial intelligence: a comparative analysis of US, EU, and UK regulatory frameworks. AI and Ethics (2021). https://doi.org/10.1007/s43681-021-00077-w.
355. Özkan, Özlem; Şahinol, Melike; Aydinoglu, Arsev Umur; Aydin Son, Yesim Reflections on Turkish Personal Data Protection Law and Genetic Data in Focus Group Discussions. NanoEthics (2022). https://doi.org/10.1007/s11569-022-00431-0.
356. Guntrum, Laura Gianna; Güldenring, Benjamin; Kuntke, Franz; Reuter, Christian Using digitally mediated methods in sensitive contexts: a threat analysis and critical reflection on data security, privacy, and ethical concerns in the case of Afghanistan. Zeitschrift für Friedens- und Konfliktforschung (2022). https://doi.org/10.1007/s42597-022-00088-2.
357. Nsengimana, Jean Philbert Reflections upon periclitations in privacy: perspectives from Rwanda's digital transformation. Health and Technology (2017). https://doi.org/10.1007/s12553-017-0196-0.
358. Weber, Philip Andreas; Zhang, Nan; Wu, Haiming A comparative analysis of personal data protection regulations between the EU and China. Electronic Commerce Research (2020). https://doi.org/10.1007/s10660-020-09422-3.
359. Bentotahewa, Vibhushinie; Hewage, Chaminda; Williams, Jason The Normative Power of the GDPR: A Case Study of Data Protection Laws of South Asian Countries. SN Computer Science (2022). https://doi.org/10.1007/s42979-022-01079-z.
360. Cervi, Giulio Vittorio Why and How Does the EU Rule Global Digital Policy: an Empirical Analysis of EU Regulatory Influence in Data Protection Laws. Digital Society (2022). https://doi.org/10.1007/s44206-022-00005-3.
361. Adams, Andrew A.; Murata, Kiyoshi; Orito, Yohko The Japanese sense of information privacy. AI & SOCIETY (2009). https://doi.org/10.1007/s00146-009-0228-z.
362. Nakagawa, Hiroshi; Orita, Akiko Using deceased people's personal data. AI & SOCIETY (2022). https://doi.org/10.1007/s00146-022-01549-1.
363. González-Pizarro, Felipe; Figueroa, Andrea; López, Claudia; Aragon, Cecilia Regional Differences in Information Privacy Concerns After the Facebook-Cambridge Analytica Data Scandal. Computer Supported Cooperative Work (CSCW) (2022). https://doi.org/10.1007/s10606-021-09422-3.
364. Canedo, Edna Dias; Calazans, Angelica Toffano Seidel; Bandeira, Ian Nery; Costa, Pedro Henrique Teixeira; Masson, Eloisa Toffano Seidel Guidelines adopted by agile teams in privacy requirements elicitation after the Brazilian general data protection law (LGPD) implementation. Requirements Engineering (2022). https://doi.org/10.1007/s00766-022-00391-7.
365. Alberton Coutinho Silva, Cecília The fundamental right of confidentiality and integrity of IT systems in Germany: a call for "IT Privacy" right in Brazil?. International Cybersecurity Law Review (2021). https://doi.org/10.1365/s43439-021-00037-4.
366. de Paiva Marques Carvalho, Túlio; de Paiva, Jailton Carlos; de Medeiros Valentim, Ricardo Alexsandro; Silva, Carlos Breno Pereira; de Lima, Diêgo Ferreira; Silva, Emerson Costa Sabiá: an authentication, authorization, and user data delivery architecture based on user consent for health information systems in Brazil. Research on Biomedical Engineering (2020). https://doi.org/10.1007/s42600-020-00058-8.

367. Adarmouch, Latifa; Felaefel, Marwan; Wachbroit, Robert; Silverman, Henry Perspectives regarding privacy in clinical research among research professionals from the Arab region: an exploratory qualitative study. BMC Medical Ethics (2020). https://doi.org/10.1186/s12910-020-0456-9.

Chapter 5
Importance of Identifying Consent

Introduction by the Editor

Certain key phrases and terms are primarily associated with the idea of consent used in the parlance of data protection and privacy. For instance, free/freely, specific, informed, unconditional, unambiguous and a clear affirmative action/act.[1] These are the smaller clusters that form the DNA of consent. Each of these terms has its own meaning and epitomises the rational mind of a data subject. They exude confidence, transparency and trust in the process that leads to the processing of personal data. An effective consent holds the data subject and data controller accountable for their activities. Even if the attributes of consent seem structured, it is often not easy to transpose these attributes into practices that data controllers should adopt. Factors like the extent of processing, the nature of the technology, and the data collection procedure make the act of harnessing consent complex. Sometimes technology would help to achieve the outcome sought by legislators.[2]It is compounded by the extent of a data subject's understanding of privacy. While the legislative intent of bringing

[1] Recital 32, Regulation (EU) 2016/679 of the European Parliament and of the Council of 27 April 2016 on the protection of natural persons with regard to the processing of personal data and on the free movement of such data, and repealing Directive 95/46/EC (General Data Protection Regulation) OJ L 119, 4.5.2016, p. 1–88. (GDPR); Section 6(1), The Digital Personal Data Protection Act, 2023 [India].

[2] Martin Bialke & others, "A FHIR has been lit on gICS: facilitating the standardised exchange of informed consent in a large network of university medicine" (2022) BMC Medical Informatics and Decision Making 22(335); Martin Bialke & others, "MAGIC: once upon a time in consent management—a FHIR® tale" (2018) Journal of Translational Medicine 16(256).

I. Gupta (ed.), *Expectations vs Realities of Information Privacy and Data Protection Measures*, https://doi.org/10.1007/978-981-99-6778-0_5

about transparency is paramount, these requirements do not always function optimally owing to reasons that are multi-faceted and have multi-variable dependencies. Consent is not merely a transaction but represents a dialogue wherein stakeholders interact, exchange emotions and sentiments and bear a certain degree of heterogenous outcome.[3]

The nature of consent varies with the nature of the personal data. As a standard practice, in most jurisdictions, personal data is categorised into sensitive and non-sensitive.[4] Sensitive data requires explicit consent from the data subject prior to processing. The explicit consent embodies a greater understanding of the purpose and objective of data processing. The process of asking for explicit consent should allow data subjects to make a reasoned and free decision. Asking for explicit consent brings about the need to opt-in data subjects before processing their personal data. There could be an opportunity for a data controller to rely upon either a data subject opting out of processing or using the soft opt-in option.[5] The standard attributes do not seem to allow the data controller to adopt the opt-out option unless there is a specific provision in the law that allows the processing of personal data that relies on data subjects voluntarily sharing data.[6] In all these cases, data controllers need not ensure that the data subject shares consent.

The timings when data controllers ask for consent are equally crucial. Ideally, it should be immediately before data processing commences, or data processing should be instantaneous to the act of consent. The data protection framework across jurisdictions would like to connect the act of notifying the purpose behind processing personal data with specific consent directed to such purpose. Under certain circumstances, consent may be delayed or deferred when the data controller cannot communicate with the subject.[7]

Implementing an effective consent mechanism assumes greater importance for processing personal data in health and allied research areas.[8] The processed data would be sensitive and thus require data controllers to adopt measured approaches to safeguard personal data and provide the necessary freedom for data subjects to exercise their free choice, including the option to opt-out. The preferred approach of using the tools of explicit and informed consent would assume importance in all these cases. Proper explicit consent can reduce the power imbalance that may exist between

---

[3] Basil Vassilicos & Marek McGann, "Qualities of consent: an enactive approach to making better sense" (2023) Phenomenology and the Cognitive Sciences.

[4] Article 9, Regulation (EU) 2016/679 of the European Parliament and of the Council of 27 April 2016 on the protection of natural persons with regard to the processing of personal data and on the free movement of such data, and repealing Directive 95/46/EC (General Data Protection Regulation) OJ L 119, 4.5.2016, p. 1–88.

[5] Paul Langford, "Telephone marketing: The importance of the opt-out" (2007) Journal of Direct, Data and Digital Marketing Practice 9, 84–97.

[6] Section 7, The Digital Personal Data Protection Act, 2023 (India).

[7] T C Jansen & others, "Deferred consent in emergency intensive care research: what if the patient dies early? Use the data or not?" (2007) Intensive Care Medicine 33, 894–900.

[8] European Society of Radiology (ESR), "The new EU General Data Protection Regulation: what the radiologist should know. Insights into Imaging" August (2017).

a controller and a data subject.[9] Similarly, informed consent relies on connecting the data controllers to data subjects.[10] Its role is critical in the case of scientific and health research activities portrayed through a series of sources included in this chapter.[11] Such connections bring about informational transparency and provide more information to the data subject, such as the purpose and the risks of processing.[12] Research has highlighted the importance of tiered consent with a granular clarity approach to fit varied research requirements,[13] thereby putting forth different models of consent that meet various ethical and legal needs.[14] Studies have also considered instances of negative impact on research due to the withdrawal of consent.[15] Using the mechanism of consent while advancing dynamic scientific research brings up a critical issue about accommodating the idea of explicit consent across changing times.[16] However, there are questions about applying specific consent in all forms of research, including genomic research.[17] The interpretation of Recital 33 provided by the Guidelines on Consent under Regulation 2016/679 (wp259rev.01) suggests:

> Recital 33 does not disapply the obligations with regard to the requirement of specific consent. This means that, in principle, scientific research projects can only include personal data on the basis of consent if they have a well-described purpose. For the cases where purposes for data processing within a scientific research project cannot be specified at the outset, Recital 33 allows as an exception that the purpose may be described at a more general level.[18]
>
> … When research purposes cannot be fully specified, a controller must seek other ways to ensure the essence of the consent requirements are served best, for example, to allow data

---

[9] Daniela Alaattinoğlu, "Rethinking Explicit Consent and Intimate Data: The Case of Menstruapps" (2022) Feminist Legal Studies 30, 157–179.

[10] Ekaterina Muravyeva & others, "Exploring solutions to the privacy paradox in the context of e-assessment: informed consent revisited" (2020) Ethics and Information Technology 22, 230–238.

[11] Elise Jacquier & others, "Facing new challenges to informed consent processes in the context of translational research: the case in CARPEM consortium" (2021) BMC Medical Ethics 22(21); DC Anderson & others, "Elements of informed consent for pharmacogenetic research; perspective of the pharmacogenetics working group" (2002) The Pharmacogenomics Journal 2, 284–292; Helena Machado & Susana Silva, "Informed Consent in Forensic DNA Databases: Volunteering, Constructions of Risk and Identity Categorization" (2009) BioSocieties 4, 335–348.

[12] Evelien De Sutter & others, "Personalized and long-term electronic informed consent in clinical research: stakeholder views" (2021) BMC Medical Ethics 22(108).

[13] Tsaone Tamuhla & others, "An e-consent framework for tiered informed consent for human genomic research in the global south, implemented as a REDCap template" (2022) BMC Medical Ethics 23(119).

[14] Svenja Wiertz & Joachim Boldt, "Evaluating models of consent in changing health research environments" (2022) Medicine, Health Care and Philosophy 2, 269–280.

[15] Stefan Eriksson & Gert Helgesson, "Potential harms, anonymization, and the right to withdraw consent to biobank research" (2005) European Journal of Human Genetics 13, 1071–1076.

[16] Mary Kirwan & others, "What GDPR and the Health Research Regulations (HRRs) mean for Ireland: "explicit consent"—a legal analysis" (2021) Irish Journal of Medical Science 190, 515–521.

[17] Dara Hallinan, "Broad consent under the GDPR: an optimistic perspective on a bright future" (2020) Life Sciences, Society and Policy 16(1); Stephanie O M Dyke & others, "Registered access: authorizing data access" (2018) European Journal of Human Genetics 26, 1721–1731.

[18] Guidelines on Consent under Regulation 2016/679 (wp259rev.01) <https://ec.europa.eu/newsroom/article29/items/623051/en> (accessed 15th July 2024).

> subjects to consent for a research purpose in more general terms and for specific stages of a research project that are already known to take place at the outset. As the research advances, consent for subsequent steps in the project can be obtained before that next stage begins. Yet, such a consent should still be in line with the applicable ethical standards for scientific research.[19]

The interpretation is arguably moving away from the framework of broad consent,[20] which may be suitable for specific kinds of health research activities[21] and promotes specificity as the way forward for all kinds of processing.[22] There is a further suggestion of introducing dynamic consent,[23] bringing all healthcare services to one single platform with data subjects having more control, thereby avoiding phenomena like consent fatigue.[24] Further, scholars have referred to moral and ethical obligations on data subjects to transfer data useful for health research and allied areas where data protection norms create impediments.[25]

The challenges in placing appropriate notice and informed consent mechanisms[26] with advancing technology like Big Data[27] and the use of AI in different situations[28] have brought about the idea of using anonymisation techniques, although there could be further impediments to medical and health research.[29] The use of

---

[19] Ibid; Etain Quigley & others, ""Data is the new oil": citizen science and informed consent in an era of researchers handling of an economically valuable resource" (2021) Life Sciences, Society and Policy 17(9).

[20] Donald J Willison & others, "Alternatives to project-specific consent for access to personal information for health research: Insights from a public dialogue" (2008) BMC Medical Ethics 9(18); Gesine Richter & others, "Secondary research use of personal medical data: attitudes from patient and population surveys in The Netherlands and Germany" (2020) European Journal of Human Genetics 29, 495–502.

[21] R Indrakusuma & others, "Context-Relative Norms Determine the Appropriate Type of Consent in Clinical Biobanks: Towards a Potential Solution for the Discrepancy between the General Data Protection Regulation and the European Data Protection Board on Requirements for Consent" (2020) Science and Engineering Ethics 26, 3271–3284.

[22] Dara Hallinan, "Broad consent under the GDPR: an optimistic perspective on a bright future" (2020) Life Sciences, Society and Policy 16(1).

[23] Deborah Mascalzoni & others, "Ten years of dynamic consent in the CHRIS study: informed consent as a dynamic process" (2022) European Journal of Human Genetics 30, 1391–1397.

[24] Harriet J A Teare & others, "Reflections on dynamic consent in biomedical research: the story so far" (2020) European Journal of Human Genetics 29, 649–656.

[25] Anton Vedder & Daniela, "Moral autonomy of patients and legal barriers to a possible duty of health related data sharing" (2023) Ethics and Information Technology 25(23); Andy Alorwu & others, "Crowdsourcing sensitive data using public displays—opportunities, challenges, and considerations" (2020) Personal and Ubiquitous Computing 26, 681–696.

[26] Luca Belli & others, "Selling your soul while negotiating the conditions: from notice and consent to data control by design" (2017) Health and Technology August 4.

[27] Adam J Andreotta & others, "AI, big data, and the future of consent" (2022) AI & SOCIETY 37, 1715–1728.

[28] Kristina Astromskė & others, "Ethical and legal challenges of informed consent applying artificial intelligence in medical diagnostic consultations" (2021) AI & SOCIETY 36, 509–520.

[29] Menno Mostert, "Big Data in medical research and EU data protection law: challenges to the consent or anonymise approach" (2015) European Journal of Human Genetics 24, 956–960.

Blockchain and other software interventions as examples have been suggested to improve anonymity, showcasing the use of technology as a way forward to cover the limitations of using conventional methods of harnessing consent in the twenty-first century.[30] This chapter puts forth the contradictions and showcases the challenges that the future holds and how the requirement of consent continues to push us to adopt innovative solutions.

Machine generated summaries

Disclaimer: The summaries in this chapter were generated from Springer Nature publications using extractive AI auto-summarization: An extraction-based summarizer aims to identify the most important sentences of a text using an algorithm and uses those original sentences to create the auto-summary (unlike generative AI). As the constituted sentences are machine selected, they may not fully reflect the body of the work, so we strongly advise that the original content is read and cited. The auto generated summaries were curated by the editor to meet Springer Nature publication standards. To cite this content, please refer to the original papers.

Machine generated keywords: consent, research, clinical, patient, participant, inform consent, genetic, sample, inform, trial, gdpr, medical, template, medical research, genomic.

5.1 Understanding Consent and Its Elements

Machine generated keywords: consent, text, message, guidance, broad consent, trial, genomic, patient, interaction, situation, send, relative, early, emergency, quality.

Broad consent under the GDPR: an optimistic perspective on a bright future [46]

This is a machine-generated summary of:

Hallinan, Dara: Broad consent under the GDPR: an optimistic perspective on a bright future [46]

Published in: Life Sciences, Society and Policy (2020)

Link to original: https://doi.org/10.1186/s40504-019-0096-3

---

[30] Giuseppe Albanese & others "Dynamic consent management for clinical trials via private blockchain technology" (2020) Journal of Ambient Intelligence and Humanized Computing 11, 4909–4926.

If you want to cite the papers, please refer to the original.

For technical reasons we could not place the page where the original quote is coming from.

Abstract-Summary

"The position of the GDPR concerning broad consent has recently been cause for concern in the genomic research community."

"This article takes an in-depth look at the situation concerning broad consent under the GDPR and—despite the understandable concern flowing from recent jurisprudence—offers a positive outlook."

"This positive outlook is argued from three perspectives, each of which is significant in defining the current, and ongoing, legitimacy and utility of broad consent under the GDPR: the principled, the legal technical, and the practical."

Introduction

"This assertion includes legislative limits imposed on the legitimate scope of consent—for example if the Article 29 Working Party's guidance were to be read as requiring specific consent rather than broad consent in genomic research."

"All elements are in place for a forceful resistance: the genomic research community did not simply accept problematic limitations on broad consent in the legislative process leading up to the GDPR [1]; the political power of the genomic research community has been proven—as discussed in Section 3, their lobbying activity was influential in the legislative process leading up to the GDPR; strong normative and legal arguments are available to the genomic research community—see Sections 4 and 5; and, finally, clear judicial avenues already exist through which to mount resistance efforts—Article 78 of the GDPR, for example, offers recourse options against DPAs' judgments."

Conclusion

"In the first instance, the GDPR is a key instrument of genomic research regulation in Europe and its position on broad consent is thus highly significant."

"The content of the Article 29 Working Party guidelines is not friendly to broad consent and the Article 29 Working Party—now European Data Protection Board—are a key entity tasked with providing authoritative interpretations of the GDPR."

"Whilst recognising the legitimacy of these concerns, this article adopted a more positive outlook and explained why broad consent in genomic research still has a bright future under the GDPR."

"There is a clear overlap between the base rationale behind consent in the GDPR and the base rationale behind broad consent in genomic research."

"Even if the EDPB were to push forward with a restrictive approach to broad consent and DPAs began to enforce the approach, the genomic research community would still have options, and have already shown their capacity, in mounting resistance."

Qualities of consent: an enactive approach to making better sense [220]

This is a machine-generated summary of:

Vassilicos, Basil; McGann, Marek: Qualities of consent: an enactive approach to making better sense [220]

Published in: Phenomenology and the Cognitive Sciences (2023)

Link to original: https://doi.org/10.1007/s11097-023-09897-z

If you want to cite the papers, please refer to the original.

For technical reasons we could not place the page where the original quote is coming from.

Abstract-Summary

"We note that the recognition of this complexity demands a theoretical framework quite different to those presently extant, and we suggest that the enactive approach is one which offers significant value in this regard."

"In consonance with arguments made by Laurie Penny about how consent is a continuous and dynamic process, we outline how an enactive approach identifies consent as temporally extended (rather than momentarily transactional), and as affected by the skilfulness of the agents involved, the fitness of community-provided resources to negotiate the consensual relationship over time, and the unfolding of circumstances in the situation in which the joint action is taking place."

"We argue that the value of an enactive perspective on consent is in highlighting these complexities, and in providing resources to articulate and theorise them in ways that are not open to other current approaches."

Introduction: of things good and better

"While there are certainly tools provided by different philosophical approaches for these questions, for the sake of space, coherence, and focus, we shall explore them by broaching the premises and implications of an enactive account of consent."

"On the basis of that sketch, we will give a preliminary justification for why an enactive account might offer an alternative notion of consent, and especially as one amenable to a concern for dynamic qualities of consent."

"(3), we will then turn to an enactive account of consent expliciter, along the way laying out some of the basic enactive concepts and principles which make it available."

"(4) with some lessons for thinking about consent within the enactive account and point to a key question for its further elaboration."

Philosophical knots of consent

"From the point of view that our behaviours and social interaction are constantly evolving from one moment to the next, consent too must be seen as a fluid human interaction subject to various forms of processual realisation in view of its context and participants—for instance, where consensual interaction between persons might (or might not) manifest qualities of type a, then type b, then type c, e.g. of uncertainty and openness and their navigability, etc. Taking this thought further, one crucial implication here is that no token or instance of consent occurs in a vacuum; no form of consent stands on its own."

"It is in this sense one can find some reference to 'quality of consent,' meaning the competence of persons at the moment that they would be given or issue a form of consent, for instance as in Matey's claim (2, 381) that "[b]oth the validity of the consent and quality of the consent can be compromised in different ways.""

Enacting enactive consent

"From an enactive perspective, agency is a continuous process of maintaining coordination between the precarious norms of the animated agent, and the complex flow of environmental events arising for that agent."

"Sensorimotor agency (and its associated sensorimotor bodies; Di Paolo and others, 3) arises in a richer domain of activities in which the autonomous systems in question are not organic, but activity-based—here one can think of how habits and skills emerge as systems of environmental coordination."

"Just as sensorimotor agency involves the continuous coping with the pull and push of multiple norms of habit, skill, and bodily need, so too is linguistic agency a process of coping with a complex normativity which exists not in, but through various sensorimotor bodies."

"The very notion of skill arises when we note that the enactive approach to agency entails that there are a range of complementarities and mutual relationships between the different aspects of what it means to be a living, sentient organism (i.e. organic, sensorimotor, social)."

Consensual lessons

"These are resources to manage and cope with the normative exposures to other agents realised in consent relations, and as we have already noted, these resources can vary widely in terms of the sensitivities they afford to partners in social interaction."

"As the interaction between the agents is being regulated through the enactment of a suite of actions, of moves that nuance or finesse the shared activity so as to address the layers of dialectical tensions, so too consent is continually improving or dis-improving."

"Particular details of the interaction, the manner in which the behaviour setting is enacted in a given instance, can be problematic, and a person can not consent to this manner of interaction, resulting in a conflict of exposures to another's agency that must be negotiated."

Conclusion

"Following the direction pointed to by Penny [4, 5], we have outlined an enactive approach to consent which shifts the description from one of transaction to one of manner."

"Within an adverbial approach, consent is a way of engaging in joint action, one in which we are sensitive not just to the level of a partner's willingness and commitment to shared direction of travel, but to its dynamics, its ebbs and flows—a dialogue more than a momentary exchange or mental state."

"It helps us understand what it is distinct about joint agency, and reinforces Penny's point that we must understand consent in a nuanced, continuously sensitive manner; this has received too little discussion in the philosophical literature."

Exploring solutions to the privacy paradox in the context of e-assessment: informed consent revisited [221]

This is a machine-generated summary of:

Muravyeva, Ekaterina; Janssen, José; Specht, Marcus; Custers, Bart: Exploring solutions to the privacy paradox in the context of e-assessment: informed consent revisited [221]

Published in: Ethics and Information Technology (2020)

Link to original: https://doi.org/10.1007/s10676-020-09531-5

If you want to cite the papers, please refer to the original.

For technical reasons we could not place the page where the original quote is coming from.

Abstract-Summary

"Informed consent for personal data use is a central instrument for ensuring the protection of personal data."

"Current informed consent practices often fail to actually inform data subjects about the use of personal data."

"This article presents the results of a requirements analysis for informed consent from both a legal and usability perspective, considering the application context of educational assessment."

"As the main outcome, the article presents a blueprint which will be the basis for the development of an informed consent template that supports data controllers in establishing an effective and efficient informed consent form."

Introduction

"This aspect brings legal and ethical challenges related to the acceptance of these technologies and the use of personal data: consenting to the use of personal data requires reading and understanding information provided to reach an informed decision."

"Obtaining informed consent poses the same challenge of ensuring that students read and understand how these instruments operate and how their (sensitive) personal data are used so that they can reach an informed decision [6]."

"The privacy paradox and the new legislation underline the need for clear guidelines for data controllers regarding informed consent for the use of personal data."

"The blueprint will be developed in the application context of e-assessment but can be integrated and applied in any context in which people may be asked to consent to the use of personal data, for instance, for online services, participation in research, and the review of such research by a research ethics committee."

Method

"The blueprint describes the total set of obligatory variables to be included in an informed consent form and possible options in connection with these variables for the data controller to decide upon in creating a template."

"The blueprint is meant to inform future development of a wizard-like tool to support data controllers in creating an informed consent template through a series of choices."

"In wait of future development of the blueprint into a wizard to support template creation, the first small scale usability study presented in this paper focusses on the experiences of data controllers using a mock-up of an informed consent template, as explained in more detail in the next section."

"In order to understand how data controllers experience working with an informed consent template derived from the blueprint developed in the current study, we adopted a mixed-methods design with concurrent collection of quantitative and qualitative data [7]."

Results

"4 GDPR, consent is "any freely given, specific, informed and unambiguous indication of the data subject's wishes by which he or she, by a statement or by a clear affirmative action, signifies agreement to the processing of personal data relating to him or her"."

"In the application context of e-assessment, for instance, if an educational institution plans to use the data collected for authorship verification for the purposes of a scientific study as well, then this information can be included in the same consent procedure, helping both students and their educational institution save time and effort in obtaining consent."

"When it comes to the usability requirements for informed consent, two parties are involved: the data controller responsible for providing information, and the data subject who is asked to consent based on the information provided."

"The blueprint aims at helping data controllers fulfil all the legal requirements in creating a template for an informed consent form, which in turn should help ensure that data subjects receive information in a way that enhances informed decision making."

Conclusions and Discussion

"A blueprint for an informed consent template form was presented with the aim to support data controllers in establishing informed consent that is both compliant with the legal requirements and responds to the data subject's needs (usability)."

"The usability study results reported in this paper in relation with an informed consent template revealed the need to provide substantial explanatory information, including examples for data controllers on how to fill in the template."

"Taking things further from here will still require considerable effort on various levels: to develop a formalised model of the blueprint to support software development for informed consent; further, usability studies with data controllers considering various contexts and various levels of complexity; and, last but not least, experts' evaluations and usability studies of the resulting informed consent forms."

Deferred consent in emergency intensive care research: what if the patient dies early? Use the data or not? [222]

This is a machine-generated summary of:

Jansen, T. C.; Kompanje, E. J. O.; Druml, C.; Menon, D. K.; Wiedermann, C. J.; Bakker, J.: Deferred consent in emergency intensive care research: what if the patient dies early? Use the data or not? [222]

Published in: Intensive Care Medicine (2007)

Link to original: https://doi.org/10.1007/s00134-007-0580-8

If you want to cite the papers, please refer to the original.

For technical reasons we could not place the page where the original quote is coming from.

Introduction

"Various approaches are used as surrogate to subject consent: waiver of consent, consent by an independent physician and deferred consent."

"Deferred consent involves randomization at the investigator's discretion according to criteria that have been explicit during ethical review of the protocol, followed by the request for patient's (deferred subject consent) or representative's (deferred proxy consent) informed consent in a later phase."

"Several emergency trials have used deferred consent [8–11]."

"During the enrolment process in an ongoing Dutch multi-centre randomized controlled trial using deferred consent the situation arose that no deferred (subject or proxy) consent was obtained from patients who died early after start of the study."

The "Early Lactate-Directed Therapy on the ICU" study as an example

"Patients eligible for inclusion are randomly allocated to either 8 h of early lactate-directed therapy or control group therapy."

"Since early timing of goal-directed therapy is essential [12, 13], patients are randomized immediately after the first available lactate level, resulting in a very short inclusion time window."

"In 11.7% of cases (13/111) consent could not be obtained due to early death (< 72 h, before relatives could be approached)."

"The Ethics Committee of the coordinating centre was asked for a judgement on the use of already obtained data of patients who died before consent could be sought."

Previous experience in patients who died before obtained deferred consent

"Some emergency studies using deferred consent have reported the use of data from patients who died early."

"In the PAC-man trial using deferred consent [9] consent was sought from patients who regained consciousness (deferred subject consent), regardless of whether relative's assent was obtained earlier."

"Fifty percent (249/500) of the patients died with obtained relative's assent (not consent) but without obtained subject deferred consent."

"In these studies [8, 9, 11], data were used if the patient died before deferred consent was obtained, and reasonable efforts were made to obtain permission from a patient representative."

Ethical considerations

"It seems that very few patients or relatives refuse consent for using already obtained data in emergency situations."

"After completion of our lactate study we plan to compare study results including or excluding data from patients who died before deferred consent could be obtained."

"If data obtained in emergency ICU situations without consent cannot be used, and selection bias is thus introduced, study results may be ruined and future patients be harmed."

"Obtaining written consent a long time after the patient has died can be impractical (telephone consent is not allowed and it is questionable whether the agreeing relative will take effort to reply a request for written consent), and, more importantly, actively approaching relatives for seeking consent in a "fatal medical research case" would still be a real burden for the relatives."

Comparison with other countries in the European Union

"Italian implementation of the European Clinical Trial Directive 2001/20/EC enables enrolment of incapacitated patients in clinical trials, and deferred consent is required according to the ethics committee's approval [14]."

"In Austrian law, which allows proxy consent in the case of mental incapacity, there is a special provision regarding "emergency" situations in which inclusion of a patient can take place without proxy consent."

"Participation of such patients must be with the prospect of a potential direct benefit, "which exceeds the risks"."

"The issue of whether data from patients who have died can be used without formal authorization from patients or representatives is addressed in neither the Dutch, British, Italian and Austrian legislation."

Synthesis

"Dutch regulatory bodies have ruled that if the patient dies shortly after randomization, and consent could not yet be obtained, this forms no reason to abandon the requirement to obtain deferred consent to use the data."

"(a) the validity of proxy consent obtained during emergency situations can be ethically questioned; (b) using data of patients who died and for whom deferred consent was not yet obtained will not harm the patient or relatives (provided that appropriate confidentiality and privacy measures have been applied); (c) not using data will probably introduce selection bias; (d) using data will benefit future patients and society; (e) confronting bereaved relatives to obtain consent is an additional burden; and (f) an individual's decision about the privacy of their medical information is not absolutely binding."

"We therefore think that it is inappropriate to enforce a strict rule that deferred consent must be obtained from bereaved relatives of deceased patients."

Recommendations

"In studies that use deferred consent, data should be used if the patient dies before written (subject or proxy) consent can be sought."

"Only if a patient dies after this period and consent is not yet obtained, should data not be used."

"If not, as a sign of respect for patient autonomy, we plea for a written notification send to the patients' general practitioner and to relatives after the early mourning phase in cases where data are being used for study analyses despite the lack of deferred consent [15]."

Telephone marketing: The importance of the opt-out [223]

This is a machine-generated summary of:
Langford, Paul: Telephone marketing: The importance of the opt-out [223]
Published in: Journal of Direct, Data and Digital Marketing Practice (2007)
Link to original: https://doi.org/10.1057/palgrave.dddmp.4350072

If you want to cite the papers, please refer to the original.

For technical reasons we could not place the page where the original quote is coming from.

The complaints

"The promotional text message campaign that led to the complaints was as follows: 'Orange customer, you may now claim your FREE CAMERA PHONE upgrade for your loyalty."

"The second complaint focused on whether World Networks had sought explicit consent before sending the message and whether the text message was misleading, because it did not tell recipients how to opt out of receiving further messages from World Networks."

"Both complainants also challenged whether the text message was misleading as it implied that it was sent by Orange."

World Networks' Arguments

"It addressed recipients as Orange customers because the message had been sent to numbers assigned to Orange."

"World Telecom said that this was explained to the complainant when he called (in this case, because he had recently received an upgrade, he would not be eligible again for another ten months)."

"Customers could call the local rate number included in the message to request to be taken off its list—the call would cost less than a standard text message, which was the standard method to opt out of receiving further messages."

"It had provided a web address as the sender ID for the message and explained that customers could contact it via the website at no cost."

The ASA's decision

"As this was not the case, it concluded that the message was misleading."

"Although the message stated 'T&C's apply', the ASA considered that the overall impression created by the statement 'you may now claim your FREE CAMERA PHONE upgrade' was that all recipients were able to claim an upgrade—not that they might be able to do so."

"It concluded that the message was misleading."

Consent

"On the issue of consent, the ASA noted that World Networks had contacted recipients with whom they had not previously negotiated a sale, without having obtained their explicit consent."

"The message had been sent without proper consent."

Opt-out

"The guidance clarifies that stated marketers should provide a postal or email address or a short code number to which recipients could send an opt-out message."

"It concluded that the message did not give recipients a clear or simple means of opting out of receiving future messages."

"It also reminded World Networks that it should provide a clear and simple means for recipients to opt out of receiving future messages, requesting it to seek guidance from the CAP Copy Advice team before advertising."

Commentary

"The Privacy Regulations apply to unsolicited electronic marketing messages sent by telephone, fax, email or text."

"Unsolicited marketing material by electronic mail (this includes texts, picture messages and emails) should only be sent if the individual has chosen to receive them, unless the email address was obtained as a result of a commercial relationship (known as the 'soft opt-in')."

"The recipient should always be given the opportunity to stop receiving the emails."

"Has the recipient consented to receiving marketing materials?"

"Has the recipient opted out from receiving this particular material?"

'Blagging'

"The Government and the Information Commissioner have been increasingly concerned about an apparent growth in the illegal trade of personal data, in particular 'blagging'."

"Blagging' is a practice whereby an individual contacts an organisation (typically by telephone) pretending to be someone else (eg the Inland Revenue) in order to extract personal data by deception."

"The acquired data is then sold on, for example to journalists or debt collection agencies."

Prosecutions

"Current penalties of a fine in the Data Protection Act 1998 (DPA) have, however, not provided a sufficiently strong deterrent and have not been viewed as strict enough."

"The DPA provides for the following penalties: on summary conviction, a fine not exceeding the statutory maximum (currently £5,000), and on conviction on indictment, an unlimited fine."

"The Government intends to amend the DPA to make the penalties below available to the Courts, in addition to the current fines: on summary conviction, up to six months imprisonment (increased to 12 months imprisonment in England and Wales when s154 of the Criminal Justice Act 2003 comes into force); and on conviction on indictment, up to two years imprisonment."

Commentary

"Gary Brooks, Head of Data Protection Unit, Business & Technology Services-gary.brooks@blplaw.com E-Commerce: Online shopping under the EU microscope Vanessa Barnett In a bold move in her first press conference, the new EU Commissioner for Consumer Affairs, Meglena Kuneva, announced in February: 'We need a root and branch review of consumer rules."

"We need to inject a new sense of consumer confidence into the e-shopping world so it becomes a trusted market space."

"The paper, however, emphasises that this potential is not being fully realised as business and consumer behaviour has not kept pace with the changes introduced by advances in technology, being restrained by 'internal market obstacles and a lack of confidence in cross-border shopping'."

"Widespread reform of EU consumer policy is therefore needed in order for consumers and businesses to fully benefit from the new opportunities which technology has created."

Complex rules

"Can the consumer open sealed packaging and still get a refund?"

"Can the online retailer withhold some of the refund money paid if the returned goods arrive in a shoddy or used condition?"

"Believe it or not, there have even been arguments about how soon a refund must arrive in the consumer's account based on the placement of a comma in the Distance Selling Regulations!"

Commentary

"According to Forrester Research, the market for online retail sales in Europe is set to more than double in under five years to reach 263bn in 2011, with the number of shoppers growing to 174 million."

"Vanessa Barnett, Senior Associate, Business & Technology Servicesvanessa. barnett@blplaw.com Sale of goods: House of Lords decision provides reminder that consumer is king Paul Langford A person returns a faulty item for repair."

"The House of Lords considered this very issue in the recent case of J&H Ritchie Ltd v Lloyd Ltd, which provides some valuable lessons for anyone who supplies goods to consumers."

Case facts

"The parties entered into a contract for sale of an industrial harrow, which was modified to suit the buyer's requirements."

"The harrow was taken back to the seller's premises for inspection, stripped down and fitted with replacement bearings, after which the buyer was told that the harrow had been repaired and was ready for collection."

"Despite the buyer's requests, the seller refused to reveal what had been done to repair the harrow and would not provide an engineer's report."

"The buyer decided to reject the equipment and brought an action for repayment of the purchase price."

House of Lords decision

"The House of Lords considered the relevant sections of the Sale of Goods Act, which provide that where a buyer is not deemed to have accepted goods, he retains the right to reject them."

"Such a right would be lost if at any time he decided to accept the goods or was deemed to have accepted them; but the buyer could not be expected to exercise the right to reject until he had the information that he needed to make an informed choice."

"The seller could not refuse to give the buyer the information that he needed to exercise it."

"The information of the kind that the buyer was asking for was obviously needed to be able to make a properly informed choice between accepting and rejecting the repaired equipment."

Commentary

"This case clarifies that a buyer will usually retain the right to reject goods until he or she is equipped with all the facts necessary to make an informed decision as to whether to reject or accept such goods."

"An important lesson for suppliers—particularly those who sell complex items—is always to provide full details of any repair that is being undertaken in order to fulfil the requirements of the Sale of Goods Act."

"It is possible to spend thousands of pounds on clever legal advice to ensure your terms and conditions cater for the bare minimum requirements under sale of goods legislation, while side-stepping the pitfalls of unfair terms regulation; but what will be the actual cost to your business of such a policy?"

"This is likely to far outweigh the occasional savings you might make by not providing a refund or replacement."

The violation

"The company's actions in selling such CDs violated European laws designed to prevent infringement of intellectual property rights (IPRs)."

"The law in this area is clear: it allows holders of IPRs to stop unlicensed companies from importing products displaying their brands, or otherwise containing their IPRs, if those products originate outside the European Economic Area (EEA)."

"CD-Wow was found to have sourced music CDs outside the EEA and to have imported them without the consent of the record companies, in direct infringement of European legislation prohibiting this."

Commentary

"Although a music industry case, it is a reminder that, even in the so-called global e-economy, territorial rights can be brought to bear by rights holders to prevent certain imports from eroding profit margins."

"Given the remedies available to rights holders for IPR infringement—injunctions, damages, accounts of profits—this is a wake-up call for all e-commerce businesses to undertake some housekeeping on their product sourcing."

"Vanessa Barnett, Senior Associate, Business & Technology Servicesvanessa.barnett@blplaw.com Copyright: Court of Appeal decides upon the holy grail of copyright protection Toby Headdon The co-authors of The Holy Blood and the Holy Grail (HBHG) have lost their appeal against the High Court decision in April last year that Dan Brown's The Da Vinci Code did not infringe copyright in HBHG."

"The Court of Appeal's decision is a clear signal that there are limits to the extent of copyright protection."

The claim

"The claim involved the alleged copying of the 'central theme' of HBHG."

"The Court of Appeal agreed with the High Court's conclusions, pointing out that what Dan Brown took from HBHG amounted only to underlying ideas not protected by copyright. (Copyright protects particular expressions of ideas, eg the specific text used to express an idea, rather than the underlying idea itself.) The Court of Appeal also provided useful guidance for considering the question of copyright infringement."

Commentary

"There have been a number of cases in which claimants have sought to push the boundaries of copyright protection in relation to a variety of subject matter such as ideas in a book (as was the case here), the 'look and feel' of computer programs, magazine layouts etc but have failed."

"Anyone producing documents with factual content or ideas will need to bear in mind that copyright protection in those documents is likely to be limited, if it exists at all."

"Toby Headdon, Associate, Intellectual Propertytoby.headdon@blplaw.com Copyright: Viacom sues YouTube for $1bn Ian De Freitas Viacom, the owner of

MTV, Nickelodeon and other entertainment brands, recently started proceedings for copyright infringement in the New York Courts against YouTube and its owner Google."

"The claim seeks damages of $1bn as well as an injunction."

Viacom's Complaint

"Viacom's complaint focuses on content posted by users of YouTube's service that comes from Viacom's own programmes such as South Park and Sponge Bob Square Pants."

"A US Supreme Court ruling in favour of the music industry effectively shut down the Grokster peer-to-peer site."

"A similar ruling in favour of the music industry was made by the Australian courts in the Sharman peer-to-peer case."

The key Issues

"They hold sites like YouTube responsible for encouraging widespread copyright infringement."

"Companies like YouTube are saying that they have created a new and popular service for users and one that advertisers are prepared to pay for and why should they not take the financial credit."

"They say they are not responsible if some people infringe copyright when using these services."

Commentary

"Some media organisations, like the BBC, have struck deals with YouTube over the licensing of content."

"The lawsuit may just be posturing in what is looking like an increasingly high stakes game of poker between Viacom and YouTube."

"The real question is likely to be how much YouTube will have to pay to make the deal."

The Case

"The case involved a dispute concerning roof tiles supplied by the claimants to the defendants, which the defendants alleged to be sub-standard."

"The defendants had obtained a copy of an internal letter written by the executive vice president of the claimant company to two directors of its holding company."

"The defendants claimed that the letter showed that the claimants were aware that the tiles were defective, despite fighting the action."

"This letter was then copied and circulated by the defendants in an attempt to force a settlement."

Commentary

"The court held that copyright subsisted in the letter and it had been infringed by the defendants copying it."

"This demonstrates the low threshold of originality required to obtain copyright protection, and that there is no need for the work to have any particular literary merit."

"It was also held that, since the letter was a private internal communication, it was protected by the law of confidence and its disclosure by the defendants was a breach of confidence."

"Ian De Freitas, Partner, Intellectual Propertyian.defreitas@blplaw.com Voice over Internet Protocol: Ofcom creeps in Kieran Healy Ofcom recently announced a new regulatory code for Voice over Internet Protocol (VoIP) service providers designed to ensure that consumers have access to key information about the capabilities of their service."

Rising impact

"VoIP services continue to have a rising impact on the UK communications sector."

"Year, a range of new services was launched and uptake has increased significantly."

"Ofcom has stressed the potential for VoIP services to offer 'significant new benefits to consumers, including more competition and choice, lower prices and new services such as second lines and nomadic services'."

Modification to General Conditions

"After considering responses to the 2006 consultation, Ofcom has decided to modify General Conditions 14 and 18, respectively, to incorporate these regulatory proposals."

"The deadline for compliance for VoIP providers is 29 May 2007."

Commentary

"We await with interest a further consultation, expected this summer, on whether certain VoIP services should be required to offer emergency services access."

"This is particularly key given that the geographical location of the user may be far away from the VoIP service provider."

"Ofcom has also revealed that it is still keeping a watchful eye on other issues relevant to VoIP providers, including naked DSL, net neutrality and the regulation of nomadic services."

"With such a rise in popularity of VoIP services, it was inevitable that Ofcom would regulate."

Greater freedom

"From 1 September 2007, industry sectors such as licensed casinos and betting operators—which have been heavily restricted with regard to where they can advertise—will be granted more freedom, including the ability to advertise on television and radio."

"The content rules for TV gambling advertisements contained in the BCAP code apply to text services and interactive TV."

"The Rules apply largely the same restrictions on content to both non-broadcast and broadcast advertisements."

Objectives and aims of the Rules

"Examples of the Rules include that gambling advertisements: must not portray, condone or encourage gambling behaviour that is socially irresponsible or could lead to financial, social or emotional harm must not suggest that gambling can be a solution to financial concerns, an alternative to employment or a way to achieve financial security must neither suggest peer pressure to gamble nor disparage abstention must not link gambling to seduction, sexual success or enhance attractiveness must not portray gambling in a context of toughness or link it to resilience or recklessness must not suggest that solitary gambling is preferable to social gambling and must not suggest gambling as a right of passage."

"The Rules also state, for example, that no child may be included in a gambling advertisement and that no one who is, or seems to be, under 25 years old may be featured gambling or playing a significant role."

Scheduling of advertisements

"Besides dealing with the content of broadcast advertisements, the new rules also police when gambling advertisements may be aired on TV or radio."

"The scheduling of most gambling adverts in a slot adjacent to children's programming or programmes directed at under 18s will be prohibited."

Spread betting

"The Rules also state that spread betting may be advertised as an investment on specialised financial channels, in specialised financial programming, or on interactive or additional TV services (including text services) only."

"Spread betting advertisements must also comply with other gambling rules."

Commentary

"Although the Rules herald a relaxation of the restrictions in relation to where gambling advertisements may be placed, the restrictions as to content remain detailed and strict."

"Operators, broadcasters and any publishers of advertisements should ensure that any representation of gambling in advertising is in line with the Rules."

"Gambling should be portrayed in a 'soft' manner, meaning its role as a responsible, fun and legitimate leisure activity should be emphasised and any sexy, reckless or 'get rich quick' style imagery or messages must be avoided."

"Tom Russell, Associate, Betting & Gaming Teamtom.russell@blplaw.com"

[Section 34]

"Telephone marketers need to be particularly careful when embarking on SMS campaigns, given the extent of legal regulation that serves to protect the privacy rights of consumers."

"World Networks Ltd recently discovered this for themselves, following two complaints to the ASA."

5.2 Consent in Different Sectors: Health Research and Allied Services

Machine generated keywords: consent, clinical, research, patient, participant, genetic, inform consent, sample, inform, template, biobank, health research, health, explicit consent, medical.

The new EU General Data Protection Regulation: what the radiologist should know [224]

This is a machine-generated summary of:

European Society of Radiology (ESR): The new EU General Data Protection Regulation: what the radiologist should know [224]

Published in: Insights into Imaging (2017)

Link to original: https://doi.org/10.1007/s13244-017-0552-7

If you want to cite the papers, please refer to the original.

For technical reasons we could not place the page where the original quote is coming from.

Abstract-Summary

"Radiologists and radiology departments should be prepared to comply with several new rules for the protection of imaging data."

"Although the new GDPR applies to all domains of the public and private sectors, some specific derogations are defined for data concerning health, aiming at protecting the rights of data subjects and confidentiality of their personal health data, whilst preserving the benefits of processing data, including digital images for research and public health purposes."

"Specific new obligations which healthcare providers (including radiologists/radiology departments) should prepare for include data access for patients, rules for data processing including explicit consent of the data subject in the absence of derogations, or technical and organisational safeguards."

"National health authorities can define exceptions and derogations from certain obligations by means of national law They will also define sanctions in the form of penalties or fines that may be applicable for organisations of the public and private sector that fail to comply with the rules of the GDPR."

"Explicit consent prior to data processing will be necessary."
"Explicit consent prior to communication of imaging data will be necessary."

Introduction

"The revised General Data Protection Regulation (GDPR) of the European Union (EU) addresses the protection of its residents (data subjects) with regard to accessing, processing and the free movement of their personal data [16]."

"The main purpose of the GDPR is to define and update several basic rights of data subjects regarding control of and access to their personal data, and to implement common rules for data protection in all member states."

"The new GDPR implies that all organisations processing personal data must be able to prove that they comply with the rules."

"The purpose of this publication is to summarise some important elements of the protection of digital data concerning health in the framework of the new GDPR, and to make radiologists aware of some key provisions of the new GDPR that should be taken into account in daily practice, clinical research and public health projects."

The new GDPR and data concerning health

"Data concerning health are defined as personal data related to the physical or mental health of a natural person (i.e. an individual), including the provision of health and care services, which reveal information about his or her health status."

"The new GDPR aims at protecting the rights of the data subject and confidentiality of personal health data as an important civil right, whilst preserving the benefits of digital image processing for research and public health purposes."

"There is also a requirement for healthcare providers to demonstrate that their procedures fulfil the conditions of lawful data processing required by the GDPR, e.g. on the basis of explicit consent of the data subject, or the use of safeguards when processing data for research or public health purposes."

GDPR versus national and local rules

"Close collaboration between healthcare providers and organisations and the national health authorities is recommended, because the framework provided by the GDPR enables individual member states to define rules for specific situations in national law Although the GDPR applies in the entire EU, it recognises the relevance of national rules and provides a certain margin of manoeuvrability for all member states to specify their rules within the general framework of the EU legislation with regard to health data."

"Both public and private hospitals and other health and care providers will need to define their lawful basis for the processing of health data and demonstrate their compliance with relevant rules."

"As long as confidentiality and data safety are guaranteed and access rules are defined with regard to the purpose of individual healthcare, there appears to be no obvious reason and no requirement to erase data such as medical images routinely after defined periods without the data subject's request."

Portability of health data

"Unless defined otherwise by national law, hospitals and other healthcare providers, such as radiology departments, may have to be prepared to provide electronic data in an appropriate format to a patient upon request, so he or she can choose to consult with another provider of care, regardless of national borders."

"Such copies of data records have to be provided free of charge."

"Further detail on data portability is provided by a recent publication from the EU working group [17]."

Personal data breach

"Personal data breach means a breach of security leading to the accidental or unlawful destruction, loss, alteration, unauthorised disclosure of, or access to, personal data transmitted, stored or otherwise processed."

"To the obligations to communicate a personal data breach to the data subject, data breaches must be notified within 72 h to the supervisory authority."

"Breaches are not notified in situations where derogations exist, records still need to be kept."

Anonymisation, pseudonymisation and encryption

"In order to facilitate the use of data in the context of research projects, public health or biobanks as well as 'big data' analytics, while protecting personal data, the new GDPR proposes technical and organisational measures such as anonymisation, pseudonymisation and encryption."

"The term anonymisation refers to removing personally identifiable information where it is not needed (e.g. the name of the patient and institution, the date and other text data on images, and perhaps even DICOM metadata from the PACS file which would allow re-identification)."

"Recital 26 (an explanatory text which is part of the legislation that sets out reasons for the provisions of Article 4) explains the following: data which have undergone pseudonymisation, which could be attributed to a natural person by the use of additional information, should be considered as information on an identifiable natural person."

Conditions of lawful data processing for research

"Owing to the potential public health interest created by the scientific use of 'big data' from biobanks and the development of 'personalised medicine', certain exemptions apply to the restriction of personal data for scientific research and storage in the public interest as outlined in Article 89 of the GDPR."

"The GDPR provides several exemptions and derogations for the use of health data, e.g. in the context of research or public health purposes under certain conditions."

"With the condition of such safeguards, the GDPR creates exemptions for researchers to process sensitive health data even beyond the purposes for which they were collected."

"The Biobanking and BioMolecular resources Research Infrastructure (BBMRI-ERIC) has recently published its position with regard to frequently asked questions in the context of data protection for biobanks under the new GDPR [18]."

Summary

"The regulation aims at protecting the confidentiality of personal health data whilst preserving the benefits of digital image processing for research and public health purposes."

"Radiologists and radiology departments should get prepared to comply with several new rules for the protection of digital imaging data, including: To obtain explicit consent from the data subject prior to processing or communication his or her data unless in situations where derogations exist."

"Rules applying in individual member states may either be specified through the GDPR itself or through national law within the framework provided by the GDPR, or defined through the European Data Protection Board."

"It is worth mentioning that the new GDPR also mentions sanctions, such as penalties and fines, to be applicable to organisations of the public and private sectors failing to comply with the rules for data protection."

What GDPR and the Health Research Regulations (HRRs) mean for Ireland: "explicit consent"—a legal Analysis [225]

This is a machine-generated summary of:

Kirwan, Mary; Mee, Blanaid; Clarke, Niamh; Tanaka, Aoife; Manaloto, Lino; Halpin, Emma; Gibbons, Una; Cullen, Ann; McGarrigle, Sarah; Connolly, Elisabeth M.; Bennett, Kathleen; Gaffney, Eoin; Flanagan, Ciaran; Tier, Laura; Flavin, Richard; McElvaney, Noel G.: What GDPR and the Health Research Regulations (HRRs) mean for Ireland: "explicit consent"—a legal analysis [225]

Published in: Irish Journal of Medical Science (2020)

Link to original: https://doi.org/10.1007/s11845-020-02331-2

If you want to cite the papers, please refer to the original.

For technical reasons we could not place the page where the original quote is coming from.

Abstract-Summary

"Irish Health Research Regulations (HRRs) were introduced following the commencement of the General Data Protection Regulation (GDPR) in 2018."

"A legal analysis presented under the auspices of the Irish Academy of Medical Sciences (IAMS) on April 8 and November 25, 2019 at the Royal College of Surgeons in Ireland welcomed the introduction of GDPR and the HRRs."

"This article aims to review academic papers, legal opinion, EU opinion and advice and data protection law in relation to research and explicit consent, in order to examine the legal burden of GDPR and the HRRs on health research in Ireland and to determine whether the analysis presented at the IAMS meetings is reflected more widely in legal text."

"Legal literature review of academic papers, legal opinion, EU opinion and advice and data protection legislation."

"Our results confirm the GDPR explicit consent requirement of the HRRs is having had a significantly negative and far-reaching impact on the conduct of health research in Ireland."

"Urgent review of the HRRs and meaningful engagement between the health research community and legislators in healthcare is required."

Introduction

"One of the requirements of the HRRs is that identified or identifiable personal data cannot be included in health research unless (a) GDPR "explicit consent" exists or (b) a consent declaration has been granted [19–25]."

"The view was expressed that the additional mandatory GDPR explicit consent requirement imposed by the HRRs is a significant impediment to conducting research in Ireland."

"A call was made to regulate informed consent in line with the common law as an achievable alternative safeguard in the Irish health research setting and would align Ireland with the approach taken in other EU Member States [26]."

"This paper sets out to examine the impact of GDPR explicit consent and the HRRs by way of literature review of academic papers, legal opinion, EU opinion, and advice and data protection law."

Results & discussion

"The HRRs in effect negate the research exemption by applying a mandatory GDPR explicit consent safeguard to the processing of personal data for the purpose of health research."

"Some proposals have included: Exempting retrospective chart review from the Privacy Rule Aligning the Privacy Rule with Common Rule informed consent Permitting data disclosure without consent, where the appropriate protections exist and, in a manner similar to direct care and treatment Ireland's GDPR explicit consent requirement remains in place at a time of national crisis with the emergence of COVID-19."

"EDPB Guidelines on the processing of data concerning health for the purpose of scientific research in the context of the COVID-19 outbreak once again draw attention to the issues around the use of GDPR explicit consent: "It has to be noted that all the conditions for explicit consent must be fulfilled….. consent must be freely given, specific, informed, and unambiguous, and it must be made by way of a statement or 'clear affirmative action'" [27]."

Conclusion

"It is important to state that the intention of the legislators was not, in the authors' opinion, to impede health research."

"Rather, GDPR presented a mechanism to regulate research and simultaneously afford greater autonomy to health research participants."

"The authors seek the regulation of consent in health research to a high, achievable standard ensuring public trust and continued support of research."

Recommendations

"Several proposed amendments were presented at both IAMS events in an attempt to rectify some of the problems created by the HRRs."

"These amendments have not been enacted, and there has been no attempt by the Department of Health to set up a consultative process with IAMS to discuss these proposals."

"Meaningful dialogue and consultation between legislators and key stakeholders including IPPOSI (Irish Platform for Patient Organizations, Science and Industry), patient advocate groups and the research community is urgently needed."

Registered access: authorizing data access [226]

This is a machine-generated summary of:

Dyke, Stephanie O. M.; Linden, Mikael; Lappalainen, Ilkka; De Argila, Jordi Rambla; Carey, Knox; Lloyd, David; Spalding, J. Dylan; Cabili, Moran N.; Kerry, Giselle; Foreman, Julia; Cutts, Tim; Shabani, Mahsa; Rodriguez, Laura L.; Haeussler, Maximilian; Walsh, Brian; Jiang, Xiaoqian; Wang, Shuang; Perrett, Daniel; Boughtwood, Tiffany; Matern, Andreas; Brookes, Anthony J.; Cupak, Miro; Fiume, Marc; Pandya, Ravi; Tulchinsky, Ilia; Scollen, Serena; Törnroos, Juha; Das, Samir; Evans, Alan C.; Malin, Bradley A.; Beck, Stephan; Brenner, Steven E.; Nyrönen, Tommi; Blomberg, Niklas; Firth, Helen V.; Hurles, Matthew; Philippakis, Anthony A.; Rätsch, Gunnar; Brudno, Michael; Boycott, Kym M.; Rehm, Heidi L.; Baudis, Michael; Sherry, Stephen T.; Kato, Kazuto; Knoppers, Bartha M.; Baker, Dixie; Flicek, Paul: Registered access: authorizing data access [226]

Published in: European Journal of Human Genetics (2018)

Link to original: https://doi.org/10.1038/s41431-018-0219-y

Copyright of the summarized publication:

The Author(s) 2018

If you want to cite the papers, please refer to the original.

For technical reasons we could not place the page where the original quote is coming from.

Abstract-Summary

"The Global Alliance for Genomics and Health (GA4GH) proposes a data access policy model—"registered access"—to increase and improve access to data requiring an agreement to basic terms and conditions, such as the use of DNA sequence and health data in research."

"In piloting registered access with the Scientific Demonstration data sharing projects of GA4GH, we provide additional ethics, policy and technical guidance to facilitate the implementation of this access model in an international setting."

Introduction

"As data sharing policies in genomics strive to keep pace with the state of data-intensive science [28, 29], current policies offer little choice for sharing genomic research data beyond the two established mechanisms of open access, when data are freely published on the World Wide Web, and controlled access (also called managed or restricted access), whereby qualified researchers apply for access on a project-by-project basis and their research plans are reviewed, often by a committee [30–32]."

"Both open and controlled access policy models have historically served the research community's needs, scientific progress and clinical care."

"Plans for greater integration of datasets and informatics platforms [33], along with ever greater sharing of health-related datasets and growing interest by clinicians and patients in also accessing genomic data, call for new streamlined models of data access that take greater advantage of the richer access-control policies current technology is capable of enforcing."

"Access-control policies, and the technology that enforces them, must enable rapid and efficient access to data that is shared only for specific purposes to a wide range of users while effectively managing ethical and legal risks."

The registered access policy model

"This model would capitalize on the well-established role-based access control (RBAC) model for information technology security enforcement [34–38] and is based on the notion that potential users could be granted online access to data according to their roles (e.g., bona fide researcher or clinical care professional) and risk analysis, rather than on the basis of a specifically described project as is normally required in the controlled access models commonly implemented for research purposes."

"While its main resource on BRCA variant interpretation is publicly available, overlaying registered access would allow enrichment of the dataset with data that cannot be shared openly: for example, patient data supporting clinical interpretations of variants may not be consented for open release but would be available to expert review teams, researchers, or clinicians."

Authentication

"This model is premised on the trust that broad categories of registered users, such as researchers and clinical care professionals, will use the data accessed with the same appropriate care as they would manage controlled access data."

"In addition to the attestation registered users will have agreed to in registering for access to data, home institutions may require researchers and clinicians—who plan to use internal or external health data—to sign up to procedures and guidance documents such as a "Code of Conduct", in order to bind them with the institutional rules and sanctions in this respect."

"In case these routes did not allow registration of atypical potential users, as an additional route to registration, any individual would also be able to apply to a standard Data Access Committee (DAC), the committees that oversee access to controlled access data, to be assessed on a case-by-case basis for registered access status."

Attestation

"One of the attestation statements refers to respecting consent-based data use permissions and restrictions, which should ideally be expressed as Consent Codes [39]."

"We aim to enable a single format for the registration process but support a model that would allow for additional attestation statements attaching extra conditions of use for some datasets or for data from some providers."

"Although our model does not include a review and approval of the user's specific research or data use plans, we considered requesting abstracts of general planned data use in lay terms that would be published to enhance transparency."

"This may be reconsidered, especially to reinforce registered users' commitments to using data for appropriate research or clinical care purposes and to further the aims of public transparency."

Authorization

"Registered access relying parties are the entities that consume OpenID authorizations and enforce access rights and privileges based on the registered access status and attributes of the users."

"To be able to use registered access claims, a relying party needs to trust one or several OpenID Providers."

"In order to establish a federation of OpenID Providers and relying parties, they need to agree on the exact semantics of registered access status and attributes; how credentials are verified by the OpenID Provider and expressed to the relying party; what technical protocols are used to share between the registry and the relying party; and how to protect the confidentiality, integrity, and availability of the communication."

"This would mean communicating only a pseudonymous identifier of the researcher (i.e., an alphanumeric code, which is needed for thwarting re-identification and other attacks on multiple relying parties simultaneously) and their registered access status (which is needed for verifying the requestor's status), including its route and provenance, i.e., which registry delivered the status."

Accessibility

"Since 2005, research and education institutions have been operating technical frameworks called identity federations that allow researchers to use their home institution's credentials (such as user accounts and passwords) to access services that are outside their home institutions."

"To register their bona fide researcher status and make the related attestations within such federations, a researcher would first need to log in at their home institution, which then delivers their fresh and validated role and affiliation information to the registration process."

"A benefit of using an identity federation for registered access is that the researcher status is not self-asserted by the researcher but instead claimed by the research institution employing the researcher."

"Additional details that could support registered access through federated identity management would be the categorization of bio/health researchers or even "registered following GA4GH standards.""

Conclusion

"The main goal of registered access is to streamline access to datasets that require acceptance of terms and conditions due to consent agreements or because of a level of ethical and legal risk, and to enable access to multiple datasets at once as well as to facilitate data discovery and use."

"We also envisage that the simplicity, and clarity, of the standard conditions of data access and use in registered access (the attestation) will both encourage greater use of the data and respect for its ethical use, as seen with licensing terms, such as GNU General Public License and Creative Commons."

"Another important aspect of improving data access is the development of ethics tools to support the assessment of data sensitivity and therefore the risk in data sharing to better determine proportionate levels of protection (e.g., open or registered)."

"Data is the new oil": citizen science and informed consent in an era of researchers handling of an economically valuable resource [227]

This is a machine-generated summary of:

Quigley, Etain; Holme, Ingrid; Doyle, David M.; Ho, Aileen K.; Ambrose, Eamonn; Kirkwood, Katie; Doyle, Gerardine: "Data is the new oil": citizen science and informed consent in an era of researchers handling of an economically valuable resource [227]

Published in: Life Sciences, Society and Policy (2021)

Link to original: https://doi.org/10.1186/s40504-021-00118-6

If you want to cite the papers, please refer to the original.

For technical reasons we could not place the page where the original quote is coming from.

Abstract-Summary

"Research methods are increasingly incorporating citizen participation in the research process, and there has been an increase in collaborative research that brings academic and industry partners together."

"This paper addresses the ethical considerations that research teams must consider when using participatory methods and/or when working with industry and outlines a novel informed consent matrix designed to maintain the high ethical standard to which academic research in the healthcare arena has traditionally adhered."

Introduction

"A typical academic research project runs by gaining explicit informed consent at the outset of the project to collect specific personal data that will be used for a specific project."

"The citizen science model includes additional ethical considerations, particularly around preventing harm to participants and obtaining consent at all stages of the research project and data usage."

"According to Article 4(5): 'pseudonymisation' means the processing of personal data in such a manner that the personal data can no longer be attributed to a specific data subject without the use of additional information, provided that such additional information is kept separately and is subject to technical and organisational measures to ensure that the personal data are not attributed to an identified or identifiable natural person [40]' This means that only data that is truly unidentifiable becomes something that can be used beyond that to which the participant consented."

Potential harms, anonymization, and the right to withdraw consent to biobank research [228]

This is a machine-generated summary of:

Eriksson, Stefan; Helgesson, Gert: Potential harms, anonymization, and the right to withdraw consent to biobank research [228]

Published in: European Journal of Human Genetics (2005)

Link to original: https://doi.org/10.1038/sj.ejhg.5201458

If you want to cite the papers, please refer to the original.

For technical reasons we could not place the page where the original quote is coming from.

Abstract-Summary

"This paper discusses the potential harms involved in biobank research and how ethical review, informed consent, withdrawals, and anonymization of samples should be handled in the light of these harms."

"Nor should a request for withdrawal necessarily stop research on identifiable samples."

"We suggest that the current view on withdrawal from research, supported by the Declaration of Helsinki and subsequent ethical guidelines, be abandoned in the context of biobank research and be replaced by an approach inspired by the Nuremberg Code."

"This approach requires those wishing to withdraw their samples from research to present sufficient reason for doing so."

"This underlines the fact that we all share a responsibility for health research and that no one should take withdrawal from biobank research lightly."

Introduction

"Further analysis is needed to clarify what negative effects biobank research may have and what recommendations regarding ethical review, informed consent, and procedures for the storage and use of biobank samples are needed in order to counter or reduce these risks."

"A presumption in much of the debate, reflected in legal regulations and ethical guidelines, is that potential problems connected with biobank research are related to the harm that may arise if sensitive personal information ends up in the wrong hands."

"We argue that anonymization should not be an automatically permissible response to requests for withdrawal; nor should a request for withdrawal necessarily stop research on identifiable samples."

"Apart from not being particularly appropriate for protecting the interests of individuals, anonymization of samples has a negative impact on research."

Risks of harm related to biobank research

"There is a certain risk of physical harm connected to the taking of samples, the amplitude depending on the kind of sample and physical state of the sample donor."

"Storing a sample in a biobank or using it for research, however, involves no risk of direct physical harm."

"[41] Although there is no risk of direct physical harm involved in biobank research, there is a palpable risk of indirect physical harm."

"Biobank research may also involve a risk of various forms of nonphysical harm."

"[42] Samples used in biobank research may harm the group to which a participant belongs, and thereby harm the individuals concerned."

"Such changes can cause economic harm, for instance more expensive insurance for group members due to expectations about their behaviour."

"Internal harm, on the other hand, is related to internal changes resulting from the spread of information from biobank research."

Moral harms

"It should be clear that the category of wrongs is relevant to the ethics of biobank research."

"In connection to medical research, as well as the procedures of informed consent usually treated as absolutely essential to ethically acceptable research, autonomy, privacy, and personal integrity are central values at stake."

"Biobank research creates moral harm when samples are used for research that the participants would object strongly to, or when researchers by means of their access to samples and personal information draw conclusions regarding paternity, look into details of eating and drinking habits, or deal with other private matters that individuals wish to keep to themselves."

"Deceased persons possess dignity (or a right to be treated with dignity) even in death, which should prevent doctors from performing certain procedures on a corpse."

"Groups are also wronged when treated unjustly, if they are systematically excluded from the benefits of research."

Conflated models of research ethics

"That ethical reviews should be carried out does not mean that biobank research should have to face the same demands as human subject research does."

"We risk conflating different models of research ethics."

"There are the ethical considerations made for epidemiology and register research."

"This differentiation is also acknowledged in some legal regulations regarding registers, in which the collection of personal information is mandatory by law The individual person's interest in self-determination is then considered to be outweighed by public health interests in allowing for research to be carried out."

"Hoeyer et al. [43] Even when they are not willing to do so, it can still be the case that the interest of the individuals concerned in exercising their autonomy regarding participation is less important than research being carried out unobstructed."

"One part of the human subjects research ethics that has been carried over to biobank research ethics is the far-reaching right to withdraw one's consent."

Anonymization as a response to request for withdrawal

"According to many regulations and guidelines, there is no need to return or destroy a sample when a participant withdraws from a study, since anonymization addresses the substantial ethical problem."

"The European Society of Human Genetics writes that the use of 'unlinked anonymized samples' secures 'absolute confidentiality' and these samples can thus be used for new purposes without the need to obtain consent."

"UNESCO [44] The German Nationaler Ethikrat maintains that if samples are anonymized, then 'donor interests calling for protection are not at issue'; therefore, withdrawal of consent need only be responded to with anonymization."

"You have the right to end your participation, but since it can be of importance to keep your sample for research, de-identification is performed."

Limited value of anonymization

"Anonymization has a limited value in protecting participants' interests: It may not decisively cut the link to a specific individual, it prevents the use of samples for purposes such as diagnostics, it may not prevent harms to groups, and it does not rule out wrongdoing."

"If demographic and clinical data accompany the anonymized sample (as is explicitly assumed in the recommendations from the ESHG), it may still be possible in some cases for individuals possessing sufficient knowledge to identify a donor."

"If someone says 'Do not use my samples', that person's request has simply not been respected by the use of anonymization."

"Apart from not being particularly appropriate for protecting the interests of participants, anonymization of biobank samples has a negative impact on research."

"A model in which continued use of identified samples is impossible if withdrawal is requested, and where a request to have one's sample destroyed results in something else (namely anonymization) is hardly satisfactory."

Restricted right to withdraw consent

"The demand for unrestricted individual rights to withdraw consent to participation in research is strongly connected to the Declaration of Helsinki from 1964."

"While the Helsinki Declaration and subsequent documents have given the individual the right to withdraw regardless of reason, the Nuremberg Code conceived of research as a common good and saw experimental subjects as participants in an important humanitarian project, which led to a different view of the right to withdraw."

"The Nuremberg Code [45] There are good reasons why this notion of consent and withdrawal gave way to the view expressed later in the Helsinki Declaration, with its greater emphasis on voluntarism and freedom of choice."

"This lends weight to the notion that there is a moral obligation to accept that one's biological samples are stored for and used in future biobank-related medical research."

Our proposal

"We propose that, instead of the present emphasis on individuals' right to withdraw consent to research on their biological samples for any reason, the clause on withdrawal in the Nuremberg Code be included in guidelines on biobank research."

"To reflect the different context, the previously cited Nuremberg clause could be rewritten as follows: During the course of the research, a participant should be at liberty to withdraw his consent if he can present sufficient reasons why it is no longer reasonable to ask for his continued participation."

"If it is well established and made public that those who strongly object to participation in research for any genuinely felt reason (not based on misconception) will be allowed to withdraw and will then have the right to decide what will happen to their samples, then this is likely to have a positive effect on trust in research."

Conclusions

"We have discussed the potential harms involved in biobank research and how informed consent, withdrawals, and anonymization of samples should be handled in the light of these potential harms."

"Since biobank research does not involve any risk of direct physical harm, and is likely to involve nonphysical harms of different magnitude as compared to research on living human subjects, it is reasonable to treat it as a special kind of research more akin to register research."

"Since it may be reasonable to treat different kinds of research differently, it is by no means clear that guidelines relevant to research on human subjects are also relevant to biobank research."

"We argue that anonymization should not be an automatically permissible response from researchers and biobank holders to requests of withdrawal."

"Donors should not take withdrawals from biobank research lightly."

Evaluating models of consent in changing health research environments [229]

This is a machine-generated summary of:

Wiertz, Svenja; Boldt, Joachim: Evaluating models of consent in changing health research environments [229]

Published in: Medicine, Health Care and Philosophy (2022)

Link to original: https://doi.org/10.1007/s11019-022-10074-3

If you want to cite the papers, please refer to the original.

For technical reasons we could not place the page where the original quote is coming from.

Abstract-Summary

"Alternative forms of consent have been suggested, based on a variety of ethical background assumptions."

"This article identifies five main ethical perspectives at stake."

"Even though Tiered Consent, Dynamic Consent and Meta Consent are designed to the demands of the self-determination perspective as well as the perspective of research as a public good, they are still also criticized from both perspectives."

"These factors include digital infrastructure and digital literacy, data safety regulation, good scientific and clinical practice, transparent debates on ethically relevant features of research, social inequalities, anti-discrimination laws and practices, trust in health care institutions and recognition of patient preferences, and consensus on unethical research."

"We argue that the role of context in determining acceptable models of consent puts the ethical importance of models of consent into perspective."

"Since altering contextual factors can help to live up to the ethical concerns at stake in debates about models of consent, opting for such a shift of focus comes without ethical loss."

Informed consent in changing medical research environments

"It is understood as study specific consent that is given by research participants after being adequately informed about the aims, benefits and burdens of a particular study."

"Insisting on study specific consent would thus severely constrain research opportunities."

"We argue that different consent models reflect different, specific ethical concerns to varying degrees."

"This overview leads us to conclude that no single model of consent can address all of the ethical concerns at stake adequately."

The GDPR and the ethical debate on models of consent

"Governance of research on sensitive personal data is not only an issue of soft law and ethics."

"Three of those conditions are potentially relevant for scientific research on sensitive personal data."

"One of them is consent, the two others are public interest and the necessity of the processing of special category data for scientific research [46]."

"With regard to consent, it is subject to scholarly controversy whether broad consent, as traditionally understood and practiced in biobank research, is acceptable under the GDPR regimen, or whether GDPR consent must rather be understood along the lines of narrower consent models such as dynamic consent [46, 47]."

"This applies all the more, given that sidestepping consent and taking resort to public interest or the GDPR research exemption implies losing the ethical and social advantages of consent."

Models of consent

"Broad Consent, in contrast to Specific Consent, asks research participants to consent to multiple future studies the nature and specificities of which are not known at the time of consenting."

"It is not in itself a model of consent, but the online platform is meant to facilitate the consent process as well as ongoing communication between researchers and participants."

"In a model of Dynamic Specific Consent, online platforms would allow research participants to view information in a format of their choice, adequate to their level of education and interest."

"Of setting a fixed form of consent for all research participants, however, it suggests to set up the platform in a way that allows individuals to give either Specific or Broad Consent, depending on personal preferences."

"As with Dynamic Specific Consent, researchers would have to provide the required information and request the system to contact potential research participants."

Ethical concerns in the debate about informed consent

"Protecting individual rights to self-determination is one of the most important ethical demands a model of consent has to live up to."

"The central question posed by all arguments addressing this concern is: Are all relevant individual rights to self-determination respected by a given model of consent?"

"Even though this is mainly a general concern about social contexts of research, models of consent have themselves occasionally been rejected or defended by reference to opportunities of participation [48, 49]."

"These properties are regarded as goods in themselves, sometimes they are considered as a means to upholding societal acceptance of research and individuals' willingness to participate: "Lastly, obtaining consent makes transparent decisions about donating and researching biospecimens."

"The relational perspective thus rightly asks how consent models shape the relationship between researchers and participants."

"It considers how different models of consent could support or undermine trust in medical research and health care institutions."

Evaluating models of consent

"From the perspective of individual self-determination, Specific Informed Consent is defended as the established standard that does justice to the human right of self-determination and shows appropriate respect for individual autonomy, since it ensures that research participants know what they are asked to consent to, and since it enables them to take an informed stance on whether to consent or not [50]."

"Taking these arguments into account, Specific Informed Consent does not appear as the best possible model to satisfy the perspective of medical progress, nor does it necessarily appear as the right answer to a perspective of self-determination in modern research contexts."

"Although consent models such as Dynamic or Meta Consent are not necessarily more cumbersome to handle than Broad Consent for individuals willing to participate in research, this argument shows that Broad Consent cannot be simply dismissed on grounds of self-determination concerns."

Balancing perspectives

"As the overview of ethical evaluations of informed consent models shows, and this is not much of a surprise, protecting individual rights to self-determination, on the one hand, while allowing for the realization of the public good of medical progress, on the other hand, is the conflict at the centre of the debate."

"From the perspective of self-determination, Specific Informed Consent is the consent model of choice."

"Besides self-determination and research as a good, other concerns play a role as well in the debate on consent models."

"Given the tension of the perspectives of self-determination and progress, and given the variety of additional ethical concerns directed at consent models, no single consent model can lay claim to be ethically best under all circumstances."

"If a participatory process for determining normatively relevant categories of research can be established, Meta Consent would offer clear advantages over Dynamic Specific Consent."

From models of consent to contexts of consent

"The GDPR and GDPR data protection rules constitute, if adhered to, a comparatively high level of data safety, which, in turn, justifies lower levels of the specificity of consent. (3) Established standards and safeguards of good scientific and clinical practice: If awareness of norms of good scientific and clinical practice is high, trustworthiness of research institutions is increased and the details of the consent procedures are of lesser importance. (4) Transparent debates on ethically relevant categories of research: If public debates on ethically problematic aspects of research exist, relevant features of research can easily be identified, and thus Tiered Consent and consent platforms can mirror these categories and go hand in hand with many individual values. (5) Social inequalities and safeguards against them: The less social inequalities exist, the lower is the risk of unequal distribution of health services and research benefit and burden."

Conclusion

"In the light of these criticisms, and given the fact that all these perspectives appear ethically relevant, it becomes impossible to declare one model ethically best under all circumstances."

"Given the tension of the perspectives of self-determination and medical progress, and given the variety of additional ethical concerns directed at consent models, no consent model can lay claim to be ethically uncontested."

"Taking these factors into account in different societies may lead to different ethical recommendations regarding models of consent."

"The role of context in determining acceptable models of consent puts the ethical importance of models of consent into perspective."

Reflections on dynamic consent in biomedical research: the story so far [230]

This is a machine-generated summary of:

Teare, Harriet J. A.; Prictor, Megan; Kaye, Jane: Reflections on dynamic consent in biomedical research: the story so far [230]

Published in: European Journal of Human Genetics (2020)

Link to original: https://doi.org/10.1038/s41431-020-00771-z

If you want to cite the papers, please refer to the original.

For technical reasons we could not place the page where the original quote is coming from.

Abstract-Summary

"There have been significant societal shifts during this time, namely in our reliance on digital tools and the use of social media, as well as a greater appreciation of the integral role of patients in biomedical research."

"This paper reflects on the development of DC to understand its importance in an age where digital health is becoming the norm and patients require greater oversight and control of how their data may be used in a range of settings."

"As looking back, it looks forwards to consider how DC could be further utilised to enhance the patient experience and address some of the inequalities caused by the digital divide in society."

Introduction

"Alongside the establishment of NHSX [51], the UK Government unit tasked with driving digital health strategy and policy, this marked a decisive moment to promote the use of digital tools to support healthcare, and by extension, biomedical research in the UK."

"Approaches such as dynamic consent (DC) could prove integral to the further adoption of digital health strategies."

"DC is an approach to informed consent that allows communication and engagement through a secure digital portal in ways that have not been possible before, with individuals being able to revisit and review consent decisions and preferences over time, as and when they choose."

"We discuss the influences of external factors on the development of DC and the importance of digital consent approaches in the future, as data become even more central to address global health challenges."

The changing context of biomedical research

"The requirement of informed consent has been significantly tested by these research agendas that have promoted a wide range of different secondary uses of samples and data [50]."

"Informed consent is regarded as central to voluntary participation in biomedical research to notify participants of the risks and benefits of taking part, and to explain what will happen during a study."

"Scholars have questioned whether participants can make a truly informed decision if they are not provided with specific details of the intended use of biological samples and data at the time of consent, which is impossible if the 'future research' is yet to be defined [52]."

"In response to the implications of these changes, the DC approach to consent was developed to allow participants to review and update their consent decisions over time and to be informed of the many research uses of their data and samples."

The development of DC

"Biobanking, with its collection of samples and data for future use, was one of three project case studies, because of the consent challenges that were arising in biomedical research [53]."

"DC was designed to allow researchers to see in real-time what permissions were associated with the data they were accessing, and enabled participants to review and update their consent decisions over time using an online portal."

"Coincident to the development of DC has been a growth in electronic consenting tools, which allow PDF copies of consent forms to be shared via email or electronic signatures to be recorded."

"These approaches tend to provide an electronic version of a paper form, maintaining a static single point consent, and an electronic receipt, rather than allowing participants to change their mind over time, and thus do not support the fundamental behaviour change that DC espouses [54]."

Criticisms of DC

"DC's development has occurred alongside the continuing reliance of biomedical research on paper-based systems for consenting, with engagement, and communication still generally restricted to face-to-face encounters."

"Concerns were raised by Steinsbekk and others that DC would be an unwanted burden for participants and researchers."

"In a DC model, participants will be asked for consent continuously, simply because each new project is a new project."

"Steinsbekk and others also argued that DC might reduce participation rates: '[b]eing confronted with the detailed complexity of biomedical research … it is likely that at least some people will struggle with feelings of falling short—that their own competence or knowledge do not suffice.'"

Participants' views of DC

"Focus groups were conducted to explore research participants' views about DC, with participants recruited from three biobanks [55]."

"Many of the focus group participants, all of whom had previously signed up to a biobank, had limited memory of the consent process."

"This led to a marked focus on communication and engagement as a key feature of DC."

Implementation of DC

"The Italian-based CHRIS study (which commenced in 2011) [56–58], implemented 'an interactive DC process for empowering participants' autonomy' and experienced notable resource benefits by adapting their recruitment approaches."

"PEER and RUDY both used a DC approach and function within a broader software solution to support their digital studies."

"Would be to develop a standalone DC tool that could be integrated with other software."

"The Oxford-based SPRAINED study (2015–17) piloted such a DC approach with a small cohort of participants, linking DC software with a clinical trials management tool [59]."

"The DC software collected consent decisions and provided information and updates to participants."

"These early adopters started demonstrating the technological realities and feasibility of DC and the opportunities and challenges it might present to research teams and participants alike."

Changing research paradigms

"Greater emphasis on large volumes of data has changed the way that research is conducted and raises specific issues that are relevant to consent."

"DC has also been considered as an innovative approach to support group decision-making—an opportunity that will become increasingly important as genomic medicine, relevant to family members, becomes more common [60, 61]."

"While these findings demonstrate the potential opportunity for DC to support patients and participants to interact with their data, there is still limited evidence surrounding how DC affects the patient experience."

"This will be a crucial step in moving beyond the promissory discourse around DC, to measure empirically the benefits and drawbacks of the tool, allowing researchers themselves to make an informed decision whether to adopt it as a viable consenting approach."

External factors that have influenced DC's development

"The development of DC was motivated by the growing need for individuals to have control of how their data (and samples) were being used, especially in the face of new technologies and research methods (although as outlined above, the opportunity for DC to support collective control in wider groups and communities is also being explored)."

"The rapid development of digital tools, following the establishment of the Internet and the unprecedented opportunity to connect and network, redefined expectations of privacy and security and emphasised the non-physical risks of research participation."

"This followed growing pressure for patients and the public to be involved in decision-making surrounding research and clinical care—as cemented by the 1996 establishment of INVOLVE [62], the National Institute of Health Research body responsible for championing patient and public involvement (PPI)."

The future of DC

"The use of health data will continue to be supported by digital technologies that enable collection, processing, and sharing on a large scale."

"If it was possible for participants to have access to a single research participation site incorporating DC, which consolidated all their health data decisions across different projects and clinical care as well, this would further mitigate against consent fatigue and help to support meaningful engagement with the decision-making process."

"While paper-based tools have not been phased out entirely and will likely always play a role, digital health is rapidly advancing in all areas of healthcare delivery, from electronic check-in at GP surgeries and hospitals to the use of tablet computers in the clinic to capture patient data."

"This assertion has been challenged, with the acceleration of digital health in response to COVID-19, as consultations with general practitioners have moved online, and digital tools have been rapidly introduced to support virtual patient care."

Conclusion

"DC goes beyond an informed consent tool for biomedical research and could help individuals and groups navigate their entire online presence, with data from a variety of different areas being shared for a range of activities."

"This is more than just being a tool for decision-making, as it raises wider societal issues about how personal information should be controlled in the digital landscape and by whom."

"It is time for the development of health dashboards that can help us to manage all aspects of our healthcare, including decisions about data relating to clinical care, involvement in research, social media, and real-world data."

"The development of DC for biomedical research, in response to broader societal concerns, demonstrates that health data cannot be treated in isolation."

Ten years of dynamic consent in the CHRIS study: informed consent as a dynamic process [231]

This is a machine-generated summary of:

Mascalzoni, Deborah; Melotti, Roberto; Pattaro, Cristian; Pramstaller, Peter Paul; Gögele, Martin; De Grandi, Alessandro; Biasiotto, Roberta: Ten years of dynamic consent in the CHRIS study: informed consent as a dynamic process [231]

Published in: European Journal of Human Genetics (2022)

Link to original: https://doi.org/10.1038/s41431-022-01160-4

If you want to cite the papers, please refer to the original.

For technical reasons we could not place the page where the original quote is coming from.

Abstract-Summary

"The Cooperative Health Research in South Tyrol (CHRIS) is a longitudinal study in Northern Italy, using dynamic consent since its inception in 2011."

"We describe the experience with the CHRIS study dynamic consent, providing an overview of its conceptualization and implementation, and of the participant-centered strategies used to assess and improve the process, directly linked to participation and communication."

"In order to comply with high ethical standards and to allow broadness in the areas of research, CHRIS dynamic consent was conceived as an interactive process: based on a strong governance and an ongoing tailored communication with participants, it aims to promote autonomy and to develop a trust-based engaged relationship with participants, also relevant for retention."

"Participants showed a high degree of participation, thus enabling the consolidation of the CHRIS resources."

"Even though a low change rate was reported in the baseline, participants valued the possibility of changing their informed consent choices."

"Dynamic consent was proven to be a flexible consent model, which allowed to meet ethical and legal standards for participation in research, and to accommodate participants' and researchers' needs."

Introduction

"Dynamic consent consists of an interactive platform and an ongoing communication approach that keeps participants informed on the development of the study over time and allows control on their end."

"Dynamic consent is considered to positively affect both recruitment and retention of participants, and the effective management of the informed consent process, particularly in biobanking and longitudinal studies where future uses of data and samples may not have been established at the time of recruitment [63, 64]."

"The logic framework for evaluation of dynamic consent [65] inspired the way we presented our experience with the CHRIS dynamic consent: the focus of the paper on participation, communication and engagement, and on what was achieved in those areas after its use for 10 years, provided an internal assessment respect to the purpose and assumptions underlying the design of the CHRIS dynamic consent."

CHRIS study and dynamic consent conceptualization

"MICROS was the forerunner of what became the CHRIS study: scientists wanted to design a longitudinal, long-lasting population study, that would allow to efficiently conduct genetic and genomic research."

"To achieve the aims of the CHRIS study, researchers designed a population study which consisted in an extensive collection of sensitive data (genetic data, genealogical information, health-related and life-style data including alcohol and drugs consumption and psychiatric information) and follow-up plans in a long-term project."

"The solution to meet both the ethical and legal requirements (transparency, specific information, specific consent and re-consent) and the needs of researchers (long-term study, extensive collection of data and samples, broad aim of research and use of data and samples) was found with the development of a consent model which was defined as "interactive" [52], consisting of an ongoing communication process, thus allowing an "open dialogue" [66, 67] between participants and researchers, and "[to incorporate, Ed] community feedback into the policy frame of continued research" [66]."

"With this participant-centered approach, the CHRIS dynamic consent was designed to develop a partnership with the study participants [52]."

The CHRIS dynamic consent model

"The conceptual basis of the CHRIS informed consent is broad as regards the study aims, that is consent within predefined research areas (cardiovascular, neurological, metabolic, and oncologic health), and very specific with regard to the policy of the project for using data and samples (such as access rules, sharing rules, information policy, oversight) and participant rights [58, 67]."

"If a participant chooses "No", the study assistant provides clarifications and explains why consent to the specific question is necessary for participating in the study (i.e., "Q1.")

"The informed consent conceived as dynamic promotes autonomy by providing participants with the possibility to easily change their perspectives over time, but also is a process that enables participants to express their preferences in accordance with changes to the research."

"Participation was extended to minors and cohabitants of CHRIS participants, and an entirely online informed consent (for the online survey study) was used."

Observations coming from 10 years of experience

"This echoes the findings of other empirical studies that found that biobank participants' preferences over samples availability changed over time, suggesting that dynamic consent, designed as an informed consent model that provides the strategy to address the possible changes in values and wishes that participants may experience through time, is relevant for participants [68]."

"Another empirical study with biobank participants investigating the experience of participating in biobanking research and of using a dynamic consent interface sample showed that the possibility of revising the consent given and the possibility of being informed about the research development through a dynamic consent tool represented for participants an opportunity for reciprocity and engagement [55]."

"What characterizes the dynamic consent of the CHRIS study is the important role that communication with participants plays in the informed consent process."

Conclusion

"We provided a description of the dynamic consent of the CHRIS study."

"With a decade-long experience with CHRIS dynamic consent, we aimed to shed light into the practical implementation of a discussed consent model, and to show the actual potentialities of dynamic consent as an ethical-legal tool for research participation in long-term research endeavors such as longitudinal studies."

"Considering that the CHRIS study has been using dynamic consent since its inception, we did not provide a formal assessment of outcomes in line with an evaluation framework."

Personalized and long-term electronic informed consent in clinical research: stakeholder views [232]

This is a machine-generated summary of:

De Sutter, Evelien; Borry, Pascal; Geerts, David; Huys, Isabelle: Personalized and long-term electronic informed consent in clinical research: stakeholder views [232]

Published in: BMC Medical Ethics (2021)

Link to original: https://doi.org/10.1186/s12910-021-00675-7

If you want to cite the papers, please refer to the original.

For technical reasons we could not place the page where the original quote is coming from.

Abstract-Summary

"A number of challenges hinder the successful and widespread deployment of eIC in clinical research."

"Interviewees identified that a powerful feature of eIC is its personalized approach as it may increase participant empowerment."

"They identified several ethical and practical challenges, such as ensuring research participants are not overloaded with information and offering the same options to research participants who would prefer a paper-based informed consent rather than eIC."

"According to the interviewees, eIC has the potential to establish efficient long-term interactions between the research participants and the research team in order to keep the participants informed during and after the study."

"Interviewees emphasized that a personal interaction with the research team is of utmost importance and this cannot be replaced by an electronic platform."

"Interviewees across the stakeholder groups supported the idea of having a harmonized eIC approach across the European Member States."

"Interviewees reported a range of design and implementation challenges which needs to be overcome to foster innovation in informing research participants and obtaining their consent electronically."

"It was considered important that the implementation of eIC runs alongside the face-to-face contact between research participants and the research team."

"If successfully implemented, eIC may facilitate the engagement of research participants in clinical research."

Background

"The informed consent (IC) process in clinical research refers to informing a potential research participant of all the pertinent aspects of a clinical study by presenting the information in a IC document [69]."

"Throughout this manuscript, eIC refers to an interactive online-based IC application which could facilitate interactions over time and could enable a personalized approach, adapted to research participants' needs."

"This Regulation is not specifically focused on obtaining participants' eIC in the context of clinical research [70, 71]."

"There is a need for structured insights regarding how an eIC platform (through which research participants are able to manage their eIC) may be effectively personalized and also on how the long-term interactions may be established."

"The insights provided by these different stakeholder groups may support the development and responsible implementation of interactive eIC approaches that will stimulate research participants to be more empowered and informed."

Methods

"Semi-structured interviews were conducted to seek the views and experiences of various stakeholder groups across Europe and the United Kingdom (UK) regarding (electronic) IC in clinical research."

"An interview guide was developed based upon a systematic literature review and the research aims [72]."

"Semi-structured interviews were conducted with people from 5 stakeholder groups: pharmaceutical industry representatives, patient organization representatives, regulator representatives, EC members, and physicians."

"Potential interviewees were identified by exploring stakeholder websites and existing literature with regard to eIC or upon referral by the research group's network."

"39 interviews were conducted with representatives of the pharmaceutical industry (n = 8), of patient organizations (n = 7), of regulators (n = 5), physicians (n = 6), and EC members (n = 13, of whom n = 4 were also physicians)."

"All interviews were conducted by the same researcher (EDS)."

Results

"A health professional can easily determine from the body language and the eye contact of study participants whether they are really ready to participate or not and whether they still have doubts or questions. (14, EC member) These interviewees believed that the paper-based IC document was the traditional method of providing information to research participants, and due to this, the paper-based form is manageable for the majority of participants."

"They have the feeling that there is one study and when the study is successful, the medication will be available tomorrow. (11, EC member) An eIC platform may also be used to return the results of the research study to the participants."

"Other interviewees indicated that information not related to serious health conditions of the research participants may be shared easily via the eIC platform."

Discussion

"Whereas previous research has mostly focused on investigating potential opportunities of eIC or assessing participants' understanding when using eIC, this study has taken a more precise approach on how an eIC platform can be effectively personalized and how the interaction over time can be guaranteed."

"A small number of interviewees were convinced that eIC would not influence the understanding of research participants compared to that of a paper-based IC."

"A review found that the implementation of digital health technologies, such as eIC, may result in increased personal contact between the research participant and physician."

"Interviewees widely agreed that eIC can support the establishment of long-term interactions between the research team and research participants."

"Interviewees mentioned that an eIC platform can also be used to provide generalized or individual results related to the research study."

"Some interviewees indicated that a personalized eIC platform could empower research participants."

Conclusions

"This qualitative study indicates that stakeholders expect several advantages to a personalized eIC platform that enables long-term contact between researchers and research participants."

"For effective and widespread eIC implementation, several challenges need to be overcome."

"Attention needs to be paid to the preferences of the research participants regarding contact with the research team and the choice between eIC or a paper-based IC."

"Interviewees expressed the wish to establish a harmonized eIC approach across the European Member States."

Rethinking Explicit Consent and Intimate Data: the case of Menstruapps [233]

This is a machine-generated summary of:

Alaattinoğlu, Daniela: Rethinking Explicit Consent and Intimate Data: The Case of Menstruapps [233]

Published in: Feminist Legal Studies (2022)

Link to original: https://doi.org/10.1007/s10691-021-09486-y

If you want to cite the papers, please refer to the original.

For technical reasons we could not place the page where the original quote is coming from.

Abstract-Summary

"This exploratory article brings menstruapps into a feminist legal debate."

"It investigates the supranational European legal standards on intimate and sensitive data processing, particularly the General Data Protection Regulation (GDPR)."

"Scrutinising explicit consent according to GDPR Article 9, this paper, through empirical examples, claims that current legal standards are not enforced."

"Of abandoning the concept, the article reimagines consent, using a contextual and communicative model where power relations are taken into consideration, building on the feminist concept of freedom to negotiate."

Period-tracking as Increased Bodily Control—But by and for Whom?

"The legal ground for intimate and sensitive data processing used by menstruapps is, as a rule, their users' consent."

"With users giving consent for menstruapps to process their personal data, the apps, in turn, often use scientific, empowering and even feminist language to appeal to customers."

"This article bridges the existing disciplinary gap between legal discussions on data privacy and consent, critical analysis of menstrual health tracking and feminist legal theory."

"It originally interrogates the European Union (EU) special legal standard of 'explicit consent' pertaining to processing sensitive and intimate data, a standard which is rarely distinguished from regular consent."

"The article is an example of how feminist theory can be used to develop the current legal standards of ethical data processing."

Menstruapps and Consent

"Consent, according to the definition in the GDPR, is a "freely given, specific, informed and unambiguous indication of the data subject's wishes by which he or she, by a statement or by a clear affirmative action, signifies agreement to the processing of personal data relating to him or her" (Article 4(11), author's emphasis)."

"A freely given consent, according to the EDPB guidelines, should not involve an imbalance of power between the data subject and the data controller (paras 16–24)."

"Valid consent in the GDPR must also be specific, which according to the EDPB means that the data subject consents separately in relation to the particular purposes of the data processing (paras 55–61)."

"The most important ground, and likely the only one applicable for menstruapps, for exceptionally processing such data is, nevertheless, the 'explicit consent' by the data subject [Article 9(2)(a)]."

Formulations of Explicit Consent in Menstruapps

"SimpleInnovation and Amila's menstruapps asked for consent to process personal data when opening the app."

"Data subjects may want to revoke their consent to future data processing (GDPR Article 7(3)), or to delete the personal user data that the apps have already collected (see GDPR Article 17)."

"When it comes to data subjects consenting to share their personal information with third parties, such as advertisers or analytics companies, the menstruapps compared provide different options."

"Users' consent to third-party information sharing was assumed upon agreeing to data processing or, alternatively, the terms and conditions of the menstruapps."

"Even though some of the apps investigated specifically asked for users' consent to personal data processing at the outset, it is also relatively common that such consent is assumed."

Reformulating Explicit Consent Through the Freedom to Negotiate

"The former section showed that there are clear examples when menstruapps fail to live up to standards on consent as determined in EU data protection law If the current legal standards, particularly those concerning explicit consent for sensitive data as special categories of personal data (GDPR Article 9), were enforced by data controllers, it might arguably increase data subjects' awareness, right to privacy and data protection."

"The validity standards of explicit consent concerning sensitive data in EU data protection law are unable to change the terms of the contract entered into by the data subject."

"Some of the app providers give users possibilities to email them when they wish to withdraw their consent, which also empowers data subjects to affect the terms of the contract."

"In a digital context, the communication between data subjects and controllers is, understandably, different and more limited in comparison to, for example, sexual encounters between natural persons, also affecting the freedom to negotiate."

Towards a Feminist Ethics of Data

"This article has theoretically, doctrinally and empirically shed light on the European legal standards pertaining to explicit consent and intimate data and investigated how they apply to menstruapps and how they are formulated in practice."

"There is, conclusively, a need to think critically about the commercial use of intimate data at all points of data processing, not merely when the data subject starts using an application and accepts its privacy policy."

"How can ethical data processing be balanced against monetisation of personal and intimate data?"

"As the lines between our physical and data bodies are blurred, intimate data processing needs normative standards and guidelines for its ethical use."

Alternatives to project-specific consent for access to personal information for health research: Insights from a public dialogue [234]

This is a machine-generated summary of:

Willison, Donald J; Swinton, Marilyn; Schwartz, Lisa; Abelson, Julia; Charles, Cathy; Northrup, David; Cheng, Ji; Thabane, Lehana: Alternatives to project-specific consent for access to personal information for health research: Insights from a public dialogue [234]

Published in: BMC Medical Ethics (2008)

Link to original: https://doi.org/10.1186/1472-6939-9-18

If you want to cite the papers, please refer to the original.

For technical reasons we could not place the page where the original quote is coming from.

Abstract-Summary

"The role of consent for research use of health information is contentious."

"We sought to elicit public values in this matter and to work toward an agreement about a common approach to consent for use of personal information for health research through deliberative public dialogues."

"Immediately before and after each dialogue, participants completed a fixed-response questionnaire rating individuals' support for 3 approaches to consent in the abstract and their consent choices for 5 health research scenarios using personal information."

"Broad opt-in consent for use of personal information garnered the greatest support in the abstract."

"When presented with specific research scenarios, no one approach to consent predominated."

"Personal controls were among the most commonly cited approaches to improving people's confidence in the responsible use of their information for research."

"Because no one approach to consent satisfied even a simple majority of dialogue participants and the importance placed on personal controls, a mechanism should be developed for documenting consent choice for different types of research, including ways for individuals to check who has accessed their medical record for purposes other than clinical care."

"Researchers and policy makers should continue to engage the public to promote greater public understanding of the research process and to look for feasible alternatives to existing approaches to project-specific consent for observational research."

Background

"Public attitudes on the need for and type of consent for research use of their health information are context-specific."

"We surveyed the Canadian public on a spectrum of alternatives to conventional project-specific consent for research use of personal information, including: no use at all, prior individual consent for each use, prior broad authorization for different types of uses, notification with an opportunity to opt-out, and use without consent or notification."

"The majority of the public was open to alternatives to conventional project-specific consent; however, there was no clear preferred approach to consent for use of personal information for health research [73]."

"Recognizing this is a complex and challenging topic, in the spring of 2005, we conducted a series of structured public dialogues across Canada as an alternative method to elicit public values and to work toward an agreement about a common approach to consent for research use of personal information."

Methods

"Participants in two of the dialogues were recruited through an invitation at the end of a public opinion survey of 1230 Canadians that we had conducted on the topic of privacy and access to personal information for health research [73]."

"The workbook presented three general approaches to the role of consent in the use of personal information for health research as a starting point for their discussion: Approach 1: consent for each research use."

"Approach 2: not requiring consent for research use of their information (called "assumed consent" in the dialogues)."

"Approach 3: broad authorization for a range of future research uses, determined by the individual (called "broad consent" during the dialogues)."

"We tabulated participants' attitudes toward the three general consent approaches to using personal information for research."

Results

"Regardless of the approach or the stated ability to opt-out at some future point, participants expressed concern over what might become of the data once released to the researcher. [Approach 1—consent for each use] "I don't think it's workable for the common good."

"That's the override that, in some situations, the researchers could simply—well, I shouldn't say "simply" 'cause they'd have to go through an REB—but they could decide to use information without, without actually having us as individuals give permission for it. (Female, Toronto) These sentiments may have arisen in response to a statement in the workbook explaining that, under Approach 1 (consent for each project), an REB can permit the researcher to use personal health information without individual consent, so long as certain established conditions were met."

Discussion and conclusion

"Dialogue participants were very supportive of their de-identified health information being used for research."

"Over half of dialogue participants (57%) felt their de-identified health information should not be used at all for marketing research (Scenario 2)."

"Given this, as Canada moves toward developing a common inter-operable health record, consideration needs to be given to developing a system for a mechanism for documenting individuals' consent choices for research and other secondary uses of their personal health information, embracing a broader array of consent options than the current dichotomous alternatives of project-specific consent or exemption from consent."

"We noted earlier that dialogue participants expressed concern over what happens to one's personal information once it is released to researchers—regardless of whether initial consent for the release of the information to the researcher was required."

A FHIR has been lit on gICS: facilitating the standardised exchange of informed consent in a large network of university medicine [235]

This is a machine-generated summary of:

Bialke, Martin; Geidel, Lars; Hampf, Christopher; Blumentritt, Arne; Penndorf, Peter; Schuldt, Ronny; Moser, Frank-Michael; Lang, Stefan; Werner, Patrick; Stäubert, Sebastian; Hund, Hauke; Albashiti, Fady; Gührer, Jürgen; Prokosch, Hans-Ulrich; Bahls, Thomas; Hoffmann, Wolfgang: A FHIR has been lit on gICS: facilitating the standardised exchange of informed consent in a large network of university medicine [235]

Published in: BMC Medical Informatics and Decision Making (2022)

Link to original: https://doi.org/10.1186/s12911-022-02081-4

If you want to cite the papers, please refer to the original.

For technical reasons we could not place the page where the original quote is coming from.

Abstract-Summary

"The "COVID-19 Data Exchange Platform" (CODEX) as part of NUM establishes a harmonised infrastructure that supports research use of COVID-19 datasets."

"All 34 participating university hospitals (NUM sites) work upon a harmonised infrastructural as well as legal basis for their data protection-compliant collection and transfer of their research dataset to the central CODEX platform."

"Each NUM site ensures that the exchanged consent information conforms to the already-balloted HL7 FHIR consent profiles and the interoperability concept of the MII Task Force "Consent Implementation" (TFCI)."

"The Independent Trusted Third-Party (TTP) of the University Medicine Greifswald supports data protection-compliant data processing and provides the consent management solutions gICS."

"A "TTP-FHIR Gateway" for the HL7 FHIR-compliant exchange of consent information using gICS was implemented."

"A last step included external integration tests and the development of a preconfigured consent template for the BC for the NUM sites."

"A FHIR-compliant gICS-release and a corresponding consent template for the BC were provided to all NUM sites in June 2021."

"The consent template simplifies the technical BC rollout and the corresponding implementation of the TFCI interoperability concept at the NUM sites."

"This article shows that a HL7 FHIR-compliant and interoperable nationwide exchange of consent information could be built using of the consent management software gICS and the provided TTP-FHIR Gateway."

Background

"The planned processing of a patient's health data at the NUM site and the standardised transfer to the central platform CODEX for research purposes requires, according to the GDPR Article 6 (1) lit."

"The NUM sites ensure that only consented data is transferred to the central research repository CODEX."

"The NUM-specific version of the MII Broad Consent (v.1.6f) is to be used at all NUM sites as a uniform basis for the data protection-compliant collection and transfer of the GECCO83 data set to the central platform CODEX."

"As part of the NUM-CODEX project, the University Medicine Greifswald supports the NUM sites to establish necessary local Trusted Third Parties and corresponding consent processes by providing the necessary software solution gICS for digital consent management."

"In order to be able to implement the clear objective in the NUM-CODEX project for cross-site data exchange via FHIR in the area of consents, current developments in the FHIR community must be considered."

Objectives

"This article describes the necessary steps for the conception and implementation of a standardised solution for the FHIR-compliant provision of consent information using gICS, which can support both existing user projects and future users in the implementation of FHIR-oriented infrastructures and processes."

"at least the FHIR-Consent Profiles as specified by the national HL7 Working Group on Consent Management (WGCM) [75] should be supported, 2."

"additional requirements resulting from the specific setting of NUM should be considered (such as the implementation situation at the sites, planned cross-site processes) 3."

"and in order to simplify the application at the sites of MII and NUM, the interoperability concept of the MII Task Force "Consent Implementation" (TFCI) based on operationalised consent policies [76] should be facilitated."

Methods

"Options for simplification of connecting the sites to the NUM infrastructure were discussed with partners from the University Medicine Erlangen, including queries to the local consent management via TTP-FHIR Gateway."

"The FHIR consent resources delivered by the TTP-FHIR Gateway were consolidated with the partner Heilbronn, as the GTH ensures that only correctly consented GECCO data sets may be transmitted to the central platform."

"This additional information is considered in the subsequent derivation of FHIR consent resources with TTP-FHIR Gateway and ensures technical conformity with the specifications of the TFCI (e.g. specifications for OIDs and permissible answer options) (cf. examples of FHIR consent resources in the Implementation Guide of the WGCM [77])."

"Pre-releases of the new and extended gICS, the TTP-FHIR gateway and the created consent template were handed over early for external evaluation to the NUM-CODEX project partner, the Medical Data Integration Center of clinical centre of the Ludwig Maximilian University of Munich (MeDIC^{LMU})."

Results

"The generic Informed Consent Management Service gICS (version 2.13.0) by default includes the implemented TTP-FHIR Gateway."

"The TTP-FHIR Gateway for gICS provides access to gICS consent content in accordance with the new and officially balloted FHIR consent profiles of the WGCM [75]."

"As soon as consents based on this template are managed in gICS, these consent contents can be exported to FHIR consent resources via the TTP-FHIR Gateway."

"Comprehensive examples of FHIR consent resources generated by the TTP-FHIR Gateway, based on the provided template and including necessary references to the TFCI-concept (as part of FHIR consent.provision) are available from Simplifier [78] accompanied by detailed descriptions within the Implementation Guide [79]."

"The NUM and MII sites were centrally provided with the approved version of gICS with FHIR support (gICS version 2.13.0 with TTP-FHIR Gateway) [80], as well as a TFCI-compliant Broad Consent templates (1.6d [81] and 1.6.f [82]) on June 29th, 2021."

Discussion

"It represents a milestone for the NUM and MII sites to achieve the set goals of exchanging consent content in FHIR format."

"The developed solution (gICS with TTP-FHIR Gateway-Extension) is explicitly not limited to an application in the context of MII and NUM."

"The provided consent template (MII BC 1.6.f) encapsulates the necessary prior technical knowledge to generate FHIR consent resources conforming to the specifications of the TFCI concepts and to consider them consistently and uniformly."

"The provision of gICS, TTP-FHIR Gateway and the consent-template are an offer to the NUM and MII sites."

"It is up to each site to accept this offer and to integrate these tools into local Trusted Third-Party processes and, thus, reduce necessary technical efforts for a HL7 FHIR- and TFCI-compliant consent management."

Conclusions

"In late 2020, the NUM and MII sites have made a binding commitment to set up local infrastructures and establish required local consent processes to enable a data protection-compliant export of GECCO83 data items (based on HL7 FHIR) to the central research platform CODEX to support research on COVID-19 and to demonstrate the necessary degree of preparedness for future pandemic situations in Germany."

"In the first funding phase of NUM-CODEX, a FHIR-based representation of informed consent, based on the MII Broad Consent (MII BC), signed by the patient and automatically enforceable, was to be transferred in addition to the patient's medical data."

"In order to support the 34 NUM sites in this challenge, this article has shown that the technical prerequisites for a FHIR-compliant and interoperable provision of consent information could be achieved with the help of the consent management gICS and the provided TTP-FHIR Gateway [80]."

MAGIC: once upon a time in consent management—a FHIR ® tale [236]

This is a machine-generated summary of:

Bialke, Martin; Bahls, Thomas; Geidel, Lars; Rau, Henriette; Blumentritt, Arne; Pasewald, Sandra; Wolff, Robert; Steinmann, Jonas; Bronsch, Tobias; Bergh, Björn; Tremper, Galina; Lablans, Martin; Ückert, Frank; Lang, Stefan; Idris, Tarik; Hoffmann, Wolfgang: MAGIC: once upon a time in consent management—a FHIR® tale [236]

Published in: Journal of Translational Medicine (2018)

Link to original: https://doi.org/10.1186/s12967-018-1631-3

If you want to cite the papers, please refer to the original.

For technical reasons we could not place the page where the original quote is coming from.

Abstract-Summary

"In the context of multi-centre research initiatives and a multitude of clinical and epidemiological studies scalable and automatable measures for digital consent management are required."

"Modular form, structure, and contents render a patient's consent reusable for varying project settings in order to effectively manage and minimise organisational and technical efforts."

"Within the DFG-funded project "MAGIC" (Grant Number HO 1937/5–1) the digital consent management service tool gICS was enhanced to comply with the recommendations published in the TMF data protection guideline for medical research."

"A structured exchange format for modular consent templates considering established standards and formats in the area of digital informed consent management was designed."

"It represents a starting point for current discussions concerning digital consent management."

"Independent of the exchange format, providing the possibility to export, modify and import templates for consents and withdrawals to be reused in similar clinical and epidemiological studies is an essential precondition for the sustainable operation of digital consent management."

Background

"If the purpose of the data processing is not specifically determined by a legal basis (GDPR Art 6 (3)), data collection for research purposes requires an informed consent (GDPR Art 6 (1) lit."

"Considering more than 35,000 inpatient cases of this hospital per year, a traditional paper-based documentation and management of the patient's informed consent conflicts with the goal to use collected data for a large number and variety of research projects."

"The formulation of participant information and informed consent documents for the study requires the description of the study's individual setting as well as planned data processing procedures."

"Based on consent templates, gICS supports to subsequently start the (B) Digital Documentation and Tracking of informed consents and withdrawals. (C) Application of the informed consent on the participant's data can be automated more fully by using an access control language to describe a consent's implications as standardized in IHE Advanced Patient Privacy Consent (APPC) [84] for a limited domain."

Objectives

"The aim of the MAGIC-partner Institute for Community Medicine of the UMG is to enable research projects with up-to-now purely paper-based consent procedures to use gICS and, thus, to benefit from the advantages of digital consent management."

"GICS will be enhanced to import and export existing consent templates."

"To facilitate these goals, this article describes the design and implementation of a structured exchange format for modular consent templates considering established standards and formats in digital informed consent management."

Methods

"The template also contains complementary information for the consent document (inter alia, the order of modules, the definition of obligatory modules, header, footer and free text fields)."

"Based on published requirements regarding informed consents [85], existing gICS source code and new requirements in current application projects (NAKO, DZHK, BFCC) necessary information for the automatic creation of domains, policies, modules, and templates were compiled to identify relevant information for the new exchange format."

"APPC provides the required mechanisms to support automated enforcement of specific policies regarding the managed research data, according to the patient's consent."

"Due to the limited granularity of FHIR Consents (regarding permissible actions and applicable rules), an automated conversion into IHE APPC for enforcement is currently hard to achieve."

"After evaluating the common profiles, formats and standards, the Consent Definition, the Consent Documentation and Tracking as well as the Consent Application currently cannot be aligned."

Results

"New functionalities were implemented in gICS to support import and export of this FHIR-based exchange format, which can be accessed directly using the provided web-service or the web-based user interface."

"Calling the import mechanism invokes validation of the specified exchange format."

"Configurations for domains, policies, modules, and templates included in this exchange format will be instantly created within gICS or updated, if permissible."

"The right digitalisation strategy is to build consents from a template database and adapt only the specifics of a study, then simply import the result into the existing consent management tools."

"Comparability of the contents of the various study consents is achieved, data quality is improved, and compliance with regulatory requirements such as the GDPR can be assured and audited."

Discussion

"Of the short presentation at the GMDS, the first version of this exchange format was introduced to the community at the periodically arranged "Forum for Interoperability of HL7 Germany, IHE Germany, German Association of Health IT Vendors (bvitg e.V.) and German Institute for Standardisation (DIN) (Forum for Interoperability)" [86]."

"An important part of the discussion with representatives of HL7 FHIR was the practical concept and benefit of using modular informed consents in cohort studies and registries."

"In a second meeting of the Forum for Interoperability, representatives of HL7, IHE, FHIR, selected MI-I consortia (MIRACUM [87], SMITH [88], HiGHmed [89]) and industry showed intent to harmonise existing standards and profiles (HL7, FHIR, IHE APPC) with regards to consent management."

"The presently separated consideration of these three aspects in existing tools (e.g. PEW, gICS) and profiles (e.g. IHE APPC, HL7 FHIR Consent) should be resolved."

Conclusions

"The format had to take common standards in the area of digital informed consent management into account."

"The technical harmonisation of HL7 FHIR (to document informed consents) and IHE APPC (for application and enforcement of the informed consent based on respective rules) shall be ensured in the long term."

"Results of the working group presumably will affect existing specifications and profiles of HL7 FHIR and IHE APPC or result in a new FHIR resource type and/or IHE profile with respect to the three use cases of consent management (as described above)."

"Sooner or later, managed informed consents and withdrawals within gICS shall be provided for simplified access using an FHIR-based REST-interface."

"Using existing tools to efficiently implement the management of informed consents and withdrawals, whether paper-based or purely digital, in research projects and healthcare can be accomplished."

Facing new challenges to informed consent processes in the context of translational research: the case in CARPEM consortium [237]

This is a machine-generated summary of:

Jacquier, Elise; Laurent-Puig, Pierre; Badoual, Cécile; Burgun, Anita; Mamzer, Marie-France: Facing new challenges to informed consent processes in the context of translational research: the case in CARPEM consortium [237]

Published in: BMC Medical Ethics (2021)

Link to original: https://doi.org/10.1186/s12910-021-00592-9

If you want to cite the papers, please refer to the original.

For technical reasons we could not place the page where the original quote is coming from.

Abstract-Summary

"In the context of translational research, researchers have increasingly been using biological samples and data in fundamental research phases."

"To explore informed consent practices, we conducted a retrospective study on informed consent documents that were used for CARPEM's translational research programs."

"This review focused on detailing their form, their informational content, and the adequacy of these documents with the international ethical principles and participants' rights."

"Informed consent forms (ICFs) were collected from CARPEM investigators."

"A content analysis focused on information related to biological samples and data treatment (context of sampling and collect, aims, reuse, consent renewal), including the type of consent."

"Most of the biological samples were collected during the healthcare routine, but the information content about secondary use of biological samples varied between ICFs."

"All documents mentioned personal data treatment but information about their reuse was not standardized in the ICFs."

"Theoretically, a digital tool allowing dynamic information and consent of participants, through an electronic interactive platform may be a good way to promote more active participation in research."

"Its feasibility in the complex environment of biological samples and data research remains to prove."

Background

"The French law provides also that informing someone and obtaining their consent to participate in research or use biological samples or data for research purposes has become a legal obligation [92–94]."

"The context of CARPEM provides the opportunity to analyse the practices related to information and consent in the translational research approach, from the time of care to the "phase 0" of collecting and using biological samples and data [95]."

"In the light of these considerations and of ethical perspectives of transparency, respect for individuals and scientific expectations, we aimed to assess the characteristics of various informed consent forms used by CARPEM researchers at the moment they collect data and biological samples for their research purposes by reviewing the current IC process of CARPEM: what kind of information current IC forms provide to research participants and how could they be improved?"

Methods

"Collected informed consent forms (ICFs), including information documents and consent forms, were studied to evaluate their readability using a specific and validated IT solution (the Flesch Score; see below) and to analyse information content delivered about the collection and the use of biological samples and data for research purposes [96]."

"After distinguishing the types of research (non-interventional or interventional), we focused on pieces of information about the collection and the use of biological samples and data for scientific purposes which were recorded in the documents."

"The following specified information about the collection of biological samples was taken from the documents: the context of the collection (care or research), storage place, place of use, the period of storage, the purpose of use, whether reuse was considered, and whether the initial consent was planned to be renewed."

Results

"Information and consent documents from 32 of 49 studies were gathered."

"Four studies required that a research participant signed two consent forms: one for the study and one specifically for genetic research."

"In all, 29 informed consent forms were analyzed, related to 25 studies."

"Opt-out consent forms were analyzed as information documents."

"Most of the biological samples used in the studies came from healthcare (16 of 25)."

"In one study, an additional sample had to be collected specifically for the research purpose if no healthcare tumor sample was available."

"The three identified types of consent were used (broad, specific, either broad or specific) whatever the type of the study (either for non-interventional and interventional ones)."

"7 of the 25 documents mentioned that data could be reused for future studies (four BC and three BSC)."

Discussion

"The lack of information about reuse shows that the scope of the consent remains unclear: according to our results, participants can't know where or to whom their biological samples or their data would be transferred in the future."

"In that case, ongoing informed consent would preserve participants' control over their data and maintain confidence in the scientific community which uses more and more personal data for research [97–99]."

"The main issue seems to be more the information content than the consent itself: qualitative studies highlighted the interest for regular information about secondary use of biological samples and personal data [98]."

"In that sense, people are not annoyed to consent only once, but they would prefer to be informed regularly about the use of their biological samples and personal data if they are looking for it [100–102]."

"Our study showed that information was too complex to expect that most participants to biological samples and data research understood it."

Conclusion

"Our survey highlights that the current translational research approach may favor the emergence of new ethical issues concerning informed consent practices."

"Such practices may be accompanied by a new informed consent process to respect the ethical right of the individuals to decide whether researchers can use their material for secondary uses."

"Given these changes in health research practices, ethical standards updated by GDPR enforcement in Europe may be a way to reconceptualize and reinvent the practice of informed consent, especially because secondary uses cannot be anticipated at the time of consent."

"IT tools could be useful for both researchers and individuals who consent to the use of their biological samples and personal data for scientific purposes."

Revisiting informed consent in forensic genomics in light of current technologies and the times [238]

This is a machine-generated summary of:

Budowle, Bruce; Sajantila, Antti: Revisiting informed consent in forensic genomics in light of current technologies and the times [238]

Published in: International Journal of Legal Medicine (2023)

Link to original: https://doi.org/10.1007/s00414-023-02,947-w

If you want to cite the papers, please refer to the original.

For technical reasons we could not place the page where the original quote is coming from.

Abstract-Summary

"While informed consent has focused on human subjects and research, the practice has been adopted willingly in the forensic science arena primarily to acquire reference samples from family members to assist in identifying missing persons."

"Given the state-of-the-art, areas in which informed consent may need to be modified and augmented are as follows: reference samples from family members in missing persons or unidentified human remains cases; targeted analysis of an individual(s) during forensic genetic genealogy cases to reduce an investigative burden; donors who provide their samples for validation studies (to include population studies and entry into databases that would be applied to forensic statistical calculations) to support implementation of procedures and operations of the forensic laboratory; family members that may contribute samples or obtain genetic information from a molecular autopsy; and use of medical and other acquired samples that could be informative for identification purposes."

"The informed consent process should cover (1) purpose for collection of samples; (2) process to analyze the samples (to include type of data); (3) benefits (to donor, target, family, community, etc. as applicable); (4) risks (to donor, target, family, community, etc. as applicable); (5) access to data/reports by the donor; (6) sample disposition; (7) removal of data process (i.e., expungement); (8) process to ask questions/assessment of comprehension; (9) follow-up processes; and (10) voluntary, signed, and dated consent."

"Addressing informed consent will allow human subjects to make decisions voluntarily and with autonomy as well as secure the use of samples for intended use."

Introduction

"There are at least four forensic scenarios in which informed consent may need to be modified and augmented based on these new capabilities: (1) reference samples from family members in missing persons or unidentified human remains cases; (2) targeted analysis of an individual(s) during forensic genetic genealogy (FGG) cases to reduce an investigative burden; (3) donors who provide their samples for validation studies (to include population studies and entry into databases that would be applied to forensic statistical calculations) to support implementation of procedures and operations of the forensic laboratory; and (4) family members that may contribute samples or obtain genetic information from a molecular autopsy [103]."

"The risks to privacy are greater (especially with the ease of genetic data analyses and that a molecular autopsy would point directly to a gene(s) and specific genetic variants) and should be included in the informed consent documentation with a greater need by medicolegal personnel to ensure the information is comprehended by sample donors and potentially impacted family members."

Conclusion

"Our goal is to raise awareness and facilitate discussion on these important and fundamental aspects of personal data, the role of informed consent, and the need to update the process to be concomitant with current technological capabilities and the times."

"While an in-depth informed consent may be perceived as cumbersome, the benefits in the short and long term are individuals will be protected and allowed to make sound decisions (i.e., autonomy) and the use of powerful technologies to investigate crime and identify missing persons and human remains will be sustained."

"Within the European Economic Area, the entering into force of the GDPR saw a distinct move away from the privileged status of informed consent as the basis for processing personal data."

"We all must hold ourselves accountable to the principles of informed consent and actions or consequences that may ensue with the samples and data obtained."

Controversies between regulations of research ethics and protection of personal data: informed consent at a cross-road [47]

This is a machine-generated summary of:

Gefenas, Eugenijus; Lekstutiene, J.; Lukaseviciene, V.; Hartlev, M.; Mourby, M.; Cathaoir, K.Ó: Controversies between regulations of research ethics and protection of personal data: informed consent at a cross-road [47]

Published in: Medicine, Health Care and Philosophy (2021)

Link to original: https://doi.org/10.1007/s11019-021-10060-1

If you want to cite the papers, please refer to the original.

For technical reasons we could not place the page where the original quote is coming from.

Abstract-Summary

"This paper explores some key discrepancies between two sets of normative requirements applicable to the research use of personal data and human biological materials: (a) the data protection regime which follows the application of the European Union General Data Protection Regulation (GDPR), and (b) the Declaration of Helsinki, CIOMS guidelines and other research ethics regulations."

"One source of this controversy is that the GDPR requires consent to process personal data to be clear, concise, specific and granular, freely given and revocable and therefore has challenged the concept of 'broad consent', which has been widely applied in the context of biobanking."

"Another source of controversy is the interplay between regulations of research ethics and protection of personal data related to the secondary use of personal data and biological materials."

"The GDPR 'research condition' provides an alternative to re-consent for the use of previously collected personal data and biological materials."

"Should consent be regarded as a priority legal basis for personal data processing in health data research?"

Consent misconception and other consequences of a 'dual' meaning of consent

"Before the entry into force of the GDPR, consent to participation in medical research was widely perceived as integrating both 'interventional consent' as well as 'informational consent' to process personal data."

"In contrast to the research ethics guidelines, the GDPR does not give consent any predefined priority for health data processing."

"Although there were alternative grounds for processing personal data for research purposes under the previous Directive 95/46 EC, the GDPR (which is a legally binding regulation directly applicable in all the EU) has made the alternatives to re-consent more explicit and therefore presented a "research-friendly approach" (Shabani and Borry 2018)."

"This also makes research without explicit consent, such as earlier mentioned 'research condition' under the Article 9(2)(j), a much less exceptional scenario for data processing as compared to what is still recommended in the current research ethics guidelines."

Diverging interpretations of consent

"A much more relaxed interpretation is provided by Recital 29 and Article 28 (2) of the Clinical Trial Regulation, which is another EU Regulation enabling universities and other research institutions to collect data from clinical trials of medicinal products to be used for future scientific research, for example for medical, natural or social sciences research purposes, if the subject gives consent to use his or her data "outside the protocol of the clinical trial"."

"By introducing the concept of "processing for purposes of general interest", the DGA may seem to also allow in certain circumstances the processing of personal data for not strictly defined research purposes which serve the general interest (e.g., scientific research) and therefore challenges the restrictive position of the Article 29 Working Party on broad consent referred to earlier."

Waivers of consent and the research condition

"To use GDPR specific terminology, in these cases we refer to the already mentioned 'research condition' scenario, which allows secondary data processing without explicit consent when this is necessary for the scientific research purposes."

"According to the EDPB, scientific research may not require a legal basis envisaged in Article 6 of the GDPR and therefore seems to allow the re-use of data for scientific research without an identified lawful basis, such as consent or even public interest [104]."

"Some countries take a more cautious approach and require taking public interest as a legal basis for data processing in health research under the 'research condition' scenario, which allows secondary data processing without explicit consent."

Implications for researchers, research institutions, RECs and DPOs

"The questions posed at the end of the previous section demonstrate the need for closer cooperation between RECs and data protection bodies when scrutinizing the ethical and legal justification of research studies using personal data and biological materials."

"This separation can easily lead to a situation where personal data assessment is being assigned to only data controllers (e.g., researcher and/or research institutions) assisted by legal units and DPOs acting at the institution, while leaving it to RECs to review studies and related informed consent issues dealing exclusively with the interventional aspects of research."

"The involvement of RECs in the assessment of research related data processing seems to be very relevant due to the need to assess the importance and social value of research, the benefit and risk ratio, as well as distress to the subjects, which are issues typically examined during the review procedure by RECs and elaborated in detail by different research ethics guidelines, such as the Declaration of Helsinki and CIOMS Guidelines specifically focusing on all the mentioned aspects of research."

Concluding remarks

"Considering that some European countries encourage the use of the 'research condition' for secondary research use of health data based on the legal basis of public interest, it is important to ensure an interaction between RECs and data protection bodies and a better integrated framework for processing of personal data, where research ethics regulations operate along with the GDPR."

"While there are still ongoing discussions on what would be the best option for data sharing: the use of dynamic consent, broad consent or data donation without consent, it seems that public preferences for a certain level of control over their data, regardless of the lawful basis relied upon under the GDPR, support a continued role for consent to maintain trust in research, shows respect to study participants, and is a default necessity under the research ethics guidelines."

Context-Relative Norms Determine the Appropriate Type of Consent in Clinical Biobanks: Towards a Potential Solution for the Discrepancy between the General Data Protection Regulation and the European Data Protection Board on Requirements for Consent [239]

This is a machine-generated summary of:

Indrakusuma, R.; Kalkman, S.; Koelemay, M. J. W.; Balm, R.; Willems, D. L.: Context-Relative Norms Determine the Appropriate Type of Consent in Clinical Biobanks: Towards a Potential Solution for the Discrepancy between the General

Data Protection Regulation and the European Data Protection Board on Requirements for Consent [239]

Published in: Science and Engineering Ethics (2020)
Link to original: https://doi.org/10.1007/s11948-020-00271-9

If you want to cite the papers, please refer to the original.

For technical reasons we could not place the page where the original quote is coming from.

Abstract-Summary

"Clinical biobanks processing data of participants in the European Union (EU) fall under the scope of the General Data Protection Regulation (GDPR), which among others includes requirements for consent."

"This discrepancy elicits the question whether clinical biobanks can fail the requirement of consent if they obtain broad consent, but not a specific consent for each biomedical study."

"We analysed this discrepancy within the framework of contextual integrity, in order to describe the context-relative informational norms that govern information flows in clinical biobanks."

"Our analysis demonstrates that there is no uniform set of norms that can be applied to all clinical biobanks."

"Neither the GDPR nor the WP29 guidance can act as a "one size fits all" approach to all clinical biobanks."

Introduction

"Explicit consent is used as the legal basis, which will be the case for most current clinical biobanks, it must meet certain requirements, which have been described in the GDPR itself and have been further specified in guidelines of the Article 29 Working Party (WP29) [105]."

"They observed that the draft WP29 guidelines "could seriously restrict the availability of broad consent to future research as a basis for processing personal data under the GDPR" and that these draft guidelines are confusing with regard to Recital

33 as "the guidelines could therefore be read to suggest that providing data subjects a description of the general areas of future research [...] might not satisfy the GDPR's requirement that consent be specific [106]."

Contextual Integrity

"These norms are referred to as context-relative informational norms."

"Contextual integrity is concerned with whether the sharing of information is appropriate for the context in which it is shared taking into account the context-relative informational norms."

"If sharing of information breaches a context-relative informational norm, this sharing then constitutes a violation of contextual integrity and therefore privacy."

"Before an information practice can be said to violate privacy, it is necessary to understand the context-relative informational norms, by describing four parameters that constitute a context-relative informational norm."

Context-relative Informational Norms for Clinical Biobanks

"Within the context of clinical biobanks, participating patients are the majority of information subjects, although data can also be collected about physicians and their care, or about relatives when genetic data is collected."

"This category of secondary recipients appears to be empirically supported, for example by a survey of 308 biobank participants (recruited via a broad consent model) in the USA, who expressed that sharing their biomaterials and data with other domestic researchers was "extremely acceptable [107]."

"Although not a biobank in name, the DIRECT study collected biomaterials and associated data from participants, set up a research database for future use and used a broad consent model and is therefore comparable to clinical biobanks."

"A relatively new, yet widely supported norm is that clinical biobanks should allow data sharing in order to achieve its original aim: to facilitate biomedical research in the pursuit of medical progress."

From Uniformity to Diversity of Norms

"Different types of clinical biobanks require different sets of context-relative informational norms."

"The underlying rationale for such an approach is based on the diversity of the contexts and contextual goals (i.e. broad scientific aims of each clinical biobank), the type of information subjects (i.e. type of patients) involved and the type of information to be collected."

"A recommendation would be to tailor at least these two norms to the type of clinical biobank."

"In the case of a clinical biobank about rare diseases, norms for broad consent and international data sharing may be justified to overcome the scientific challenge of small sample sizes."

"The role of a medical-ethical and/or biobank committee would be to assess whether sufficient ethical justification exists for the proposed norms and whether they are appropriate for the specific context."

Conclusion

"Although clinical biobanks are united by the overarching goal of enabling future biomedical research, there is no uniform set of context-relative informational norms that could govern information flows for all clinical biobanks."

"The discussion needs to progress beyond choosing exclusively for either the GDPR or the WP29 guidelines."

"The diversity among clinical biobanks, specifically with regard to the contextual goals and patient populations, warrants context-relative norms that inform the appropriate type of consent."

An E-Consent framework for tiered informed consent for human genomic research in the global south, implemented as a REDCap template [240]

This is a machine-generated summary of:

Tamuhla, Tsaone; Tiffin, Nicki; Allie, Taryn: An e-consent framework for tiered informed consent for human genomic research in the global south, implemented as a REDCap template [240]

Published in: BMC Medical Ethics (2022)

Link to original: https://doi.org/10.1186/s12910-022-00860-2

If you want to cite the papers, please refer to the original.

For technical reasons we could not place the page where the original quote is coming from.

Abstract-Summary

"Research involving human participants requires their consent, and it is common practice to capture consent information on paper and store those hard copies, presenting issues such as long-term storage requirements, inefficient retrieval of consent forms for reference or future use, and the potential for transcription errors when transcribing captured informed consent."

"There have been calls to move to electronic capture of the consent provided by research participants (e-consent) as a way of addressing these issues."

"The branching logic feature streamlines the consent process by making follow-up information available depending on participant response, the 'wet signature' feature enables a timestamped electronic signature to be appended to the e-consent documents and the 'auto-archiver' allows for PDF copies of the e-consent documents to be stored in the database."

"Building informed consent protocols and forms without prior experience can be daunting, so we have provided researchers with a REDCap template that can be directly incorporated into REDCap databases."

"It prompts researchers about the types of consent they can request for genomics studies and assists them with suggestions for the language they might use for participant information and consent questions."

"The use of this tiered e-consent module can ensure the accurate and efficient electronic capture and storage of the consents given by participants in a format that can be easily queried and can thus facilitate ethical and effective onward sharing of data and samples whilst upholding individual participant preferences."

Background

"For tiered informed consent where participants answer a variety of questions about allowed data or specimen use, paper-based consents are inefficient and impractical for determining whose data or which specific data elements have consent for onward sharing, meta-analyses or sharing in aggregated form [113]."

"The framework is designed to improve the informed consent process for both participants and researchers involved in human genomic research, firstly by providing a comprehensive list of information for researchers to include in the consent documents, thus providing a tool which they can use as a check list to ensure that all essential information is available; and secondly providing researchers with ready-to-use, downloadable template consent documents which have been written in straightforward genomic research language that is easier for participants to understand."

"The REDCap-based tiered e-consent module presented here facilitates the electronic capture of participants' consent choices so no additional data entry is required and errors are kept to a minimum."

Construction and content

"All the online tiered e-consent documents can be downloaded and submitted as PDFs to the institutional review board (IRB) and if required, a link can be set-up to give the IRB access to consent surveys online, on the REDCap platform."

"The tiered e-consent module has three data capture instruments documents namely, the main consent and withdrawal of consent, which are both surveys, and an optional study meta data form."

"The content of the main consent was adapted from the tiered informed consent framework of Nembaware and others [115] with some modifications, most notable of which was the addition of consent for the use of participant genomic data in population and ancestry studies which was excluded from that framework."

"When designing the content layout of the main consent document in the tiered e-consent framework, emphasis was placed on the flow of information to optimise participant understanding."

Utility of the platform

"One of the objectives of this tiered e-consent framework was to improve participant understanding of human genomics research so that they could make truly informed consent."

"While e-consent modules are commonly designed to be participant self-administered [116–120], the main consent document in the tiered e-consent framework will be administered face-to-face by a trained member of the research team."

"The 'Add/Edit records' function under the data collection page on the e-consent framework is used to initiate the consent process and launch the consent documents as surveys."

"The PID is central to data retrieval, because once assigned it is linked to all data collection documents in the e-consent module for each participant."

"To ensure data security and privacy, the consent documents in the tiered e-consent framework are strictly for collecting consent information and do not collect sensitive participant information such as demographic data or contact details."

Discussion

"The use of this e-consent module can thus facilitate efficient and ethical data- and sample-sharing, whilst respecting the specific preferences and choices of each participant."

"Whilst this REDCap template utilises a digital approach to presenting and capturing the informed consent process, which comes with the described advantages such as improved data fidelity and streamlined databasing of participant choices, the fundamental process and material content of tiered informed consent remains consistent with current paper-based tiered informed consent processes for health genomics research [115, 121]."

"Other important inputs include training researchers in the use of the digital informed consent process to ensure high quality data collection as well as to ensure participants understand how the digital tool is being used."

Variant data sharing by clinical laboratories through public databases: consent, privacy and further contact for research policies [241]

This is a machine-generated summary of:

Shabani, Mahsa; Dyke, Stephanie O. M.; Marelli, Luca; Borry, Pascal: Variant data sharing by clinical laboratories through public databases: consent, privacy and further contact for research policies [241]

Published in: Genetics in Medicine (2018)

Link to original: https://doi.org/10.1038/s41436-018-0316-x

If you want to cite the papers, please refer to the original.

For technical reasons we could not place the page where the original quote is coming from.

Abstract-Summary

"Clinical laboratories can play a significant role in sharing patients' data through public variant databases."

"Despite clear benefits to sharing, questions may arise about the adequate form of consent to be obtained from patients when sharing data from their clinical tests through public databases."

"We provide an overview and critical analysis of the relevant consent policies of the major public databases, and of the consent forms of clinical laboratories that share variant data via ClinVar."

Introduction

"To facilitate access to variant interpretation data, the role of clinical laboratories in sharing patients' data through public variant databases has been recognized by professional communities."

"[122] In a recent statement, the American College of Medical Genetics and Genomics stressed the importance of clinical and laboratory data sharing to improve health care: "Responsible sharing of genomic variant and phenotype data will provide the robust information necessary to improve clinical care and empower device and drug manufacturers that are developing tests and treatments for patients.""

"[123] Similarly, a survey study with genetic counselors in the United States has concluded: "Genetic counselors have the ability to support laboratories who publically share their variant data by choosing to order their patients' testing from those laboratories.""

"[124] To facilitate data sharing, various public databases that accept data submission from laboratories, clinicians, researchers, and patients, have been established."

Consent for data sharing

"The first approach includes explicitly informing patients about sharing their data with other clinical laboratories and/or through public variant databases such as ClinVar, and providing an opt-out option."

"Only ARUP followed this approach and has prepared a Genetic Testing and Data Sharing document, which provides information about data sharing to patients and allows them to opt out of the sharing."

"The second approach is informing the patients about data sharing but not providing an option for opting out."

"Five laboratories, such as Partners HealthCare and GeneDX, inform patients about sharing de-identified data with other laboratories or through public databases such as ClinVar, but do not provide an option for opting out."

"In the fourth approach, in the consent forms of laboratories such as Athena Diagnostics, data sharing is not mentioned directly; only the use of specimens, clinical information and data in a de-identified way for research, educational studies, commercial purposes."

Privacy and the risk of re-identification

"According to the ClinVar guidance, the risks of re-identification of de-identified variant-level data are very low, owing to the fact that full genomic datasets are not shared and multiple variants from the same individual are not linked together in the database."

"The consent forms of clinical laboratories inform patients that only their de-identified data will be shared."

"The GeneDX consent form is one example, where the patients are informed that although the risk of re-identification is generally low, the risk is greater if they have already shared their genetic or health information with public resources, such as genealogy websites."

"Some laboratories also assure patients that data will be shared by federal Health Information Privacy and Accountability Act of 1996 (HIPAA)-compliant databases, or explicitly refer to ClinVar as an example of public databases that data will be shared with."

Further contact with the patients for research purposes

"Sharing variant data through public databases may result in a request for contact with the patient or access to further information about the patient by other laboratories, researchers, or clinicians."

"DECIPHER's "an introduction for families" leaflet includes the possibility of requests for sharing further information to share with other clinicians who are interested in similar clinical cases: "If your clinician is contacted by another clinician who through DECIPHER has identified other individuals with the same/similar genomic variant and the same/similar clinical features, you may be contacted and asked whether you wish to give permission for further details to be exchanged with a view to furthering understanding of this genomic variant.''"

"The consent forms of six laboratories, such as ARUP, inform patients that they may be contacted later on for participation in research, and that their identifying information will be revealed to interested researchers/clinicians only after they explicitly agree."

Are the current policies towards consent, policy, and further contacts adequate?

"Public variant databases strive to facilitate variant data sharing by hosting data submissions through online platforms, and providing guidance on how to adequately address issues related to consent, privacy, and secondary uses of data."

"Although public variant databases provide general guidance regarding consent and privacy, questions remain about the most adequate policies regarding opt-out options, the amount of information to be communicated to the patients regarding data sharing, and the adopted privacy-preserving methods."

"Although implicit or opt-out options for consent are adopted for variant-level sharing, the current guidance provided by variant databases stresses the importance of obtaining explicit consent once more detailed information is to be shared publicly (tiered data sharing), as both the risk of re-identification and the probable use of the data beyond the immediate clinical applications are increased."

Monitoring advances including consent: learning from COVID-19 trials and other trials running in UKCRC registered clinical trials units during the pandemic [242]

This is a machine-generated summary of:

Love, Sharon B.; Armstrong, Emma; Bayliss, Carrie; Boulter, Melanie; Fox, Lisa; Grumett, Joanne; Rafferty, Patricia; Temesi, Barbara; Wills, Krista; Corkhill, Andrea: Monitoring advances including consent: learning from COVID-19 trials and other trials running in UKCRC registered clinical trials units during the pandemic [242]

Published in: Trials (2021)

Link to original: https://doi.org/10.1186/s13063-021-05225-5

If you want to cite the papers, please refer to the original.

For technical reasons we could not place the page where the original quote is coming from.

Abstract-Summary

"The COVID-19 pandemic has affected how clinical trials are managed, both within existing portfolios and for the rapidly developed COVID-19 trials."

"Sponsors or delegated organisations responsible for monitoring trials have needed to consider and implement alternative ways of working due to the national infection risk necessitating restricted movement of staff and public, reduced clinical staff resource as research staff moved to clinical areas, and amended working arrangements for sponsor and sponsor delegates as staff moved to working from home."

"Organisations have often worked in isolation to fast track mitigations required for the conduct of clinical trials during the pandemic; this paper describes many of the learnings from a group of monitoring leads based in United Kingdom Clinical Research Collaboration (UKCRC) Clinical Trials Unit (CTUs) within the UK."

"The UKCRC Monitoring Task and Finish Group, comprising monitoring leads from 9 CTUs, met repeatedly to identify how COVID-19 had affected clinical trial monitoring."

"Monitoring is defined as involving on-site monitoring, central monitoring or/and remote monitoring."

"Monitoring, required to protect the safety of the patients and the integrity of the trial and ensure the protocol is followed, is often best done by a combination of central, remote and on-site monitoring."

"If on-site monitoring is not possible, workable solutions can be found using only central or central and remote monitoring."

"The arrival of COVID-19 in the UK has forced consideration of and changes to how clinical trials are conducted in relation to monitoring."

Background

"The type of monitoring conducted by CTUs for a particular trial is determined by a risk assessment [125] and summarised in a trial monitoring plan."

"Rather than monitoring routinely throughout the trial, the monitoring is directed at pre-defined risks to the trial, and to risks which become apparent during the active phase."

"As other areas of trial conduct have had to adapt to the pandemic, so has monitoring."

"We report the experience of the United Kingdom Clinical Research Collaboration (UKCRC) Monitoring Task and Finish Group in undertaking monitoring both within COVID-19 trials and non-COVID-19 trials being conducted during the pandemic."

"Some of the learning is specific to COVID-19 trials, or to trials of similar infection/mode of transmission, some to non-COVID-19 trials operating during a pandemic and some ways of working could be used as routine post the COVID-19 pandemic."

Consent

"In an infectious disease setting, the patient may have the capacity to consent but there are practical limitations to capturing written evidence of this."

"A hard copy of the patient information sheet and consent form can enter the ward, but cannot leave the ward once signed, due to the risk of infection transmission on paper documents."

"There are established procedures that can be borrowed from entering adults lacking the capacity to consent [126] and minors [127] onto a clinical trial without consent."

Monitoring without on-site visit option

"The risk assessment and/or monitoring plan may need to be adapted with the acceptance of some risks which would under usual circumstances be mitigated by on-site monitoring review."

"Direct access to site electronic health record (EHR) provided to the monitor away from the site enables monitoring with little site input."

"This method is highly effective and does not cause an increased burden for trial sites as monitors can access the data and perform review without site staff input, other than the site providing access and for addressing critical issues at the end of the review."

"Some trial sites are considering this option where it is not possible to provide the monitor with direct access to the electronic health record."

"These visits will not be possible, alternative actions should be outlined such as telephone contact with site staff to assess compliance/issues and/or other types of remote monitoring activities as detailed above."

Monitoring without on-site or remote monitoring option

"The focus should be on patient safety, primary outcome measure data and any other data considered critical."

"An example from one CTU was to focus on data relating to patient safety."

"Data relating to adverse events and protocol deviations were collated and reviewed at monthly Trial Management Group (TMG) meetings where the frequency and type of events as well as the quality of the data submitted were considered."

"A Data Monitoring Committee (DMC) was also convened monthly with the aim of independently reviewing safety data."

Minimising data

"The collection of non-essential data has long been known to increase the burden on both participating sites and sponsors, and this is now starting to be quantified."

"If the data collected can be reduced to that which is considered critical, this will reduce the burden on sites for CRF completion and sponsors for monitoring."

"The reduction in data collection also has wide-reaching benefits for the CTU including simplified and more rapid CRF and database development, more efficient data management and statistical analysis and, most pertinent to this review, fewer data points for which monitoring is required."

"If the COVID-19 pandemic can bring one benefit to clinical research in the long-term, minimising data collection would be a wide-reaching, mutually beneficial process improvement."

Altered data collection and retention

"Limiting paper contamination within COVID-19 areas, as well as clinical trial burden due to the reduced staff resource at participating sites, has required other trial management adaptations."

"Examples from COVID-19 trials have included the completion of validated patient-reported outcome measures (PROMs) tools via the phone rather than patients completing the forms directly."

"This could be done by research staff at sites, or by CTU staff if patients have given consent for this."

"COVID-19 trials also presented additional consideration when discussing where and how patients were identified and confirmed as COVID-19 positive."

Site setup changes

"In the RECOVERY-RS trial, a large site-setup team was established, and SIVs were conducted every weekday at three time slots per day."

"This was cost- and time-efficient for the trial team and for site staff and enabled multiple sites to receive a SIV simultaneously."

"It also enabled multiple site staff to attend different SIV slots, from numerous locations, fitting in with workload and availability."

"Risk-proportionate approaches should always be used to ensure oversight of adequate experience and training of site staff, and the current situation can provide an opportunity to tailor the oversight more specifically to each individual trial."

"Equally, a hierarchy of study-specific training requirements may be implemented to request a core set of study training to be completed and signed off by the PI only, with PI oversight to confirm site staff have appropriate training to perform their role within the trial."

Speed of trial development

"Many CTUs redirected resource to COVID-19 trial development and staff have worked at pace, contributing many hours in a short time, collaborating with many individuals and organisations (pharma, laboratories, clinicians, government) to pull together and fast track trial ideas into solid trial proposals, creating protocols in just a few weeks."

"Due to the speed of setup of COVID-19 studies, monitors have been required to fast track trial risk assessments and the subsequent trial monitoring plans, including considerations of mitigations due to site access restrictions and minimised site resource requirements."

"Accelerated protocol development and trial setup processes have required a substantial resource allocation, which in most cases has been delivered by experienced staff working evenings and weekends for prolonged periods."

"Once approved, the trials have required frequent amendment to account for the changing pandemic landscape, emerging safety information and differing processes across healthcare settings."

Impact on Non-COVID-19 portfolio

"These actions have required changes to central monitoring of data, for example a metric alarm threshold for deviations would need to be increased."

"As sponsors and CTUs plan for trial restart, they should consider COVID-19 impact trial viability assessments which address if/how the trial population, interventions and outcomes have been impacted to inform if trials are able to re-open or not, and whether any changes are required."

"Trial opening and change decisions may require trial risk assessments and the data monitoring plans to be updated."

"Sites need to commit to the necessary monitoring, and if the site cannot, then consideration may need to be given to keeping the site on hold to recruitment; this capacity review will often form part of the site-level risk assessment, which are likely to be required prior to re-opening [128]."

Discussion

"With the arrival of COVID-19 in the UK and the need for COVID-19 trials to be rapidly created and open to recruitment, monitoring aspects of trials had to develop and adapt at speed."

"As electronic patient record systems evolve, if sites could ensure that their electronic patient record systems are able to offer remote access to trial staff, restricted to specific trial participants, monitoring would be enabled."

"Some efficiency improvements could continue; examples include efficient document creation using learnings from COVID-19 trials (e.g. consent options), minimising data collection, development of database library documents and better understanding of SIV options."

"We hope it helps others setting up COVID-19 or managing non-COVID trials during the pandemic and gives some material for consideration of changes to the conduct of clinical trials more broadly."

Informed Consent—suggested procedures for informed consent for unrelated haematopoietic stem cell donors at various stages of recruitment, donor evaluation, and donor workup [243]

This is a machine-generated summary of:

Rosenmayr, A; Hartwell, L; Egeland, T: Informed Consent—suggested procedures for informed consent for unrelated haematopoietic stem cell donors at various stages of recruitment, donor evaluation, and donor workup [243]

Published in: Bone Marrow Transplantation (2003)
Link to original: https://doi.org/10.1038/sj.bmt.1703851

If you want to cite the papers, please refer to the original.

For technical reasons we could not place the page where the original quote is coming from.

Abstract-Summary

"This paper considers the topic of informed donor consent, but recognises that the recommendations contained within the paper may be subject to cultural variances in interpretation, and to adjustment to meet the legal requirements of individual countries."

Main

"This paper contains recommendations and requirements for the appropriate information as well as the mode of transmission to be given to volunteer haematopoietic stem cell donors at all stages, from initial recruitment, through final selection for donation, to post-donation."

"These stages are: 1 The public domain 2 Initial recruitment on to a registry of volunteer donors 3 Additional histocompatibility testing 4 Confirmatory (final stage) histocompatibility testing 5 Haematopoietic stem cell donation 6 Post donation The volunteer must be given literature pertinent to each of the above stages at the appropriate time."

1. **The public domain**

"Haematopoietic stem cell donation is: Undertaken voluntarily, and withdrawal of the volunteer is permitted at all times Not remunerated but incurred expenses are met. (This will be documented within the policy of the responsible Registry and may include loss of earnings.) Guaranteed to be anonymous for both patient and donor Health restricted including a requirement that the individual volunteer must not be at high risk of transmittable diseases Age restricted."

"The minimum age for a volunteer must be determined by the legal requirement covering general anaesthesia and blood donation in the country in which the donor resides."

"A haematopoietic stem cell donor may be asked to donate peripheral blood cells (whole blood, lymphocytes or platelets) for the treatment of relapse or other clinical conditions—usually for a patient who has already undergone a haematopoietic stem cell transplantation and predominantly to whom the volunteer has already donated (patient-directed donation)."

2. **Initial recruitment onto the registry of volunteer donors**

"The volunteer must be provided with information on the principles, general procedures, restrictions and risks of providing blood samples and haematopoietic stem cells (either via marrow or peripheral blood donation)."

"A simple, comprehensible consent form must be signed by the volunteer, which confirms that he/she understands and agrees to the following: Join the Registry as a potential haematopoietic stem cell donor and provide relevant samples for HLA typing and IDM testing."

"The volunteer must understand that certain diseases in a donor preclude haematopoietic stem cell donation."

"The volunteer must be advised that blood samples or buccal swabs will be used primarily for histocompatibility testing, blood grouping and IDM testing relevant for the registry file, donor search processes, and haematopoietic stem cell donation and transplantation, and the volunteer must sign a written consent for these purposes."

3. **Confirmatory (final stage) histocompatibility testing**

"Final stage histocompatibility testing is undertaken by the transplant centre (or associated HLA laboratory) to validate the histocompatibility type of a volunteer donor."

"An information session between volunteer and a qualified representative of the donor centre must take place."

"At no time must the volunteer be told that he/she is the only match for a patient."

"Final stage confirmatory testing has determined the volunteer's likely selection as a donor."

"A medical examination is necessary to assess the volunteer's fitness for the donation procedure."

"A further medical questionnaire must be completed by the volunteer at the time of final stage confirmatory testing."

"The updated medical information should identify any volunteers whose current medical condition now precludes them from donating, in the best interests of either the volunteer or the patient."

4. **Haematopoietic Stem Cell Donation**

"It is the responsibility of the donor centre to ensure that volunteers selected to talk to potential donors in this way have a balanced view of the experience and that ample time is provided for the donor to consider the donation."

"The advocate must be an independent third party who is not a member of the Registry, harvest, donor or transplant centre; the advocate must be conversant with haematopoietic stem cell donation."

"The volunteer must be informed/reminded of the following: He/she is considered a match for a particular patient and is now requested to donate haematopoietic stem cells."

"The volunteer has discussed, with a suitably qualified person, the collection procedures permissible, and understands the related risks. (Including anaesthesia, transfusion reaction, infection, bone, nerve or other tissue damage at the needle insertion site and the possibility of short- and long-term adverse effects of haematopoietic growth factors.) 4.3.2."

5. **Post Donation**

"For those volunteers still eligible and wishing to remain on the donor Registry, the following must be considered."

"The volunteer must be reminded that he/she may be approached for a further donation (haematopoietic stem cell or other cell therapy) for the same patient and, where possible, given an indication of any timescale(s)."

"The volunteer should be asked to consider seriously their willingness, in principle, to be approached for further donation for the same patient."

"Following haematopoietic stem cell donation, the volunteer must be asked of his/her willingness to remain on the Registry with a view to being approached for further donations to other patients. (WMDA guidelines or Registry policy will determine the frequency with which a donor may be approached again)."

Elements of informed consent for pharmacogenetic research; perspective of the pharmacogenetics working group [244]

This is a machine-generated summary of:

Anderson, DC; Gomez-Mancilla, B; Spear, BB; Barnes, DM; Cheeseman, K; Shaw, PM; Friedman, J; McCarthy, A; Brazell, C; Ray, SC; McHale, D; Hashimoto, L; Sandbrink, R; Watson, ML; Salerno, RA; Cohen, N; Lister, CE: Elements of informed consent for pharmacogenetic research; perspective of the pharmacogenetics working group [244]

Published in: The Pharmacogenomics Journal (2002)

Link to original: https://doi.org/10.1038/sj.tpj.6500131

If you want to cite the papers, please refer to the original.

For technical reasons we could not place the page where the original quote is coming from.

Introduction

"The explosion of new human genetic data and an increasing awareness that genetic variation influences drug responsiveness has signaled a new era of pharmacogenetic research."

"Pharmacogenetic investigations should provide important advances in the design, development and delivery of safe and efficacious pharmaceuticals, including opportunities to provide customized drugs for specific patient populations as defined by their unique genotypes."

"Pharmacogenetic research in drug development involves special considerations and disclosures in the informed consent process."

"References [129–133] The Pharmacogenetics Working Group (PWG) is a voluntary association of pharmaceutical companies involved in clinical drug trials and genotyping whose goal is to advance the understanding and development of pharmacogenetics by addressing non-competitive ethical, regulatory, and legal issues."

"The major goals of this communication are to assist researchers, IRBs/IECs and regulatory agencies in better understanding issues specific to pharmaceutical company-sponsored pharmacogenetic research so that they can best assure protection of subjects, while at the same time facilitating timely review, approval and implementation of pharmacogenetic trials."

Informed consent process for pharmacogenetic investigations

"Informed consent can be defined as both the document used to obtain consent as well as the process utilized to communicate the intended disclosures and to ensure accountabilities for the consequences of obtaining consent."

"When appropriate to enable clear communication, consent materials should be given to subjects prior to the research visit so there is sufficient time for review."

"In order to facilitate understanding of a pharmacogenetic study, some sponsors have created educational pamphlets and videos as an adjunct to the consent form to provide additional information to assist the subject in making decisions about participation."

"It is noteworthy that some background information a subject may have received prior to being asked to participate in a pharmacogenetic trial is likely to have come from media or a historical perspective suggesting (to the subject) that a pharmacogenetic trial involves cloning, reproductive choices or risks for serious monogenic disorders."

Key elements for disclosure and consent in pharmacogenetic studies

"Comparable issues are present with respect to unintended results in non-genetic studies as with the pharmacogenetic results described above, ie, the quality and veracity of individual research results and need for appropriate clinical context for result sharing vs the need for and benefit of disclosure of information to the subject."

"An issue generally perceived to be of major relevance for clinical genetic research is the possibility of informational risks associated with intentional or inadvertent disclosure of genetic data to third parties (or to study subjects themselves)."

"These procedures should provide well thought-out plans for the handling of subject's withdrawal of specimen, the use of database firewalls, internal stringent data access control, consent for secondary use of samples in additional genetic studies, plans for archiving or destruction of pharmacogenetic samples, and possible options, if any, for future contact with study subjects to discuss genetic results or participation in future studies."

Commercialization

"In the IC process associated with pharmacogenetic studies, the prospective subject should be informed that the contribution of his/her pharmacogenetic sample might result in commercial gains or intellectual property for the sponsoring pharmaceutical company (who should be named) or other designated parties."

"Disclosing commercial objectives in the IC process for pharmacogenetic research is a transparent and honest approach and allows study subjects to make an informed choice as to whether to contribute genetic materials."

"The process of informed consent for pharmacogenetic research should distinguish the issues of compensation associated with the reimbursement of expenses incurred by subjects who participate in a study from those issues of financial compensation that would be an inducement for subjects to participate in pharmacogenetic research."

"As for investigational drug trials, it is generally considered unacceptable to provide financial compensation to induce research subjects to participate in pharmacogenetic studies."

Conclusions

"Some important or unique considerations for pharmacogenetic trials include: unfamiliar terminology; study purpose that may be difficult for subjects to understand; perceived informational risks that may result in discrimination or psychological distress; potential societal benefits such as a better understanding of the underlying causes of variable drug response and discoveries of safer and more effective drugs; possible extended durations of genetic studies that far exceed the subject's participation (and that genetic research can continue indefinitely); potential commercial benefit that the sponsor may derive; and the issue of compensating study subjects."

"Only by assuring an appropriate IC process and document can subjects fully understand the implications of a decision to participate in pharmacogenetic research."

"The 'points to consider' discussed in this communication are intended to highlight elements of pharmacogenetic research encountered by pharmaceutical companies that are of key relevance to the IC process and a subject's participation."

International web-based survey of informed consent procedures in genetic epidemiological studies: towards the establishment of a research coordinator accreditation system [245]

This is a machine-generated summary of:

Yamamoto, Midori; Funamizu, Makiko; Muto, Kaori; Hata, Akira: International web-based survey of informed consent procedures in genetic epidemiological studies: towards the establishment of a research coordinator accreditation system [245]

Published in: Journal of Human Genetics (2009)

Link to original: https://doi.org/10.1038/jhg.2009.103

If you want to cite the papers, please refer to the original.

For technical reasons we could not place the page where the original quote is coming from.

Abstract-Summary

"We conducted an international web-based survey to obtain information about the informed consent (IC) procedures used in genetic epidemiological studies of multifactorial diseases."

"Although most support personnel had documented and/or verbal confidentiality agreements, and all had undertaken IC training, accreditation examinations were only conducted in 60% of studies."

"On the basis of the survey results obtained, support personnel dedicated to completing the IC procedures are reported to be beneficial and indispensable in alleviating the burden on medical doctors, helping participants to make autonomous decisions and promoting genome research."

"The establishment of a training program and accreditation system for such personnel is warranted, especially in Japan, where medical staff usually finds it difficult to conduct IC procedures in clinical settings due to heavy workloads."

Introduction

"Results of genomic analyses must be verified using an independent human sample set and, therefore, collection of human specimens from a large number of volunteers in medical facilities and/or communities is indispensable in conducting this kind of study."

"It is rational to suggest that support personnel specializing in IC are necessary for the collection of large numbers of human specimens in accordance with ethical guidelines."

"In Japan, it is noted by the Japanese Ministry in the 'Ethics Guidelines for Human Genome/Gene Analysis Research' [134] (revised in 2005) that the principal investigator can have support personnel perform the necessary practice regarding IC, and should provide opportunities for on-the-job training to the support personnel."

"Rather than carrying out a simple international comparison of IC procedures, the purpose of this survey was to gather information about the use of support personnel to establish a new qualification system for genome medical research coordinators in Japan."

Materials and methods

"Respondents were selected from studies that aimed at genetic analyses of multifactorial diseases and which were published in a medical journal with a considerably high impact factor (>5.0) from January 2005 to January 2008."

"The invitation to participate was sent by e-mail and responders accessed the questionnaire form by connecting to the designated URL."

"A wide range of diseases was represented in the 53 studies, including cardiovascular diseases, malignant tumors, immune abnormalities, neurological diseases, ophthalmological diseases, skeletal diseases and skin diseases."

"Additional questionnaires regarding the personnel who carried out IC procedures were sent to 20 respondents by e-mail, of whom nine responded."

Results

"In all, 40 of 53 respondents reported that doctors or nurses directly recruited participants for their studies."

"The support personnel included the research coordinator, study recruiter, interviewer, research assistant, and student or professionals other than medical doctors, nurses, genetic counselors or PhD doctorates."

"It was suggested that well-trained coordinators have a better understanding of IC procedures and study projects, and hence they can answer to the participant's questions more thoroughly than medical doctors, nurses and genetic counselors."

"In 45 of the 53 responses, the peripheral blood was collected from study subjects regardless of personnel qualifications of those in charge of obtaining IC."

Discussion

"This international survey was conducted to comprehend the actual state of IC procedures in other countries so as to provide basic information for the establishment of training programs and accreditation systems for research coordinators."

"Well-trained support personnel not only reduce the workload of hospital medical staff, but also understand privacy policies, IC procedures and research projects better than most medical staff, and thus are better able to answer to participant's questions and handle their anxiety and lack of understanding."

"An accreditation system for highly qualified research coordinators could also be helpful in gaining the cooperation of hospital staff and study subjects."

"We gathered sufficient information regarding the recruitment of subjects performed in other countries, IC procedures, and the collection of specimen and personal data during recent clinical genome research on multifactorial diseases."

Comparison of participant information and informed consent forms of five European studies in genetic isolated populations [246]

This is a machine-generated summary of:

Mascalzoni, Deborah; Janssens, A Cecile JW; Stewart, Alison; Pramstaller, Peter; Gyllensten, Ulf; Rudan, Igor; van Duijn, Cornelia M; Wilson, James F; Campbell, Harry; Quillan, Ruth Mc: Comparison of participant information and informed consent forms of five European studies in genetic isolated populations [246]

Published in: European Journal of Human Genetics (2009)

Link to original: https://doi.org/10.1038/ejhg.2009.155

If you want to cite the papers, please refer to the original.

For technical reasons we could not place the page where the original quote is coming from.

Abstract-Summary

"In common with all studies in humans, those in genetically isolated populations need ethical approval; however, existing ethical frameworks may be inadequate to protect participant privacy and confidentiality and to address participants' information needs in such populations."

"Using the ethical–legal guidelines of the Council for International Organizations of Medical Sciences (CIOMS) as a template, we compared the participant information leaflets and consent forms of studies in five European genetically isolated populations to identify additional information that should be incorporated into information leaflets and consent forms to guarantee satisfactorily informed consent."

"We highlight the additional information that participants require on the research purpose and the reasons why their population was chosen; on the potential risks and benefits of participation; on the opportunities for benefit sharing; on privacy; on the withdrawal of consent and on the disclosure of genetic data."

Introduction

"Family-based genetic research in isolated populations is an effective methodology for identifying loci influencing variation in disease traits."

"These issues apply partly to every kind of genetic study, but although these issues are also applicable to family-based studies performed in genetically isolated communities, the special characteristics of such communities provide additional challenges to ethical approaches designed for large population-based biobanks."

"As in other genetic studies, family-based studies conducted in genetically isolated populations can potentially uncover information about a family's health that may have wider implications for individual members of that family, even if the individuals concerned were non-participants."

"The aim of this paper is to identify specific issues that should be addressed and incorporated into participant information leaflets and consent forms for family-based studies in genetically isolated populations."

"We compared the participant information and informed consent forms of five European studies in genetically isolated populations to identify what additionally needed to be addressed."

Methods

"The Erasmus Rucphen Family study [135] (ERF; the Netherlands) is a family-based study that investigates the genetic origins of complex diseases in a genetically isolated community in the south-west of the Netherlands."

"MICROisolates in South Tirol (MICROS; Italy) [136] is a family-based population study that investigates the genetic origins of complex diseases."

"The Orkney Complex Disease Study (ORCADES; UK) is a family-based study that seeks to identify the genetic factors influencing cardiovascular and other disease risks in the population isolate of Orkney, an archipelago off the north coast of Scotland."

"The Northern Sweden Population Health Study (NSPHS; Sweden) is a family-based population study that aims to identify genetic and environmental risk factors for common, mainly non-communicable disease in populations living in the arctic and sub-arctic areas."

"About 60% of those invited (740 individuals) participated in the study and underwent a comprehensive health examination and sampling for serum, plasma and DNA analyses."

Results

"References [137, 138] The MICROS information leaflet states that although no commercial benefits are anticipated, any that are forthcoming will be reinvested in research and health facilities in the region but that there will be no individual benefits to participants."

"Article 11 of CIOMS G specifies that participants should be informed about the expected benefits of the research to the community or society in general or to scientific knowledge."

"The only information leaflet to mention the risk of identification of participants in publications is that of the ORCADES study and this states that individuals will not be identifiable in any way in research publications; however, this may be a problematic point as it is hard to guarantee this 100% in studies of this kind."

"Article 16 of the CIOMS G states that researchers must inform participants about their policies regarding the disclosure of personal and familial genetic data to third parties."

Discussion

"References [139–141] Second, most topics that were not addressed in the information leaflets may have been communicated in other ways by the studies, for example, in information meetings and brochures and in face-to-face discussions between individual participants and research staff."

"The implications for and rights of non-participants in this type of study are dimensions that are missing in current research guidelines, but it is essential that researchers working in family-based genetic studies in isolated populations engage with the wider community so that non-participants are also aware of the possible implications of the study and given an opportunity to make their voices heard."

"It is important to reiterate that although we have concentrated fairly narrowly here on the information given to participants through the information and consent stage of the project, many of the issues raised here were addressed by all our projects at other stages, starting from discussions with research ethics committees at the project inception stage, through to public meetings and one-to-one communications with individual participants later on."

Informed Consent in Forensic DNA Databases: Volunteering, Constructions of Risk and Identity Categorization [247]

This is a machine-generated summary of:

Machado, Helena; Silva, Susana: Informed Consent in Forensic DNA Databases: Volunteering, Constructions of Risk and Identity Categorization [247]

Published in: BioSocieties (2009)

Link to original: https://doi.org/10.1017/s1745855209990329

If you want to cite the papers, please refer to the original.

For technical reasons we could not place the page where the original quote is coming from.

Abstract-Summary

"We aim to expand the traditional ethical issues related to the use and storage of genetic information in forensic DNA databases by exploring the Portuguese practices of informed consent in the context of DNA sample collection for forensic processing."

"We argue that the Portuguese forensic DNA database is an interesting case study to re-focus the ethical debate on the implications of collecting genetic information, elucidating how this activity reconfigures interpersonal relations and social hierarchy, the power of legal medicine and criminal justice, and the social construction of personal autonomy, privacy and individual rights."

DNA sample collection for the Portuguese forensic DNA database

"Some implications of these legal provisions will be explored in this article, with the aim of raising a number of ethical issues regarding practices of informed consent in the context of the Portuguese forensic DNA database."

"According to 146, the DNA profiles obtained from the analysis of samples, and the corresponding personal data, are only integrated into the database with the free, written and informed consent of volunteers, relatives of missing persons or forensic technicians who collect and analyse DNA samples (no)."

"We will critically examine some ethical issues related to informed consent in the context of collecting biological samples and DNA profiling in the Portuguese national forensic DNA database."

Consent forms for DNA sample collection

"The Portuguese forensic DNA database contains profiles from volunteers (i.e. non-criminals who want to give samples on their own initiative) which can be cross-referenced with profiles from criminal suspects and convicted individuals."

"This means that DNA samples are 'given' by volunteers to the Portuguese forensic DNA database (as is the case in the context of medical DNA databases), but 'taken' from criminal suspects or convicted offenders."

"There is no specific informed consent form to be signed by forensic technicians who collect and analyse DNA samples, although the law mentions the need for free, written and informed consent in such cases."

"In the case of volunteers and relatives of missing persons, the consent forms include the following statement: 'I declare that I give my free and informed consent to collection of biological samples and to obtaining my DNA profile, as well as to insertion, communication and interconnection, according to the 146 of 12 February.'"

Volunteers for DNA profile inclusion in the forensic database

"We examine some specific concerns regarding informed consent which are raised by the written request provided by volunteers to apply for his/her DNA profile to be included in the database."

"Regarding profile entry criteria, 146 establishes that DNA profiles obtained from the analysis of samples, and the corresponding personal data, are only integrated into the database with the free, written and informed consent of volunteers, relatives of missing persons and forensic technicians who collect and analyse DNA samples (no)."

"Forensic technicians are required to provide their DNA profile in order to be admitted to work in the national forensic DNA database, but volunteers and relatives of missing persons, on their own initiative, must send a written sample collection request to the Laboratory of the Scientific Police and the National Institute of Legal Medicine (no)."

The information sheet

"It was considered necessary to produce an information sheet that would protect and respect fundamental ethical values and citizens' individual rights."

"Take the Portuguese information sheet, for example: first, although individuals from whom DNA samples are collected might have different needs, values and expectations, the provision of information is based on general principles; second, there is an underlying assumption that the provision of information is an essential way of legitimizing the collection of human biological material."

"How do we make the provision of information compatible with individual requirements, values and expectations?"

"How do we provide information that takes into account, in the foreseeable future, uncertain and potential uses of personal data and individual genetic profiles incorporated into the forensic DNA database?"

"This is especially important because genetic information may have far-reaching repercussions, not only for the individual but also for the individual's immediate and extended family, and because uncertainties about DNA profiling and the uses and applications of this information still exist (147)."

General principles of information and communication of uncertainties

"Some probable risks related to possible laboratory errors and to the specific nature of genetic information are omitted or at least clearly reduced."

"The Portuguese forensic DNA database information sheet does not contain any safeguard against the fact that a DNA genetic profile could also be used to establish biological kinship relations."

"Some uncertainties about the DNA profile entry and the uses and applications of the information obtained through the collection of biological samples still exist and, in the future, genetic information could be used not only to identify people, but also to prevent some kind of 'danger' (148)."

"The inherently complex relationships and concerns that emerge in the context of forensic genetic testing and profiling, including those regarding privacy and discrimination, suggest a need for extra requirements in the informed consent process, that should exceed the requirements of informed consent to undergo traditional medical tests (149)."

Conclusion

"Along with the socio-political concern with the identity of populations, the inclusion of the ethnic group category in the personal data obtained from individuals in the context of the Portuguese forensic DNA database forms an alignment of science and state bureaucracy that could have significant consequences (150) for the management, labelling and surveillance of citizens by the state."

"The practice of informed consent in the context of the Portuguese forensic DNA database might thus be an interesting case study to re-focus the ethical debate on the implications of collecting genetic information, elucidating how this activity reconfigures interpersonal relations and the social hierarchization of individuals, as well as the power of legal medicine and criminal justice, and the cultural and social constraints in the framing of personal autonomy, privacy and individual rights."

[Section 7]

"We aim to expand the traditional ethical issues related to the use and storage of genetic information in forensic DNA databases by exploring the emerging Portuguese practices of informed consent in the context of DNA sample collection for forensic processing."

"This discussion re-focuses the ethical debate on the consent process undergone during DNA sample donation and collection for forensic purposes, elucidating how this activity both transforms and reproduces interpersonal relations and social hierarchization of individuals, the power of legal medicine and criminal justice, and the cultural and social constraints in the framing of the concepts of personal autonomy, privacy and individual rights."

"Our empirical approach will be to analyse three different documents which are included in the legislation for the Portuguese forensic DNA database: the mandatory written request provided by volunteers or relatives of missing persons to request that his/her DNA profile be included in the Portuguese forensic DNA database; the information sheet to be shown to individuals before they sign the informed consent form; and the personal data for identification of donors included in the consent form."

Can a 'consent to contact' community help research teams overcome barriers to recruitment? The development and impact of the 'Research for the Future' community [248]

This is a machine-generated summary of:

Grady, Katherine; Gibson, Martin; Bower, Peter: Can a 'consent to contact' community help research teams overcome barriers to recruitment? The development and impact of the 'Research for the Future' community [248]

Published in: BMC Medical Research Methodology (2019)

Link to original: https://doi.org/10.1186/s12874-019-0843-4

If you want to cite the papers, please refer to the original.

For technical reasons we could not place the page where the original quote is coming from.

Abstract-Summary

"Innovation is required to meet policy commitments to help patients take part in health research."

"One innovation that may help meet those policy goals is the development of 'consent to contact' systems, where people give generic consent to be contacted about research opportunities."

"Despite their potential, there are few empirical assessments of different ways of recruiting patients to such communities, or of the value of such communities to local research teams."

"We describe the development of the 'Research for the Future' consent to contact community, outline the recruitment of patients to the community, and present data on their participation in research."

"The 'Research for the Future' consent to contact community has proven feasible and useful for local research teams."

"Further evaluation is needed to assess the cost-effectiveness of different recruitment strategies, explore patient and researcher experience of its advantages and disadvantages, and explore how the community can be more reflective of the wider population."

Background

"The NHS White paper set out a proposal to ensure that patients are made aware of research studies relevant to them [147], and the NHS Constitution outlined guiding principles for the promotion, conduct and use of research to improve population health [148]."

"One innovation that could have major implications for recruitment to health research is the use of 'consent to contact' models."

"Using information in medical records to find suitable patients may offer a solution to recruitment difficulties—a view supported by NHS England mandate 2016/17 which recommended better use of technology to support research, innovation and growth [149]."

"One way of doing this is to develop registers of patients who have given prior consent, who can be approached by researchers with appropriate studies."

"was recently identified as one of the top 20 research questions relating to how best to improve the process of recruiting people to randomised trials [150]."

"The aim is to make a contribution to the developing evidence base around consent to contact models and their role in health research."

Main text

"Each 'Help BEAT' campaign invites people with a particular health condition to register their details and research interests as part of a research community."

"We aimed to describe the 'Research for the Future' consent to contact community, how patients are recruited into the community, and the effectiveness of different methods of recruitment."

"The objectives of the pilot were: To set up a database so that people who register their interest via the Help BEAT Diabetes campaign can be contacted directly about specific studies they were eligible to take part in Develop a communications strategy to maintain volunteer engagement and interest in the campaign through the development of a website and promotional materials Demonstrate that the contact centre and database can be used to recruit volunteers to diabetes portfolio studies."

"5587 people are registered on the Research for the Future community, and 3981 (71%) live in the target area (corresponding to the NIHR CRN: Greater Manchester footprint)."

Conclusions

"Our consent to contact community has registered over 5000 patients in three disease areas, and has facilitated the recruitment of over 1500 patients to a variety of research studies."

"If there was sufficient interest in the potential of systems such as Research for the Future, formal research could provide more accurate data on these issues, as well as exploring the experience of patients and professionals through detailed qualitative research."

"Some important data of interest to the development of consent to contact communities (such as the proportions of patients who engage with research in any time period) cannot be gathered through our systems at present."

"Engaging with patients in the community may be crucial to ensuring ongoing registration, both in terms of providing plenty of opportunities for research, but also providing other benefits such as information on health and research more generally, as well as opportunities to engage with the wider community."

Moral autonomy of patients and legal barriers to a possible duty of health related data sharing [249]

This is a machine-generated summary of:

Vedder, Anton; Spajić, Daniela: Moral autonomy of patients and legal barriers to a possible duty of health related data sharing [249]

Published in: Ethics and Information Technology (2023)

Link to original: https://doi.org/10.1007/s10676-023-09697-8

If you want to cite the papers, please refer to the original.

For technical reasons we could not place the page where the original quote is coming from.

Abstract-Summary

"Informed consent bears significant relevance as a legal basis for the processing of personal data and health data in the current privacy, data protection and confidentiality legislations."

"This paper discusses the compatibility of a moral duty to share data for the sake of the improvement of healthcare, research, and public health with autonomy in the field of data protection, privacy and medical confidentiality."

"It explores several ethical-theoretical justifications for a duty of data sharing, and then reflects on how existing privacy, data protection, and confidentiality legislations could obstruct such a duty."

"Consent, as currently defined in the General Data Protection Regulation—a key legislative framework providing rules on the processing of personal data and data concerning health—and in the recommendation of the Council of Europe on the protection of health-related data—explored here as soft-law—turns out not to be indispensable from various ethical perspectives, while the requirement of consent in the General Data Protection Regulation and the recommendation nonetheless curtails the full potential of a duty to share medical data."

Introduction

"In current privacy, data protection and confidentiality legislation, (informed) consent by the individuals involved plays an important role as one of the legitimising grounds for processing personal data, including personal health-related data."

"We will discuss several possible ethical-theoretical justifications for a moral duty of personal health related data sharing."

"We will indicate which forms the implementations of the different justifications might take, and identify the impediments that existing relevant and applicable privacy, data protection and confidentiality laws might raise against these."

"We will pay special attention to the consent requirement with its latent normative autonomy theory in the background, but we will also discuss other legal bases as possible alternatives for the grounds for lawful processing of personal data than consent."

"We are first and foremost interested in whether and in what ways the existing legal framework in the field of privacy, data protection, and confidentiality impedes the implementation of such duties."

A data sharing duty and personal autonomy

"Questions concerning moral responsibilities or moral duties of patients to transfer data concerning health can be expected to arise much more often in the near future."

"Various ethical theories (in particular, deontological, utilitarian, communitarian ones, and ethics of care) may deliver different approaches to the justification of a moral duty of patients to transfer their health data."

"These professionals will act either on the basis of a direct duty in its own rights—e.g. a duty based on the expected professional benefit from data transfers such as improving the accuracy of research—or on the basis of a derivative duty, i.e. one that is logically building on the patients' responsibilities."

"Rather, we will ask: how would a moral duty of patients to transfer (health) data for the benefit of health care, research, and public health improvement in the health sector sit within the currently existing confidentiality, privacy, and data protection legislations?"

Transferring data

"In the context of all of these eHealth services, the processing of health data relies on the sharing of information and thus implies the exchange of data between different actors (e.g. patients, physicians) or organisations (e.g. research institutes, governmental authorities)."

"The data collected may then, in a subsequent step, be transferred further to other stakeholders for the purpose of either the patients' own treatment, research and/or public health (or a combination of these)."

"Patients or volunteering citizens who wish to partake in scientific research may make their health data available for the investigation of diseases, treatments or drugs by providing access to their medical records for the purpose of medical research and public health."

"Researchers may exchange the patients' data with third parties (e.g. government agencies) to facilitate the re-use of the data for the purpose of medical research, public health or the improvement of healthcare overall."

[Section 4]

"We may differentiate between two forms in which a duty of patients to share data could exist: one being a basic duty and the other being a supererogatory duty."

"The latter can set higher standards for a moral duty to share data where individuals impose an obligation based on their personal moral ideals, but supererogation comes with restrictions."

"The different moral outlooks referred to in the previous section may require different implementations of an alleged duty to transfer health data."

"The utilitarian approach does not so much rely on consent as a legal basis for health data transfers but rather requires clearly defined legal grounds that justify the transfer of health data for the benefit of society."

"A moral outlook other than the utilitarian requires voluntariness as a pre-requisite for an action, a moral duty of patients to share data may persist in form of a supererogatory duty instead of a basic duty, and vice versa."

The current legal context

"Both frameworks introduce provisions on the processing of health data (data concerning health as per article 4(15), recital 35 General Data Protection Regulation and health-related data as defined per article 3 Council of Europe recommendation, respectively), thereby governing the transmission of health data."

"This provision covers the further sharing of health-related data with recipients who are authorized by law The recommendation does not clarify the terms 'exchange', 'data sharing' or 'communication', but, similarly to the General Data Protection Regulation, introduces a non-exhaustive list of examples in article 3 which indicates that the exchange, data sharing or the communication of data can be considered as a data processing activity, such as the disclosure or making available of data."

"The General Data Protection Regulation foresees a ban with permit reservation allowing only the exceptional processing of data concerning health (articles 9(2–4), 6 General Data Protection Regulation), while the Council of Europe recommendation legitimises the processing of health-related data only if one of the legitimate bases set down in article 5 apply."

Is the current legal framework an impediment to the duty to transfer?

"The use and re-use of health-related data, as we have attempted to describe above, can take place in different forms, but often requires researchers to obtain approval from patients and research participants for the processing of their health data for research for other purposes than the original one."

"Recital 33 of the General Data Protection Regulation acknowledges that it may not always be possible to clearly identify the purpose at the time of data collection and, therefore, preserves some flexibility regarding the specification of consent, allowing participants to agree to have their data processed for certain areas of scientific research."

"This approach possibly leads to a tension between the consent requirement and the need for a more flexible further use of health-related data that was originally collected on the basis of consent."

[Section 7]

"The current predominant consent conception pervading the Regulation on the processing of personal data is reaching its limits when it comes to facilitating health related data sharing on the basis of an individual's agreement."

"The complexity of data sharing operations create a lack of control for individuals that the current conceptualization of consent cannot sufficiently address when it comes to the re-/use of the patients' health data for the improvement of healthcare, research, and public health."

"Despite the existing legal impediments for the sharing of health data, various accounts of a moral duty to transfer health data present some advantages."

"All of these ethical outlooks suggest that the concept of autonomy is not strictly tied to the subjective will of the patient or research participant, and can provide arguments for data sharing based on altruistic reasons beyond what the current legal framework foresees."

What GDPR and the Health Research Regulations (HRRs) mean for Ireland: a research perspective [250]

This is a machine-generated summary of:

Mee, Blanaid; Kirwan, Mary; Clarke, Niamh; Tanaka, Aoife; Manaloto, Lino; Halpin, Emma; Gibbons, Una; Cullen, Ann; McGarrigle, Sarah; Connolly, Elisabeth M; Bennett, Kathleen; Gaffney, Eoin; Flanagan, Ciaran; Tier, Laura; Flavin, Richard; McElvaney, Noel G.: What GDPR and the Health Research Regulations (HRRs) mean for Ireland: a research perspective [250]

Published in: Irish Journal of Medical Science (2020)

If you want to cite the papers, please refer to the original.

For technical reasons we could not place the page where the original quote is coming from.

Abstract-Summary

"Irish Health Research Regulations (HRRs) were introduced following the European Union (EU) General Data Protection Regulation (GDPR) in 2018."

"The HRRs described specific supplementary regulatory requirements for research regarding governance, processes and procedure that impact on several facets of research."

"The numerous problems that the HRRs and particularly "explicit consent" inadvertently created were presented under the auspices of the Irish Academy of Medical Sciences (IAMS) on November 25, 2019, at the Royal College of Surgeons in Ireland."

"The objective of this review was to obtain feedback and to examine the impact of GDPR and the HRRs on health research in Ireland in order to determine whether the preliminary feedback, presented at the IAMS meetings, was reflected at a national level."

"Individuals from the research community were invited to provide feedback on the impact, if any, of the HRRs on health research."

"Our results confirm that the HRRs have had a significantly negative impact on health research in Ireland."

Introduction

"One of the requirements of the HRRs is that identified or identifiable personal data cannot be included in health research unless (a) explicit consent exists or (b) a consent declaration has been granted [19–25]."

"Presented were preliminary findings that the HRRs had precipitated [151] (a) huge site to site variation, because institutions had interpreted and applied the legislation differently; (b) serious concerns regarding Ireland's ability to attract clinical trials now and into the future, due to site to site variability; (c) a go-slow in research, with some studies halted altogether due to (i) protracted decision-making on the part of institutions, (ii) immense outlay in both time and funds, required to implement GDPR and the HRRs, and (iii) difficulties navigating a new regulatory system for research."

"An example of retrospective patient recruitment and consent for health research, outside a hospital setting, is included."

Methods

"Individuals were invited to provide information via email on any of the following: (a) general understanding of the HRRs, (b) time to obtain Research Ethics Committee (REC) approval, (c) RECs shutting down to become compliant, (d) RECs reluctance to accept external templates due to lack of nationally approved templates, e.g. patient information leaflets, (e) time outlay related to Health Research Consent Declaration Committee (HRCDC) applications and (f) Data Protection Officer (DPO) guidance."

"This study sought to obtain consent retrospectively from previous breast cancer patients diagnosed at St. James's Hospital between 2004 and 2014 to take part in a Breast Predict study and also consent to their inclusion in the St James's Hospital Histopathology Biobank (SJHHB)."

"The Breast Predict Consent Form and Information Leaflet were modelled on the SJHHB documentation which underwent robust review, namely [152]: i. Review by department of Legal and Insurance."

Results and discussion

"The inclusion of GPs (general practitioners) as gatekeeper to contacting patients or their next of kin was important in attempting to establish whether clinical changes had occurred in the interim, which might affect the participants' capacity to consent."

"It is not clear whether a written record of telephone conversations, where respondents clarified their wishes, would have satisfied the requirements of "explicit consent". (c) The prospect of recontacting patients or their next of kin, for a second time, raised concerns."

"Towards the end of the research nurse's contract, it became unclear whether the retrospective consent was GDPR and HRR compliant and even were funds available, whether it would be ethical to return to patients and next of kin again."

"Participant consent has long existed as an ethical and legal requirement for the conduct of health research in Ireland [153, 154]."

Conclusion

"The introduction of legislation and standards to afford greater protection to health research participants will always be welcome."

"It is important to state that the intention of the legislators was not, in the authors' opinion, to impede health research."

"Rather GDPR presented a mechanism to regulate research and simultaneously afford greater autonomy to health research participants."

"The authors seek the regulation of consent in health research to a high standard."

Recommendations

"Several proposed amendments were presented at both IAMS events in an attempt to rectify some of the problems created by the HRRs."

"These amendments have not been enacted, and there has been no attempt by the Department of Health to set up a consultative process with IAMS to discuss these proposals."

"Meaningful dialogue and consultation between legislators and key stakeholders including IPPOSI (Irish platform for Patient Organisations, Science and Industry), patient advocate groups and the research community are urgently needed."

Clinicians' knowledge and practice of data protection legislation and information management [251]

This is a machine-generated summary of:

Ryan, F. S.; Cedro, M. K.; Pabari, S.; Davenport-Jones, L.; Noar, J. H.: Clinicians' knowledge and practice of data protection legislation and information management [251]

Published in: British Dental Journal (2009)

Link to original: https://doi.org/10.1038/sj.bdj.2009.53

If you want to cite the papers, please refer to the original.

For technical reasons we could not place the page where the original quote is coming from.

Abstract-Summary

"Information governance is an integral part of clinical governance and it is incumbent on all clinicians to be aware of their duties toward their patients."

Introduction

"Patient information can be defined as anything that is used to identify a patient either directly or indirectly and is bound by legal and ethical obligations of confidentiality."

"Reference [155] Information offered in confidence should not be utilised or disclosed in any way that might identify a patient without his or her explicit consent."

"Reference [156] The management of confidential patient information is central to the principles of clinical governance and its importance is highlighted by the fact that it is outlined twice in the Department of Health's ten point list on the main components of clinical governance."

"Reference [156] The Caldicott Committee was set up by the Department of Health as a result of concerns that had arisen about the security of patient-identifiable information."

Materials, subjects and methods

"Following a review of current guidelines regarding information management within the NHS, a questionnaire was developed to audit clinicians' knowledge of these guidelines."

"Reference [155–164] The questionnaire was piloted on five members of staff for readability and ease of administration and minor wording changes were made following this."

"The questionnaire was distributed to all NHS clinical members of staff in the Orthodontic Department at Eastman Dental Hospital during a meeting, and participants were given 30 min to complete the questionnaire under exam conditions."

Results

"Reference [155] You are going on holiday outside the EU and would like to take your laptop with you (containing patient information) and you are data protected."

"Correct answer: yes, personal data should not be transferred outside the EU without the assurance of adequate data protection, compliance with the act and that the personal data is registered for processing."

"Correct answer: yes it is, but there are specific requirements in the local information governance policy and the Data Protection Act (1998)."

"Confidentiality: NHS code of practice supports this answer and states that 'staff should not normally take patient records home'."

"Correct answer: it is and disposal should be done securely in accordance with the Data Protection Act."

"Keeping confidential patient information secure is: Correct answer: a legal, ethical and NHS contractual obligation."

"Reference [160] The Department of Health also states that 'patient information is generally held under legal and ethical obligations of confidentiality'."

Discussion

"This audit was carried out as it was apparent that there was confusion among clinicians regarding the correct protocol with respect to information governance in our department."

"A review of current Department of Health, NHS and General Dental Council guidelines, together with local Trust policy, revealed that there are many areas where absolute guidance cannot be given and a combination of policy and clinical judgment must be exercised."

"Having said that, knowledge of information governance exhibited by the clinicians within this department was quite good."

"This was perhaps more to do with good clinical judgement rather than explicit knowledge or understanding of published guidelines, as the responses to Question 19 reveal."

"To improve local knowledge and due to the limitations of the published guidance on information governance, locally agreed, concise guidelines are being devised for our department."

Conclusions

"Clinicians' knowledge and practice of information governance principles in our department was good, with an overall correct response rate of 73%."

"There is scope for improvement and as dentists, we are continually being trusted with confidential patient information."

"It is imperative that all practitioners are aware of their ethical, legal and contractual obligations towards their patients."

Towards a comprehensive and interoperable representation of consent-based data usage permissions in the german medical informatics initiative [252]

This is a machine-generated summary of:

Bild, Raffael; Bialke, Martin; Buckow, Karoline; Ganslandt, Thomas; Ihrig, Kristina; Jahns, Roland; Merzweiler, Angela; Roschka, Sybille; Schreiweis, Björn; Stäubert, Sebastian; Zenker, Sven; Prasser, Fabian: Towards a comprehensive and interoperable representation of consent-based data usage permissions in the German medical informatics initiative [252]

Published in: BMC Medical Informatics and Decision Making (2020)

Link to original: https://doi.org/10.1186/s12911-020-01138-6

If you want to cite the papers, please refer to the original.

For technical reasons we could not place the page where the original quote is coming from.

Abstract-Summary

"To this, Data Integration Centers are set up at university medical centers, which address data harmonization, information security and data protection."

"To capture patient consent, a common informed consent template has been developed."

"On the technical level, a common digital representation of information from signed consent templates is needed."

"As the partners in the initiative are free to adopt different solutions for managing consent information (e.g. IHE BPPC or HL7 FHIR Consent Resources), we had to develop an interoperability layer."

"We compiled an overview of data items required to reflect the information from the MII consent template as well as patient preferences and derived permissions."

"We then compared this data model to conceptual models describing representations of consent information using different interoperability standards."

"We used the result of this comparison to derive an interoperable representation that can be mapped to common standards."

"The digital representation needs to capture the following information: (1) version of the consent, (2) consent status for each module, and (3) period of validity of the status."

"We found that there is no generally accepted solution to represent status information in a manner interoperable with all relevant standards."

"The proposed digital representation is (1) generic enough to capture relevant information from a wide range of consent documents and data use regulations and (2) interoperable with common technical standards."

Background

"On the technical level, an interoperable digital representation of information encoded in signed consent forms is needed to facilitate common data use and sharing."

"An additional requirement was that the solution should not only enable the representation of status information from signed MII Informed Consent Templates but also for further consent forms and other data use policies (e.g. derogations relating to processing for scientific purposes)."

"The developed digital representation of consent information and data use permissions is described in this article."

"An essential requirement for the digital representation presented in this article was the ability to capture relevant information from the MII Informed Consent Template, which consists of different sections addressing different use permissions."

"We emphasize that the objective of the work described in this article was to develop an interoperable digital representation of consent information and permissions related to the use of data and/or biosamples."

Methods

"We compiled an overview of data items required to reflect the information from consent templates comprising several modules as well as patient preferences and derived permissions with respect to each module."

"We then compared this data model to conceptual data models describing representations of consent information using relevant standards to analyze similarities and differences between different implementations."

"Of this comparison, we developed a digital representation of consent information which is interoperable with all standards used within the consortia."

Results

"In these code systems the codes are defined in such a way that they clearly represent policies in a specific version of the consent template."

"Common value sets combine all codes that refer to a specific policy of the consent template."

"If consent to a policy is withdrawn, this can be documented by creating a corresponding instance of the digital documentation referencing the code representing the status label "not valid" for this policy."

"The proposed digital representation can not only be used to represent information from consent templates, but also from other contexts, such as regulatory policies."

"Similar to statements from consent templates, such statements from regulatory policies can also be combined with status codes, and those combinations can then be referenced via unique identifiers."

"In HL7 FHIR, status information for policies can be implemented with the Consent resource [165]."

Discussion

"The primary aim of the digital representation described in this article is to provide a cross-site interoperability layer for representing the validity of data use policies derived from signed informed consent templates and regulatory frameworks."

"All four consortia plan to use gICS to manage the respective consent templates and to map OIDs to the original textual representation."

"The use of MII consent status codes (valid, not valid, unknown) enables the establishment of an interoperability layer across all consortia."

"Due to the minimalistic approach of the common representation, the mapping of data from more complex implementations (e.g. withdrawn, invalidated, pending, etc.) into the interoperability layer is associated with loss of information."

"We plan to support additional status codes to further facilitate harmonization of consent management processes, e.g. by explicitly representing states such as "withdrawn" or "decision pending" instead of summarizing them under a common state "not valid"."

Conclusions

"We have presented a digital representation of information from signed consent templates developed in the German MII."

"Our solution is generic, as it supports representing information from signed instances of the MII Informed Consent Template as well as other consent templates and policies from data use regulations."

"The proposed solution for an interoperable representation of the information on the consent status is based on individual codes per distinct combination of a specified policy and its validity."

"The coordination of such modifications will require additional organizational efforts, which will be implemented by utilizing the governance framework already set up within the MII."

Towards a comprehensive and interoperable representation of consent-based data usage permissions in the German medical informatics initiative [252]

This is a machine-generated summary of:

Bild, Raffael; Bialke, Martin; Buckow, Karoline; Ganslandt, Thomas; Ihrig, Kristina; Jahns, Roland; Merzweiler, Angela; Roschka, Sybille; Schreiweis, Björn; Stäubert, Sebastian; Zenker, Sven; Prasser, Fabian: Towards a comprehensive and interoperable representation of consent-based data usage permissions in the German medical informatics initiative [252]

Published in: BMC Medical Informatics and Decision Making (2020)

Link to original: https://doi.org/10.1186/s12911-020-01138-6

If you want to cite the papers, please refer to the original.

For technical reasons we could not place the page where the original quote is coming from.

Abstract-Summary

"To this, Data Integration Centers are set up at university medical centers, which address data harmonization, information security and data protection."

"To capture patient consent, a common informed consent template has been developed."

"On the technical level, a common digital representation of information from signed consent templates is needed."

"As the partners in the initiative are free to adopt different solutions for managing consent information (e.g. IHE BPPC or HL7 FHIR Consent Resources), we had to develop an interoperability layer."

"We compiled an overview of data items required to reflect the information from the MII consent template as well as patient preferences and derived permissions."

"We then compared this data model to conceptual models describing representations of consent information using different interoperability standards."

"We used the result of this comparison to derive an interoperable representation that can be mapped to common standards."

"The digital representation needs to capture the following information: (1) version of the consent, (2) consent status for each module, and (3) period of validity of the status."

"We found that there is no generally accepted solution to represent status information in a manner interoperable with all relevant standards."

"The proposed digital representation is (1) generic enough to capture relevant information from a wide range of consent documents and data use regulations and (2) interoperable with common technical standards."

Background

"On the technical level, an interoperable digital representation of information encoded in signed consent forms is needed to facilitate common data use and sharing."

"An additional requirement was that the solution should not only enable the representation of status information from signed MII Informed Consent Templates but also for further consent forms and other data use policies (e.g. derogations relating to processing for scientific purposes)."

"The developed digital representation of consent information and data use permissions is described in this article."

"An essential requirement for the digital representation presented in this article was the ability to capture relevant information from the MII Informed Consent Template, which consists of different sections addressing different use permissions."

"We emphasize that the objective of the work described in this article was to develop an interoperable digital representation of consent information and permissions related to the use of data and/or biosamples."

Methods

"We compiled an overview of data items required to reflect the information from consent templates comprising several modules as well as patient preferences and derived permissions with respect to each module."

"We then compared this data model to conceptual data models describing representations of consent information using relevant standards to analyze similarities and differences between different implementations."

"Of this comparison, we developed a digital representation of consent information which is interoperable with all standards used within the consortia."

Results

"In these code systems the codes are defined in such a way that they clearly represent policies in a specific version of the consent template."

"Common value sets combine all codes that refer to a specific policy of the consent template."

"If consent to a policy is withdrawn, this can be documented by creating a corresponding instance of the digital documentation referencing the code representing the status label "not valid" for this policy."

"The proposed digital representation can not only be used to represent information from consent templates, but also from other contexts, such as regulatory policies."

"Similar to statements from consent templates, such statements from regulatory policies can also be combined with status codes, and those combinations can then be referenced via unique identifiers."

"In HL7 FHIR, status information for policies can be implemented with the Consent resource [165]."

Discussion

"The primary aim of the digital representation described in this article is to provide a cross-site interoperability layer for representing the validity of data use policies derived from signed informed consent templates and regulatory frameworks."

"All four consortia plan to use gICS to manage the respective consent templates and to map OIDs to the original textual representation."

"The use of MII consent status codes (valid, not valid, unknown) enables the establishment of an interoperability layer across all consortia."

"Due to the minimalistic approach of the common representation, the mapping of data from more complex implementations (e.g. withdrawn, invalidated, pending, etc.) into the interoperability layer is associated with loss of information."

"We plan to support additional status codes to further facilitate harmonization of consent management processes, e.g. by explicitly representing states such as "withdrawn" or "decision pending" instead of summarizing them under a common state "not valid"."

Conclusions

"We have presented a digital representation of information from signed consent templates developed in the German MII."

"Our solution is generic, as it supports representing information from signed instances of the MII Informed Consent Template as well as other consent templates and policies from data use regulations."

"The proposed solution for an interoperable representation of the information on the consent status is based on individual codes per distinct combination of a specified policy and its validity."

"The coordination of such modifications will require additional organizational efforts, which will be implemented by utilizing the governance framework already set up within the MII."

Secondary research use of personal medical data: attitudes from patient and population surveys in the Netherlands and Germany [253]

This is a machine-generated summary of:

Richter, Gesine; Borzikowsky, Christoph; Lesch, Wiebke; Semler, Sebastian C.; Bunnik, Eline M.; Buyx, Alena; Krawczak, Michael: Secondary research use of personal medical data: attitudes from patient and population surveys in The Netherlands and Germany [253]

Published in: European Journal of Human Genetics (2020)

Link to original: https://doi.org/10.1038/s41431-020-00735-3

If you want to cite the papers, please refer to the original.

For technical reasons we could not place the page where the original quote is coming from.

Abstract-Summary

"Making routine clinical-care-data available for medical research requires adequate consent to legitimize use and exchange."

"To broad consent, the idea of (consent-free) data donation has been brought into play as another way to legitimize secondary research use of medial data."

"Flanking the implementation of broad consent policies or data donation, the attitude of patients, and the general public toward different aspects of these approaches needs to be assessed."

"28.8% of the Dutch patients generally approved secondary data-use for non-academic research, 42.3% would make their decision dependent upon the type of institution in question."

"In the German survey addressing the general population, 78.8% approved data donation without explicit consent as an alternative model of legitimization, the majority of those who approved (96.7%) would allow donated data to be used by universities and public research institutions."

"Our findings thus not only add empirical evidence to the debate about broad consent and data donation, but also suggest that widespread public discussion and education about the role of industry in medical research is necessary in that context."

Introduction

"This approach is difficult to translate from prospective clinical studies to the secondary research use of medical data collected in the course of clinical care, where not all potential purposes of the future use of the data can be anticipated at the time of collection."

"According to Article 6 of the EU-GDPR, processing of personal medical data is forbidden unless (a) the data subject (i.e., the patient) has given consent or (b) the processing (i.e., the research) is carried out in the public interest and under conditions laid down by EU or member state law [167]."

"Against the above background, it appears worthwhile exploring whether data donation (as defined, for example, by the German Ethics Council) represents an acceptable and adequate alternative form of consent, arguing that with its intention to actively support medical research, data donation clearly has the common good as its main foundational ethical principle [168–170]."

Methods

"The present work comprises two surveys, carried out in Germany and The Netherlands in 2019, that addressed the attitude of different societal groups toward the secondary research use of personal medical data under certain consent frameworks."

"The second study, from 2018, was conducted under the same conditions and assessed the level of acceptance of a no-consent policy to legitimize secondary data use in medical research."

"Both studies revealed that German patients were not only willing to give broad consent but were also likely, under certain circumstances, to accept use of their medical data for research without explicit consent [171]."

"In a second, representative population survey, we asked members of the general public in Germany for their attitude toward data donation, i.e., toward the practical implementation of a no-consent policy ("Attitude toward data donation without explicit consent in Germany" section)."

Acceptance of broad consent in the Netherlands

"The only notable difference with regard to content concerned the policy of reporting incidental findings, which had to be re-phrased for the Dutch study in order to comply with national law In contrast to the two German surveys, where the attitude toward broad consent was evaluated in the clinic, the goal of the Dutch study was to find out whether patients would accept broad consent even if only affecting them sometime in the future."

"Of the 22,000 Dutch patients approached, 7430 participated in the web-based survey (33.8%) and 5258 (70.8% of those who participated) also read the (hypothetical) broad consent documents, followed by completion of the corresponding part of the survey."

"Some 3422 of these (92.3%) expressed subsequent willingness to give broad consent, suggesting an approval rate among Dutch patients that is comparable to that noted in the German surveys [171, 172]."

Attitude toward data donation without explicit consent in Germany

"The main outcome of the survey, namely that a considerable majority of participants (75.7%) approved of such a consent-free approach, brought into play the possibility of data donation as an alternative form of legitimizing the secondary research use of medical data."

"The aim of the 2019 TMF survey reported here was to evaluate the attitude of the general population in Germany toward medical data donation for research purposes."

"The 2019 survey on data donation also marked the beginning of a comprehensive dialogue between the German university hospitals, represented by the MII, and patient organizations as well as self-help groups, which are well organized in Germany and have a strong interest in medical research."

"Of the 1006 participants, a majority of 793 (78.8%) approved donation of data from their digital health records as well as sharing of these data with third parties for medical research, anonymously and free of charge."

Limitations

"We observed significantly better understanding of the broad consent issues in the Dutch survey, compared to the 2016 study in Germany [172], likely due to the overall higher educational level in the Dutch sample, which may point toward a lack of comparability as a limitation of our current study."

"The different levels of comprehension could also be due to the fact that the topic has received more public attention over the last three years in The Netherlands, thereby resulting in better prior knowledge among the Dutch patients."

"It is not likely that the Dutch had better prior knowledge because, of those Dutch patients who did not change their opinion (n = 3562) upon reading the broad consent brochure, only 21.1% (n = 751) claimed to have known the information beforehand."

Discussion

"As a means to legitimize the secondary research use of medical data, broad consent policies are becoming adopted more and more widely in Germany and The Netherlands."

"The 2018 German patient survey and, even more so, the recent population survey by TMF revealed a generally positive attitude towards data donation for research by way of an opt-out policy, in line with other reports [173]."

"Both surveys also revealed that, whereas universities and public research institutions seem to enjoy great public trust, data use for commercial research of data is seen more critically by patients and the general population alike."

"We observed a generally positive public attitude in Germany and The Netherlands towards data donation for medical research, even without explicit consent."

Crowdsourcing sensitive data using public displays—opportunities, challenges, and considerations [254]

This is a machine-generated summary of:

Alorwu, Andy; van Berkel, Niels; Goncalves, Jorge; Oppenlaender, Jonas; López, Miguel Bordallo; Seetharaman, Mahalakshmy; Hosio, Simo: Crowdsourcing sensitive data using public displays—opportunities, challenges, and considerations [254]

Published in: Personal and Ubiquitous Computing (2020)

Link to original: https://doi.org/10.1007/s00779-020-01375-6

If you want to cite the papers, please refer to the original.

For technical reasons we could not place the page where the original quote is coming from.

Abstract-Summary

"Little work has exploited public displays for collecting tagged data that might be useful beyond a single application."

"We set to fill this gap and present two studies: (1) a field study where we investigated collecting biometrically tagged video-selfies using public kiosk-sized screens, and (2) an online narrative transportation study that further elicited rich qualitative insights on key emerging aspects from the first study."

"The field study indicates that people are willing to donate even highly sensitive data about themselves in public."

"The two studies combined in this article pave the way forward towards a vision where volunteers can, should they so choose, ethically and serendipitously help unleash advances in data-driven areas such as computer vision and machine learning in health care."

Introduction

"One interesting research direction where data generally perceived as sensitive is of paramount importance is digital health and, more specifically, automated approaches to detect or monitor health conditions using media (images, video)."

"We investigate the potential of interactive public displays in collecting sensitive data, in the form of tagged media, from passersby."

"Related to the challenge described above, computer vision depends on such high-quality training data and metadata, and public displays may well offer an avenue worth exploring for the collection of ecologically valid and accurately tagged media."

"The key contributions of our work are as follows: A contribution to crowdsourcing literature, by investigating the use of public display interactive technologies to collect potentially sensitive data."

"In the first study (presented in [174]), our display setup was deployed in the wild for a period of 61 days where we collected 199 15-s video selfies, 78 of which had voluntarily labelled metadata attached to them by participants."

Related work

"An attribute making public displays particularly suitable for data collection for a variety of use cases is their serendipitous nature of attracting passersby and their attention [175–178]."

"Several types of data can be collected using public displays."

"Public displays have been used to collect large-scale civic polling data through open-ended long format answers [179] or even classified ads directly using the screen [180]."

"Playfulness plays a crucial role in public display interactions and can be used in designing engaging applications [181]."

"Coined as a way to distribute large tasks as smaller chunks to online crowds, crowdsourcing (CS) is now the primary means to collect high-quality data from human subjects at scale [182]."

"These digital data collection ecosystems are, as an unfortunate side effect, rapidly antiquating the concept of informed consent [183, 184]—the very ethical cornerstone of any research dealing with human contributions."

System design

"The screen contains two text areas that can be dynamically configured to provide different task instructions at different times."

"In the metadata entry screen, users are asked to provide basic demographic data (age, gender) and, for this study, we configured the screen to ask height, weight, and a self-reported truthfulness value about the reported height and weight (using a scale from 1 to 5)."

"The different configurable text fields are injected in the notifications as additional payload, and the moment that the device where VideoSourcing is installed receives the notification, i.e. the user does not have to even open the notification, VideoSourcing is configured with the new dynamic values."

"The task description (title + instructions) in the welcome screen can be configured to encourage the user to perform a specific action or task while taking the video."

Study 1: A field study with videosourcing

"Before commencing a longer deployment, we validated the technical solution and feasibility of data collection in two short studies: (1) as an example of a technology probe during a lecture on Human–Computer Interaction, and (2) as a standalone installation in an open laboratory space within our own research premises."

"During these studies, we discovered technical problems such as loss of data due to the devices not staying connected to the public WiFi network and minor usability problems with the screen flow of the application itself (e.g. ambiguous wording in the final submit button)."

"Field studies with public displays are particularly important, since only with field studies can researchers really delve into matters that manifest only in natural settings such as privacy considerations of authentic responses to, e.g. questionnaires that are parts of the deployment [185]."

Study 2: Online narrative transportation

"After this, the second section of the questionnaire contained another single item, a text field on 'first thoughts,' to elicit open-ended ideas, commentary, and questions about the use of such situated tech for collecting data."

"After these initial thoughts were recorded, the subsequent section of the questionnaire contained two open-ended items: (1) what type of data would the participants be happy to donate with the deployment, and (2) what kind of control would the participants want to exercise, as articulated by themselves."

"In the last section, we specifically elicited open-ended ideas on five of the issues that we deemed emerging as the most interesting ideas from Study 1 and that in our opinion are key to unlocking the future potential of using situated technologies for collecting potentially sensitive data: Privacy and data control Compensation Institutional trust Role of surrounding context in data collection Thoughts on pervasive data collection infrastructures"

Results

"Participants largely agree that the sharing of data which is already 'publicly' observable and cannot be used to identify an individual is acceptable to collect; 'I would be happy disclosing 'personal' data such as age, height, weight etc., because all those things are publicly available anyway—you only have to look at me to work out more or less what those things are.'"

"The inclusion of video footage (as in our described scenario) limited some of the data types which participants were willing to 'link' to their identifiable video; 'If it was anonymous I would have no issues, but as I can be identified I think my limit would be age, height, gender, relationship status, employment status, etc.' Consent in the age of digital data collection is no longer a simple matter [183]."

Discussion

"Our work is, in a way, a provocation aimed at probing a new use case: collecting sensitive data for research use through public interactive displays."

"Further, the users of our deployment could not even proceed to the video capture stage before agreeing to our simple but, as far as research goes, all-encompassing disclaimer that we may use all of their data for our own non-profit purposes."

"The concerns raised by our participants regarding our intentions with the data as well as ethics concerns in general alert us to rethink the user-facing introduction to the concept, the invitation to donate data."

"This would in all likelihood increase participation, reduce concerns about data use, and make the collection procedure more ethical and less suspicious in general."

"We plan to extend the deployment first by ensuring granular data use rights, by letting the users indicate their preferences prior to taking a video and donating metadata."

Conclusion

"In the second study, we showed how willing people are to donate different data types and present vast amount of qualitative insights around different aspects of such public data collection systems, such as the perceptions of private vs. public organisations as data collectors, role of context, and reuse of the donated data."

"All this presents provocative ways of looking at pervasive data collection: on the one hand, people are willing to donate data, but on the other hand, the systems should be designed to collect granular data and have clear data management policies attached to them."

"The findings in the article contribute not only a novel application concept but also a great deal of considerations for the future design of public data collection tools."

5.3 Placing Consent in the Midst of Future Technology

Machine generated keywords: consent, medical research, dataset, medical, healthcare, compliance, participant, trust, transparency, management, artificial, artificial intelligence, freely, user, problem.

Big Data in medical research and EU data protection law: challenges to the consent or anonymise approach [255]

This is a machine-generated summary of:

Mostert, Menno; Bredenoord, Annelien L; Biesaart, Monique C I H; van Delden, Johannes J M: Big Data in medical research and EU data protection law: challenges to the consent or anonymise approach [255]

Published in: European Journal of Human Genetics (2015)

Link to original: https://doi.org/10.1038/ejhg.2015.239

If you want to cite the papers, please refer to the original.

For technical reasons we could not place the page where the original quote is coming from.

Abstract-Summary

"The current EU data protection law reform has led to an intense debate about its potential effect on this processing of data in medical research."

"To contribute to this evolving debate, this paper reviews how the dominant 'consent or anonymise approach' is challenged in a data intensive medical research context, and discusses possible ways forwards within the EU legal framework on data protection."

"We underline the necessity of an appropriate research exemption from consent for the use of sensitive personal data in medical research to take account of all legitimate interests."

"Further deliberation is required to determine when a shift away from consent as a legal basis is necessary and proportional in a data-intensive medical research context, and what safeguards should be put in place when such a research exemption from consent is provided."

Introduction

"The EU data protection law reform has led to an intense debate about its potential effect on medical research."

"The discussion is about where limits should be drawn to the use of sensitive personal data in medical research."

"Key issues are related to the scope and limitations of consent as a legal basis for the use of sensitive personal data in medical research and its possible alternatives."

"A dominant approach in some EU member states is that the conventional or only alternative to obtain consent is anonymising these data."

"References [186, 187] Even so, derogations to this approach can be laid down in data protection law in so-called 'research exemptions'."

"References [188–194] To contribute to this evolving debate, this paper reviews how the consent or anonymise approach is challenged in a data-intensive medical research context, and discusses possible ways forward within the EU legal framework on data protection."

The context of data-intensive medical research

"Large worlds of complex health-related and genomic data, often referred to as 'Big Data', are becoming available to medical researchers."

"The UK Biobank provides access to medical researchers from all around the world to a wide variety of health-related data and human samples from more than 500 000 participants."

"Reference [195] Also, many initiatives exist to promote or facilitate the large-scale re-use and linkage of health-related and genomic data, such as the Global Alliance for Genomics and Health (Global Alliance)."

"Reference [196] These developments illustrate how medical research is increasingly becoming a data-intensive activity, in which health-related, genomic and other data are being collected, re-used and linked on a large scale."

EU legal framework on data protection

"At the heart of the current principal EU data protection law, the DPD, are the principles of fairness and lawfulness."

"The principle of lawfulness demands that each processing of personal data must be based on consent or another legitimate basis laid down by law, as is also enshrined in Article 8 of the Charter."

"When it comes to the processing of sensitive personal data, such as health-related data, a more restrictive set of legal bases is provided by EU data protection law Genetic data will be explicitly recognised as sensitive in the forthcoming GDPR, without granting this type of data a status different from other categories of sensitive personal data."

"Reference [197] At present, the legal base provided by Article 8 (2a) of the DPD for the processing of sensitive personal data, in any context, is explicit consent."

Ways forward within the consent or anonymise paradigm

"The difficulties in obtaining consent, when personal data are to be available for linkage, re-use and analysis for largely undetermined future research purposes, have been discussed extensively in the literature."

"Reference [198] Others argue that broad consent is an ethically sound alternative for specific consent, although individuals are not given specific information about future research projects."

"Some indicate that broad consent may not meet the conditions on consent as defined in the Parliament's draft GDPR, regarding the information that must be given to the individual."

"Reference [194] This position is reflected in Recital 25aa of the Council's draft GDPR, which states that 'data subjects can give their consent to certain areas of scientific research when in keeping with recognised ethical standards for scientific research.'"

"References [64, 199] A conventional method to protect data and avoid consent or other legal requirements is anonymisation."

Ways forward outside the consent or anonymise paradigm

"An alternative approach is to search for ways forward outside the consent or anonymise paradigm, by creating another legal basis than consent for the processing of sensitive personal data for medical research purposes."

"According to Article 81 (2a) of the Parliament's draft GDPR, such a research exemption from consent should be provided by national law, for 'research that serves a high public interest'."

"Article 9 (2i) of the Council's draft GDPR indicates that consent is not required when the processing is necessary for scientific purposes, subject to certain conditions and safeguards laid down in law Differing positions on the appropriate scope of research exemptions are also reflected in the literature."

"Technological and organisational or governance measures have also been proposed in the literature to justify alternative legal bases to consent, such as opt-out registration, [200] authorisation by an ethics committee, [187] limiting data access and use, and engaging in public participation."

Discussion

"In the debate on how to deal with the challenges to the consent or anonymise approach in the context of data-intensive medical research, within the EU legal framework on data protection, we suggest that the following considerations should be taken into account."

"This necessity may increase even further, dependent on what definitions on consent and anonymisation will be provided by the forthcoming GDPR, which need to be clear to reduce legal uncertainty and prevent the erosion of data protection law Then, we recommend that further debate should focus on two issues related to research exemptions in data protection law First, we do not expect that a high level of harmonisation on the conditions of research exemptions will be provided by the forthcoming GDPR."

"Further interdisciplinary research is however needed to determine when a shift away from consent as a legal basis is necessary and proportionate in a data-intensive medical research context, and what technological and governance measures should be put in place when such a research exemption from consent is provided."

AI, big data, and the future of consent [256]

This is a machine-generated summary of:

Andreotta, Adam J.; Kirkham, Nin; Rizzi, Marco: AI, big data, and the future of consent [256]

Published in: AI & SOCIETY (2021)

Link to original: https://doi.org/10.1007/s00146-021-01262-5

If you want to cite the papers, please refer to the original.

For technical reasons we could not place the page where the original quote is coming from.

Abstract-Summary

"We discuss several problems with current Big data practices which, we claim, seriously erode the role of informed consent as it pertains to the use of personal information."

"To illustrate these problems, we consider how the notion of informed consent has been understood and operationalised in the ethical regulation of biomedical research (and medical practices, more broadly) and compare this with current Big data practices."

"We do so by first discussing three types of problems that can impede informed consent with respect to Big data use."

"We discuss the re-repurposed data problem."

"We discuss the meaningful alternatives problem."

Introduction

"Google owned YouTube were fined $US170 million (Singer and Conger 212) for extracting personal information, without parental consent, from children using the platform, and then using the data to target advertisements towards them."

"The opacity of the terms of this collaboration have raised a number of pressing questions regarding the protection of privacy, the regulation of data sharing and use, and the structural difficulties to implement a measure of meaningful individual consent and control in the face of increasingly common transfers of population-derived datasets to large private companies (Powles and Hodson 213)."

"These cases, and others like them, show that express consent (consent pertaining specifically to the activity to which the consent is given) has not always been sought, let alone received, by companies who use, share, and/or sell, peoples' personal data."

"We discuss several problems with current Big data practices which, we claim, seriously erode the significance of informed consent."

Informed consent

"In biomedical contexts, for example, such information allows a patient to become aware of the potential risks that may arise if they consent to a procedure or determine whether consenting to a procedure may conflict with their values or preferences."

"Before laying out three of the main problems with current Big data practices, as we see them, it is worth reflecting on the asymmetry between the seriousness in which informed consent is dealt with in the medical context; and the perfunctory manner in which it is dealt with by big technology companies."

"It is important to secure an individual's informed consent with respect to Big data use in our view because, just like in the medical context, failing to do so can cause harm to individuals."

Big data and informed consent

"If users have a statutory 'right to an explanation' about how their data are being used [as they do, for instance, under the EU's GDPR], and this explanation is a necessary part of their ability to make informed decisions, then the problem of procuring meaningful 'informed consent' becomes extremely challenging in these contexts."

"The general problem of re-purposed data, then, is that data users (e.g., companies and institutions) have not always limited their use of personal data to the purpose for which the subjects' original consent was applicable."

"Cutting edge medical research has also managed to address both the re-purposed data problem and the meaningful alternatives problem through the mechanism of 'dynamic consent', which allows participants to constantly recalibrate their initial decision and provide (or refuse) fresh consent to new and emerging uses of their data beyond what they initially consented to (Kaye and others 65)."

Big data and the future of informed consent

"We propose that some uses of personal data for commercial and administrative objectives could be subject to a 'soft governance' ethical regulation, akin to the way that any research involving human participants (spanning the collection of human medical data and tissue use, to participation in interviews and focus groups) are regulated in many countries through the Human Research Ethics Committees (HRECs)."

"The use of supercookies to gather browser data without user consent would have been unlikely to pass through a HREC-like review board, on the grounds that users were not informed about the consequences of using the services."

"We think that the reasons Newson and Lipworth provide for the justification of ethics approvals, with respect to research involving human subjects, can also be used to justify our suggested ethics approval component for certain Big data practices."

Conclusion

"Alternative ways of securing informed consent exist; it is worth exploring the ethical implications of applying them."

"Since this approach only addresses one half of the problem, we also focused on improving ways in which consent is actually sought."

"These proposals are by no means all that is required to solve the problems of Big data use that we have identified here."

"Big data technologies and their applications are new and will continue to evolve and bring with them new ethical problems (consider recent developments in AI emotion capturing technology, for example)."

"It is worth attempting to understand that history, so that we might minimize the harm that current Big data use is causing."

Dynamic consent management for clinical trials via *private blockchain technology* [257]

This is a machine-generated summary of:

Albanese, Giuseppe; Calbimonte, Jean-Paul; Schumacher, Michael; Calvaresi, Davide: Dynamic consent management for clinical trials via private blockchain technology [257]

Published in: Journal of Ambient Intelligence and Humanized Computing (2020)

Link to original: https://doi.org/10.1007/s12652-020-01761-1

If you want to cite the papers, please refer to the original.

For technical reasons we could not place the page where the original quote is coming from.

Abstract-Summary

"An essential factor for the success of a CT is the appropriate management of its participants and their personal data."

"Healthcare institutes need to obtain freely given, specific, informed, and unambiguous consent before being able to collect the data."

"We present SCoDES, an approach for trusted and decentralized management of dynamic consent in clinical trials, based on blockchain technology (BCT)."

"The usage of blockchain provides a set of features that allow maintaining consent information with trust guarantees while avoiding the need for a dedicated or centralized third trusted party."

Introduction

"In this scope, some of the key aspects required for trusted and secure management of consent in CTs include: (i) certified authentication for access to data and resources, (ii) decentralized and trusted consent data access across different healthcare institutions and participants, (iii) verifiable, unfalsifiable and accountable registry of operations, transactions and consent authorizations, (iv) secure and consistent modification of dynamic consent, and (v) participant-centered control over her own consent preferences, approvals and revocations."

"We propose an approach for trusted and decentralized management of dynamic consent in clinical trials, based on blockchain technology (BCT)."

"The most relevant features of BCT implemented in SCoDES include: Trust management trust is not centralized on a single actor, but it is distributed among the peers/participants of the network (in this case healthcare institutions and research facilities); Immutability the ledger cannot be modified, nor any transaction stored in it can be deleted."

State of the art

"We explore existing BCT-based works in the health and the biomedical domain, focusing on consent management, clinical trial management systems (CTMS), and the usage of blockchain for ensuring privacy and trust in such a domain."

"In the health domain, biomedical data-types involved in BCT-based applications are mostly medical electronic health records (EHR), personal health records, consent forms, drug information, environmental data, location, or medical evidence data (Drosatos and Kaldoudi 216)."

"The consent related to a set of sensitive data is implemented through standard database fields representing status (e.g., unopened, viewed, and signed) reflecting the conditions of actual (physical) documents still part of the clinical trial initialization process."

"Beyond blockchain-based approaches, other prototypes such as the mobile based data acquisition application by Beierle et al. (217) explored ways of protecting data privacy, showing what types of data a participant is willing to share with researchers."

System design

"Such a high-level objective can be decomposed in several structured system goals: G1 to improve the current management of consent for CTs; G2 to allow data-transfer between medical platforms, enabling to provide the information required in a consistent form ensuring the patient confidentiality; G3 to enforce the trust by ensuring the authenticity and integrity of participants' data and consent."

"The data model contains the design and definition of the assets (Crosby et al. 218), the participants, the transactions, and the events that are part of the system."

"While fabric allows defining the concrete components of the network (i.e., peers, certification authorities (CAs) and orderers), hyperledger composer allows to abstract the process of defining the actual network."

"In the SCoDES project, there is a single model file defining the structure and the transactions related to the assets of the network (i.e., contracts and trials)."

System implementation

"There are: a dashboard, in which the users can see a summary of a set of data (user-role dependent); a trials page in which all available trials are listed; a contracts page that shows the consent status; and the patient profile displaying his/her medical data (in this case stored on REDCap)."

"On this page, the doctors can issue new contracts to possible participants to register the consent to access their medical data."

"The medical-data page is available for doctors to manage the biometric data related to a given CT, and for patients, to insert here their biometric data (which are stored in REDCap)."

"Contract: represents the official agreement bounding the participant (and his/her data) with a clinical trial (and its supervisors/observers)."

"Such functions are used respectively by Patients and Doctors: Patients: to insert medical data related to a trial to which they give consent."

System validation and discussion

"This paper proposed the integration and combination of BCT (Hyperledger Composer and Fabric) and web technologies (React, REST Server, and REDCap) to dynamically handle participants' informed consent in CTs (requirement R1.1)."

"The information composing the Trial goes through all the layers of the network until it is stored in the world-state DB of the blockchain (requirements R3.1 and R3.2), and the respective transaction is submitted on the ledger, guaranteeing authenticity, integrity and trust (requirements R3.3, R3.4, and R3.5)."

"When a doctor goes to his Medical Data page, the system retrieves (from REDCap—requirement R4.3) and displays the data relative to all the patients that accepted the doctor's contracts (possible to filter)."

"An alternative approach could be to develop a blockchain service, providing a consent management module possible to be plugged into any existing or future platform with a lightweight integration process."

Conclusions

"We have presented a full implementation of a consent management application for clinical trials, based on blockchain technology."

"The resulting proof-of-concept addresses the following challenges related to consent management: Improve current procedures for managing consent while ensuring confidentiality; Enable sensitive data sharing between medical platforms as electronic data capture software (e.g., REDCap); Ensure trust in the data sources identities; Ensure traceability and integrity of data through a tamper-proof system."

"This study shows that the use of blockchain technology within a network of researchers and healthcare institutions/practitioners can facilitate both the process and collection of patient data and ensure the level of anonymity required for clinical trials and the reproducibility of the research."

"Thanks to the BCT, the control data sharing for scientific research is simplified, while privacy is ensured in cross-source."

The generic Informed Consent Service gICS®: implementation and benefits of a modular consent software tool to master the challenge of electronic consent management in research [258]

This is a machine-generated summary of:

Rau, Henriette; Geidel, Lars; Bialke, Martin; Blumentritt, Arne; Langanke, Martin; Liedtke, Wenke; Pasewald, Sandra; Stahl, Dana; Bahls, Thomas; Maier, Christian; Prokosch, Hans-Ulrich; Hoffmann, Wolfgang: The generic Informed

Consent Service gICS®: implementation and benefits of a modular consent software tool to master the challenge of electronic consent management in research [258]

Published in: Journal of Translational Medicine (2020)

Link to original: https://doi.org/10.1186/s12967-020-02457-y

If you want to cite the papers, please refer to the original.

For technical reasons we could not place the page where the original quote is coming from.

Abstract-Summary

"According to these regulations, the Informed Consent (IC) is an inevitable element of research with human subjects."

"In the era of "big data medicine", aspects of IC become even more relevant since research becomes more complex rendering compliance with legal and ethical regulations increasingly difficult."

"To address the requirements, the free-of-charge, open-source software "generic Informed Consent Service" (gICS®) was developed by ICM to provide a tool to facilitate and enhance usage of digital ICs for the international research community covering various scenarios."

"gICS facilitates IC management based on IC modularisation and supports various workflows within research, including (1) electronic depiction of paper-based consents and (2) fully electronic consents."

"Since the consent's content is a prerequisite for securing participants' rights, application of gICS is no guarantee for legal compliance."

"gICS simplifies and supports sustained IC management as a major key to successfully conduct studies and build trust in research with human subjects."

Background

"According to MITRE [208] "[...] consent management is a system, process, or set of policies that enables [...]" participants to decide what healthcare providers and researchers are allowed to do with their health information, and to extend this to all kinds of specific personal information."

"The participant's permissions for usage together with his/her choice of restrictions including partial and complete withdrawals can be handled based on automated algorithms by an Informed Consent Management System [208]."

"The University Medicine Greifswald developed a scalable generic– thus, portable and adaptable—approach to implement a generic Informed Consent Service (gICS) to facilitate consent management for all possible study types and settings, with a focus on epidemiological research studies and registries."

"This paper discusses the organisational development and technical implementation of the software solution gICS for the creation, management and modularisation of informed consents as well as its support of policy-specific automated queries and withdrawals."

Methods

"Only if all policies of this module are consented, data collection is worthwhile."

"This approach eradicates the possibility of contradictory consent states: If a policy appears in different modules within the same consent template, a participant might first decline this policy in the first module and, afterwards, consent to the same policy in the second module leading to inconsistencies."

"Subsequent changes of contents and wording in already consented policies, modules and templates lead automatically to a new version, respectively."

"After introducing the concept of policies and modules, consent states are used to depict the participant's will."

"Based on research studies, the following states are used: Accepted—the participant answered with "yes" to a policy, e. g. "Do you consent to data capture for this research project?""

"Declined—the participant answered with "no" to a policy, e. g. "Do you consent to data storage for this research project?""

Results

"gICS' SOAP interface (gicsService, v.2.9.1) provides functionalities regarding consent management, queries, checks and validation, workflows and support functions."

"Using templates for consents and versioning them as supported by gICS simplifies processes and can reduce the burden on study staff [209]."

"Policies and consent templates tend to be added, edited or even deleted during the duration of long-term studies, e. g. the NAKO project."

"By managing different consent versions as supported by the gICS' versioning, the burden on study staff as well as the central data management is further reduced."

"To support various workflows within a research study, different ways of collecting ICs with gICS are possible: (1) the electronic depiction of a paper-based consent and (2) fully electronic consent using the web-based service architecture of gICS."

"It includes the paper-based collection of consent, which is then scanned and stored in a digital format as a participant's consent within gICS."

Discussion

"To ensure reusability in new projects as well as a widespread provision of a consent management tool for the scientific community, gICS is provided as open-source, free of charge software tool."

"Due to Docker, gICS is easy to install and supports research projects with limited resources regarding IT experts and finances."

"GICS is suited for research projects with short as well as long project durations and has no practical limits regarding number of consents or consent template versions."

"Using gICS does not automatically guarantee a high degree of legal compliance, since the consent's content is responsible for securing participants' rights."

"gICS only supports the management of consent documents as well as permissions and prohibitions (regarding policies) stated by the participant."

"A future gICS-feature could also include providing the consent electronically to the participant, e.g. via e-mail."

Conclusions

"The introduced approach of the generic Informed Consent Service gICS supports the automated processing of ICs, and use cases of a broad range of research projects to collect and manage ICs in compliance with their workflows as well as legal and ethical requirements."

"It will be used and further enhanced in the complex multi-site network MIRACUM (Medical Informatics in Research and Care in University Medicine; grant number: FKZ 01ZZ1801M) [87] leading most likely to further use cases."

"Interested researchers are invited to use gICS [210] and provide feedback for further improvements."

Ethical and legal challenges of informed consent applying artificial intelligence in medical diagnostic consultations [259]

This is a machine-generated summary of:

Astromskė, Kristina; Peičius, Eimantas; Astromskis, Paulius: Ethical and legal challenges of informed consent applying artificial intelligence in medical diagnostic consultations [259]

Published in: AI & SOCIETY (2020)

Link to original: https://doi.org/10.1007/s00146-020-01008-9

If you want to cite the papers, please refer to the original.

For technical reasons we could not place the page where the original quote is coming from.

Abstract-Summary

"This paper inquiries into the complex issue of informed consent applying artificial intelligence in medical diagnostic consultations."

"To achieve this objective, the first part of the paper analyzes ethical aspects of the alleged right to explanation, privacy, and informed consent, applying artificial intelligence in medical diagnostic consultations."

"This analysis is followed by a legal analysis of the limits and requirements for the explainability of artificial intelligence."

Introduction

"Artificial intelligence systems, accompanied by increasing computational powers, can analyze personal characteristics, medical records, large amounts of literature, and other medical data with high-accuracy and using just a fraction of time and cost required for the same task to be completed by a human counterpart."

"This article contributes to the ongoing scientific discussion on the risks and benefits of artificial intelligence use in healthcare and aims to expose contemporary legal and ethical challenges of informed consent applying artificial intelligence in medical diagnostic consultations."

"The article exposes ethical aspects of the right to explanation, privacy, and informed consent, applying artificial intelligence in medical diagnostic consultations, followed with a legal analysis of the limits and requirements for it."

Ethical challenges of applying artificial intelligence in medical diagnostic consultations

"Even if artificial intelligence system is modeled in an explainable way, the interpretability of the outcomes it provides may require assessment of risks related to possible disparities in the raw data that system has been trained on, especially in the high stakes—life or death—situations.. Fairness appears as another important concern while dealing with disproportionally inaccurate medical decision-making on minorities (Schönberger 222)."

"To make decisions on a treatment plan, physicians should not merely consider quality of data, collected and used in the process of training artificial intelligence systems (no matter how detailed) but should also employ their experience (to conclude results according to individual situation) and a deeper understanding of illness to conclude everything in the certain contextual picture (to take into account patient's psychological, emotional, social, or even religious aspects)."

Legal challenges of applying artificial intelligence in medical diagnostic consultations

"The legal problem of interpretation does not revolve around whether a right to an explanation in the process of informed consent to medical diagnostic procedures exists, especially under the EU law Under the generally accepted principles of contract and tort law, complemented with data protection law in the context of artificial intelligence applications, physician has the duty to explain and patient has the right to demand such explanation."

"The process of informed consent in diagnostics requires the physician to provide the patient with meaningful information before or during that process-when no medical treatment decisions are made yet."

"The legal standard of care, applied to the physician's professional duties in the process of informed consent, requires a full understanding of the medical treatment and its options that must be individually and comprehensibly explained to every patient."

Conclusions

"The process of informed consent is one of the major ethical concerns of the New Health phenomenon, especially in the context of technological complexity and opacity of artificial intelligence systems."

"Legal analysis of the informed consent process revealed that the right to an explanation does not require exposure of the artificial intelligence model and it may be technically feasible to shift the explanation burden from physician to the technology developers."

"An explanation is a professional duty of the physician and, thus, the information provided by the developers of artificial intelligence systems will be incomplete in the context of meaningful physician intervention and oversight in the diagnostic process."

"Despite the possibility to shift at least partially the explanation burden, a physician still needs to understand the complexity of artificial intelligence algorithms used in medical diagnostics."

Selling your soul while negotiating the conditions: from notice and consent to data control by design [260]

This is a machine-generated summary of:

Belli, Luca; Schwartz, Molly; Louzada, Luiza: Selling your soul while negotiating the conditions: from notice and consent to data control by design [260]

Published in: Health and Technology (2017)

Link to original: https://doi.org/10.1007/s12553-017-0185-3

If you want to cite the papers, please refer to the original.

For technical reasons we could not place the page where the original quote is coming from.

Abstract-Summary

"This article claims that the Notice and Consent (N&C) approach is not efficient to protect the privacy of personal data."

"N&C could be seen as a license to freely exploit the individual's personal data."

"Through this paper, we aim at uncovering evidences of the inefficiency of the N&C approach, as well as the possibility to further enhance PbD, in order to provide the individual with increased control on her personal data."

"We are putting forth the Data Control by Design (DCD) concept, which we see as an essential complement to N&C and PbD approaches advocated by data-protection regulators."

"Argue that data protection frameworks should foster the adoption of DCD mechanisms in conjunction with PbD approaches, and privacy protections should be designed in a way that allows every individual to utilise interoperable DCD tools to efficiently manage the privacy of her personal data."

"After having scrutinised the N&C, PbD and DCD approaches we discuss the specificities of health and genetic data, and the role of DCD in this context, stressing that the sensitivity of genetic and health data requires special scrutiny from regulators and developers alike."

Notice and consent: good intentions with questionable outcomes

"Rather than allowing individuals to manage their personal-data privacy, the N&C approach has increasingly become an instrument for individuals to pay for online services with the consent to access, process, and disclose their personal data."

"It is important to acknowledge that, despite the stated intention to attribute informational self-determination to individuals, the N&C mechanism has de facto evolved into a payment model, consisting in trading consent to personal-data exploitation for the possibility to access online services."

"Virtually every mobile application can be used only under the previous acceptance of the ToS, which typically foresee the application provider right to collect and process a wide range of personal information spanning form the user telephone number and IP address to geo-localisation data, list of contacts, etc. In this regard, it has been argued that the N&C failure to protect privacy and simultaneous success in easing the exploitation of personal data may be the fruit of specific ideological choices."

Designing privacy into architectures

"The aim of both PETs and PbD is to prompt a proactive approach to data privacy, in which the user can play an active role thanks to the privacy protections that are embedded ab origine into the services, processes, networks, etc. The goal of PbD is to lead engineers, providers and administrators to consider carefully the interests of the individuals, including privacy protections, when they design or redesignspecific information systems."

"PbD aims at utilising both technical and governance strategies, considering privacy within organisations' risk management and establishing user-friendly features that allow individuals to manage the privacy of their data."

"Although the implementation of appropriate technical and organisational measures can improve individuals' capacity to protect their data privacy, it would be overconfident to argue that PbD guarantees individuals' full control over how their data are utilised."

Data control by design

"While the prospect of individual people successfully managing their reputations within fluid digital contexts seems challenging, technology and standards experts have been working to create novel approaches that give users simple, centralised control over the dissemination of their data, such as the Kantara Initiative's User-Managed Access (UMA) protocol [212]. By basing data control around standardised, user-driven protocols that give users the ability to explicitly consent to how personal data is disseminated and accessed, the UMA protocol has provided the technological standard to facilitate a new model of data control."

Specificities on consent for health and genetic data

"Precision medicine is based on mining datasets encompassing the individual's genetic, health and lifestyle data, with the aim of recognising specific patterns of illness and nurturing research in prevention techniques."

"The possibilities of collecting and processing genetic and health data, besides data on individuals' lifestyle is increasingly "total.""

"The processing of health and genetic data may reveal highly sensitive information for which traditional N&C schemes and many existing PbD approaches may fail to provide appropriate guarantees."

"The complexity of the assessments to be made in order to express consent and define preferences with regard to health and genetic data collection and processing brings to attention the growing need for trust in both the data processor and the health professional that should accompany the data subject in her choices."

"This situation becomes more complicated when it involves people who need to provide consent concerning their health or genetic data for research, as they are generally unable to identify the nature of the information that may emerge."

Conclusion

"In such contexts, the element of trust becomes increasingly important to construct responsible mechanism and institutions that "only earn their designations as trustworthy if there are feasible procedures by which an individual can check on what is done." (Hanen, 2009) Building trust through a user-centric DCD approach, implemented by initiatives such as MyData, plays a key role in allowing individuals to exert control over their personal data and take autonomous decisions about their lives, thus and freely developing their personalities."

"The availability of intelligible information is necessary to carefully consider and to reflect on the pros and cons of giving consent to the collection and processing one's personal data—and, even more importantly, one's health and genetic data—but does not guarantee per se individuals' control on personal data."

[Section 6]

"The methodology of this paper is based on both a literature review and an analysis of concrete implementations of technical tools designed to allow users to assert control over their personal data management, with a particular focus on the MyData initiative [213]. The first section of this paper will provide an overview of the "traditional" N&C mechanism, used to protect individuals' data privacy online and will emphasise the failures of such mechanism."

"PbD still does not fully empower users to take control over their privacy because, as in the N&C context, individuals are only given the option to choose whether or not their data will be collected and processed rather than being able to exert full control over how their personal data are used and by whom."

"Just-in-time" generation of datasets by considering structured representations of given consent for GDPR compliance [261]

This is a machine-generated summary of:

Debruyne, Christophe; Pandit, Harshvardhan J.; Lewis, Dave; O'Sullivan, Declan: "Just-in-time" generation of datasets by considering structured representations of given consent for GDPR compliance [261]

Published in: Knowledge and Information Systems (2020)

Link to original: https://doi.org/10.1007/s10115-020-01468-x

If you want to cite the papers, please refer to the original.

For technical reasons we could not place the page where the original quote is coming from.

Abstract-Summary

"Data processing is increasingly becoming the subject of various policies and regulations, such as the European General Data Protection Regulation (GDPR) that came into effect in May 2018."

"One important aspect of GDPR is informed consent, which captures one's permission for using one's personal information for specific data processing purposes."

"The state-of-the-art primarily focuses on, for instance, the analysis of prescriptive models and posthoc analysis on logs to check whether data processing is compliant to GDPR."

"We argue that GDPR compliance can be facilitated by ensuring datasets used in processing activities are compliant with consent from the very start."

"We propose RDF and OWL ontologies to represent the consent that an organization has collected and its relationship with data processing purposes."

"We use this ontology to annotate schemas, allowing us to generate declarative mappings that transform (relational) data into RDF driven by the annotations."

"We demonstrate the approach with a synthetic dataset simulating users (re-) giving, withdrawing, and rejecting their consent on data processing purposes of systems."

"It is argued that the approach facilitates transparency and compliance verification from the start, reducing the need for posthoc compliance analysis common in the state-of-the-art."

Introduction

"Rather than analysing the processes (before or after execution), we want to investigate whether an organization can also use this information for the creation of a compliant dataset, therefore, guaranteeing a compliant execution of a process."

"This would be useful for data processing activities (e.g. sending newsletters, using one's data and history for a recommender system, etc.) that require one's informed consent."

"Organizations often hold data in various heterogeneous data sources, motivating the need for a common graph-based data model, and for standardized technologies for transforming non-RDF data in RDF (called "uplift")."

"In [214], we demonstrated the feasibility of generating RDF Data Cube datasets that consider the consent an organization has gathered."

"This paper demonstrates how one can use Semantic Web technologies (which are open and standardized) to generate compliant datasets from a knowledge base of gathered consent information in a declarative manner."

Objectives, scope, and assumptions

"The objective of this study is to propose a method for generating datasets in a "GDPR-aware" manner, with a focus on consent and fit for particular data processing purposes."

"To design and develop a model for representing the consent gathered by an organization and its relations to a system's data processing purposes."

"We assume the consent (and the information thereof) an organization has stored is valid."

"We also note that we are primarily concerned with the generation of datasets for data processing purposes."

Design

"We decompose the design and development of our solution in different components: an ontology for representing consent, the generation of uplift mappings, and the component that uses these two for the generation of compliant datasets."

"Our consent ontology is used to annotate a dataset schema description with a data processing purpose as well as to populate a knowledge base with previously gathered consent."

"To generate those by the mapping engine, we adopt vocabularies to represent datasets and annotate those with references to tables, columns, and so on."

"The mapping engine will process both the schema and the annotations to generate an R2RML [215] mapping, a W3C Recommendation for declaring how relational data should be transformed into RDF."

"We stated in the introduction that prior work, presented in [214, 216], looked into the generation of R2RML from annotated Data Structure Definitions (DSDs) and the generation of compliant datasets from these R2RML mappings."

Representing consent information

"The ontology we developed for prior work (in [214]) was built from the perspective of an organization and only concerned with whether consent was given by data subjects."

"The concepts and relations we needed to include in our consent ontology are mostly related to the appropriate DPV ones and also linking dataset schemas with an organization's data."

"Although DPV provides most of the classes and relations we need (i.e. around Data Subject, Consent, and Purpose), we needed to introduce the link between a dataset (schema) and a data processing purpose."

"In DPV, the class "Consent" represents the consent given by a Data Subject (in most cases) or their legal representative."

"The DPV also does not provide information regarding modelling of information regarding the evolution of consent over time—such as a Data Subject providing consent that has been expired—was previously withdrawn, or was refused."

Generating uplift mappings in R2RML

"When creating a dataset for data processing purposes, a location (and thus, also, an absolute URI) will be generated We start off with the (re)use of a schema describing the tabular data, which we will annotate with information on where to fetch the data."

"We have chosen to adopt a declarative approach to generating the R2RML mapping via a sequence of SPARQL CONSTRUCT queries."

"The various queries can be summarized as follows: (1) create the triples maps (for mapping tables, views or queries to RDF); (2) use the columns to create subject maps; (3) create predicate object maps for the columns, and (4) connect the dataset that will be generated with its schema."

"With these mappings—which are declarative and implemented as SPARQL CONSTRUCT queries—we are able to generate an executable R2RML mapping that will be used in the generation of the "just-in-time" compliant dataset in the next process."

Creating compliant datasets

"(3.1); and (3) how we combine (1) and (2) to generate a consent-compliant dataset for use in data processing operations."

"We retrieve consent information for a particular purpose using a SPARQL SELECT query (see Listing 9) upon the Consent Information Knowledge Base (whose information is structured according to the Consent Ontology)."

"The synthetic dataset contains, for purpose #30 "sending newsletters", consent information about three users."

"With the query in Listing 9, we fetch the latest consent information of users."

"This information is used to construct a list of users that have given their consent."

"That list is then used to create and execute a SPARQL DESCRIBE query returning to us a dataset that is fit for the purpose for which consent has been agreed."

"The DESCRIBE query is shown in Listing 12 and is applied to the dataset to obtain a subset with only the information of those that have given their consent."

On generalizing the approach

"We provide the motivation for supporting tabular representations next to RDF Data Cubes as well as the differences in the implementation (i.e. the SPARQL queries)."

"With the Data Cube Vocabulary, one can only declare one unit of measurement per observation."

"This article has demonstrated how an initial approach to generating compliant datasets can be generalized to cater for different types of data; tabular data using an RDF representation of CSV files (which can be translated into CSV files according to the CSVW specification), and RDF Data Cubes based on the aforementioned prior work."

"In RDF Data Cube, each observation is identified by its dimensions."

"In RDF Data Cubes, there are dimensions and measures."

"As for validating the datasets, RDF Data Cube prescribes a set of SPARQL ASK queries implementing so-called "integrity constraints" that test whether a dataset adheres to a Data Structure Definition."

Design of the service

"We assume that one maps the consent information stored in some non-RDF format to our consent knowledge base (by using, for example, R2RML) and expose that information via a SPARQL endpoint."

"This assumption is motivated by the fact that organizations oftentimes store information in various (non-RDF) data sources."

"The tool also requires information about the non-RDF datastore (location or connection URL, username and password if needed, etc.)."

"Our approach can be easily extended to store such information in separate graphs and instead point to the URI of a datastore description in RDF."

"All (intermediate) files are to be stored in a secure location with references (URLs) in the provenance graph."

"Although an appropriate manager should for these governance strategies, we currently store these files in RDF files in a separate folder."

Conclusions and future work

"The question we addressed in this article is: "How can we generate datasets for a specific purpose "just-in-time" that complies with given consent?""

"The third objective O3 "generating compliant datasets" proposed a declarative approach to manipulating the resulting datasets to exclude data for which no consent is given."

"We have, thus, demonstrated that the use of Semantic Technologies for creating given consent-compliant datasets is not only feasible but also facilitates compliance verification processes and provides organizations with a compliance-by-design approach for their operations."

"While it does so automatically, our approach allows one to look up why someone's personal data was included in a particular dataset via the consent knowledge base, the sequence of SPARQL queries, and the timestamps captured in the provenance model."

"Our model relates dataset schemas with consent, GConsent and DPV have more fine-grained representations for personal data and data categories."

User consent modeling for ensuring transparency and compliance in smart cities [262]

This is a machine-generated summary of:

Fernández, Javier D.; Sabou, Marta; Kirrane, Sabrina; Kiesling, Elmar; Ekaputra, Fajar J.; Azzam, Amr; Wenning, Rigo: User consent modeling for ensuring transparency and compliance in smart cities [262]

Published in: Personal and Ubiquitous Computing (2020)

Link to original: https://doi.org/10.1007/s00779-019-01330-0

Copyright of the summarized publication:

If you want to cite the papers, please refer to the original.

For technical reasons we could not place the page where the original quote is coming from.

Abstract-Summary

"Smart city infrastructures such as transportation and energy networks are evolving into so-called cyber physical social systems (CPSSs), which collect and leverage citizens' data in order to adapt services to citizens' needs."

"Current systems either try to escape the privacy challenge via anonymization or use very rigid, hard-coded workflows that have been agreed with a data protection authority."

"We focus on a concrete setting where we extend the domain-agnostic SPECIAL policy language for a smart mobility use case supplied by Vienna's largest utility provider."

"To that end, (1) we create an extension of SPECIAL in terms of a core CPSS vocabulary that lowers the semantic gap between the domain agnostic terms of SPECIAL and the vocabulary of the use case; (2) we propose a workflow that supports defining domain-specific vocabularies for complex CPSSs; and (3) show that these two contributions allow successfully achieving the goals of our setting."

Introduction

"The EU H2020 SPECIAL project strives to enable companies to work with the data subjects to sustain trust in complex systems, such as CPSSs."

"If the data subject sets a preference in the CPSS, e.g., via his mobile device, the CPSS needs to make sure said preference are followed in the subsequent complex workflows that sometimes even transcend organizational borders."

"In a nutshell, the novelty of this paper lies in a non-trivial extension of the semantics used within cyber physical systems in order to prepare a much more sophisticated approach to data protection and GDPR compliance for CPSSs."

"Concretely, we make the following contributions: The SPECIAL-CPSS core vocabulary, which serves as a means for describing usage constraints across a variety of CPSS domains and is usable within a SPECIAL-like system; The practical workflow, which enables CPSS preference, constraint, and consent modeling in general; A practical use case to show how these techniques can be applied in a CPSS setting."

Related work

"Formal policy languages are designed to unambiguously represent usage policies, which makes it possible to automatically verify whether data processing is covered by data subjects' given consent."

"Since the GDPR has come into effect, data controllers must provide transparency to data subjects with respect to the processing of personal data and compliance, i.e., the CPSS data controller must demonstrate that the usage of personal data complies with data subjects' consent (it respects/does not violate any requests)."

"As for transparency about data processing, relevant work primarily focuses on the re-purposing of existing logging mechanisms as the basis for personal data processing transparency and compliance [217]."

"To existing approaches, in this paper, we focus on vocabularies that can be used to record both usage policies and data processing and sharing events in a manner that supports automatic compliance checking."

Background and methodology

"In order to enable companies to comply with consent and transparency requirements stipulated by the GDPR, SPECIAL provides a policy language, vocabularies, and a consent, transparency, and compliance framework, which can be adapted and extended specifically for CPSS needs."

"We rely on the output of Step 1, i.e., the SPECIAL-CPSS core ontology, and exemplify its usage to derive a domain ontology."

"We validate both the SPECIAL-CPSS core ontology and the UC consent vocabularies by exemplifying their use to create usage policies required to support our smart mobility use cases."

"The major, reusable outputs of our methodology are: The SPECIAL-CPSS core ontology, which can also serve as a starting point for describing usage policies vocabularies in other CPSS domains (output of Step1)."

"The workflow to define CPSS data subjects' consent and data usage policies (output of Step2); The overall methodology itself, can be followed whenever adapting SPECIAL to new domains."

CitySPIN smart mobility use case

"We present four scenarios of the CitySPIN use case, exemplified with a generic WienMobil APP user, Doris, and a WStW transportation network planner, Eva."

"Doris installs the WienMobil APP, provided by a subsidiary of WStW, which allows her to obtain real-time public transport routing information in Vienna."

"This can be integrated with external non-personal data sources (e.g., traffic congestion and city events) and will be used to analyze her mobility patterns, including the attendance to events in the city (e.g., concerts, sport events), in order to recommend the best routes and notify about delays in the future."

"The day of the event, the APP suggests to keep her current GPS active in order to receive live updates of museum attendance, routes, and sub-event recommendations."

Deriving the SPECIAL-CPSS core vocabulary

"The study aimed to answer the following research questions in order to provide an in-depth understanding of CPSS as described in the literature: RQ1: What is an overarching definition of CPSS?"

"EC3: Studies where a CPSS is only mentioned as a side topic, e.g., this term appears only in the title or a reference or as an example."

"Besides bibliographic information, we collected data items relevant to our research questions, e.g., CPSS definition, application domain, CPSS purposes, CPSS process steps/activities, involvement of human actors, data sources, collected data."

"OpinionsAndFeedback provided by users are important sources of information for CPSSs that aim to adjust recommendations according to the users' perception of some service (e.g., a restaurant)."

A practical workflow for conceptualizing CPSS data usage policies

"When it comes to representing CPSS data usage policies using the SPECIAL model, a first step involves a deeper analysis of the potential data usage policies for the use case."

"A second step consists of (i) identifying standard vocabularies to represent such terms (within the SPECIAL auxiliary vocabularies or others existing and reusable vocabularies) or (ii) extending the SPECIAL auxiliary vocabularies to cover the CPSS use case needs."

"The category of the final relevant personal data in the CPSS (i.e., data that cannot be further anonymized) needs to be represented in the SPECIAL policy language."

"Given the inherent complexity of CPSSs, where multiple components are often organized in a "pipeline" architecture, the first step is to analyze the information flow and the components identified in the planning phase, and to describe the CPSS components processing personal data."

Validation: user policy representation in CitySPIN use cases

"The data category can be represented with a union (ObjectUnionOf) of three use case-specific terms (wm: SmartMobilityHistory, wm: MobilityPattern and wm: WienMobileGPSData) that accurately reflect the personal data involved in the scenario."

"The storage (location and duration) and recipient (just ours) directly follow from the use case description and are encoded according to the SPECIAL policy language (e.g., the 2-year period is represented with a xsd: maxinclusive restriction)."

"That both concepts are represented using specific use case terms; hence, they should include a human-readable comprehensive definition of the actual type of data considered in each case (e.g., cultural and heritage events in the case of the partner recommendations)."

"To the previous cases, the policy represents the data with a union (ObjectUnionOf) of terms to capture the real-time use case location data history (WienMobile GPSDataRealTime) and the planned event data (in this case, using the CPSS upper level ontology via svd-cpss: PlannedEventData)."

Summary and future work

"In this paper, we exemplify extending the SPECIAL domain-agnostic policy language for describing user policies in a smart mobility use case provided by Vienna's largest utility provider."

"We relied on an approach which aims to support CPSS owners in general, and WStW in particular, in adapting the policy language for the needs of their own use cases in two ways: (1) by providing the SPECIAL-CPSS core vocabulary that already extends the domain-agnostic SPECIAL terms toward the domain of CPSS; (2) by proposing a novel practical workflow that can be used to elicit vocabularies for defining CPSS data subjects' consent and data usage policies."

"We validate the resulting vocabularies (both core and use case-specific) by demonstrating that they can be used successfully to construct usage policies according to the SPECIAL specification."

"A current limitation of this work is that the SPECIAL-CPSS core vocabulary and the proposed workflow have been tested on a mobility use case only."

Adapting HTML5 Web applications to user privacy preferences [263]

This is a machine-generated summary of:

Kapitsaki, Georgia M.; Charalambous, Theodoros: Adapting HTML5 Web applications to user privacy preferences [263]

Published in: World Wide Web (2018)

Link to original: https://doi.org/10.1007/s11280-018-0628-4

If you want to cite the papers, please refer to the original.

For technical reasons we could not place the page where the original quote is coming from.

Abstract-Summary

"Users can give their consent on the use of this sensitive information, but should have the right to express their privacy preferences, so that Web applications can adapt themselves accordingly."

"We address the above by specifying a privacy preferences language for users tailored to HTML5 Web applications employing the eXtensible Access Control Markup Language, whereas we introduce a mechanism that adapts the Web application considering these user preferences."

"We describe the process followed for the creation of the privacy preferences, the process of application adaptation and the benefits this approach provides to end-users via a demonstration and evaluation of the use of the extension."

Introduction

"Users should be able to specify their privacy preferences in the framework of HTML5 Web applications, so that access to data is allowed only with user's consent."

"Having as motivation the above, in this paper we are presenting our work on the adaptation of Web applications considering user's privacy preferences using a simple enforcement mechanism, whereas our approach is accompanied by a tool implemented for popular Web browsers (Firefox and Chrome)."

"The contribution of this work is threefold: (1) we introduce the structure of a privacy preferences language for Web users in the framework of HTML5 Web applications adopting for this purpose the eXtensible Access Control Markup Language (XACML) 3.0, (2) we propose a mechanism for adapting Web applications on-the-fly based on blocking and obfuscation mechanisms, and (3) we introduce context as an important parameter for HTML5 Web application adaptation, when accessing user-relevant resources."

Related work

"Privacy policies may refer to service provider or end-user preferences on privacy settings."

"It gives providers the opportunity to define their privacy policies in P3P format, so that Web browsers can compare them with user's privacy settings, and take appropriate actions, if needed (e.g. alert the user)."

"It is a context-dependent privacy policy language, where user's context is used as an additional condition for deciding upon granting or denying access to the requested resource."

"It uses for this reason different constructs that are not required in HTML5 Web applications, but considers additionally contextual elements (e.g. user location)."

"Previous works have focused on obfuscation mechanisms for user location and address relevant attacks."

"Location Guard offers an approach that adds noise to user's location or provides a fake address and is available as a browser extension [218]."

User privacy policies language

"Regarding the context elements supported by PrivacySafer we are considering the following that have not been included in a taxonomy, since they are few: Day Time Battery status Website domain Specific Web location We have added the possibility of users choosing the time of access to private information (Day and Time elements), as they may wish to allow access to sensitive data when at work but not at other times."

"The main categories of the data access taxonomy form the user policy set: {Device ∪ Communication ∪ Data And Files} Items in each set are represented as 2-tuples that indicate the specific element (e.g. bluetooth) and a value of true or false showing respectively that access is either allowed or denied."

"When specific context conditions are also applied (e.g. for a specific Web domain or Web application), each item contains more information: the resource, the user (of the resource), the day and time access to the resource is granted and any additional condition that may exist."

Application adaptation mechanism

"This may lead to less personalized application experience, but, on the other hand, it limits the execution of code that accesses data and helps in storing user relevant information in Web cookies providing an opportunity for personalized privacy."

"We apply the following measures to obfuscation that are captured in different choices provided to the user: Specific coordinates: the user may specify the coordinates that will be returned to any Web application requesting geolocation access."

"If the user selects this option, but does not provide specific coordinates, coordinates that point to a specific city in the world are returned to any Web application that asks for user's location."

"Auto: PrivacySafer applies obfuscation by providing to any Web application a location within a specific location radius (currently set to 20 km) from the user's actual location."

PrivacySafer implementation

"The extension can function similarly on Firefox due to recent changes to the extension implementation mechanism (add-ons versus Web extensions)."

"PrivacySafer does not affect browsers actions related with asking users about access to their data (e.g. Will you allow website to access the location?)."

"We have relied on page mode, simple storage and Firefox workers and have moved the user interfaces to separate Web locations, where the users can set and edit their settings."

"The Web locations for editing user settings correspond to the two panels presented for the desktop extension (basic and advanced settings)."

"These settings are subsequently stored in the browser's local storage, since the applications are only used for providing a presentation to users."

"After retrieving user settings from the local storage, page mode communicates with the corresponding webpage in order to provide their settings, allow their editing and store them back, if changes are made."

Demonstration and evaluation

"We have evaluated the efficiency of PrivacySafer by measuring the page loading time (all transfers finished and page fully loaded) for different cases comparing loading time with and without the use of the extension."

"Some of the pages we employed for loading time measurements use specific HTML5 features that are relevant to sensitive data, whereas others do not perform any data request."

"Using PrivacySafer leads to lower loading times (e.g. in the first two Web locations that request access to user's location)."

"When the user blocks access to specific resources, loading page elements may take less time, since these elements do not need to be loaded."

"Loading time is higher for many Web locations on Firefox, but this observation is irrelevant to the use of PrivacySafer."

Limitations

"Although HTML5 features supported in each browser are considered in the extension implementation, changes between browser versions may cause some features not to function."

"PrivacySafer functionality is restricted to access to information from HTML5 features."

"Other means used by applications to receive user data are not considered (e.g. third-party trackers, IP address), whereas other measures are required in order to prevent attacks (e.g. code injection) [219]."

"A further limitation can be found in the fact that PrivacySafer needs access to sensitive information (e.g. battery level), in order to decide about applying a privacy rule."

"If the user is, for instance, in the only city close to a lake and the obfuscation type she chooses (e.g. radius or auto) places her in the lake, the application may be able to infer that the real location is on the shore."

Conclusions

"We have described our work on PrivacySafer, a mechanism to adapt HTML5 Web applications to user's privacy policies, along with an XACML notation for expressing user privacy preferences."

"PrivacySafer permits or blocks access to different features of HTML5, when requested by Web applications, whereas this selective access considers also user's context (e.g. day and time, battery level)."

"As future work, we intend to extend the support of HTML5 features in PrivacySafer, as browser support expands."

"Regarding location obfuscation, we intend to work on optimization techniques for Web applications using source code analysis, where the type of obfuscation supported will rely on the level of data granularity required for the Web application to function properly."

References

1. Thompson, Beth. 2016. Data protection: how medical researchers persuaded the European Parliament to compromise. http://blogs.lse.ac.uk/brexit/2016/02/11/data-protection-how-medical-researchers-persuaded-the-european-parliament-to-compromise/. Accessed 31 Oct 2019.
2. Matey, J. (2021). Sexual consent and lying about one's self. Philosophy and Phenomenological Research, 102(2), 380–400.
3. Di Paolo, E., Cuffari, E. C., & De Jaegher, H. (2018). Linguistic bodies. MIT Press.
4. Penny, L. (2017, October 10). The horizon of desire. Longreads. https://longreads.com/2017/10/10/the-horizon-of-desire/.
5. Penny, L. (2022). Sexual revolution: Modern fascism and the feminist fightback. Bloomsbury.
6. Drachsler, H., & Greller, W. (2016). Privacy and analytics—it's a DELICATE Issue. A Checklist for Trusted Learning Analytics. LAK Conference Proceedings. https://doi.org/10.1145/2883851.2883893.
7. Creswell, J. W. (2013). Educational research: Planning, conducting, and evaluating quantitative and qualitative research (4th ed.). London: Pearson Education Limited.
8. Annane D, Sebille V, Charpentier C, Bollaert PE, Francois B, Korach JM, Capellier G, Cohen Y, Azoulay E, Troche G, Chaumet-Riffaut P, Bellissant E (2002) Effect of treatment with low doses of hydrocortisone and fludrocortisone on mortality in patients with septic shock. JAMA 288:862–871.

9. Harvey S, Harrison DA, Singer M, Ashcroft J, Jones CM, Elbourne D, Brampton W, Williams D, Young D, Rowan K (2005) Assessment of the clinical effectiveness of pulmonary artery catheters in management of patients in intensive care (PAC-Man): a randomised controlled trial. Lancet 366:472–477.
10. Marshall LF, Maas AI, Marshall SB, Bricolo A, Fearnside M, Iannotti F, Klauber MR, Lagarrigue J, Lobato R, Persson L, Pickard JD, Piek J, Servadei F, Wellis GN, Morris GF, Means ED, Musch B (1998) A multicenter trial on the efficacy of using tirilazad mesylate in cases of head injury. J Neurosurg 89:519–525.
11. Young B, Runge JW, Waxman KS, Harrington T, Wilberger J, Muizelaar JP, Boddy A, Kupiec JW (1996) Effects of pegorgotein on neurologic outcome of patients with severe head injury. A multicenter, randomized controlled trial. JAMA 276:538–543.
12. Kern JW, Shoemaker WC (2002) Meta-analysis of hemodynamic optimization in high-risk patients. Crit Care Med 30:1686–1692.
13. Rivers E, Nguyen B, Havstad S, Ressler J, Muzzin A, Knoblich B, Peterson E, Tomlanovich M (2001) Early goal-directed therapy in the treatment of severe sepsis and septic shock. N Engl J Med 345:1368–1377.
14. Wiedermann CJ, Almici M, Mangione S, Giarratano A, Mayr O (2006) Clinical research in Italy in adult patients unable to consent: after implementation of the European Union's Directive 2001/20/CE. Intensive Care Med 33:316–318.
15. Spence JM, Notarangelo V, Frank J, Long J, Morrison LJ (2005) Responses to written notification during out-of-hospital care trials using waiver of informed consent. Acad Emerg Med 12:1099–1103.
16. New General Data Protection Regulation published in EU Official Journal on 4 May 2016. http://eur-lex.europa.eu/legalcontent/EN/TXT/?uri=uriserv:OJ.L_2016.119.01.0001.01.ENG&toc=OJ:L:2016:119:TOC.
17. European Commission (2016) Data protection: Article 29 Working Party. http://ec.europa.eu/newsroom/just/item-detail.cfm?item_id=50083.
18. BBMRI- ERIC. http://www.bbmri.nl/wp-content/uploads/2017/03/20170314-BBMRIERIC_FAQs_on_the_GDPR_V2.0.pdf.
19. The EU General Data Protection Regulation (GDPR) is the most important change in data privacy regulation in 20 years.eugdpr.org. EU: c2019 [cited 20 April 2020]. Available from: https://eugdpr.org/.
20. General Data Protection Regulation (GDPR). Recitals. Key Issues. gdpr-info.eu. EU: c2019 [cited 20 April 2020]. Available from: https://gdpr-info.eu/.
21. Clarke N, Vale G, Reeves EP, Kirwan M, Smith D, Farrell M, Hurl G, McElvaney NG (2019) GDPR: an impediment to research? Ir J Med Sci 188:1129–1135. https://doi.org/10.1007/s11845-019-01980-2.
22. Data Protection Act (2018) (Section 36(2)) (Health Research) Regulations 2018. irishstatutebook.ie. Ireland:c2019 [cited 20 April 2020]. Available from: http://www.irishstatutebook.ie/eli/2018/si/314/made/en/pdf.
23. Health Research Consent Declaration Committee. Hrcdc.ie. Ireland: c2019 [cited 20 April 2020]. Available from: https://hrcdc.ie/.
24. Health Research Board Decision Tree. hrb.ie. Ireland: c2019 [cited 20 April 2020]. Available from: https://www.hrb.ie/fileadmin/user_upload/Decision_Tree_30072018.pdf.
25. Guidance notes to support and application to process or further process personal data for the purposes of health research commencing before 8 August 2018 (re-consenting). hrcdc.ie. Ireland. c2019 [cited 20 April 2020]. Available from: https://hrcdc.ie/wp-content/uploads/2018/12/HRCDC-Guidance_Re-consenting.pdf.
26. Woods K (2015) Researchers: consent rules put future medical trials at risk. Sunday Business Post (Ireland). Sect. A:6 (col. 1).
27. Guidelines 03/2020 on the processing of data concerning health for the purpose of scientific research in the context of the COVID-19 outbreak. https://edpb.europa.eu/our-work-tools/our-documents/guidelines/guidelines-032020-processing-data-concerning-health-purpose_en. Accessed 27 Apr 2020.

28. Bobrow M. Funders must encourage scientists to share. Nature. 2015;522:129.
29. Auffray C, Balling R, Barroso I, Bencze L, Benson M, Bergeron J, et al. Making sense of big data in health research: towards an EU action plan. Genome Med. 2016;8:71.
30. Toronto International Data Release Workshop Authors, Birney E, Hudson TJ, Green ED, Gunter C, Eddy S, et al. Prepublication data sharing. Nature. 2009; 461:168–70.
31. Ramos EM, Din-Lovinescu C, Bookman EB, McNeil LJ, Baker CC, Godynskiy G, et al. A mechanism for controlled access to GWAS data: experience of the GAIN Data Access Committee. Am J Human Genet. 2013;92:479–88.
32. Brenner SE. Be prepared for the big genome leak. Nature. 2013;498:139.
33. Manolio TA, Fowler DM, Starita LM, Haendel MA, MacArthur DG, Biesecker LG, et al. Bedside back to bench: building bridges between basic and clinical genomic research. Cell. 2017;169:6–12.
34. Ferraiole D, Kuhn, R. Role-based access control. Proceedings of the 19th National Computer Security Conference. 1992.
35. Ferraiole D, Cugini, J, Kuhn, R. Role-based access control: features and motivations. Proceedings of the 11th Conference in Computer Security Applications. 1995.
36. Sandhu R, Coyne E, Feinstein H, Youman C. Role-based access control models. IEEE Comput. 1996;29:38–47.
37. Barkley J, Cincotta, A, Ferraiolo, D, Gavrila, S, Kuhn, D. Role based access control for the world wide web. 20th National Computer Security Conference. 1997.
38. Ferraiole D, Sandhu, R, Gavrila, S, Kuhn, D, Chandramouli, R. A proposed standard for role-based access control. National Institute of Standards and Technology. 2000. ACM Transactions on Information and System Security, Vol. 4, No. 3, August 2001, pp. 224–274.
39. Dyke SO, Philippakis AA, Rambla De Argila J, Paltoo DN, Luetkemeier ES, Knoppers BM, et al. Consent codes: upholding standard data use conditions. PLoS Genet. 2016;12:e1005772.
40. Art. 4(5)GDPR—Conditions for consent | General Data Protection Regulation (GDPR) [Internet]. 2018 General Data Protection Regulation (GDPR). [cited 2 May 2019]. Available from: https://gdpr-info.eu/art-4-gdpr/.
41. World Medical Association. Declaration of Helsinki, Ethical Principles for Medical Research Involving Human Subjects (2004), latest version available electronically at http://www.wma.net/e/policy/b3.htm.
42. Clayton EW: Panel comment: why the use of anonymous samples for research matters. J Law Med Ethics 1995; 23: 375–377.
43. Hoeyer K, Olofsson B-O, Mjörndal T, Lynöe N: Informed consent and biobanks: a population-based study of attitudes towards tissue donation for genetic research. Scand J Public Health 2004; 32: 224–229.
44. UNESCO: International declaration on human genetic data. Eur J Health Law 2004; 11: 93–107.
45. The Nuremberg Code. (1947). BMJ 1996; 313: 1448.
46. Hallinan, Dara. 2020. Broad consent under the GDPR: an optimistic perspective on a bright future. Life sciences, society and policy 16 (1): 1. https://doi.org/10.1186/s40504-019-0096-3.
47. Gefenas, Eugenijus, J., V. Lekstutiene, M. Lukaseviciene, M. Hartlev, Mourby, and K. Ó. Cathaoir. 2021. Controversies between regulations of research ethics and protection of personal data: informed consent at a cross-road. Medicine, health care, and philosophy. 25 (2): 23–30. https://doi.org/10.1007/s11019-021-10060-1.
48. Arnason, Vilhjalmur. 2004. Coding and consent: moral challenges of the database project in Iceland. Bioethics 18 (1): 27–49. https://doi.org/10.1111/j.1467-8519.2004.00377.x.
49. Gould, Carol C. 2019. How democracy can inform consent: cases of the Internet and bioethics. Journal of Applied Philosophy 36 (2): 173–191.
50. Caulfield, Timothy, Jane Kaye. 2009. Broad Consent in Biobanking: Reflections on Seemingly Insurmountable Dilemmas. Medical Law International 10 (2): 85–100. https://doi.org/10.1177/096853320901000201.
51. NHSX is leading the largest digital health and social care transformation programme in the world. NHSX. 2019. https://www.nhsx.nhs.uk/. Accessed 26 Jun 2020.

52. Kaye J, Curren L, Anderson N, Edwards E, Fullerton SM, Kanellopoulou N, et al. From patients to partners: participant-centric initiatives in biomedical research. Nat Rev Genet. 2012;13:371–6. https://doi.org/10.1038/nrg3218.
53. Hofmann, B. 2009. Broadening consent–and diluting ethics? Journal of medical ethics 35 (2): 125–129. https://doi.org/10.1136/jme.2008.024851.
54. Prictor M, Lewis MA, Newson AJ, Hass M, Baba S, Kim H, et al. Dynamic consent: an evaluation and reporting framework. J Empir Res Hum Res Ethics. 2019;15:175–86. https://doi.org/10.1177/1556264619887073.
55. Teare HJ, Morrison M, Whitley EA, Kaye J. Towards 'Engagement 2.0': insights from a study of dynamic consent with biobank participants. Digit Health. 2015;1:1–13. https://doi.org/10.1177/2055207615605644.
56. Eurac Research. CHRIS Gesundheitsstudie. Vinschgau Südtirol: Eurac Research. https://de.chris.eurac.edu/. Accessed 26 Jun 2018.
57. Piciocchi C, Ducato R, Martinelli L, Perra S, Tomasi M, Zuddas C, et al. Legal issues in governing genetic biobanks: the Italian framework as a case study for the implications for citizen's health through public-private initiatives. J Community Genet. 2018;9:177–90. https://doi.org/10.1007/s12687-017-0328-2.
58. Pattaro C, Gögele M, Mascalzoni D, Melotti R, Schwienbacher C, De Grandi A, et al. The Cooperative Health Research in South Tyrol (CHRIS) study: rationale, objectives, and preliminary results. J Transl Med. 2015;13:348. https://doi.org/10.1186/s12967-015-0704-9.
59. Keene DJ, Schlüssel MM, Thompson J, Hagan DA, Williams MA, Byrne C, et al. Prognostic models for identifying risk of poor outcome in people with acute ankle sprains: the SPRAINED development and external validation study. Health Technol Assess. 2018;22:1–112. https://doi.org/10.3310/hta22640.
60. Minari J, Teare H, Mitchell C, Kaye J, Kato K. The emerging need for family-centric initiatives for obtaining consent in personal genome research. Genome Med. 2014;6:118. https://doi.org/10.1186/s13073-014-0118-y.
61. Knoppers BM, Kekesi-Lafrance K. The genetic family as patient? Am J Bioeth. 2020;20:77–80. https://doi.org/10.1080/15265161.2020.1754505.
62. NIHR INVOLVE. INVOLVE Supporting public involvement in NHS, public health and social care research. NIHR INVOLVE. 1996. https://www.invo.org.uk/. Accessed 26 Jun 2020.
63. Budin-Ljøsne, Isabelle, Harriet J. A. Teare, Stephan Kaye, Heidi Beate Beck, Luciana Bentzen, Clive Caenazzo, Flavio D Collett'Heike Abramo, and Felzmann, Teresa Finlay, Muhammad Kassim Javaid, Erica Jones, Višnja Katić, Amy Simpson, and Deborah Mascalzoni. 2017. Dynamic Consent: a potential solution to some of the challenges of modern biomedical research. BMC medical ethics 18 (1): 4. https://doi.org/10.1186/s12910-016-0162-9.
64. Kaye, J., Whitley, E. A., Lund, D., Morrison, M., Teare, H., & Melham, K. (2015). Dynamic consent: A patient interface for twenty-first century research networks. European Journal of Human Genetics, 23, 141–146. https://doi.org/10.1038/ejhg.2014.71.
65. Prictor M, Lewis MA, Newson AJ, Haas M, Baba S, Kim H, et al. Dynamic consent: an evaluation and reporting framework. J Empir Res Hum Res Ethics. 2019;15:175–186. https://doi.org/10.1177/1556264619887073.
66. Mascalzoni D, Hicks, A, Pramstaller, PP. Consenting in population genomics as an open communication process. Stud Ethics Law Technol. 2009;3. https://doi.org/10.2202/1941-6008.1085.
67. Mascalzoni D, Hicks A, Pramstaller P, Wjst M. Informed consent in the genomics era. PLoS Med. 2008;5:e192. https://doi.org/10.1101/cshperspect.a036582.
68. Pacyna JE, McCormick JB, Olson JE, Winkler EM, Bublitz JT, Hathcock MA, et al. Assessing the stability of biobank donor preferences regarding sample use: evidence supporting the value of dynamic consent. Eur J Hum Genet. 2020;28:1168–77. https://doi.org/10.1038/s41431-020-0625-9.
69. International Council for Harmonisation of Technical Requirements for Pharmaceuticals for Human Use. Integrated addendum to ICH E6 (R1): guideline for Good Clinical Practice E6 (R2). 2016. https://database.ich.org/sites/default/files/E6_R2_Addendum.pdf.

70. Regulation (EU) No 910/2014 of the European Parliament and of the Council of 23 July 2014 on electronic identification and trust services for electronic transactions in the internal market and repealing Directive 1999/93/EC. https://ec.europa.eu/futurium/en/system/files/ged/eidas_regulation.pdf.
71. Derks S. eConsent readiness in 9 countries; 2020. https://www.castoredc.com/whitepaper/econsent-readiness-by-country-guide/.
72. De Sutter E, Zaçe D, Boccia S, Di Pietro ML, Geerts D, Borry P, et al. Implementation of electronic informed consent in biomedical research and stakeholders' perspectives: systematic review. J Med Internet Res. 2020;22(10): e19129. https://doi.org/10.2196/19129.
73. Willison DJ, Schwartz L, Abelson J, Charles C, Swinton M, Northrup D, Thabane L: Alternatives to project-specific consent for access to personal information for health research: What is the opinion of the Canadian public?. Journal of the American Medical Informatics Association. 2007, 14: 706-712. https://doi.org/10.1197/jamia.M2457.
74. Bialke M, Penndorf P, Wegner T, Bahls T, Havemann C, Piegsa J, et al. A workflow-driven approach to integrate generic software modules in a Trusted Third Party. J Transl Med. 2015. https://doi.org/10.1186/s12967-015-0545-6.
75. HL7 Germany. Simplifier-Project of the HL7 Working Group Consent Management (German). https://simplifier.net/Einwilligungsmanagement (2021). Accessed 01 May 2021.
76. MII Task Force "Consent Implementation". ART-DECOR MII Consent Policy ValueSet. https://art-decor.org/decor/services/RetrieveValueSet?id=2.16.840.1.113883.3.1937.777.24.11.36&language=en-US (2021). Accessed 05 May 2021.
77. HL7 Germany. Implementation Guide of the HL7 Working Group Consent Management—Examples for FHIR Consent (German). https://ig.fhir.de/einwilligungsmanagement/stable/Consent.html (2022). Accessed 24 Nov 2022.
78. Independent Trusted Third Party of the University Medicine Greifswald. Simplifier Implementation Guide TTP-FHIR Gateway—Examples Consent Bundle (German). https://simplifier.net/ths-greifswald/~resources?category=Example&exampletype=Bundle (2021). Accessed 29 Jun 2021.
79. Independent Trusted Third Party of the University Medicine Greifswald. Implementation Guide TTP-FHIR Gateway for Consent Management (German). https://ths-greifswald.de/gics/fhir (2022). Accessed 24 Nov 2022.
80. Independent Trusted Third Party of the University Medicine Greifswald. Consent Management gICS—Download. https://www.ths-greifswald.de/forscher/gics/#download (2021). Accessed 29 Sep 2021.
81. Independent Trusted Third Party of the University Medicine Greifswald. MII Broad Consent Version 1.6d—gICS-Template (JSON). https://www.ths-greifswald.de/gics/templates/mii/1.6.d (2021). Accessed 03 Mar 2021.
82. Independent Trusted Third Party of the University Medicine Greifswald. MII Broad Consent Version 1.6f for NUM-CODEX—gICS-Template (JSON). https://www.ths-greifswald.de/gics/templates/mii/1.6.f (2021). Accessed 29 Jun 2021.
83. Michalik C, Naziyok TP, Schlünder I, Harnischmacher U, Röhrig R. Informed Consent—Update des TMF-Wizards zur Erstellung von Patienteninformation und Einwilligung für medizinische Forschungsvorhaben. In Proceedings of HEC 2016: Health-Exploring Complexity. Joint Conference of GMDS, DGEpi, IEA-EEF, EFMI; 2016; München. https://doi.org/10.3205/16gmds137.
84. Idris T. "Patient has consented"—the IHE APPC Profile (presentation slides) [GERMAN]; 2016 http://download.hl7.de/veranstaltungen/jahrestagungen/2016/14-idris.pdf. Accessed 01 Aug 2017.
85. Bahls T, Liedtke W, Geidel L, Langanke M. Ethics meets IT: aspects and elements of computer-based informed consent processing. In: Fischer T, Langanke M, Marschall P, et al., editors. Individualized medicine, ethical, economical and historical perspectives. Springer: Berlin; 2015. p. 209–29.
86. HL7 Deutschland e.V. Forum for Interoperability of HL7 Germany, IHE Germany, bvitg and DIN; 2018. http://interoperabilitaetsforum.de/. Accessed 01 Jan 2018.

87. Prokosch HU, Acker T, Bernarding J, Binder H, Boeker M, Boerries M, et al. MIRACUM: medical informatics in research and care in university medicine. Methods Inf Med. 2018;57(S01):e82–91. https://doi.org/10.3414/ME17-02-0025.
88. Winter A, Stäubert S, Ammon D, Aiche S, Beyan O, Bischoff V, et al. Smart medical information technology for healthcare (SMITH). Methods Inf Med. 2018; 57(S 01): e92-e105. https://doi.org/10.3414/ME18-02-0004.
89. Haarbrandt B, Schreiweis B, Rey S, Sax U, Scheithauer S, Rienhoff O, et al. (2018) HiGHmed—an open platform approach to enhance care and research across institutional boundaries. Methods Inf Med 57(S 01): e66-e81. https://doi.org/10.3414/ME18-02-0002.
90. Beskow LM. Lessons from hela cells: the ethics and policy of biospecimens. Annu Rev Genom Hum Genet août. 2016;17:395–417.
91. Cumyn A, Barton A, Dault R, Cloutier A, Jalbert R, Ethier J. Informed consent within a learning health system: A scoping review. Learn Health Syst [Internet]. avr 2020 [cité 1 juill 2020];4(2). Disponible sur: https://onlinelibrary.wiley.com/doi/abs/, https://doi.org/10.1002/lrh2.10206
92. Article L1126–1 [Internet]. Code de Santé publique, Loi n°2012–300 mars 5, 2012. Disponible sur: https://www.legifrance.gouv.fr/affichCodeArticle.do;jsessionid=40175270A1F85098A5747145AD426107.tpdila18v_1?cidTexte=LEGITEXT000006072665&idArticle=LEGIARTI000025454365&dateTexte=20170209&categorieLien=id#LEGIARTI000025454365.
93. Article 7 [Internet]. Loi n°78–17 relative à l'informatique, aux fichiers et libertés juin 1, 1978. Disponible sur: https://www.legifrance.gouv.fr/affichTexteArticle.do;jsessionid=5E06597CB3A2171C22AC02115FE50850.tpdila16v_2?idArticle=LEGIARTI000006528071&cidTexte=LEGITEXT000006068624&dateTexte=20170412.
94. Article L1211–2 [Internet]. Code de la Santé publique, Loi n°2004–800 août 6, 2004. Disponible sur: https://www.legifrance.gouv.fr/affichCode.do?idArticle=LEGIARTI000006686056&idSectionTA=LEGISCTA000006171017&cidTexte=LEGITEXT000006072665&dateTexte=20170324.
95. Hostiuc S, Moldoveanu A, Dascălu M-I, Unnthorsson R, Jóhannesson ÓI, Marcus I. Translational research-the need of a new bioethics approach. J Transl Med. 2016;14:16.
96. Ménoni V, Lucas N, Leforestier JF, Dimet J, Doz F, Chatellier G, et al. The readability of information and consent forms in clinical research in France. PloS One. 2010;5(5):e10576.
97. Prictor M, Teare H, Bell J, Taylor MJ, Kaye J. Consent for Data Processing Under the General Data Protection Regulation: Could "Dynamic Consent" Assist Researchers? SSRN Electron J [Internet]. 2018 [cité 22 mars 2019]; Disponible sur: https://www.ssrn.com/abstract=3222394.
98. Aitken M, de St Jorre J, Pagliari C, Jepson R, Cunningham-Burley S. Public responses to the sharing and linkage of health data for research purposes: a systematic review and thematic synthesis of qualitative studies. BMC Med Ethics. 2016;17(1):73.
99. Kondylakis H, Koumakis L, Hänold S, Nwankwo I, Forgó N, Marias K, et al. Donor's support tool: Enabling informed secondary use of patient's biomaterial and personal data. Int J Med Inf janv. 2017;97:282–92.
100. O'Doherty KC, Burgess MM. Engaging the public on biobanks: outcomes of the BC biobank deliberation. Public Health Genom. 2009;12(4):203–15.
101. Hoeyer K. Donors perceptions of consent to and feedback from biobank research: time to acknowledge diversity? Public Health Genom. 2010;13(6):345–52.
102. Hansson, M. G. 2009. Ethics and biobanks. British journal of cancer 100 (1): 8–12. https://doi.org/10.1038/sj.bjc.6604795.
103. Sajantila A, Budowle B (2016) Postmortem medico-legal genetic diagnostics also require reporting guidance. Eur J Hum Genet 24(3):329–330.
104. EDPB (European Data Protection Board). 2019. Opinion 3/2019 concerning the Questions and Answers on the interplay between the Clinical Trials Regulation (CTR) and the General Data Protection regulation (GDPR) (art. 70.1.b)). https://edpb.europa.eu/sites/edpb/files/files/file1/edpb_opinionctrq_a_final_en.pdf. Accessed June 7, 2021.

105. Article 29 Working Party. (2018b). Guidelines on consent under regulation 2016/679. Retrieved June 27, 2020, from https://ec.europa.eu/newsroom/article29/item-detail.cfm?item_id=623051.
106. Secretary's advisory committee on human research protections. (2018). Attachment B: European Union's general data protection regulations. Retrieved June 27, 2020, from https://www.hhs.gov/ohrp/sachrp-committee/recommendations/attachment-b-implementation-of-the-european-unions-general-data-protection-regulation-and-its-impact-on-human-subjects-research/index.html.
107. Warner, T. D., Weil, C. J., Andry, C., Degenholtz, H. B., Parker, L., Carithers, L. J., et al. (2018). Broad consent for research on biospecimens: The views of actual donors at four U.S. medical centers. Journal of Empirical Research on Human Research Ethics, 13(2), 115–124. https://doi.org/10.1177/1556264617751204.
108. Artal R, Rubenfeld S. Ethical issues in research. Best Pract Res Clin Obstet Gynaecol. 2017;43:107–14.
109. Dankar FK, Gergely M, Dankar SK. Informed consent in biomedical research. Comput Struct Biotechnol J. 2019;17:463–74.
110. Pranati. Informed consent: Are we doing enough? Perspect Clin Res. 2010;1:124–7.
111. Lentz, J., Kennett, M., Perlmutter, J., & Forrest, A. (2016). Paving the way to a more effective informed consent process: Recommendations from the Clinical Trials Transformation Initiative. Contemporary Clinical Trials, 49, 65–69. https://doi.org/10.1016/j.cct.2016.06.005.
112. Vanaken H, Masand SN. Awareness and collaboration across stakeholder groups important for eConsent achieving value-driven adoption. Ther Innov Regul Sci. 2019;53:724–35.
113. Chalil Madathil K, Koikkara R, Obeid J, Greenstein JS, Sanderson IC, Fryar K, et al. An investigation of the efficacy of electronic consenting interfaces of research permissions management system in a hospital setting. Int J Med Inf sept. 2013;82(9):854–63.
114. Tiffin, Nicki. 2018. Tiered informed consent: respecting autonomy, agency and individuality in Africa. BMJ global health 3 (6): e001249. https://doi.org/10.1136/bmjgh-2018-001249.
115. Nembaware, Victoria, Katherine Johnston, Alpha A. Diallo, J. Maritha, Alice Kotze, Keymanthri Matimba, Godfrey B. Moodley, and Tangwa, Rispah Torrorey-Sawe, and Nicki Tiffin. 2019. A framework for tiered informed consent for health genomic research in Africa. Nature genetics 51 (11): 1566–1571. https://doi.org/10.1038/s41588-019-0520-x.
116. Haussen DC, Doppelheuer S, Schindler K, Grossberg JA, Bouslama M, Schultz M, et al. Utilization of a smartphone platform for electronic informed consent in acute stroke trials. Stroke Am Heart Assoc. 2017;48:3156–60.
117. Haas, M.A., H. Teare, M. Prictor, et al. 2021. 'CTRL': An online, Dynamic Consent and participant engagement platform working towards solving the complexities of consent in genomic research. European Journal of Human Genetics 29: 687–698. https://doi.org/10.1038/s41431-020-00782-w.
118. Doerr M, Suver C, Wilbanks J. Developing a transparent, participant-navigated electronic informed consent for mobile-mediated research [Internet]. Rochester, NY: Social Science Research Network; 2016 Apr. Report No.: ID 2769129. https://papers.ssrn.com/abstract=2769129.
119. Wilbanks J. Design issues in e-consent. J Law Med Ethics. 2018;46:110–8.
120. Kim H, Bell E, Kim J, Sitapati A, Ramsdell J, Farcas C, et al. iCONCUR: informed consent for clinical data and bio-sample use for research. J Am Med Inform Assoc JAMIA. 2017;24(2):380–7.
121. Nembaware V, Munung NS, Matimba A, Tiffin N. Patient-centric research in the time of COVID-19 conducting ethical COVID-19 research in Africa. BMJ Global Health BMJ Spec J. 2020;5:e003035.
122. Harrison SM, et al. Clinical laboratories collaborate to resolve differences in variant interpretations submitted to ClinVar. Genet Med. 2017;19:1096.
123. Raza S, et al. Data sharing to support UK clinical genetics and genomics services. Workshop report. PHG Foundation. 2015. http://www.phgfoundation.org/report/data-sharing-to-support-uk-clinical-genetics-and-genomics-services . Accessed 18 September 2018.

124. Colleen A, Brokamp E. The utility of genomic variant databases in genetic counseling. Human Genetics Theses and Capstones. Paper 15. 2016.
125. Beever D, Swaby L. An evaluation of risk-based monitoring in pragmatic trials in UK Clinical Trials Units. Trials. 2019;20;556. https://doi.org/10.1186/s13063-019-3619-6.
126. The medicines for human use (clinical trials) and blood safety and quality amendment 2008 [Available from: http://www.legislation.gov.uk/uksi/2008/941/contents/made] Accessed 25_Jun_2020.
127. The medicines for human use (clinical trials) (amendment no. 2) regulation 2006 2006 [Available from: http://www.legislation.gov.uk/uksi/2006/2984/contents/made] Accessed 14_Jul_2020.
128. A framework for restarting NIHR research activities which have been paused due to COVID-19 [Available from: https://www.nihr.ac.uk/documents/restart-framework/24886] Accessed 18_Jun_2020.
129. The American Society of Human Genetics. Statement on informed consent for genetic research Am J Hum Genet 1996 59: 471–474.
130. Reilly PR, Boshar MF, Holtzman SH. Ethical issues in genetic research: disclosure and informed consent Nat Genet 1997 15: 16–20.
131. Clayton E. Informed consent for genetic research on stored tissue samples JAMA 1995 274: 1786–1792.
132. Deschenes M, Cardinal G, Knoppers BM, Glass KC. Human genetic research, DNA banking and consent: a question of 'form' Clin Genet 2001 59: 221–239.
133. Beskow MB, Burke W, Merz JF, Barr PA, Rerry S, Penchaszadeh VB et al. Informed consent for population-based research involving genetics JAMA 2001 286: 2315–2326.
134. Ministry of Economy, Trade and Industry, Ministry of Education, Culture, Sports, Science and Technology, and Ministry of Health, Labour and Welfare—Japan. Ethics Guidelines for Human Genome/Gene Analysis Research (2005) http://www.mext.go.jp/a_menu/shinkou/seimei/genome/04122801.htm.
135. Pardo LM, MacKay I, Oostra B, van Duijn CM, Aulchenko YS : The effect of genetic drift in a young genetically isolated population. Ann Hum Genet 2005; 69: 288–295.
136. Pattaro C, Marroni F, Riegler A, Mascalzoni D, Pichler I, Volpato CB, et al. The genetic study of three population microisolates in South Tyrol (MICROS): study design and epidemiological perspectives. BMC Med Genet. 2007;8:29. https://doi.org/10.1186/1471-2350-8-29.
137. Haddow G, Laurie G, Cunningham-Burley S, Hunter KG : Tackling community concerns about commercialisation and genetic research: a modest interdisciplinary proposal. Soc Sci Med 2007; 64: 272–282.
138. Kettis-Lindblad A, Ring L, Viberth E, Hansson MG : Genetic research and donation of tissue samples to biobanks. What do potential sample donors in the Swedish general public think? Eur J Public Health 2006; 16: 433–440.
139. Gibbons SM, Kaye J : Governing genetic databases: collection, storage and use. Kings Law J 2007; 18: 201–208.
140. Foster MW, Sharp RR : Ethical issues in medical-sequencing research: implications of genotype-phenotype studies for individuals and populations. Hum Mol Genet 2006; 15 (Spec no. 1): R45–R49.
141. Gulcher JR, Stefansson K : Ethics of population genomics research. Nature 1999; 400: 307–308.
142. Law 5/2008 of 12 February 2008. URL (accessed October 2009): www.mj.gov.pt/sections/pessoas-e-bens/base-de-dados-geneticos8948/proposta-de-leique/downloadFile/attachedFile_f0/Lei_5_de_2008_de_12_de_Fevereiro.pdf?nocache=1206710619.26.
143. Nuffield Council on Bioethics (2007). The forensic use of bioinformation: Ethical issues. URL (accessed October 2009): www.nuffieldbioethics.org/fileLibrary/pdf/The_forensic_use_of_bioinformation_-_ethical_issues.pdf.
144. Aas K.F. (2006). 'The body does not lie': Identity, risk and trust in technoculture. Crime, Media, Culture, 2, 143–158.

145. Nisker J., & Daar A.S. (2006). Moral presentation of genetics-based narratives for public understanding of genetic science and its implications. Public Understanding of Science, 15, 113–123.
146. Smart A., Tutton R., Martin P., Ellison G., & Ashcroft R. (2008). The standardization of race and ethnicity in biomedical science editorials and UK biobanks. Social Studies of Science, 38, 407–423.
147. Department of Health. Equity and excellence: liberating the NHS. London: Department of Health; 2010.
148. NHS: Handbook to the NHS Constitution. 2009.
149. Department of Health. The Government's mandate to NHS England 2016–17. London: Department of Health; 2017. https://www.gov.uk/government/uploads/system/uploads/attachment_data/file/600604/NHSE_Mandate_2016-17.pdf, Accessed 15 Jan 2018.
150. Healy P, Galvin S, Williamson PR, Treweek S, Whiting C, Maeso B, Bray C, Brocklehurst P, Moloney MC, Douiri A, et al. Identifying trial recruitment uncertainties using a James Lind Alliance priority setting partnership—the PRioRiTy (Prioritising recruitment in randomised trials) study. Trials. 2018;19(1):147.
151. Woods K (2019) Researchers: consent rules put future medical trials at risk. Sunday Business Post (Ireland). Sect. A:6 (col. 1).
152. Mee B, Gaffney E, Glynn SA et al (2013) Development and progress of Ireland's biobank network: ethical, legal, and social implications (ELSI), standardized documentation, sample and data release, and international perspective. Biopreserv Biobank 11(1):3–11. https://doi.org/10.1089/bio.2012.0028.
153. (1950) Declaration of Helsinki, Article 40.3 Bunreacht Na hEireann, the common law, European Convention for the Protection of Human Rights and Fundamental Freedoms, European Convention on Human.
154. Health Service Executive (2019) National Quality Improvement Division. National Consent Policy—Quality Improvement Programmes. HSE. https://www.hse.ie/eng/about/who/qid/other-quality-improvement-programmes/consent/. Accessed 20 April 2020.
155. Department of Health. Confidentiality: NHS code of practice. London: Department of Health, 2003.
156. Department of Health. Clinical governance: in the new NHS. London: Department of Health, 1999. HSC 1999/065.
157. Department of Health. Patient confidentiality webpage. http://www.dh.gov.uk/en/Managingyourorganisation/Informationpolicy/Patientconfidentialityandcaldicottguardians/DH_4084181 (accessed 30 October 2008).
158. Department of Health, NHS Executive, Quality and Consumers Branch. Implementing the Caldicott Report: consultation document. London: Department of Health, 1998.
159. Office of Public Sector Information. The Data Protection Act, 1998. London: The Office of Public Sector Information, 1998.
160. General Dental Council. Principles of patient confidentiality. London: General Dental Council, 2005.
161. Office of Public Sector Information. Health and Social Care Act, 2001. London: The Office of Public Sector Information, 2001.
162. Office of Public Sector Information. Children Act, 2004. London: The Office of Public Sector Information, 2004.
163. Office of Public Sector Information. Computer Misuse Act, 1990. London: The Office of Public Sector Information, 1990.
164. British Dental Association. Data protection. London: British Dental Association, 2000. Advice Sheet B2.
165. FHIR Release 4: Resource Consent [cited 2019 Nov 5]. Available from: URL: https://www.hl7.org/fhir/consent.html.
166. Meszaros J, C Ho C. Building trust and transparency? Challenges of the opt-out system and the secondary use of health data in England. Med Law Int. 2019;19:159–81.

167. Regulation (EU) 2016/679 of the European Parliament and of the Council of 27 April 2016 on the protection of natural persons with regard to the processing of personal data and on the free movement of such data, and repealing Directive 95/46/EC (General Data Protection Regulation), http://data.europa.eu/eli/reg/2016/679/2016-05-04.
168. Prainsack B, Buyx A. Solidarity in biomedicine and beyond. Cambridge, UK: Cambridge University Press; 2017.
169. Krutzinna J, Taddeo M, Floridi L. Enabling posthumous medical data donation: an appeal for the ethical utilisation of personal health data. Sci Eng Ethics. 2019;25:1357–87.
170. Prainsack B, Data Donation: how to resist the iLeviathan. In: Krutzinna J, Floridi L, editors. The ethics of medical data donation. Philosophical Studies Series. Cham: Springer, Cham; 2019. vol 137. 12:22.
171. Richter G, Borzikowsky C, Lieb W, Schreiber S, Krawczak M, Buyx A. Patient views on research use of clinical data without consent: legal, but also acceptable? Eur J Hum Genet. 2019;27:841–7.
172. Richter G, Krawczak M, Lieb W, Wolff L, Buyx A. Broad consent for health care-embedded biobanking: understanding and reasons to donate in a large patient sample. Genet Med. 2018;20:76–82.
173. Boulos D, Morand E, Foo M, Trivedi JD, Lai R, Huntersmith R, et al. Acceptability of opt-out consent in a hospital patient population. Int Med J. 2018;48:84–87.
174. Hosio S, Alorwu A, van Berkel N, López MB, Seetharaman M, Oppenlaender J, Goncalves J (2019) Fueling ai with public displays?: a feasibility study of collecting biometrically tagged consensual data on a university campus. In: Proceedings of the 8th ACM international symposium on pervasive displays, PerDis '19. ACM, New York, pp 14:1–14:7, https://doi.org/10.1145/3321335.3324943.
175. Goncalves J, Hosio S, Liu Y, Kostakos V (2014) Eliciting situated feedback: a comparison of paper, web forms and public displays. Displays 35(1):27–37.
176. Hosio S, Goncalves J, Kostakos V (2013) Application discoverability on multipurpose public displays: popularity comes at a price. In: Proceedings of the 2nd ACM international symposium on pervasive displays. ACM, pp 31–36.
177. Kostakos V, Kukka H, Goncalves J, Tselios N, Ojala T (2013) Multipurpose public displays: how shortcut menus affect usage. IEEE Comput Graph Appl 33(2):56–63.
178. Müller J, Alt F, Michelis D, Schmidt A (2010) Requirements and design space for interactive public displays. In: Proceedings of the 18th ACM international conference on multimedia. ACM, pp 1285–1294.
179. Hosio S, Goncalves J, Kostakos V, Riekki J (2015) Crowdsourcing public opinion using urban pervasive technologies: lessons from real-life experiments in Oulu. Polic Internet 7(2):203–222.
180. Alt F, Kubitza T, Bial D, Zaidan F, Ortel M, Zurmaar B, Lewen T, Shirazi AS, Schmidt A (2011) Digifieds: insights into deploying digital public notice areas in the wild. In: Proceedings of the 10th international conference on mobile and ubiquitous multimedia. ACM, pp 165–174.
181. Tomitsch M, Ackad C, Dawson O, Hespanhol L, Kay J (2014) Who cares about the content? An analysis of playful behaviour at a public display. In: Proceedings of the international symposium on pervasive displays. ACM, p 160.
182. Parameswaran A, Sarma AD, Venkataraman V (2016) Optimizing open-ended crowdsourcing: the next frontier in crowdsourced data management. Bull Tech Committee Data Eng 39(4):26.
183. Christen M, Domingo-Ferrer J, Herrmann D, van den Hoven J (2017) Beyond informed consent—investigating ethical justifications for disclosing, donating or sharing personal data in research. In: Philosophy and computing. Springer, pp 193–207.
184. O'Neill O (2003) Some limits of informed consent. J Med Ethics 29(1):4–7.
185. Alt F, Schneegaß S, Schmidt A, Müller J, Memarovic N (2012) How to evaluate public displays. In: Proceedings of the 2012 international symposium on pervasive displays. ACM, p 17.
186. Academy of Medical Sciences: Personal Data for Public Good: Using Health Information in Medical Research, 2006. Available at http://www.acmedsci.ac.uk/download.php?f=file&i=13206.

187. Sethi N, Laurie G : Delivering proportionate governance in the era of eHealth: making linkage and privacy work together. Med Law Int 2013; 13: 168–204.
188. Ploem MC, Essink-Bot ML, Stronks K : Proposed EU data protection regulation is a threat to medical research. BMJ 2013; 346: f3534.
189. Di Lorio CT, Carinci F, Oderkirk J : Health research and systems' governance are at risk: should the right to data protection override health? J Med Ethics 2014; 40: 488–492.
190. Kerr DJ : Policy: EU data protection regulation-harming cancer research. Nat Rev Clin Oncol 2014; 11: 563–564.
191. Dolgin E : New data protection rules could harm research, science groups say. Nat Med 2014; 20: 224.
192. Nyrén O, Stenbeck M, Grönberg H : The European Parliament proposal for the new EU General Data Protection Regulation may severely restrict European epidemiological research. Eur J Epidemiol 2014; 29: 227–230.
193. Dove ES, Townend D, Knoppers BM : Data protection and consent to biomedical research: a step forward? Lancet 2014; 384: 855.
194. Coppen R, van Veen EB, Groenewegen PP et al: Will the trilogue on the EU Data Protection Regulation recognise the importance of health research? Eur J Public Health 2015; 25: 757–758.
195. Van Ommen G-JB, Törnwall O, Bréchot C et al: BBMRI-ERIC as a resource for pharmaceutical and life science industries: the development of biobank-based Expert Centres. Eur J Hum Genet 2015; 23: 893–900.
196. Knoppers BM : International ethics harmonization and the global alliance for genomics and health. Genome Med 2014; 6: 13.
197. Hallinan, D., Friedewald, M., & De Hert, P. (2013). Genetic data and the data protection regulation: Anonymity, multiple subjects, sensitivity and a prohibitionary logic regarding genetic data? Computer Law & Security Review, 29(4), 317–329.
198. Kaye J : The tension between data sharing and the protection of privacy in genomics research. Annu Rev Genomics Hum Genet 2012; 13: 415–431.
199. Williams H, Spencer K, Sanders C et al: Dynamic consent: a possible solution to improve patient confidence and trust in how electronic patient records are used in medical research. JMIR Med Informatics 2015; 3: e3.
200. Ruyter KW, LOuk K, Jorqui M, Kvalheim V, Cekanauskaite A, Townend D: From research exemption to research norm: recognising an alternative to consent for large scale biobank research. Med Law Int 2010; 10: 287–313.
201. Singer N, Conger K (2019) Google is fined $170 million for violating children's privacy on YouTube. The New York Times. Retrieved from https://www.nytimes.com/2019/09/04/technology/google-youtube-fine-ftc.html. Accessed 23 Aug 2021.
202. Powles J, Hodson H (2017) Google DeepMind and healthcare in an age of algorithms. Heal Technol 7:351–367. https://doi.org/10.1007/s12553-017-0179-1.
203. Nakamoto S (2008) Bitcoin: a peer-to-peer electronic cash system. http://bitcoin.org/bitcoin.pdf.
204. Agbo CC, Mahmoud QH, Eklund JM (2019) Blockchain technology in healthcare: a systematic review. In: Healthcare, vol 7. MDPI, Basel, p 56.
205. Drosatos G, Kaldoudi E (2019) Blockchain applications in the biomedical domain: a scoping review. Comput Struct Biotechnol J. https://doi.org/10.1016/j.csbj.2019.01.010.
206. Beierle F, Tran VT, Allemand M, Neff P, Schlee W, Probst T, Zimmermann J, Pryss R (2019) What data are smartphone users willing to share with researchers? J Ambient Intell Human Comput. https://doi.org/10.1007/s12652-019-01355-6.
207. Crosby M, Pattanayak P, Verma S, Kalyanaraman V et al (2016) Blockchain technology: beyond bitcoin. Appl Innov 2(6–10):71.
208. MITRE. Electronic Consent Management: Landscape Assessment, Challenges, and Technology. McLean2014 29/10/2014 Contract No.: 1.
209. Bialke M, Bahls T, Geidel L, Rau H, Blumentritt A, Pasewald S, et al. MAGIC: Once Upon a time in consent management—a FHIR tale. J Transl Med. 2018;16:256. https://doi.org/10.1186/s12967-018-1631-3.

210. Institute for Community Medicine of the University Medicine Greifswald. A Generic Informed Consent Service gICS (GitHub). 2018. https://github.com/mosaic-hgw/gICS. Accessed 03 Dec 2018.
211. Schönberger D (2019) Artificial intelligence in healthcare: a critical analysis of the legal and ethical implications. Int J Law Info Tech 27(2):171–203.
212. Machulak, M. P., Maler, E. L., Catalano, D., & Van Moorsel, A. (2010). User-managed access to web resources. In Proceedings of the 6th ACM workshop on Digital identity management (pp. 35–44). ACM.
213. Poikola, A., Kuikkaniemi, K. and Honko, H. (2015). "MyData: A Nordic Model for human-centered personal data management and processing." Finnish Ministry of Transport and Communications. http://www.lvm.fi/documents/20181/859937/MyData-nordic-model/2e9b4eb0-68d7-463b-9460-821493449a63?version=1.0.
214. Debruyne C, Pandit HJ, Lewis D, O'Sullivan D (2019) Towards generating policy-compliant datasets. In: 13th IEEE international conference on semantic computing, ICSC 2019, Newport Beach, CA, USA, 30 Jan–1 February 2019. IEEE, pp 199–203. https://doi.org/10.1109/ICOSC.2019.8665631.
215. Das S, Cyganiak R, Sundara S R2RML: RDB to RDF mapping language. https://www.w3.org/TR/r2rml/. Accessed 20 Dec 2019.
216. Debruyne C, Lewis D, O'Sullivan D (2018) Generating executable mappings from RDF data cube data structure definitions. In: On the move to meaningful internet systems: OTM 2018 conferences—confederated international conferences: CoopIS, CT&C, and ODBASE 2018, Valletta, Malta, 22–26 Oct 2018. Proceedings. Springer, pp 333–350. https://doi.org/10.1007/978-3-030-02671-4_21.
217. Bonatti P, Kirrane S, Polleres A, Wenning R (2017) Transparent personal data processing: the road ahead. In: Proc. of TELERISE, pp 337–349.
218. Karjoth, G., Schunter, M., Waidner, M.: Privacy-enabled services for enterprises. In: IEEE 13th Int. Workshop on Databases and Expert Systems Applications, pp. 483–487 (2002).
219. Voss, W. G.: European Union Data Privacy Law Reform: General Data Protection Regulation, Privacy Shield, and the Right to Delisting (2017).
220. Vassilicos, Basil; McGann, Marek Qualities of consent: an enactive approach to making better sense. Phenomenology and the Cognitive Sciences (2023). https://doi.org/10.1007/s11097-023-09897-z.
221. Muravyeva, Ekaterina; Janssen, José; Specht, Marcus; Custers, Bart Exploring solutions to the privacy paradox in the context of e-assessment: informed consent revisited. Ethics and Information Technology (2020). https://doi.org/10.1007/s10676-020-09531-5.
222. Jansen, T. C.; Kompanje, E. J. O.; Druml, C.; Menon, D. K.; Wiedermann, C. J.; Bakker, J. Deferred consent in emergency intensive care research: what if the patient dies early? Use the data or not?. Intensive Care Medicine (2007). https://doi.org/10.1007/s00134-007-0580-8.
223. Langford, Paul Telephone marketing: The importance of the opt-out. Journal of Direct, Data and Digital Marketing Practice (2007). https://doi.org/10.1057/palgrave.dddmp.4350072.
224. European Society of Radiology (ESR) The new EU General Data Protection Regulation: what the radiologist should know. Insights into Imaging (2017). https://doi.org/10.1007/s13244-017-0552-7.
225. Kirwan, Mary; Mee, Blanaid; Clarke, Niamh; Tanaka, Aoife; Manaloto, Lino; Halpin, Emma; Gibbons, Una; Cullen, Ann; McGarrigle, Sarah; Connolly, Elisabeth M.; Bennett, Kathleen; Gaffney, Eoin; Flanagan, Ciaran; Tier, Laura; Flavin, Richard; McElvaney, Noel G. What GDPR and the Health Research Regulations (HRRs) mean for Ireland: "explicit consent"—a legal analysis. Irish Journal of Medical Science (2020). https://doi.org/10.1007/s11845-020-02331-2.
226. Dyke, Stephanie O. M.; Linden, Mikael; Lappalainen, Ilkka; De Argila, Jordi Rambla; Carey, Knox; Lloyd, David; Spalding, J. Dylan; Cabili, Moran N.; Kerry, Giselle; Foreman, Julia; Cutts, Tim; Shabani, Mahsa; Rodriguez, Laura L.; Haeussler, Maximilian; Walsh, Brian; Jiang, Xiaoqian; Wang, Shuang; Perrett, Daniel; Boughtwood, Tiffany; Matern, Andreas; Brookes, Anthony J.; Cupak, Miro; Fiume, Marc; Pandya, Ravi; Tulchinsky, Ilia; Scollen,

Serena; Törnroos, Juha; Das, Samir; Evans, Alan C.; Malin, Bradley A.; Beck, Stephan; Brenner, Steven E.; Nyrönen, Tommi; Blomberg, Niklas; Firth, Helen V.; Hurles, Matthew; Philippakis, Anthony A.; Rätsch, Gunnar; Brudno, Michael; Boycott, Kym M.; Rehm, Heidi L.; Baudis, Michael; Sherry, Stephen T.; Kato, Kazuto; Knoppers, Bartha M.; Baker, Dixie; Flicek, Paul Registered access: authorizing data access. European Journal of Human Genetics (2018). https://doi.org/10.1038/s41431-018-0219-y.

227. Quigley, Etain; Holme, Ingrid; Doyle, David M.; Ho, Aileen K.; Ambrose, Eamonn; Kirkwood, Katie; Doyle, Gerardine "Data is the new oil": citizen science and informed consent in an era of researchers handling of an economically valuable resource. Life Sciences, Society and Policy (2021). https://doi.org/10.1186/s40504-021-00118-6.
228. Eriksson, Stefan; Helgesson, Gert Potential harms, anonymization, and the right to withdraw consent to biobank research. European Journal of Human Genetics (2005). https://doi.org/10.1038/sj.ejhg.5201458.
229. Wiertz, Svenja; Boldt, Joachim Evaluating models of consent in changing health research environments. Medicine, Health Care and Philosophy (2022). https://doi.org/10.1007/s11019-022-10074-3.
230. Teare, Harriet J. A.; Prictor, Megan; Kaye, Jane Reflections on dynamic consent in biomedical research: the story so far. European Journal of Human Genetics (2020). https://doi.org/10.1038/s41431-020-00771-z.
231. Mascalzoni, Deborah; Melotti, Roberto; Pattaro, Cristian; Pramstaller, Peter Paul; Gögele, Martin; De Grandi, Alessandro; Biasiotto, Roberta Ten years of dynamic consent in the CHRIS study: informed consent as a dynamic process. European Journal of Human Genetics (2022). https://doi.org/10.1038/s41431-022-01160-4.
232. De Sutter, Evelien; Borry, Pascal; Geerts, David; Huys, Isabelle Personalized and long-term electronic informed consent in clinical research: stakeholder views. BMC Medical Ethics (2021). https://doi.org/10.1186/s12910-021-00675-7.
233. Alaattinoğlu, Daniela Rethinking Explicit Consent and Intimate Data: The Case of Menstruapps. Feminist Legal Studies (2022). https://doi.org/10.1007/s10691-021-09486-y.
234. Willison, Donald J; Swinton, Marilyn; Schwartz, Lisa; Abelson, Julia; Charles, Cathy; Northrup, David; Cheng, Ji; Thabane, Lehana Alternatives to project-specific consent for access to personal information for health research: Insights from a public dialogue. BMC Medical Ethics (2008). https://doi.org/10.1186/1472-6939-9-18.
235. Bialke, Martin; Geidel, Lars; Hampf, Christopher; Blumentritt, Arne; Penndorf, Peter; Schuldt, Ronny; Moser, Frank-Michael; Lang, Stefan; Werner, Patrick; Stäubert, Sebastian; Hund, Hauke; Albashiti, Fady; Gührer, Jürgen; Prokosch, Hans-Ulrich; Bahls, Thomas; Hoffmann, Wolfgang A FHIR has been lit on gICS: facilitating the standardised exchange of informed consent in a large network of university medicine. BMC Medical Informatics and Decision Making (2022). https://doi.org/10.1186/s12911-022-02081-4.
236. Bialke, Martin; Bahls, Thomas; Geidel, Lars; Rau, Henriette; Blumentritt, Arne; Pasewald, Sandra; Wolff, Robert; Steinmann, Jonas; Bronsch, Tobias; Bergh, Björn; Tremper, Galina; Lablans, Martin; Ückert, Frank; Lang, Stefan; Idris, Tarik; Hoffmann, Wolfgang MAGIC: once upon a time in consent management—a FHIR® tale. Journal of Translational Medicine (2018). https://doi.org/10.1186/s12967-018-1631-3.
237. Jacquier, Elise; Laurent-Puig, Pierre; Badoual, Cécile; Burgun, Anita; Mamzer, Marie-France Facing new challenges to informed consent processes in the context of translational research: the case in CARPEM consortium. BMC Medical Ethics (2021). https://doi.org/10.1186/s12910-021-00592-9.
238. Budowle, Bruce; Sajantila, Antti Revisiting informed consent in forensic genomics in light of current technologies and the times. International Journal of Legal Medicine (2023). https://doi.org/10.1007/s00414-023-02947-w.
239. Indrakusuma, R.; Kalkman, S.; Koelemay, M. J. W.; Balm, R.; Willems, D. L. Context-Relative Norms Determine the Appropriate Type of Consent in Clinical Biobanks: Towards a Potential Solution for the Discrepancy between the General Data Protection Regulation and

the European Data Protection Board on Requirements for Consent. Science and Engineering Ethics (2020). https://doi.org/10.1007/s11948-020-00271-9.

240. Tamuhla, Tsaone; Tiffin, Nicki; Allie, Taryn An e-consent framework for tiered informed consent for human genomic research in the global south, implemented as a REDCap template. BMC Medical Ethics (2022). https://doi.org/10.1186/s12910-022-00860-2.
241. Shabani, Mahsa; Dyke, Stephanie O. M.; Marelli, Luca; Borry, Pascal Variant data sharing by clinical laboratories through public databases: consent, privacy and further contact for research policies. Genetics in Medicine (2018). https://doi.org/10.1038/s41436-018-0316-x.
242. Love, Sharon B.; Armstrong, Emma; Bayliss, Carrie; Boulter, Melanie; Fox, Lisa; Grumett, Joanne; Rafferty, Patricia; Temesi, Barbara; Wills, Krista; Corkhill, Andrea Monitoring advances including consent: learning from COVID-19 trials and other trials running in UKCRC registered clinical trials units during the pandemic. Trials (2021). https://doi.org/10.1186/s13063-021-05225-5.
243. Rosenmayr, A; Hartwell, L; Egeland, T Informed Consent—suggested procedures for informed consent for unrelated haematopoietic stem cell donors at various stages of recruitment, donor evaluation, and donor workup. Bone Marrow Transplantation (2003). https://doi.org/10.1038/sj.bmt.1703851.
244. Anderson, DC; Gomez-Mancilla, B; Spear, BB; Barnes, DM; Cheeseman, K; Shaw, PM; Friedman, J; McCarthy, A; Brazell, C; Ray, SC; McHale, D; Hashimoto, L; Sandbrink, R; Watson, ML; Salerno, RA; Cohen, N; Lister, CE Elements of informed consent for pharmacogenetic research; perspective of the pharmacogenetics working group. The Pharmacogenomics Journal (2002). https://doi.org/10.1038/sj.tpj.6500131.
245. Yamamoto, Midori; Funamizu, Makiko; Muto, Kaori; Hata, Akira International web-based survey of informed consent procedures in genetic epidemiological studies: towards the establishment of a research coordinator accreditation system. Journal of Human Genetics (2009). https://doi.org/10.1038/jhg.2009.103.
246. Mascalzoni, Deborah; Janssens, A Cecile JW; Stewart, Alison; Pramstaller, Peter; Gyllensten, Ulf; Rudan, Igor; van Duijn, Cornelia M; Wilson, James F; Campbell, Harry; Quillan, Ruth Mc Comparison of participant information and informed consent forms of five European studies in genetic isolated populations. European Journal of Human Genetics (2009). https://doi.org/10.1038/ejhg.2009.155.
247. Machado, Helena; Silva, Susana Informed Consent in Forensic DNA Databases: Volunteering, Constructions of Risk and Identity Categorization. BioSocieties (2009). https://doi.org/10.1017/s1745855209990329.
248. Grady, Katherine; Gibson, Martin; Bower, Peter Can a 'consent to contact' community help research teams overcome barriers to recruitment? The development and impact of the 'Research for the Future' community. BMC Medical Research Methodology (2019). https://doi.org/10.1186/s12874-019-0843-4.
249. Vedder, Anton; Spajić, Daniela Moral autonomy of patients and legal barriers to a possible duty of health related data sharing. Ethics and Information Technology (2023). https://doi.org/10.1007/s10676-023-09697-8.
250. Mee, Blanaid; Kirwan, Mary; Clarke, Niamh; Tanaka, Aoife; Manaloto, Lino; Halpin, Emma; Gibbons, Una; Cullen, Ann; McGarrigle, Sarah; Connolly, Elisabeth M; Bennett, Kathleen; Gaffney, Eoin; Flanagan, Ciaran; Tier, Laura; Flavin, Richard; McElvaney, Noel G. What GDPR and the Health Research Regulations (HRRs) mean for Ireland: a research perspective. Irish Journal of Medical Science (2020). https://doi.org/10.1007/s11845-020-02330-3.
251. Ryan, F. S.; Cedro, M. K.; Pabari, S.; Davenport-Jones, L.; Noar, J. H. Clinicians' knowledge and practice of data protection legislation and information management. British Dental Journal (2009). https://doi.org/10.1038/sj.bdj.2009.53.
252. Bild, Raffael; Bialke, Martin; Buckow, Karoline; Ganslandt, Thomas; Ihrig, Kristina; Jahns, Roland; Merzweiler, Angela; Roschka, Sybille; Schreiweis, Björn; Stäubert, Sebastian; Zenker, Sven; Prasser, Fabian Towards a comprehensive and interoperable representation of consent-based data usage permissions in the German medical informatics initiative. BMC

Medical Informatics and Decision Making (2020). https://doi.org/10.1186/s12911-020-01138-6.
253. Richter, Gesine; Borzikowsky, Christoph; Lesch, Wiebke; Semler, Sebastian C.; Bunnik, Eline M.; Buyx, Alena; Krawczak, Michael Secondary research use of personal medical data: attitudes from patient and population surveys in The Netherlands and Germany. European Journal of Human Genetics (2020). https://doi.org/10.1038/s41431-020-00735-3.
254. Alorwu, Andy; van Berkel, Niels; Goncalves, Jorge; Oppenlaender, Jonas; López, Miguel Bordallo; Seetharaman, Mahalakshmy; Hosio, Simo Crowdsourcing sensitive data using public displays—opportunities, challenges, and considerations. Personal and Ubiquitous Computing (2020). https://doi.org/10.1007/s00779-020-01375-6.
255. Mostert, Menno; Bredenoord, Annelien L; Biesaart, Monique C I H; van Delden, Johannes J M Big Data in medical research and EU data protection law: challenges to the consent or anonymise approach. European Journal of Human Genetics (2015). https://doi.org/10.1038/ejhg.2015.239.
256. Andreotta, Adam J.; Kirkham, Nin; Rizzi, Marco AI, big data, and the future of consent. AI & SOCIETY (2021). https://doi.org/10.1007/s00146-021-01262-5.
257. Albanese, Giuseppe; Calbimonte, Jean-Paul; Schumacher, Michael; Calvaresi, Davide Dynamic consent management for clinical trials via private blockchain technology. Journal of Ambient Intelligence and Humanized Computing (2020). https://doi.org/10.1007/s12652-020-01761-1.
258. Rau, Henriette; Geidel, Lars; Bialke, Martin; Blumentritt, Arne; Langanke, Martin; Liedtke, Wenke; Pasewald, Sandra; Stahl, Dana; Bahls, Thomas; Maier, Christian; Prokosch, Hans-Ulrich; Hoffmann, Wolfgang The generic Informed Consent Service gICS®: implementation and benefits of a modular consent software tool to master the challenge of electronic consent management in research. Journal of Translational Medicine (2020). https://doi.org/10.1186/s12967-020-02457-y.
259. Astromskė, Kristina; Peičius, Eimantas; Astromskis, Paulius Ethical and legal challenges of informed consent applying artificial intelligence in medical diagnostic consultations. AI & SOCIETY (2020). https://doi.org/10.1007/s00146-020-01008-9.
260. Belli, Luca; Schwartz, Molly; Louzada, Luiza Selling your soul while negotiating the conditions: from notice and consent to data control by design. Health and Technology (2017). https://doi.org/10.1007/s12553-017-0185-3.
261. Debruyne, Christophe; Pandit, Harshvardhan J.; Lewis, Dave; O'Sullivan, Declan "Just-in-time" generation of datasets by considering structured representations of given consent for GDPR compliance. Knowledge and Information Systems (2020). https://doi.org/10.1007/s10115-020-01468-x.
262. Fernández, Javier D.; Sabou, Marta; Kirrane, Sabrina; Kiesling, Elmar; Ekaputra, Fajar J.; Azzam, Amr; Wenning, Rigo User consent modeling for ensuring transparency and compliance in smart cities. Personal and Ubiquitous Computing (2020). https://doi.org/10.1007/s00779-019-01330-0.
263. Kapitsaki, Georgia M.; Charalambous, Theodoros Adapting HTML5 Web applications to user privacy preferences. World Wide Web (2018). https://doi.org/10.1007/s11280-018-0628-4.

Chapter 6
Interface of Data Protection and Other Areas of Law

Introduction by the Editor

This chapter highlights how privacy and data protection issues are intertwined with other laws that seek to protect consumer interests. Online business models constantly change with the exponential need for data that furthers every change. At all costs, the problem of information symmetry must be addressed because consumers and users tend to miss out on information that would help them make informed decisions. These information asymmetries can be exacerbated due to the shared information, the design layout in which such information is presented, and the nature and content of such information.[1] The previous chapters have highlighted that data subjects must have opportunities to understand and connect with the reasons behind processing personal data. Data protection regimes have underlined the need for information clarity and improved notification systems. The strength of data protection laws and data protection principles can be used to prevent similar situations where there are possibilities to manipulate consumer decisions through advanced techniques. There are instances of data protection legislation creating safeguard mechanisms in cases of profiling consumer interests and practices.[2] These preventive measures will bring about transformational changes and transparency in how online businesses communicate with customers. Transparency as a core concept refers to multiple attainable value propositions. As explained under Recital 58:

> ...requires that any information addressed to the ... data subject be concise, easily accessible and easy to understand, and that clear and plain language and, additionally, where appropriate, visualisation be used. Such information could be provided in electronic form, for example, when addressed to the public, through a website. This is of particular relevance in situations where the proliferation of actors and the technological complexity of practice make it difficult

[1] Jan Trzaskowski, "Data-driven value extraction and human well-being under EU law" (2022) Electronic Markets 32, 447–458.

[2] Article 22, Regulation (EU) 2016/679 of the European Parliament and of the Council of 27 April 2016 on the protection of natural persons with regard to the processing of personal data and on the free movement of such data, and repealing Directive 95/46/EC (General Data Protection Regulation) OJ L 119, 4.5.2016, pp. 1–88. (GDPR).

I. Gupta (ed.), *Expectations vs Realities of Information Privacy and Data Protection Measures*, https://doi.org/10.1007/978-981-99-6778-0_6

> for the data subject to know and understand whether, by whom and for what purpose personal data relating to him or her are being collected, such as in the case of online advertising. Given that children merit specific protection, any information and communication, where processing is addressed to a child, should be in such a clear and plain language that the child can easily understand.[3]

While the first part of the recital refers to the presentation and the content shared, the latter half refers to the complexities resulting from technological advancements. The network of data controllers using such advanced technology leaves the data subjects in a quandary about the nature of data collection and their use. The discussion about consumers, consumer protection law, and its cross-section with data protection law would be incomplete without considering the vulnerable group needing additional data protection and consumer protection support. The first evaluation report of the GDPR points to the concern of a fragmented approach when dealing with the data of such vulnerable individuals.[4]

It is not just about identifying personal data but also about the rights that follow if data processing is established. The issue of aligning all forms of data processing to the objective criteria is often a problem, and the challenges increase with different sectoral applications. The debate should not be about whether GDPR will apply and extend to all forms of processing, which is also the set method we tend to adopt in all possible situations, but about whether we can ensure that expectations, as a result, are equally implemented. There are questions about data generated as an outcome of scientific research, the value they generate, the ownership of such data and the impact of the data subjects exercising their rights. Further questions are about how other rights position themselves with existing data protection frameworks.[5]

On the same note, there have been discussions about personalised pricing systems for particular goods and services in online business models. A personalised pricing system primarily works through the use of cookies and the online presence of customers.[6] Taking a leaf from data protection legislation, specifically about transparency requirements in overall personal data processing, can help address pricing issues. Any form of price discrimination and unfair practices in online businesses can be reduced with greater responsibilities assigned to providers by bringing them within the purview of data protection legislation. For that to happen, greater clarity would be required to ascertain the extent of such legislation and the use of its tools

---

[3] Regulation (EU) 2016/679 of the European Parliament and of the Council of 27 April 2016 on the protection of natural persons with regard to the processing of personal data and on the free movement of such data, and repealing Directive 95/46/EC (General Data Protection Regulation) OJ L 119, 4.5.2016, pp. 1–88. (GDPR).

[4] Data protection as a pillar of citizens' empowerment and the EU's approach to the digital transition - two years of application of the General Data Protection Regulation. < https://eur-lex.europa.eu/legal-content/EN/TXT/PDF/?uri=CELEX:52020DC0264> (accessed 15th July 2024).

[5] Jasper A Bovenberg & Mara Almeida, "Patients v. Myriad or the GDPR Access Right v. the EU Database Right" (2018) European Journal of Human Genetics 27, 211–215.

[6] Frederik Zuiderveen Borgesius & Joost Poort, "Online Price Discrimination and EU Data Privacy Law" (2017) Journal of Consumer Policy 40, 347–366.

like informed consent in these matters. With consent becoming the legal basis of processing, customers would have greater clarity.[7]

The chapter also captures the contradictions a data protection legislative framework may have with the other laws. The contradictions relate to the nature of personal data in question, the scope of personal data, ownership of data and its further use. There are other issues with applying data protection law vis-à-vis incentive structure promoting the application of business models that primarily rely on data to offer services. Owing to the complexity of processes in a digital environment, there could be multiple legislative interventions to protect the interests of consumers.[8] For instance, the Directive on "... certain aspects concerning contracts for the supply of digital content and digital services."[9] The Directive recognised that personal data, being a fundamental right, is not a commodity. Recital 24 broadly refers to the "[d]igital content or digital services [that] are often supplied also where the consumer does not pay a price but provides personal data to the trader."[10] The possible overlap with GDPR means that further complexities could exist with the independent enforcement procedures under these legislative interventions.[11]The scope of legislative reach is a question where multiple pieces of legislation are converging together.[12]As a result, contradictory positions could prolong the enforcement and remedial measures adopted.

Regarding protecting personal data, researchers suggest intellectual property as an alternative to other remedial means. The right emerges from the possible ownership of personal data and the value one attaches to such data.[13] Intellectual property (IP) rights give better control and autonomy to personal data and provide additional support to the data protection framework.[14]It stems from the idea of ownership that one associate with personal data. There are two further considerations in this regard. The impact of possible value addition to such an IP by a third party over different transactions and a clear segregation of IP rights amongst multiple stakeholders. Second, is the impact and level of understanding a data subject must have about exercising such IP rights. A further consideration is the general idea of extending IP rights to all kinds of personal data and its impact.

---

[7] Ibid.

[8] Romain Robert & Lara Smit, "The proposal for a directive on digital content: a complex relationship with data protection law" (2018) ERA Forum 19, 159–177.

[9] Directive (EU) 2019/770 of the European Parliament and of the Council of 20 May 2019 on certain aspects concerning contracts for the supply of digital content and digital services OJ L 136, 22.5.2019, pp. 1–27.

[10] Recital 24, ibid.

[11] Paolisa Nebbia, "The interaction of competition, consumer and data protection laws: a few comments inspired by the recent case law of the Court of Justice of the European Union" (2023) ERA Forum 23, 515–527.

[12] Ibid.

[13] Leon Trakman & others, "Is Privacy and Personal Data Set to Become the New Intellectual Property?" (2019) IIC—International Review of Intellectual Property and Competition Law 50, 937–970.

[14] Ibid.

Thus, the chapter captures a glimpse of the possible relation that data protection legislation has with other legislation that closely relates to consumer-centric online transactions.

Machine Generated Summaries

Disclaimer: The summaries in this chapter were generated from Springer Nature publications using extractive AI auto-summarization: An extraction-based summarizer aims to identify the most important sentences of a text using an algorithm and uses those original sentences to create the auto-summary (unlike generative AI). As the constituted sentences are machine selected, they may not fully reflect the body of the work, so we strongly advise that the original content is read and cited. The auto generated summaries were curated by the editor to meet Springer Nature publication standards. To cite this content, please refer to the original papers.

Machine generated keywords: consumer, business model, protection law, contract, personalize, property, commercial, economic, pay, charter, abuse, law, video, test, business

Data-driven value extraction and human well-being under EU law [28]

This is a machine-generated summary of:

Trzaskowski, Jan: Data-driven value extraction and human well-being under EU law [28].

Published in: Electronic Markets (2022).

Link to original: https://doi.org/https://doi.org/10.1007/s12525-022-00528-0.

If you want to cite the papers, please refer to the original.

For technical reasons we could not place the page where the original quote is coming from.

Abstract-Summary

“We explore the intersection between data protection law and consumer protection law in their application to data-driven business models that monetise attention and personal data.”

“We approach European Union law from a constitutional perspective, as in addition to economic harms some aspects of predatory business models also pose a risk for human dignity, human well-being, social welfare and democracy.”

Introduction

We address how new online business models create new forms of information asymmetry, placing consumers at greater disadvantage relative to retailers, requiring changes to consumer protection laws and regulations.

Consumer protection laws have always been aimed at reducing this information asymmetry.

"Consumer protection law, including the protection of privacy and personal data, must be adapted to restore some sense of equity in online commercial interactions."

"The intent of Consumer Protection and Marketing Law is to ensure that consumers can become adequately informed about the quality of products that they consider purchasing."

"Consumer Protection and Marketing Law need to adapt, to ensure that consumers continue to have the ability to make informed decisions."

Literature review—information economics, information systems, large platforms, and the results of changes in information asymmetry

"In online markets, information asymmetry increases as the platform owners such as Google, Facebook, or Netflix know more about consumers."

"In the Internet, an enormous amount of information is available to consumers—this includes not only product information but also online reviews of other consumers."

"They tap into the long tail of products for niche consumers, and as a result, more consumers can access products that address their individual needs and create delight. (Clemons & Gao, 1; Clemons and others, 2; Markopoulos & Clemons, 3; Clemons and others, 4) While in theory, the Internet makes an ever-increasing amount of information available to consumers, the dominance of Big Tech companies such as Google and Facebook has created a situation where few companies control what consumers see."

"Research also shows that online platforms can increase information asymmetry, and increase sellers' ability to manipulate and to abuse consumers."

Monetization of Attention as a Mechanism for Value Extraction

"The monetisation of attention is at the core of data-driven business models."

"The novelty lies in the scale of the underlying surveillance of behaviour and in how big data—including personal data—and behavioural sciences together with artificial intelligence can be used to manufacture consumer behaviour, including increased engagement and 'consent' to the processing of personal data."

"Since the amount and value of attention may be increased by using personal data and sophisticated persuasion, the providers of data-driven business models have an (economic) incentive to optimise surveillance and manipulation."

"Quite simply, the better a data-driven business understands its users, the more accurately it can target material in order to increase their engagement and to directly manipulate their behaviour (Ghose and Todri-Adamopoulos 5)."

"Data-driven business models exist in both online and offline environments and usually rely on the monetisation of (1) personal data that is collected from surveillance of behaviour and/or (2) attention, by means of personalised marketing, including targeted advertising (Trzaskowski, 6; Selz, 7; Spiekermann and others, 8)."

Law, human dignity and human well-being

"Human well-being and human dignity are at the centre of the aims of the EU treaties, and in the following we will use these terms to include the mentioned values such as freedom, democracy, equality and the rule of law Thirdly, the market is also not an end in itself but a means to ensure sustainable development and 'balanced' economic growth, the efficient allocation of goods and services, the development of pricing signals to guide the allocation investment funds and the allocation of labour, etc In the early stages of the development of EU consumer policy, it was also emphasised that 'the improvement of the quality of life is one of the tasks of the Community and as such implies protecting the health, safety and economic interests of the consumer'. (Council Resolution of 14 April 1975 on a preliminary programme of the European Economic Community for a consumer protection and information policy and Preliminary programme of the European Economic Community for a consumer protection and information policy, OJ, C 92, 25 April 1975, pp."

Consumer protection law, including the protection of personal data

"The impacts of addictive and persuasive design of technology in the form of user interfaces may be more a matter of human well-being and product safety(see, e.g., European Commission, 9) than marketing law and data protection law However, the awareness of its existence and the implications thereof may be important in a more holistic assessment of the legality of data-driven business models, including the extent to which the use of personal data is appropriate, considering the requirements of legitimacy, proportionality and accountability."

"Transparency is a prerequisite for empowerment (see, e.g., Case C-54/17, Wind Tre, ECLI:EU:C:2018:710, paragraph 45: A 'free choice' supposes, in particular, that information provided is 'clear and adequate'), and information requirements permeate both consumer protection law and data protection law In a market economy, consumers are assumed to exercise due care and read information, but we must understand how personalisation and choice architecture may impact agency and undermine empowerment."

The information paradigm and information asymmetry

"21-22) relies primarily on two assumptions: (1) information is recognised as the least intrusive (and thus, the preferred) form of market intervention, and (2) consumers are able to read, understand and consider available information so as to make informed ('rational') decisions."

"In both consumer protection law (see Case C-628/17, Orange Polska, ECLI:EU:C:2019:480) and data protection law (see Case C-673/17, Planet49, ECLI:EU:C:2019:801), it seems sufficient that the consumer has been able to read information in order to make an informed decision, probably assuming that the decision of not reading information is a sufficiently informed decision in itself."

"Information is power, and information asymmetry is important to the rationale for protecting consumers."

"In order to discuss data-driven business models from a consumer protection perspective, it may be helpful to focus on three tiers of information asymmetry which relate to (1) the offer, (2) human behaviour and (3) the user (see Trzaskowski, 10)."

Modifying consumer protection and marketing law to level information asymmetries

"It may, e.g., be straightforward to explain that personal data are collected—and possibly shared with partners—for marketing purposes and that the user experience is adjusted to 'personal preferences', but it may not be clear what information will be filtered out and how the personalisation is likely to affect the individual consumer's decision-making (her 'economic interests')."

"It should be obvious that the information paradigm has its limitations and that there is a need to explore other means for levelling information asymmetries, including by means of marketing law and data protection law To the extent information asymmetries cannot be sufficiently levelled, it may be necessary to pass legislation that restrict the use or abuse of these asymmetries, including the underlying processing of personal data."

"Of proportionality in the GDPR, transparency in the context of 'automated decision-making, including profiling' (Article 22), entails providing 'meaningful information about the logic involved', as well as information about 'the significance and the envisaged consequences of such processing for the data subject' (emphasis added)."

Conclusions: A role for social welfare computing

"There is much good to be said about modern information technology, and as demonstrated above there are some significant caveats to consider in the context of data-driven business models."

"Consumer protection law, including privacy and data protection, seems like a suitable legal tool to ensure empowerment through transparency and the absence of manipulation, including subliminal techniques in a broad sense."

"It is important to understand not only which data are being used for personalised marketing, but also when, how and why, so that the consumer may better understand the logic involved and possible consequences."

"Even though data-driven business models are guided by economic incentives, the value extraction has consequences beyond the economic interests of consumers."

"If people at large do not understand the societal implications and personal consequences of data-driven business models, there is less economic incentive to abandon predatory business models."

Patients v. Myriad or the GDPR Access Right v. The EU Database Right [29]

This is a machine-generated summary of:

Bovenberg, Jasper A.; Almeida, Mara: Patients v. Myriad or the GDPR Access Right v. the EU Database Right [29]

Published in: European Journal of Human Genetics (2018).

Link to original: https://doi.org/https://doi.org/10.1038/s41431-018-0258-4.

If you want to cite the papers, please refer to the original.

For technical reasons we could not place the page where the original quote is coming from.

Abstract-Summary

"Asserting that Myriad's refusal to provide them with this information violated the HIPAA Privacy Rule, the patients sought a determination of a right to access all their genetic information from testing laboratories."

"It would put the GDPR right of access to personal data against Myriad's database right under the EU Database Right Directive."

"The outcome of this case could impact the fate of personalized medicine, which depends on the one hand on patients' having control over their genetic data, and on the other hand on incentives for genetic testing companies to generate these data."

"We then analyse how GDPR rights could play out in the context of clinical genetic testing and conclude that the GDPR access right stops short of granting unconditional access to all data generated in the process of testing, to the extent that its exercise would result in the violation of medical-professional norms, expose the testing company to potential liability, or compromise normal exploitation of the database of which the personal data form part."

Background

"As the patents had allowed Myriad to monopolize clinical testing of these genes and to assemble an enormous amount of information on variants of a gene associated with breast cancer (BRCA) from the patients it tested [12], Myriad could continue to base its business model on claiming proprietary control over these data, calling itself a 'genetic information business' [11, 13]."

"The patients had obtained testing from Myriad to determine their hereditary risk for breast, ovarian and other cancers and to guide potential treatment options."

"The legal dispute was initiated when Myriad refused to provide access to the additional, underlying information it had obtained during the testing process, but not included in the test reports."

"Asserting that Myriad's refusal violated the Health Insurance Portability and Accountability Act (HIPAA) Privacy Rule, the patients sought a determination by the HHS that patients have a right to access all their genetic information from testing laboratories and adoption by Myriad of a policy providing for patients' access to their own genetic information [11]."

Does the GDPR apply to medical records?

"The scope of the GDPR is laid down in article 2 which, inter alia, explicitly states that the GDPR does not apply to activities which fall outside the scope of Union law That raises the question of whether processing in the context of medical care is within or without the scope of Union law As a piece of EU legislation, the GDPR must be compatible with the EU Treaty under the principle that legislative competence not conferred upon the EU remains with the Member States."

"Since the medical record is a mandatory part of the delivery of care, it could be argued that, since the EU has no competence over the practice of medicine, the application of the GDPR to medical records infringes the EU Treaty [14]."

A case against Myriad or against the doctor?

"That raises the question of who controls the genetic data produced in a clinical setting by a genetic testing company?"

"In such a setting, it is the clinician and not the testing company, who decides whether or not to order a genetic test and so determines the purposes and the means of the processing of the genetic data."

"The testing company would qualify as the processor of this data, acting on behalf of, and at the instruction of, the clinician as the controller."

"The GDPR requires that when a controller (in casu: the clinician) outsources a certain processing (in casu genetic testing) to a third party (in casu: the testing company), such outsourcing must be governed by a contract between the controller and the processor."

"A clinician ordering a genetic test from Myriad must stipulate in the agreement that the latter is to assist the clinician in meeting such requests."

The scope of the access right

"Under the GDPR, Myriad must provide modalities to facilitate the exercise of this access right, and where possible, provide remote access to a secure system providing the data subject with direct access to her personal information."

"This opens up the additional question of whether the GDPR access right could trump statutory and professional limitations on providing certain types of data to patients?"

"Applied to clinical genetic testing, must sequence data which, for medical–professional reasons, are not part of the medical record and not clinically actionable, nevertheless be provided to patients on the basis of the GDPR access right?"

"The fact that the GDPR access right as applied in the context of clinical information leaves no room for the above clinical considerations and statutory limitations illustrates the point made earlier that the GDPR is not fit for application to medical records in the first place."

The right to the algorithm

"This raises the question of whether these algorithms are individually identifiable to the patients and hence constitute personal data."

"While pure algorithms do not identify individuals, data related to the applications of these algorithms to specific individuals could qualify as such when they include personal data [15]."

"Whether that means that this data become subject to the access right is an open question."

The right to data portability

"The GDPR provides for a novel right that seeks to bestow that kind of control over one's personal data: the right to 'data portability'."

"This right comprises both the right to receive, in a structured, commonly used and machine-readable format, the personal data which a data subject has provided to the controller, and the right to have these data transmitted directly to a third party."

"It is expressly limited, however, to the personal data which the data subject has provided to the controller."

"As the genomic data produced by Myriad, such as the raw genomic reads and even the genetic test results, are not provided by the data subject but inferred by Myriad from the data (sample) provided by the patient, these data are not subject to the right to data portability."

Access right versus database right

"Then, a class action of multiple patients invoking their access right would enable them to reconstruct, base by base, Myriad's databases and make them available for the research community or other parties."

"To ward off such a class action, Myriad could claim that its databases are eligible for legal protection under the EU Database Right."

"The right to extract or reutilize an insubstantial part of the database only applies to databases which have been made available to the public, in 'whatever manner'."

"The database right does not preclude, for example, anyone from exercising his/her individual data protection access right with respect to personal data contained in the database [16]."

"The database right does not allow repeated and systematic extraction and/or reutilization of insubstantial parts which conflict with a normal exploitation of the database."

Is research none of Myriad's business?

"The first question then is what, exactly, amounts to a 'normal exploitation' of a genetic database generated by a genetic testing company?"

"Is 'normal exploitation' limited to selling the test results, or does it include, if not require, the conduct of scientific research on the data the company has generated?"

"The second question is whether the exercise of the GDPR access right would conflict with this normal exploitation?"

"Using the access right to claim data beyond the tests results in order to make these data also public, is equally obvious to conflict with Myriad's normal exploitation of these data, as it would be hard to prevent Myriad's competitorsfrom having a free ride."

Conclusions

"Pursuant to EU law, the GDPR does not apply to the practice of medicine and hence the medical record."

"For data outside the medical record, exercise of the GDPR access right may be limited to the extent that its exercise would result in the violation of medical-professional norms, expose the testing company to potential liability, or compromise normal exploitation of the database of which the personal data form part."

The proposal for a directive on digital content: a complex relationship with data protection law [30]

This is a machine-generated summary of:

Robert, Romain; Smit, Lara: The proposal for a directive on digital content: a complex relationship with data protection law [30].

Published in: ERA Forum (2018).

Link to original: https://doi.org/https://doi.org/10.1007/s12027-018-0506-7.

If you want to cite the papers, please refer to the original.

For technical reasons we could not place the page where the original quote is coming from.

Abstract-Summary

"While the EU is in the process of comprehensively reforming its data protection legal framework, another legislative proposal intends to address certain aspects of the digital economy, with a potential legal shift from the balance reached by the General Data Protection Regulation: the Proposal for a Digital Content Directive."

Introduction

"This is obviously an important piece of legislation that will affect the European digital economy and the conditions under which personal data can be used in the digital world to protect individuals against any illegal use of their data."

"While the GDPR is not even yet applicable, another legislative proposal aims at regulating certain aspects of the digital economy and interfere with the data protection legislation: the Proposal for a Digital Content Directive (hereinafter the DCD or the Proposal)."

"At face value, the proposed DCD tabled by the Commission is not directly related to data protection."

"It aims at harmonising some key aspects concerning contracts for the supply of digital content with the intention to benefit both consumers and businesses and to boost the EU Digital Single Market."

The main issue at stake: the acceptance of the idea that one can pay with personal data

"As regards its material scope, the Proposal intends to cover all contracts for the provision of digital goods (e.g. films, music, computer programs) or digital services (e.g. cloud computing, social media, video sharing platforms) not only when consumers pay a price in exchange for such goods or services, but also—and this is what has sparked some debates—when consumers do not pay a price but 'actively provide personal data or other data as counter-performance'."

"By including contracts for the provision of digital content in exchange for non-pecuniary counter-performance, the intention of the Commission is to include in the scope of the DCD the so-called 'free services' (e.g. Facebook, YouTube, Instagram) and ensure that they are subject to the same requirements in terms of consumer protection."

"The idea that, in the digital economy, services can be supplied against counter-performance other than money (e.g. personal data) is not something new in EU law However, if the Proposal were to be adopted as such, it would become the first EU legislation to explicitly provide a legal framework for the use of personal data as counter-performance."

The interplay between the DCD, the GDPR and the ePrivacy legislation

"Some processing of personal data by suppliers of digital content will be subject to the requirements of the DCD, the GDPR and the ePrivacy, meanwhile other processing will only be subject to the requirements of the GDPR and the ePrivacy, with the latter potentially requiring a consent for all types of data (passively or actively provided)."

"In the context of contracts for the supply of digital content, the legal basis for processing personal data under the GDPR would, at first glance, appear to be either the consent of the person concerned (Article 6(1)(a)), the legitimate interest of the controller (Article 6(1)(f)), the compliance with a legal obligation (e.g. compliance with obligations of conformity or data retention obligations) (Article 6(1)(c)) or the strict performance of the contract (Article 6(1)(b))."

The Proposal at the European Parliament and the Council

"The question of the legal characterisation of the relationship between the consumer and the supplier is of the essence since the protection granted by the DCD will only operate when a contract exists under national law As regards its relationship with the EU data protection framework, the Council position specifically refers to the applicability of EU law on the protection of personal data to any personal data processed in connexion with contracts for the supply of digital content, with further references to the GDPR, the ePrivacy Directive and Article 8 of the EU Charter in the recitals."

"In the event of termination of the contract, the Council position provides that suppliers must comply with the provisions of the GDPR in relation to the personal data of consumers, and does no longer refer to an additional right for consumers to retrieve their data."

Conclusion: Still a lot to clarify down the road

"The proposed DCD is a perfect example of the difficult dialogue between both areas of law the recent ruling by the Berlin Court against Facebook confirms that consumer law can help to enforce data protection legislation."

"While European consumer law is built on the protection of consumers in the context of a contract, data protection law does not."

"Speaking, both the Council and the European Parliament have made great efforts to align the DCD with the GDPR and ensure coherence between these texts, while still covering contracts for the supply of digital goods or services in exchange for which a price is paid or data is provided (or collected)."

"Let's hope that the legislator will seize this opportunity to articulate consumer law and data protection law in an efficient and compatible manner to protect individuals in the Digital Single Market."

Online Price Discrimination and EU Data Privacy Law [31]

This is a machine-generated summary of:

Zuiderveen Borgesius, Frederik; Poort, Joost: Online Price Discrimination and EU Data Privacy Law [31].

Published in: Journal of Consumer Policy (2017).

Link to original: https://doi.org/https://doi.org/10.1007/s10603-017-9354-z.

If you want to cite the papers, please refer to the original.

For technical reasons we could not place the page where the original quote is coming from.

Abstract-Summary

"This paper explores personalized pricing from a legal and an economic perspective."

"The paper analyses how this dislike of personalized pricing may be linked to economic analysis and to other norms or values."

"The paper examines whether European data protection law applies to personalized pricing."

"Data protection law applies if personal data are processed, and this paper argues that that is generally the case when prices are personalized."

"Data protection law requires companies to be transparent about the purpose of personal data processing, which implies that they must inform customers if they personalize prices."

"If enforced, data protection law could thereby play a significant role in mitigating any adverse effects of personalized pricing."

"It could help to unearth how prevalent personalized pricing is and how people respond to transparency about it."

Personalized Pricing

"Online price discrimination or personalized pricing can be described as differentiating the online price for identical products or services partly based on information a company has about a potential customer."

"The latter suggest that smart sellers use online price discrimination: "Just as it's easy for customers to compare prices on the Internet, so is it easy for companies to track customers' behaviour and adjust prices accordingly.""

"The report adds that "[t]he increased availability of behavioural data has also encouraged a shift from (…) price discrimination based on broad demographic categories towards personalized pricing." (Executive Office of the President of the United States 17, pp. 2–4)."

"When a company does personalize prices online, it can do this on the basis of information from a wide variety of sources, containing different sorts of data: Information can be voluntarily and knowingly provided by a customer, e.g., when he or she created a customer account for a previous purchase."

Price Discrimination: Economics

"First-degree price discrimination enables the sellers to extract all consumer surpluses."

"Without price discrimination, the monopolist will set a price of €15, sell 50 units and leave low spenders unsupplied, even though they are willing to pay more than the marginal costs of production."

"When a non-discriminating price is raised above €8, welfare is destroyed because customers who have a willingness to pay above marginal costs are not served by the market which entails so-called dead-weight-losses."

"In the stylized case above, price discrimination can lead to a welfare improvement for both consumers and sellers."

"Even then, price discrimination will lead to a loss of welfare for some consumers."

"Price discrimination can lead to a transfer of welfare from consumers to sellers or even to a reduction of total welfare."

People's Attitude Towards Personalized Pricing

"Many people regard personalized pricing generally as unfair or manipulative."

"Many people seem to dislike online price discrimination, even if it leads to receiving discounts."

"People may dislike personalized pricing for various reasons."

"Although there would only be cause for regret when personalization leads to higher prices, loss aversion implies that people are inclined to give this scenario more weight."

"The mere fear or suspicion of paying a premium could cause people to dislike personalized pricing."

"There can also be more generic reasons for people's dislike of personalized pricing."

"To enable personalized pricing, companies might collect more information about internet users (Turow 18, pp. 108–110)."

"Many questions are still open regarding people's aversion to personalized pricing."

"Do people just need time to get used to this new kind of price discrimination?"

Data Protection Law Usually Applies to Personalized Pricing

"As Nile.com recognizes Alice and Bob on the basis of their log-in information (name, email address, etc), Nile.com processes personal data."

"Nile.com knows that the person behind cookie xyz often visits price comparison websites and buys mostly cheap or second-hand products."

"Nile.com concludes that the person behind cookie xyz is price-sensitive and in some financial trouble."

"When the person behind cookie xyz visits the Nile.com shopping website, Nile.com shows that person prices with a 10% discount but disables the option to buy now and pay later."

"We argue that the Carol and Dave examples (uniquely identifying cookies) also concern personal data processing, in line with the GDPR and most likely also in line with the CJEU."

"EU data protection law applies to most if not all types of personalized pricing, because most personalized pricing entails the processing of personal data."

Data Protection Law and Transparency

"A company must inform customers (data subjects) if it processes personal data with the purpose of personalising prices."

"The ePrivacy Directive (Art 5, 3) only allows storing a cookie on somebody's computer, after that person has given consent for the cookie, "having been provided with clear and comprehensive information, in accordance with [general data protection law], inter alia, about the purposes of the processing."

"There seems to be no online shops that inform their customers that they personalize prices, even though the 1995 Data Protection Directive (Art 10 and 11) requires, like the GDPR, transparency."

"Another possibility is that some of these companies do personalize prices but do not comply with data protection law's transparency requirements."

"The transparency requirement in data protection law requires companies to inform customers about personalising prices."

Data Protection Law and a Legal Basis for Processing

"Another core principle of data protection law is the requirement that companies that process personal data have a legal basis for doing so."

"A company can have a legal basis for personal data processing if "the data subject has given consent to the processing of his or her personal data for one or more specific purposes" (Art 6(a), GDPR)."

"Generally, the only legal basis for personal data processing is the data subject's consent (Steppe (19) arrives at a similar conclusion)."

"A company could legally process special categories of personal data if the data subject gives explicit consent for processing (Art 9(2)(a) GDPR)."

"The data subject's consent is generally required for processing personal data for price discrimination."

"If the consumer does not consent (and the company cannot rely on another legal basis), personal data processing is not allowed."

Data Protection Law and Automated Decisions

"The main rule of the GDPR's provision on "automated individual decision-making, including profiling" (Art 22, GDPR) reads as follows: The data subject shall have the right not to be subject to a decision based solely on automated processing, including profiling, which produces legal effects concerning him or her or similarly significantly affects him or her."

"Slightly rephrasing Mendoza and Bygrave (20), four conditions must be met for the provision to apply: (i) There is a decision, which is that decision is based (ii) solely (iii) on automated data processing, including profiling; (iv) the decision has legal or similarly significant effects for the person."

"To the general transparency requirements of the GDPR, [T]he controller [the company] shall provide the data subject with the following information necessary to ensure fair and transparent processing in respect of the data subject: (...) the existence of automated decision-making, including profiling (...) and, at least in those cases, meaningful information about the logic involved, as well as the significance and the envisaged consequences of such processing for the data subject."

Some Caveats and Suggestions for Further Research

"We argued that European law generally requires transparency and informed consent for personalized pricing."

"It is too early to conclude that data protection law is the most appropriate instrument to regulate online price discrimination, even when assuming compliance and enforcement to be adequate."

"The transparency requirements could help to unearth how prevalent personalized pricing is and how people respond to transparency about it."

"Regulators could use data protection law's transparency requirements to obtain information about price discrimination."

"Some public concerns could be mitigated if the law required that online personalized pricing could only lead to lower prices compared to some reference price based on anonymous browsing."

"If personalized pricing leads (indirectly) to discrimination on the basis of race or gender, it may be prohibited under non-discrimination law (see, e.g., Art 21, Charter of Fundamental Rights of the European Union; European Agency for Fundamental Rights 21, 22, pp. 29–31).”

Conclusion

“An online shop can recognize a customer, for instance through a cookie, and categorize the customer as a price-sensitive or a price-insensitive person.”

“While such price discrimination could lead to a situation in which large amounts of consumer surplus are extracted to the benefit of companies that apply personalized pricing, the effect on total welfare will be ambiguous.”

“This paper argued that European data protection law generally applies to personalized pricing.”

“Data protection law requires a company to inform people about the specific purpose of processing their personal data.”

“Data protection law generally requires companies to obtain the consumer’s consent for such pricing.”

[Section 10]

“On the internet, companies are able to personalize prices based on information about consumers.”

“This paper explores personalized online price discrimination (“personalized pricing”) from both a legal and an economic perspective.”

“The aim is neither to contribute to economic scholarship of price discrimination itself nor to deal with all areas of law that may be relevant to personalized pricing.”

“This paper discusses whether EU data privacy law applies to online price discrimination and if so, what implications that has.”

“We analyse how a widespread dislike of personalised pricing can be understood from an economic perspective.”

“We turn to data protection law we argue that personalised pricing generally entails the processing of personal data.”

“Companies must generally obtain the customer’s informed consent before they can legally use personalised pricing.”

“We mention some caveats about the usefulness of data protection law in the context of personalised pricing.”

The interaction of competition, consumer and data protection laws: a few comments inspired by the recent case law of the Court of Justice of the European Union [32]

This is a machine-generated summary of:

Nebbia, Paolisa: The interaction of competition, consumer and data protection laws: a few comments inspired by the recent case law of the Court of Justice of the European Union [32].

Published in: ERA Forum (2023).

Link to original: https://doi.org/https://doi.org/10.1007/s12027-023-00,733-0.

If you want to cite the papers, please refer to the original.

For technical reasons we could not place the page where the original quote is coming from.

Abstract-Summary

"This article reviews some recent judgments of the Court of Justice of the European Union (the CJEU) with a view to identifying areas of potential interaction, including conflict, among competition, consumer and data protection laws."

Introduction: Three different goals

"Policy and academic debate on the interplay between competition, consumer and data protection laws has now been intense for at least a decade, sparked by the ambition to achieve the synergies necessary to ensure a more integrated form of protection of consumers in the digital environment and to ensure well-balanced, efficient forms of enforcement; more recently, an increasing number of cases, some of which will be described in this article, has also started keeping the European Union (EU) judiciary busy."

"While data protection can on some occasions play a complementary role vis-à-vis consumer and competition law, it may also create areas of inevitable conflict which may ultimately be adjudicated by the CJEU, and which this article aims to explore with the purpose of drawing broader inferences as to the interplay among them."

First Scenario: Interaction by Internalisation of Exogenous Values

"The case essentially concerned, inter alia, the competence of a national competition authority such as the German Federal Cartel Office to examine the conduct of an undertaking under Art 102 TFEU in the light of certain provisions of Regulation (EU) 2016/679, commonly known as the General Data Protection Regulation (GDPR)."

"De la Tour AG noted that the legal basis of GDPR, namely Art 16 TFEU (right to protection of personal data) precludes the view that in adopting it the EU would have pre-empted all the ramifications which the protection of personal data may have in other areas—interestingly, as if preparing the soil for the subsequent opinion of Rantos AG, he noted that the "other areas" of law included not only consumer but also competition law In addition, while acknowledging the existence of several overspills from data protection into other areas of law, de la Tour AG emphasised that the harmonisation implemented by the GDPR is limited to the aspects specifically covered by that regulation in that area and that, apart from those aspects, the Member States remain free to legislate (and, one would add, interpret legislation), provided that they do not undermine the content and the objectives of GDPR (para)."

Second scenario: Separate and parallel tracks

"At least in theory (and on the basis of a very limited knowledge of the arguments made by the parties before the AGCM) the case highlights an important potential conflict between data protection rules and competition law In fact, under Art 13 GDPR, where personal data are collected, the controller shall inform the relevant data subject not only of the purposes of the processing for which the personal data are intended, but also of the recipients or categories of recipients of the personal data."

"One, last (but not least) reason for clinging to parallel and separate enforcement routes, at least as far as competition and consumer authorities are concerned, may reside in their lack of significant experience and knowledge of data protection law So, for example, the Commission decision in the Google/Fitbit merger may have raised eyebrows within the data protection community for allegedly dismissing with nonchalance data protection concerns; however, it may have been too much to expect that an authority, institutionally devoted to competition enforcement, engages in complex prospective analysis that would require them to predict and assess not only how the market power ensuing from holding large amount of data has been used, but how it will be used in a post-merger scenario."

Conclusion

"A similar solution, going far beyond what is currently provided for by Art 14 of the Digital Markets Act, would require that proposed mergers, capable of having a significant impact on the processing of personal data, are notified, whether on a voluntary or compulsory basis, to the data protection supervisory authorities for the assessment of the data-protection aspects."

"As far as areas of potential enforcement conflict are concerned, national authorities and courts may be able to rely of a twofold source of inspiration in their search for solutions that may reconcile, rather than keep apart, questions raising potential conflict."

"The CJEU has not, so far, refrained from providing resourceful and sometimes elaborate solutions in adjoining areas, which presented marked conflicts between opposing needs and rationales—in this respect, examples include avoiding that the enforcement of standard essential patents becoming a bottleneck to innovation, or preventing the use of sensitive nature of personal data for the purpose of restraining the Commission's investigatory powers."

Is privacy and personal data set to become the new intellectual property? [33]

This is a machine-generated summary of:

Trakman, Leon; Walters, Robert; Zeller, Bruno: Is Privacy and Personal Data Set to Become the New Intellectual Property? [33]

Published in: IIC—International Review of Intellectual Property and Competition Law (2019).

Link to original: https://doi.org/https://doi.org/10.1007/s40319-019-00,859-0.

If you want to cite the papers, please refer to the original.

for technical reasons we could not place the page where the original quote is coming from.

Abstract-Summary

"A pressing concern today is whether the rationale underlying the protection of personal data is itself a meaningful foundation for according intellectual property (IP) rights in personal data to data subjects."

"A further question is how IP law can provide them with the requisite protection of their private space, or whether other means of protecting personal data, such as through general contract rights, render IP protections redundant, or at least, less necessary."

"This paper maintains that lawmakers often fail to distinguish between general property and IP protection of personal data; that IP protection encompasses important attributes of both property and contract law; and that laws that implement IP protection in light of its sui generis attributes are more fitting means of protecting personal data than the alternatives."

"The paper demonstrates that one of the benefits of providing IP rights in personal data goes some way to strengthening data subjects' control and protection over their personal data and strengthening data protection law more generally."

Introduction

"This paper argues that characterizing personal data that is protected by data protection law as IP provides data subjects with a higher level of protection of their personal data."

"In arguing for the protection of data as intellectual property, this paper makes a distinction between personal data as property, and personal data as IP."

"Treating personal data as IP rather than general property, the paper highlights that general property rights overly rigidify the protection of personal data."

"Focusing on limitations in the legal protection of personal data, the paper examines how laws in the EU, Australia and Singapore, regulate the misuse of personal data, including through IP rights."

"It is our view that both general and sensitive personal data ought to be protected through IP rights."

"IP rights in personal data is appropriately protected as a "public good", as distinct from a general property right in that data."

The Significance of a Right to Privacy

"The second question is how IP can provide better protection than alternative measures, such as through the law of privacy or consent to contract, either of which could conceivably render a property right in personal data unnecessary."

"He contends that personal data and information have intrinsic value and therefore are eligible for protection as IP rights."

"What is less readily appreciated is that regulating the use of personal data through IP rights in that data is different from regulating through general principles of property law A property right, as distinct from an IP right, provides legal protection to whosever owns that property."

"Relying on contract rights to protect the personal information of data subjects is also deficient."

"Establishing rules of IP that are directed specifically at protecting personal data would serve to balance the rights of data subjects, while not stifling public access to that information."

Emerging Case Law

"There is emerging case law in jurisdictions beyond the EU, Australia and Singapore, that demonstrate the propensity of courts to protect personal data through an IP right."

"In quoting from an earlier U.S. case, Price v. Hal Roach Studios, the court reasoned that: [t]here is no single generic term that satisfactorily covers all rights which comprise intellectual property rights."

"The position taken by the court is that as long as the personal information is confidential and of commercial value it has a property right and can be treated as intellectual property."

"The court in Price v. Hal Roach Studios did not focus on whether Laurel and Hardy's property rights in their names were analogous to sui generis IP rights."

"The court did demonstrate that, while they retained an inalienable property right in their names, they could waive some of their rights to privacy during their lifetimes."

A Pathway Ahead

"Regulators and courts have also opened the door to such IP protection through the data subject's consent to use his/or personal data."

"A requirement that the data subject consent to the use of personal data is distinct from that subject having an IP right in that data."

"The result is that, if the consent of data subjects to the collection, processing and transmission of their personal data is deficient, the case for a property right, and arguably, an IP right in personal data, becomes more legally defensible."

"A third and tempered response is that data subjects are entitled to protection in the use of their personal data, such as through privacy or contract rights."

"An important purpose in developing transnational data protection law is to address the extent to which personal information and data ought to be treated as property, and as IP rights in particular."

Conclusion

"Regulators and courts have also demonstrated, albeit incrementally, a greater willingness to provide data subjects with some level of ownership and control, including through IP rights, over the collection and use of their personal data."

"They have recognized that IP rights can reconcile limitations in both the contract and general property rights of data subjects, providing those subjects with a staging platform on which to protect their personal information."

"A conception of IP that includes privacy, consent and property rights can provide data subjects with effective and fairer protection of their personal data."

"In responding to the extent to which individuals are induced or deceived into consenting to wide access and use of their personal data, EU regulators have sought to protect that personal data from commercial or other uses that violate the privacy of data subjects, and arguably, undermine their IP rights."

Intellectual property and data ownership in the age of video recording in the operating room [34]

This is a machine-generated summary of:

Filicori, Filippo; Addison, Poppy: Intellectual property and data ownership in the age of video recording in the operating room [34].

Published in: Surgical Endoscopy (2021).

Link to original: https://doi.org/https://doi.org/10.1007/s00464-021-08,692-8.

If you want to cite the papers, please refer to the original.

For technical reasons we could not place the page where the original quote is coming from.

Abstract-Summary

"These videos can be both manually and automatically analyzed for performance analysis."

"We performed a qualitative survey of surgeons, hospital administrators, lawyers and commercial entities offering video recording capabilities for serious issues that the average surgeon who records their videos should consider."

"We identified several key medico-legal constraints, including data ownership and storage, FDA compliance, privacy and potential for use in litigation, present the legal background and potential solutions."

"Given the availability of surgical recording and the future of video-based performance analysis, surgeons need to become comfortable with the medico-legal issues and the potential solutions available with national physician-led lobbying."

Auxiliary recording devices

"While traditional laparoscopic towers provide video recording capabilities, a number of external auxiliary recording devices has been recently popularized."

"Once uploaded to the surgeon's portal the software then allows for video editing, automated and manual video segmentation, temporal and spatial annotation, and automated or manual performance analysis [23]."

"To transition to automated video analysis, images need to be annotated to generate machine-readable data out of a surgical video."

"Annotations refer to either the specific phase of the surgical procedure or anatomical structures, instruments, or other spatial features present in the video."

"The process of automatically recognizing patterns, movements, anatomical structures and temporal flow of events where a machine uses manual annotations to generate automated predictions is termed computer vision (CV)."

Intellectual property and data ownership

"Congress passed legislation 35 U.S.C. 287(c) in 1997 allowing medical procedures to be patented but made the pursuit of patent infringements against physicians nearly impossible [24]."

"Despite there being no claim of overall ownership of a surgical procedure, the gestures of the surgery represent the culmination of a surgeon's education and experience."

"Unlike sharing surgical videos with colleagues, if an external entity is involved in VBA, such videos could be used to generate commercial products; therefore surgeons should know what is transferred and what the ownership structure of these data is."

"Institutions can negotiate contracts to maintain surgeon ownership of the data and to control their use in the development of commercial products."

Credentialing

"The Supreme Court in Taylor v. Intuitive Surgical, Inc. established the burden to ensure competency with new surgical technologies (including surgical robots) with the institution and the surgeon."

"While some surgeons may be concerned about unwarranted scrutiny into their operative techniques, this push towards objective metrics is an evolution that mirrors the shift to evidence-based medical practice."

"Once standardized VBA tools are available and endorsed by the major surgical societies, they will provide useful tools to train residents and an evidence-based resource to determine operative proficiency for credentialing."

Consent and the Scope of the Recording

"While surgical videos provide incredible value for the development of new technologies, most surgical consent forms fail to address the purpose of recordings, who will see it, and for how long the recording will be stored [25]."

"Informed consent for surgical videos is not formally required by any governing body."

"Uncertainty, even though recordings for quality improvement, education, and research with no financial return are generally included in standard surgical consent forms, any commercial use of video data should be disclosed to the patient and explicit consent obtained and documented."

Privacy

"If the videos were recorded for the purpose of obtaining a performance metric score (e.g., OSATS score), after the score is assigned, the videos should be deleted."

"Intra-abdominal videos have a low risk of patient identification and removing unnecessary footage (e.g., when the laparoscope is removed from the abdomen), further limits the potential for including identifying features."

"Research groups that need to associate performance analysis with surgical outcomes have created a two-way patient identifier hashing mechanism to encrypt medical record numbers and allow the association between specific patient videos and their outcomes without sharing any sensitive data with a third party [26]."

"Data from a video recording used for quality improvement, education, or research should not be included in the medical record."

Litigation

"Few surgical videos have been summoned as evidence in a legal proceeding, but the number could grow as plaintiff's attorneys become increasingly aware of the popularity of recording surgical videos among surgeons."

"Any information stored in the electronic medical record is discoverable, and surgical videos meant for quality improvement or research should be stored in a separate database."

"While videos collected for quality improvement render them inaccessible for medical litigation, if patients are made aware of surgical videos during the consent process, those videos may now be discoverable, even if stored on a surgeon's personal computer or hard drive [27]."

"Quality improvement is limited to video review and performance analysis, which might curtail the application of such valuable data for research and development of applications that can result in better surgical care."

Conclusion

"Surgical video recording is a growing reality."

"Surgeons should familiarize themselves with this technology and its medicolegal implications, including intellectual property and data ownership, consent and scope of videos, privacy and the potential for litigation."

"Some of these concerns should be addressed by physician lobbying groups to allow for a legislative resolution at the state and national levels."

"Multiple commercial platforms are already available to aid in educational efforts and performance analysis, but the most interesting applications of such technology will flourish over the next decade."

[Section 8]

"As recording technologies have become ubiquitous in the operating rooms over the recent years, surgeons have witnessed an increase in the number of intraoperative videos collected."

"While recording for self-review or to showcase surgical techniques has been a longstanding practice, multiple companies have entered the market in the recent years to provide automated video recording and performance evaluations."

"We provide a background of the currently available technology for video recording and data processing and address its implications on intellectual property and data ownership; credentialing; consent and scope of recordings; patient privacy; and potential litigation."

References

1. Clemons, E. K., & Gao, G. G. (2008). Consumer informedness and diverse consumer purchasing behaviors: Traditional mass-market, trading down, and trading out into the long tail. Electronic Commerce Research and Applications, 7(1), 3–17.
2. Clemons, E. K., Gao, G. G., & Hitt, L. M. (2006). When online reviews meet hyperdifferentiation: A study of the craft beer industry. Journal of Management Information Systems, 23(2), 149–171.
3. Markopoulos, P. M., & Clemons, E. K. (2013). Reducing buyers' uncertainty about taste-related product attributes. Journal of Management Information Systems, 30(2), 269–299.
4. Clemons, E. K., Spitler, R., Gu, B., & Markopoulos, P. (2005). Information, hyperdifferentiation, and delight: The value of being different. The Broadband Explosion.
5. Ghose, A., & Todri-Adamopoulos, V. (2016). Towards a digital attribution model: Measuring the impact of display advertising on online consumer behavior. MIS Quarterly, 40(4), 889–910.
6. Trzaskowski, J. (2021). Data-driven business models. In E. Kosta & R. Leenes (Eds.), Research Handbook on EU Data Protection Law. Edward Elgar Publishing.
7. Selz, D. (2020). From electronic markets to data driven insights. Electronic Markets, 30(1), 57–59.
8. Spiekermann, S., Acquisti, A., Böhme, R., & Hui, K. L. (2015). The challenges of personal data markets and privacy. Electronic Markets, 25(2), 161–167. https://doi.org/10.1007/s12525-015-0191-0 .
9. European Commission. (2020b). Report on the safety and liability implications of Artificial Intelligence, the Internet of Things and robotics (COM(2020b) 64 final).
10. Trzaskowski, J. (2021). Your privacy is important to us!—Restoring human dignity in data-driven marketing. Ex Tuto Publishing.
11. Health Information Privacy Complaint against Myriad, 19 May 2016. https://Www.Aclu.Org/Legal-Document/Aclu-Hipaa-Complaint.
12. US Supreme Court. Association for Molecular Pathology v. Myriad Genetics, 133 S. Ct. 2107, 2013.

13. Kevin Davies & Michael White, quoted in Health Information Privacy Complaint; Turna Ray, GenomeWeb, 5 April 2016.
14. The Treaty on EU. E.g., Court of Justice of the EU, Philipp Morris, C-547/14, EU: C: 2016: 325, 4 May 2016.
15. Guerrini CJ, McGuire AL, Majunder MA. Myriad take two: can genomic databases remain secret? Science. 2017;356:586–7.
16. Article 13 of the Directive 96191 of the European Parliament and of the Council on the Legal Protection of Databases, 11 March 1996.
17. Executive Office of the President of the United States (Council of Economic Advisors) (2015). Big Data and Differential Pricing. https://www.whitehouse.gov/sites/default/files/whitehouse_files/docs/Big_Data_Report_Nonembargo_v2.pdf. Accessed 6 June 2017.
18. Turow J. (2011). The Daily You: How the new advertising industry is defining your identity and your worth. Yale University Press.
19. Steppe, R. (2017). Prijsdiscriminatie in het digitale tijdperk: Beschouwingen over de nieuwe algemene verordening gegevensbescherming. In M. E. Storme & F. Helsen (Eds.), Innovatie en disruptie in het economisch recht. Antwerpen: Intersentia.
20. Mendoza, I., & Bygrave, L. A. (2017). The Right Not to Be Subject to Automated Decisions Based on Profiling. In T. Synodinou, P. Jougleux, C. Markou, & T. Prastitou (eds.), EU Internet Law: Regulation and Enforcement. Springer, 2017, Forthcoming; University of Oslo Faculty of Law Research Paper No. 2017–20. https://ssrn.com/abstract=2964855. Accessed 6 June 2017.
21. European Agency for Fundamental Rights (2010a). Handbook on European non-discrimination law. Publications Office of the European Union.
22. European Agency for Fundamental Rights (2010b). Data Protection in the European Union: the Role of National Data Protection Authorities. Publications Office of the European Union.
23. Shah NA, Jue J, Mackey TK (2020) Surgical data recording technology. Ann Surg 271(3):431–433. https://doi.org/10.1097/SLA.0000000000003510
24. Reisman J (1995) Physicians and surgeons as inventors: reconciling medical process patents and medical ethics. High Tech Law J 10(2):355–403
25. Makary MA (2013) The power of video recording taking quality to the next level. JAMA. https://doi.org/10.1001/jama.2013.595
26. Addison P, Yoo A, Duarte-Ramos J et al (2020) Correlation between operative time and crowd-sourced skills assessment for robotic bariatric surgery. Surg Endosc. https://doi.org/10.1007/s00464-020-08019-z
27. Jue J, Shah NA, Mackey TK (2020) An interdisciplinary review of surgical data recording technology features and legal considerations. Surg Innov. https://doi.org/10.1177/1553350619891379
28. Trzaskowski, Jan Data-driven value extraction and human well-being under EU law. Electronic Markets (2022). https://doi.org/10.1007/s12525-022-00528-0
29. Bovenberg, Jasper A.; Almeida, Mara Patients v. Myriad or the GDPR Access Right v. the EU Database Right. European Journal of Human Genetics (2018). https://doi.org/10.1038/s41431-018-0258-4
30. Robert, Romain; Smit, Lara The proposal for a directive on digital content: a complex relationship with data protection law. ERA Forum (2018). https://doi.org/10.1007/s12027-018-0506-7
31. Zuiderveen Borgesius, Frederik; Poort, Joost Online Price Discrimination and EU Data Privacy Law. Journal of Consumer Policy (2017). https://doi.org/10.1007/s10603-017-9354-z
32. Nebbia, Paolisa The interaction of competition, consumer and data protection laws: a few comments inspired by the recent case law of the Court of Justice of the European Union. ERA Forum (2023). https://doi.org/10.1007/s12027-023-00733-0
33. Trakman, Leon; Walters, Robert; Zeller, Bruno Is Privacy and Personal Data Set to Become the New Intellectual Property? IIC—International Review of Intellectual Property and Competition Law (2019). https://doi.org/10.1007/s40319-019-00859-0
34. Filicori, Filippo; Addison, Poppy Intellectual property and data ownership in the age of video recording in the operating room. Surgical Endoscopy (2021). https://doi.org/10.1007/s00464-021-08692-8

The manufacturer's authorised representative in the EU is Springer Nature Customer Service Centre GmbH, Europaplatz 3, 69115 Heidelberg, Germany. If you have any concerns regarding our products, please contact ProductSafety@springernature.com

Printed and bound by CPI Group (UK) Ltd, Croydon, CR0 4YY
01/10/2026
02225963-0008